Berechenbarkeit der Welt?

Wolfgang Pietsch · Jörg Wernecke
Maximilian Ott
(Hrsg.)

Berechenbarkeit der Welt?

Philosophie und Wissenschaft
im Zeitalter von Big Data

 Springer VS

Herausgeber
Wolfgang Pietsch
München, Deutschland

Maximilian Ott
München, Deutschland

Jörg Wernecke
München, Deutschland

ISBN 978-3-658-12152-5   ISBN 978-3-658-12153-2   (eBook)
DOI 10.1007/978-3-658-12153-2

Die Deutsche Nationalbibliothek verzeichnet diese Publikation in der Deutschen National-
bibliografie; detaillierte bibliografische Daten sind im Internet über http://dnb.d-nb.de abrufbar.

Springer VS

Gedruckt auf säurefreiem und chlorfrei gebleichtem Papier

Springer VS ist Teil von Springer Nature
Die eingetragene Gesellschaft ist Springer Fachmedien Wiesbaden GmbH
Die Anschrift der Gesellschaft ist: Abraham-Lincoln-Str. 46, 65189 Wiesbaden, Germany

Festschrift für Klaus Mainzer zum Anlass seiner Emeritierung

Inhalt

II   Berechenbarkeit

Vorwort

Es gibt Traditionen des universitären Lebens, die es wert sind, fortgeführt und gepflegt zu werden. Zu ihnen gehört die Herausgabe einer Festschrift, die mit der Emeritierung einer besonders verdienten Gelehrtenpersönlichkeit noch einmal deren im akademischen Leben bearbeitete wissenschaftliche Forschungsfelder in den publizistischen Fokus rückt.

Die in diesem Festband versammelten Autorinnen und Autoren vermitteln eindrücklich die große Reichweite im Wirken des Münchner Philosophen Klaus Mainzer, der national und international als Forscherpersönlichkeit sehr präsent war und stets auch den internen Austausch innerhalb seiner Wirkungsstätte, der TU München suchte. Zudem dokumentiert die Breite der unterschiedlichen Themenfelder dieser Festschrift, von der Informatik über die Mathematik oder Physik bis hin zu gesellschaftlichen und politischen Fragestellungen, die intellektuelle Spannkraft in einem langjährigen Wirken.

Es war ein großes Glück für unsere Universität, mit Klaus Mainzer im Jahre 2008 einen Philosophen und Wissenschaftstheoretiker gewinnen zu können, der sich den Forschungsthemen einer Technischen Universität verpflichtet sah und die interdisziplinäre Zusammenarbeit als zentralen Bestandteil seines Wissenschaftsprogramms tagtäglich aktiv, ja vorbildhaft lebte.

Viele Forschungsthemen der Zukunft sind auf eine interdisziplinäre Bearbeitung angewiesen. Die TU München hat mit der Berufung von Klaus Mainzer auf den Lehrstuhl für Philosophie und Wissenschaftstheorie bereits früh auf die Herausforderung einer interdisziplinären Vernetzung reagiert. So gelang es Klaus Mainzer, mit seinen Forschungsprojekten überaus erfolgreich in die Universität hineinzuwirken und den Forschungsdiskurs zu beflügeln. Hiervon zeugen nicht zuletzt die in diesem Band versammelten Fachbeiträge, etwa von Nadine Gissibl, Claudia Klüppelberg und Johanna Mager (TU München, Mathematik) oder von Christoph Lütge (TU München, Wirtschaftsethik). Deutlich wird dabei auch die Breite der Forschungsperspektiven, die sich von Grundlagenfragen der Berechenbarkeit und Komplexitätstheorie, über deren Anwendung in aktuellen Big Data-Analysen, bis hin zu politischen und ethischen Herausforderungen erstrecken.

Angesichts seines auf Inter- und Transdisziplinarität ausgerichteten Erkenntnisinteresses war es naheliegend, auch auf institutioneller Ebene im Rahmen der zweiten Exzellenzinitiative Herrn Mainzer mit der Gründung des „Munich Center for Technology in Society" (MCTS) zu beauftragen, als dessen Direktor er bis 2014 in höchst verdienstvoller Weise wirkte. Bereits in seiner Funktion als Direktor der

Carl von Linde-Akademie seit 2008 war es ihm gelungen, den Gedanken der Interdisziplinarität mit einem fachübergreifenden Lehrangebot für BA- und MA-Studi-
engänge TUM-weit sowie mit einem eigens konzipierten Masterstudiengang die
Wissenschafts- und Technikphilosophie auch in der Ausbildung von begabtem
Nachwuchs erfolgreich zu verankern. Immer deutlicher wurde aber auch: eine auf
Exzellenz in der Forschung ausgerichtete Technische Universität bedarf angesichts
der heutigen Reichweite wissenschaftlich-technologischer Innovationen der Einbeziehung des gesellschaftlichen Diskurses. Dies ist eine zentrale Aufgabe des MCTS
und auch Gegenstand von Beiträgen dieser Festschrift, wie etwa die Beiträge von
Jürgen Mittelstraß (Universität Konstanz, Philosophie) oder Naoshi Yamawaki
(Tokyo University, Social Philosophy) prominent zeigen.

Schließlich vermittelt der Festband einen Ausblick auf innovative, von Inter-
und Transdisziplinarität geprägte Forschungsstrategien, die nicht zuletzt den Anspruch verfolgen, den technologischen und gesellschaftlichen Herausforderungen
der Gegenwart und Zukunft gerecht zu werden.

Klaus Mainzer hat an der TU München das intellektuelle Erbe eines Spiritus
Rector hinterlassen, das weiterhin vor Ort seine Wirkung entfalten wird. Seine Ide-
en und Ansätze haben, wie der Festband eindrücklich vermittelt, zur Entwicklung
neuer Forschungsthemen angeregt, die weit über die TU München hinaus die wissenschaftlichen Diskurse begleiten werden.

Klaus Mainzers Berufung an unsere Universität war ein Glücksfall ganz beson-
derer Art. Wir haben damals nicht nur einen international hochrenommierten Philosophen für uns gewinnen können, sondern auch eine Gelehrtenpersönlichkeit,
die eigenständig und loyal, zielgerichtet und integrativ den Wissenschaftskosmos
der TU München zukunftsorientiert erweitert und mit seinem eigenen Genius be-
fruchtet hat. Davon gibt die vorliegende Festschrift, der ich eine rasche Verbreitung
wünsche, ein lebendiges Zeugnis ab.

Wolfgang A. Herrmann
Präsident
Technische Universität München

1. Einführung: Zehn Thesen zu Big Data und Berechenbarkeit

Wolfgang Pietsch, Jörg Wernecke

1.1 Ein Versuch in Interdisziplinarität

Der vorliegende Band ist ein Experiment. In einer thematischen und methodischen Breite, wie sie einst von Francis Bacon zu Anfang der modernen westlichen Wissenschaft eingefordert wurde, wie sie aber in der heutigen Wissenschaftslandschaft mit ihrer immer weiter fortschreitenden Spezialisierung kaum noch praktiziert wird, widmet er sich dem Thema Berechenbarkeit. Zu Anfang des dritten Jahrtausends stellt sich die Frage nach der Berechenbarkeit der physischen, psychischen und sozialen Welt gänzlich neu – vor allem aufgrund von Fortschritten im Bereich Informations- und Kommunikationstechnologie sowie aufgrund der damit einhergehenden Datenflut nicht nur in der Wissenschaft, sondern auch in vielen anderen Lebensbereichen, der Wirtschaft und unserem Privatleben.

Der Band ist *interdisziplinär* angelegt, er versucht möglichst viele Blickwinkel auf die Themen Big Data und Berechenbarkeit zu vereinen, von den formalen Wissenschaften wie Mathematik, Statistik oder Informatik über Natur- und Sozialwissenschaften bis hin zu Geisteswissenschaften einschließlich der Philosophie. Des Weiteren ist der Band *grundlagenorientiert*, indem viele der hier diskutierten Fragen vor allem fundamentale Konzepte und Methoden betreffen, beispielsweise Berechenbarkeit, Information oder Komplexität. Und schließlich ist er *anwendungs-*

bezogen, untersucht eine aktuelle Fragestellung, nämlich inwieweit die Datenflut des Informationszeitalters wissenschaftliche Konzepte und Methodik beeinflusst.

Die genannten drei Aspekte zeichnen auch das Werk des Münchner Wissenschaftstheoretikers Klaus Mainzer aus, dem wir den Band zu seiner Emeritierung als Festschrift widmen. Die Themen des Sammelbandes gehören zu den zentralen Forschungsanliegen in Mainzers Werk, der immer wieder ein ausgeprägtes Gespür für innovative Fragestellungen im Spannungsfeld zwischen Philosophie und Wissenschaft bewiesen hat. So gehörte er zu den ersten Wissenschaftstheoretikern, die Entwicklungen in der Komplexitätsforschung aus einer Grundlagenperspektive kritisch reflektierten. Auch erfasste er bereits zu einem frühen Zeitpunkt die wissenschaftstheoretische Relevanz von Big Data und hinterfragte gängige Ansichten in der öffentlichen Debatte.

Dabei betrachtet Mainzer solche Themen nie ausschließlich aus philosophischer und wissenschaftstheoretischer Perspektive, sondern hat sich in seiner vielfältigen wissenschaftlichen Laufbahn immer wieder für die Belange interdisziplinärer Forschung eingesetzt. Diese Verankerung in der wissenschaftlichen Praxis ist eine entscheidende Voraussetzung, dass auch abstrakte Überlegungen über die engen Zirkel philosophischer Fachdebatten hinaus relevant bleiben können. Und schließlich war Mainzer immer auch ein großer Kommunikator nicht nur mit anderen Disziplinen, sondern hat wissenschaftliche und philosophische Themen in vielen Vorträgen, Interviews und Büchern auch in eine breitere Öffentlichkeit getragen.

Liest man die Wissenschaftsseiten und Feuilletons der großen Tageszeitungen, so könnte für Außenstehende schnell der Eindruck entstehen, dass interdisziplinäre Forschung heutzutage eher die Norm als die Ausnahme ist. Die akademische Praxis ist jedoch eine andere. Der Weg zu akademischem Erfolg und Professur ist nach wie vor fast ausschließlich disziplinär organisiert. Die Probleme von Interdisziplinarität kann zudem nur verstehen, wer sich bewusst macht, dass sich auch in den Wissenschaften unterschiedliche Kulturen gegenüberstehen, die von unterschiedlichen Wissensbeständen ausgehen und vor allem auch unterschiedliche Erkenntnisinteressen und Wertesysteme besitzen.

Trotz all dieser Schwierigkeiten gibt es viele Querschnittsthemen, die letztlich nur interdisziplinär angegangen werden können, weil relevantes Wissen und Verständnis zu verschiedenen Fachgebieten gehören. Eine veränderte Sicht auf Berechenbarkeit im Zeitalter von Big Data ist sicher eines dieser Querschnittsthemen.

1.2 Zehn Thesen zu Berechenbarkeit und datenintensiver Wissenschaft

Die Frage der Berechenbarkeit der Natur ist so alt wie die Wissenschaft selber, wie Bemühungen des Menschen seine physische und soziale Umgebung vorherzusagen, um zu überleben. Dennoch haben sich die Perspektiven auf dieses Thema über die Jahrhunderte immer wieder geändert, häufig im Gleichschritt mit bedeutenden Entwicklungen sowohl im Bereich wissenschaftliche Methodik als auch in der Verfügbarkeit von Daten. Der Induktivismus eines Francis Bacon und der neue Fokus aufs Experimentieren und Beobachten ermöglichten in der Renaissance den Durchbruch der modernen empirischen Wissenschaft, die sich von den Dogmen der Scholastik befreien konnte. Die Anfänge der modernen Statistik, um ein weiteres Beispiel zu nennen, liegen im Bestreben mit den ‚Big Data' des 18. und 19. Jahrhunderts fertig zu werden, als unter anderem viele staatliche Statistikbüros entstanden, die eine wahre Flut gedruckter Zahlen hervorbrachten (Hacking 1990, Ch. 4). Alles deutet darauf hin, dass wir auch heute wieder an einem Wendepunkt stehen, an dem moderne Informationstechnologien in Kombination mit riesigen Datensätzen die Frage der Berechenbarkeit der physischen und der sozialen Welt neu aufwerfen.

Wie so oft in der Wissenschaftsgeschichte konnte sich Big Data als Schlagwort und Leitmotiv auch deswegen durchsetzen, weil der Begriff so vielschichtig und mehrdeutig ist und in verschiedenen Gebieten von der Informatik bis zur Wirtschaft ganz unterschiedlich gedeutet werden kann. Während in den Wissenschaften nicht selten konzeptionelle Unschärfen fruchtbar genutzt werden können, verlangt die Philosophie eine größere begriffliche Exaktheit. Deswegen muss im Folgenden zuerst einmal geklärt werden, was genau unter Big Data und datenintensiver Wissenschaft zu verstehen ist.

In der einschlägigen Literatur wird Big Data häufig mit Verweis auf die sogenannten 3 Vs charakterisiert, also Datenmenge (Volume), Datenrate (Velocity) und schließlich die heterogene Struktur der Daten (Variety) (vgl. Laney 2001). Dieser Ansatz ist verschiedentlich erweitert worden auf 5 oder 7 Vs. Während eine solche Definition aus Sicht der technischen Herausforderungen im Umgang mit großen Datensätzen angemessen scheint, erweist sie sich für weitergehende erkenntnistheoretische und methodische Fragestellungen als eher ungeeignet. Das größte Problem ist der ausschließliche Fokus auf die Daten selber und ihre Struktur, wobei die Algorithmen und die Methoden der Datenverarbeitung in den Hintergrund treten, die aber gerade eine erkenntnistheoretische Neuheit begründen könnten.

Im Folgenden sollen daher zwei methodische Aspekte datenintensiver Wissenschaft herausgestellt werden. Da wäre zuerst die vielzitierte Behauptung, dass Big Data in der Lage sei, alle Individuen einer Population zu erfassen, was oft in der griffigen Formulierung „N=alle" zusammengefasst wird (bspw. Mayer-Schönberger & Cukier 2013, S. 26-31). Würde man diese Eigenschaft als Bestandteil einer Definition von Big Data akzeptieren, so ließe sich der zuvor genannte Einwand entkräften, weil „N=alle" klare methodische Implikationen hat. Zum Beispiel gehen konventionelle statistische Ansätze grundsätzlich von Stichproben aus, die mit Blick auf eine gegebene Fragestellung möglichst repräsentativ sein sollen. Aufgrund der Stichprobe können dann Wahrscheinlichkeitsaussagen über weitere Individuen der untersuchten Population getroffen werden. Ein derartiger methodisch anspruchsvoller Extrapolationsprozess findet im Fall einer über „N=alle" definierten datenintensiven Wissenschaft offenbar nicht mehr statt. Die Herangehensweise scheint viel simpler, im Extremfall kann einfach in einer tabellarischen Zusammenstellung aller Möglichkeiten nachgeschlagen werden, was für ein bestimmtes Individuum der Fall ist.

Andererseits ist es unrealistisch, wirklich alle Individuen einer Population in einem Datensatz zu erfassen, beispielsweise können zukünftige Fälle kaum berücksichtigt werden. Es ist außerdem wenig plausibel, dass überhaupt keine Form von Komplexitätsreduktion mehr stattfindet, zum Beispiel indem völlig identische Fälle eliminiert werden. Eine weitere unerwünschte Konsequenz von „N=alle" ist, dass bei einfachen Systemen bereits eine geringe Anzahl von Daten alle Möglichkeiten abdeckt, zum Beispiel enthalten die Datenpunkte (Schalter an, Licht an), (Schalter aus, Licht aus) die vollständige Variation eines Systems aus Schalter und Glühbirne. Es scheint nicht sinnvoll, in solchen Fällen von Big Data zu sprechen. Um dieses Problem zu umgehen, könnte man die Definition explizit auf ausreichend komplexe Phänomene beschränken, die dann notwendigerweise große Datenmengen erfordern.

Ein Vorschlag hinsichtlich einer Präzisierung von „N=alle" wäre demnach:

Datenintensive Wissenschaft bezeichnet die systematische Erfassung und Analyse von Datensätzen, welche einen Großteil möglicher Variationen eines komplexen Phänomens[1] wiedergeben, so dass ohne weitere theoretische Hintergrundannahmen die für eine gegebene Fragestellung relevante kausale Struktur algorithmisch erschlossen werden kann.

---

1   Man spricht in diesem Zusammenhang auch von *variationaler Evidenz.*

Damit ist datenintensive Wissenschaft vornehmlich ein induktivistisches Forschungsprogramm. Auf einige Details dieser Definition, insbesondere die Theoriefreiheit und den Fokus auf Kausalität, werden wir später noch eingehen. Dabei wird sich insbesondere eine enge Verwandtschaft zwischen datenintensiver Wissenschaft und explorativem Experimentieren zeigen.

Die zweite zentrale methodische Neuerung betrifft die vollständige Automatisierung des wissenschaftlichen Prozesses (vgl. etwa Leonelli 2012). Diese hat zur Folge, dass Wissenschaft unter epistemischen Rahmenbedingungen stattfindet, die sich von denen des menschlichen Erkenntnisapparates grundlegend unterscheiden, insbesondere was die Speicherfähigkeit und die Verarbeitungsgeschwindigkeit der Daten betrifft. In automatisierten Prozessen kann somit Wissen generiert werden, welches dem Menschen grundsätzlich nicht mehr zugänglich ist. Vor allem können Phänomene behandelt werden, die zu komplex sind um im Rahmen einer ‚menschlichen Wissenschaft' verlässliche Vorhersagen zu treffen oder einen effektiven Eingriff zu begründen, zum Beispiel weil diese Phänomene von zu vielen Variablen abhängen oder weil die Abhängigkeiten nicht durch einfache Funktionen darstellbar sind. Man könnte also weiter präzisieren, dass die Phänomene, die in datenintensiver Wissenschaft untersucht werden, gemeinhin so komplex sind, dass automatisierte Methoden genauere Vorhersagen liefern als herkömmliche Ansätze im Rahmen einer ‚menschlichen Wissenschaft'.

In zehn Thesen soll im Folgenden das Thema Berechenbarkeit aus der Perspektive datenintensiver Wissenschaft betrachtet werden. Beginnen wir mit dem großen Versprechen:

Erste These: Big Data führt zur zunehmenden Berechenbarkeit komplexer Phänomene, vor allem zu verlässlicheren kurzfristigen Vorhersagen.

Diese Eigenschaft lässt sich aus den zuvor beschriebenen beiden Charakteristika ableiten, dass einerseits verlässlichere Vorhersagen möglich sind, wenn Daten ausreichend viele Systemzustände komplexer Phänomene widerspiegeln, andererseits die Veränderung in den epistemischen Rahmenbedingungen, insbesondere die verbesserten Möglichkeiten der Speicherung und Verarbeitung von Daten, neue Ansätze zur Analyse von Komplexität bietet.

Die Struktur vieler Anwendungen von Big Data ist die folgende: Gegeben ist ein Datensatz, der eine größere Anzahl von Prädiktorvariablen mit einer meist überschaubaren Zahl von Zielvariablen verknüpft. Anhand eines Teils des Datensatzes, der sogenannten Trainingsmenge, wird algorithmisch ein Modell der Daten entwickelt, welches mit Hilfe der verbliebenen Daten, der Testmenge, überprüft wird. Beispielsweise können so die Ergebnisse einer Internetsuche aufgrund der Suchhistorie bestimmt werden oder Kaufempfehlungen bei Internethändlern mithilfe von Nutzerprofilen und Kaufverhalten ausgesprochen werden. Die entscheidende Neuerung ist, dass durch die genannte Automatisierung von Datenerhebung und -analyse viel mehr Variablen und viel mehr Datenpunkte berücksichtigt werden können als in klassischen statistischen Ansätzen. In diesem zweifachen Sinne haben wir es also mit Big Data zu tun.

Im Folgenden soll kurz auf den Begriff der Komplexität eingegangen werden. Zuerst einmal lässt sich eine Komplexität in den Beschreibungen (*epistemische Komplexität*) von einer Komplexität in den Phänomenen unterscheiden (*phänomenologische Komplexität*). Der bekannteste Ansatz für den ersten Typus ist die sogenannte Kolmogorow-Komplexität, nach welcher Komplexität als „minimale Länge eines Programms [aufgefasst wird], durch das man ein Objekt x mit der Programmiermethode S darstellen kann" (Kolmogorow 1983, S. 33, unsere Übersetzung). Kolmogorow zufolge gibt es einen engen Zusammenhang zwischen Komplexität und Zufälligkeit, zufällige Objekte sind demnach solche mit maximaler Komplexität. Eine geringe Komplexität haben hingegen Objekte, die aus wenigen grundsätzlichen Annahmen ableitbar sind. Weiterhin ist eine solche epistemische Komplexität beschreibungsabhängig, hängt insbesondere von der verwendeten Programmiermethode ab, von den getroffenen Annahmen sowie von den erlaubten Ableitungsschritten.

Hierin liegt auch der wichtigste Unterschied im Vergleich zur phänomenologischen Komplexität, die in den Phänomenen selber verortet ist, also im Idealfall objektiv und beschreibungsunabhängig sein sollte. Im Fall der phänomenologischen Komplexität liegt es nahe, eine weitere Unterscheidung zwischen *emergenter* und *irreduzibler* Komplexität einzuführen.[2] Im ersten Fall liegt dem komplexen Phänomen ein relativ einfaches System an Gesetzen oder Anweisungen zugrunde, also einfache Basis-Phänomene, die durch wenige Variablen und wohldefinierte Abhängigkeiten zwischen den Variablen dargestellt werden können. Die komplexe Dynamik ist dann emergent, sie resultiert üblicherweise aus Nicht-Linearitäten, die eine starke Sensibilität des Systemverhaltens gegenüber kleinen Variationen in den Anfangsbedingungen zur Folge haben. Typische Systeme mit emergenter Komplexität werden in der Physik des Nicht-Gleichgewichts und in der Chaostheorie behandelt, zum Beispiel das Doppelpendel, einfache Räuber-Beute-Systeme oder die Rayleigh-Bénard-Konvektion.

Im Gegensatz dazu zeichnet sich irreduzible Komplexität gerade dadurch aus, dass sich entsprechende Phänomene nicht auf wenige Gesetze mit wenigen Varia-

---

2   Eine eng verwandte Unterscheidung zwischen kompositorischer und dynamischer Komplexität formuliert Meinard Kuhlmann (2011, vgl. auch Mitchell 2008).

blen und klar definierten Abhängigkeiten zurückführen lassen. Vielmehr hat man es mit einer komplizierten Anordnung einer großen Anzahl oft sehr unterschiedlicher Systembestandteile zu tun, die gemäß komplizierten kausalen Abhängigkeiten untereinander wechselwirken. Die dynamische Komplexität ist hier also nicht in dem Sinne emergent, dass sie aus einem zugrundeliegenden einfachen Modell ableitbar ist. Vielmehr ist das System selber irreduzibel komplex. Selbstverständlich können alle Eigenschaften emergenter Komplexität wie Nicht-Linearitäten, Rückkopplungsschleifen oder das Fehlen von Additivität trotzdem auftreten. Hinzu kommt aber die fehlende Rückführbarkeit auf wenige einfache Phänomene. Damit zusammenhängend lässt sich bei irreduzibler Komplexität auch nicht mehr eine streng hierarchische Struktur verschiedener Beschreibungsebenen einführen, wie sie beispielsweise in der Physik realisiert ist: Quarks, Elementarteilchen, Atome, Moleküle etc. Vielfältige Wechselwirkungen zwischen den Ebenen sind möglich, was letztlich de facto zu einem Auflösen der Beschreibungsebenen führt.

Es ist insbesondere diese letztgenannte Form von Komplexität, für welche datenintensive Wissenschaft einen großen Vorteil bedeutet. Denn um die komplexen Abhängigkeiten, die durch den menschlichen Erkenntnisapparat nicht mehr vollständig fassbar sind, korrekt abbilden zu können, benötigt man notwendigerweise eine große Anzahl an Daten. Nach allem, was man heute weiß, treten solche komplexen Phänomene in fast allen anwendungsnahen Wissenschaften auf, insbesondere den Sozialwissenschaften, der Medizin oder der Biologie. Phänomene in den theoretischen Grundlagenwissenschaften, vor allem in der Physik, sind hingegen zumeist nur in einem emergenten Sinne komplex, was vor allem an der bewussten Auswahl der Problemstellungen in der Physik liegt (s. Diskussion zur zweiten These).

Mit Blick auf das Thema des Sammelbandes sollte noch ein verwandter, ähnlich vielschichtiger Begriff kurz besprochen werden, der der Berechenbarkeit. In der Informatik wird Berechenbarkeit vor allem im Zusammenhang mit der Lösbarkeit mathematischer Probleme diskutiert. Wenn für ein bestimmtes Feld, beispielsweise die Arithmetik, allgemeine Aussagen oder Axiome gegeben sind, lässt sich dann für jeden Satz aus dem zugehörigen Bereich bestimmen, ob dieser wahr oder falsch ist? Bei dieser Form von Berechenbarkeit geht es offenbar vor allem um die Frage, was sich aus wenigen allgemeingültigen Aussagen sicher folgern lässt.

Gerade bei Phänomenen irreduzibler Komplexität ist die beschriebene Auffassung von Berechenbarkeit aber nicht zielführend, eben weil es die genannten allgemeinen Axiome oder Grundgesetze per definitionem nicht gibt. Es geht insgesamt nicht so sehr um das Herleiten gültiger Aussagen aus einer beschränkten Anzahl allgemeingültiger Sätze, sondern vielmehr um Vorhersage, ob etwas in der Welt der Fall ist oder nicht, auf der Basis eines im Prinzip beliebig erweiterbaren Datensatzes von Einzelereignissen. Man könnte diese beiden Sichtweisen wiederum grob als epistemische und als phänomenologische Berechenbarkeit einordnen. Offenbar tritt im Fall der phänomenologischen Berechenbarkeit nicht mehr das vieldiskutierte Problem unentscheidbarer Sätze auf, aus dem einfachen Grund, weil es hier nur um Aussagen geht, die entweder in der Welt realisiert sind oder nicht. Es wird damit auch ein anderer Wahrheitsbegriff zugrunde gelegt.

Die Tatsache, dass datenintensive Wissenschaft irreduzibel komplexe Phänomene analysiert, ist nun ein wichtiger Grund dafür, dass Änderungen in der Modellierung auftreten (s. auch Norvig, dieser Band):

Zweite These: Es findet ein Wandel in der Modellierung statt von stark theoriebeladenen Ansätzen mit wenig Daten hin zu einfachen Modellen mit vielen Daten.

Unter Berücksichtigung der Unterscheidung zwischen emergenter und irreduzibler Komplexität ist diese Entwicklung wenig überraschend. Hat man es mit emergenter Komplexität zu tun, so lässt sich über theoretisches Hintergrundwissen die grundlegende Struktur der Abhängigkeiten zwischen den einzelnen Variablen bestimmen. Beispielsweise lassen sich die Grundgleichungen eines Doppelpendels aus den Gesetzen der Mechanik ableiten. Es genügen dann meist wenige möglichst genaue und gut gewählte Datenpunkte, um die zusätzlichen Parameter für das Systemverhalten festzulegen. Zum Beispiel hat ein gewöhnliches Doppelpendel vier Parameter, die beiden Massen und die Länge der beiden Arme.

Im Fall irreduzibler Komplexität hingegen kann die Struktur der Gleichungen gleichsam per definitionem nicht durch theoretische Hintergrundannahmen festgelegt werden, sondern muss stattdessen aus den Daten erschlossen werden, falls diese in ausreichendem Maße vorhanden sind. Daraus ergibt sich der induktivistische und weitgehend theoriefreie Ansatz datenintensiver Wissenschaft.

Das klassische Beispiel für den in der zweiten These beschriebenen Wandel in der Modellierung stammt aus der Linguistik (Halevy et al. 2009). Stark vereinfacht gesagt, gab es im Bereich maschineller Übersetzung historisch zwei unterschiedliche Ansätze, der eine regelbasiert, der andere datengetrieben. Die Herangehensweise im ersten Fall ist die verschiedenen Sprachen mit all ihren grammatikalischen Regeln in einem Computerprogramm zu modellieren. Ein Übersetzungsprogramm muss demnach zuerst mit Hilfe dieser Regeln die grammatikalische Struktur eines Satzes in der Quellsprache erkennen, dann die einzelnen Begriffe des Satzes anhand eines Wörterbuchs in die Zielsprache übersetzen, und schließlich den Satz in die grammatikalische Struktur der Zielsprache übertragen. Dieser historisch frühere Ansatz stieß jedoch schnell an seine Grenzen. Der Hauptgrund für das Scheitern liegt letztlich in der (irreduziblen) Komplexität von Sprache. Es gibt einfach zu viele grammatikalische Regeln und wiederum zu viele Ausnahmen von diesen Regeln als dass es möglich wäre, alle händisch in ein Computerprogramm einzupflegen. Was sich überraschenderweise als viel erfolgreicher herausstellte, war ein datenbasierter Ansatz, der auf eine explizite Modellierung grammatikalischer Regeln verzichtet. Stattdessen wird mit riesigen Textkorpora gearbeitet, die beispielsweise aus dem Google Books Projekt oder aus dem Internet stammen. Die vorgegebene Modellstruktur ist nun weit weniger komplex als im regelbasierten Ansatz. Statt der Vielzahl grammatikalischer Regeln wird ein einfacher Bayesscher Algorithmus verwendet, der aufgrund der relativen Häufigkeiten von Wortfolgen in den verschiedenen Korpora die wahrscheinlichste Übersetzung bestimmt.[3] Übersetzungsprogramme haben nach wie vor ihre Schwächen, aber der datenbasierte Ansatz ermöglicht heute zumindest die Grundgedanken fremdsprachiger Texte zu erfassen. Es zeigt sich also, genau wie in These zwei behauptet wird, dass einfache Modellstrukturen mit vielen Daten manchmal erfolgreicher sind als komplexe Modellierungsansätze mit wenigen Daten.

Diese Veränderung in der Modellierung lässt sich gut anhand einer Unterscheidung zwischen phänomenologischer und theoretischer Wissenschaft einordnen, die sich bei einigen Autoren findet, zum Beispiel bei Pierre Duhem (1954) oder Nancy Cartwright (1983). Ein einschlägiges Beispiel für einen phänomenologischen Zugang sind die Ingenieurwissenschaften, die Physik hingegen ist eine klassische theoretische Wissenschaft.

Unterschiede finden sich auf fast allen Ebenen wissenschaftstheoretischer Begrifflichkeiten. Während beispielsweise die Gesetze in phänomenologischen Wissenschaften kausal und kontextuell sind, zielen theoretische Wissenschaften vor allem auf abstrakte, universelle Zusammenhänge ab (s. auch These sechs weiter unten). So sind die Ingenieurwissenschaften, um auf das erwähnte Beispiel zurückzukommen, interessiert an vielfältigem und hochspezialisiertem kausalem Wissen, um technische Probleme lösen zu können, während die Physik vor allem die wenigen grundlegenden Gesetze der Natur erkennen und erforschen will[4], von den Newtonschen Axiomen bis zu den Grundgesetzen der Quantenmechanik. Damit zusammenhängend untersuchen phänomenologische Wissenschaften eine viel

---

3   Im Kern sieht die Modellstruktur folgendermaßen aus: $Pr(e|f) = \mathrm{argmax}_e\, Pr(e)\, Pr(f|e)$, wobei $Pr(e)$ die relative Häufigkeit einer Wortfolge e in der Zielsprache bedeutet und $Pr(f|e)$ die relative Häufigkeit f in der Ursprungssprache, gegeben die Wortfolge e findet sich an der entsprechenden Stelle in der Zielsprache (vgl. „The Unreasonable Effectiveness of Data", Vortrag von Peter Norvig an der UBC, 23.9.2010. http://www.youtube.com/watch?v=yvDCzhbjYWs bei 38:00.).

4   Diese Axiome sind nicht mehr in einem starken Sinne kausal, eine Diskussion dieses Punktes würde hier aber zu weit führen.

größere Breite von Phänomenen als theoretische Wissenschaften. Die Ingenieurwissenschaften beispielsweise schaffen eine Vielzahl von Artefakten, die unter den
unterschiedlichsten Bedingungen ihren jeweiligen Zweck erfüllen sollen. Hingegen
interessiert sich die Physik vor allem für einen sehr beschränkten Bereich exempla-
rischer und paradigmatischer Phänomene wie beispielsweise das Pendel, die schie-
fe Ebene oder in neuerer Zeit Teilchenbeschleuniger. Das geschieht offenbar unter
der Annahme, dass solche paradigmatischen Phänomene die Zusammenhänge in
der Natur in möglichst unverfälschter Form aufzeigen und damit helfen können,
vereinheitlichende Prinzipien in der Natur aufzudecken. Der Anspruch ist, dass
sich mit diesen Prinzipien die Welt verstehen lässt. In theoretischen Wissenschaften geht es daher stark um Erklärung und Vereinheitlichung, in phänomenologischen Wissenschaften um verlässliche Vorhersage und erfolgreichen Eingriff in die
Welt (vgl. These acht unten).

Mit Bezug auf diese Unterscheidung findet datenintensive Wissenschaft vor allem auf phänomenologischer Ebene statt, wie sich am Beispiel maschineller Übersetzung gut erläutern lässt. Der datengetriebene Ansatz verzichtet auf das explizite
Modellieren grammatikalischer Regeln, die Zusammenhänge erklären könnten,
und konzentriert sich stattdessen auf die ‚Vorhersage' brauchbarer Übersetzungen.
In den Textkorpora wird sprachliche Praxis in ihrer kompletten Breite aufgezeichnet, statt sich auf einschlägige Beispiele zu konzentrieren, die konkrete Grammatikregeln gut illustrieren. Und schließlich geht es eben nicht darum, allgemeingültige
Grammatikregeln aufzustellen, sondern Regeln zu finden, die in sehr spezifischen
Kontexten die richtige Übersetzung liefern.

Was die maschinelle Übersetzung angeht, erwies sich aus wissenschaftstheoretischer Sicht offenbar ein induktivistischer Ansatz als erfolgreicher als ein hy-
pothetisch-deduktives Vorgehen, das auf theoretischem Hintergrundwissen und
komplexen Modellannahmen beruht. Das ist insofern bemerkenswert, als induktivistische Ansätze im 20. Jahrhundert generell eher auf Ablehnung stießen – nicht
zuletzt in der Statistik. Viele Statistiker betrachten bis zum heutigen Tag die statistische Herangehensweise als notwendigerweise modellbasiert, was viele induktive
Methoden zum Beispiel aus dem maschinellen Lernen von vornherein ausschließt.
Das führt uns zur dritten These:

*Dritte These: Die klassische Statistik ist der Datenflut nur beschränkt gewachsen,
neue induktive Ansätze sind erforderlich.*

Diese Aussage wird hervorragend illustriert durch einen vieldiskutierten Aufsatz
von Leo Breiman „Statistical Modeling: The Two Cultures" (2001; s.a. Norvig, dieser Band), in welchem er eine Kultur der Datenmodellierung von einer Kultur al-

gorithmischer Modellierung unterscheidet. Im ersten Fall handelt es sich um den klassischen Ansatz in der Statistik, bei dem ein parametrisches stochastisches Modell postuliert wird, das eine funktionale Abhängigkeit zwischen Prädiktor- und Zielvariablen beschreibt. In der Folge wird überprüft, wie gut das Modell den Daten entspricht, ob beispielsweise die Abweichungen von der postulierten Funktion einer Normalverteilung genügen. Lineare Regressionsverfahren sind ein typisches Beispiel für einen solchen Ansatz.

Gemäß der zweiten Herangehensweise, die vor allem auch außerhalb der akademischen Statistik von Informatikern oder Physikern entwickelt wurde, soll ein Algorithmus aus den Prädiktorvariablen die Zielvariable berechnen. Es gibt nun kein Modell mehr im Sinne eines fest eingegrenzten funktionalen Zusammenhangs, dessen Parameter zu bestimmen sind. Vielmehr bleibt der Zusammenhang zwischen Prädiktor- und Zielvariablen weitgehend eine Black Box und die Güte des Algorithmus wird einfach danach bemessen, wie gut die Vorhersagen sind. Beispiele für diese Art der Modellierung sind Entscheidungsbäume oder neuronale Netze. Breiman zufolge entwickeln im Jahr 2001 ungefähr 98% aller professionellen Statistiker Datenmodelle, nur 2% verwenden algorithmische Modellierung. Diese Zahlen haben sich seither sicher geändert.

Ein enger Zusammenhang besteht zu einer weiteren und etwas älteren Unterscheidung in der Statistik zwischen parametrischer und nicht-parametrischer Modellierung.[5] Parametrische Modelle sind dabei durch eine begrenzte Anzahl freier Parameter festgelegt. Bei nicht-parametrischer Modellierung hingegen ist die Modellstruktur soweit offen, dass es keine Beschränkung in der Anzahl relevanter Parameter gibt, wie es beispielsweise bei neuronalen Netzen oder auch Entscheidungsbäumen der Fall ist. Offenbar gibt es eine relativ eindeutige Zuordnung von parametrischen Modellen zu Breimans Datenmodellen, die beide in beträchtlichem Maße auf theoretisches Hintergrundwissen zurückgreifen, und von nicht-parametrischen Modellen zu algorithmischen Modellen, die jeweils viel stärker explorativ und induktiv arbeiten.

Verschiedene Autoren haben darauf verwiesen, dass nicht-parametrische Modellierung erst mit dem Aufkommen der ersten Computer zu einem wissenschaftlich interessanten und erfolgversprechenden Ansatz avancierte. Diese Art der Modellierung ist nämlich fast immer datenintensiv, eben weil nicht mehr nur eine geringe Anzahl von Parametern einer vorgegebenen Modellstruktur bestimmt wer-

---

5 Eine Anfrage beim Google Books Ngram Viewer zeigt, dass Begriffe wie „parametric statistics" und „parametric modeling" erst seit den 50er und 60er Jahren des vergangenen Jahrhunderts verwendet werden, also genau seit der Zeit als die ersten Computer in die Forschung Einzug erhielten.

den muss, sondern die Modellstruktur selber in beträchtlichem Maße unbestimmt ist. Folglich ergeben sich auch unterschiedliche Ansprüche an die Daten. Die möglichst exakte Bestimmung einiger weniger Parameter in der parametrischen Modellierung erfordert gewöhnlich eine überschaubare Anzahl möglichst genauer Datenpunkte, während in nicht-parametrischen Ansätzen die Quantität der Daten im Vergleich zur Qualität wichtiger wird. Zusammenfassend lässt sich also sagen, dass wenn irreduzibel komplexe Phänomene auf der Basis von großen Datensätzen modelliert werden sollen, man es fast immer mit algorithmischer und nicht-parametrischer Modellierung zu tun hat, wie etwa beim datengetriebenen Ansatz im Beispiel maschineller Übersetzung.

Offenbar lassen sich solche nicht-parametrischen Modelle nicht mehr im Rahmen von gekoppelten Gleichungssystemen darstellen, dem mathematischen Paradigma, das die Wissenschaft seit Jahrhunderten beherrscht. In vielen Fällen, insbesondere in der Physik, waren es vor allem partielle Differentialgleichungen, die die Dynamik der Systemvariablen wiedergeben, beispielsweise die Hamiltonschen Gleichungen in der klassischen Mechanik oder die Schrödinger-Gleichung der Quantenmechanik. Solche gekoppelten Gleichungssysteme sind immer parametrische Modelle. Es folgt:

Vierte These: Neue Arten der formalen Repräsentation werden bei der Modellierung irreduzibel komplexer Phänomene notwendig.

Hinzu kommen insbesondere diskrete, beliebig skalierbare Modellierungsstrukturen, vor allem Netzwerke (z.B. Entscheidungsbäume, neuronale Netzwerke, oder Bayessche Netzwerke). Diese Modelle sind häufig so umfangreich, dass sie nicht mehr ohne weiteres händisch darstellbar sind. Damit verändert sich auch das Medium für die Repräsentation. Das Blatt Papier, das Buch oder das menschliche Gehirn werden zunehmend durch den Computer ersetzt, der alleine solche Modelle entwickeln kann.

Zentral für die beiden vorangegangen Thesen ist die Unterscheidung zwischen theoriegeleiteten und explorativen Ansätzen in der Statistik. Interessanterweise gibt es eine verwandte Unterscheidung im Bereich des klassischen Experimentierens, was uns zur nächsten These führt:

Fünfte These: Es gibt starke Analogien zwischen explorativem Experimentieren und datenintensiver Wissenschaft.

Zuerst einmal muss die wichtige Unterscheidung zwischen explorativem und theoriegeleitetem Experimentieren besprochen werden, die erstaunlicherweise erst vor ein paar Jahren explizit in die wissenschaftstheoretische Literatur Eingang gefunden hat, und zwar unabhängig voneinander durch Richard Burian und Friedrich Steinle im Jahr 1997. Dieser späte Zeitpunkt überrascht vor allem deswegen, weil es sich um die grundlegendste Kategorisierung experimenteller Forschung handelt, die vorstellbar ist.

Nichtsdestotrotz finden sich beide Ansichten zur Rolle von Experimenten bereits in den Werken früherer Autoren, es fehlt aber die vergleichende Gegenüberstellung. Karl Popper beispielsweise charakterisiert Experimente ausschließlich als theoriegeleitet: „Der Experimentator wird durch den Theoretiker vor ganz bestimmte Fragen gestellt und sucht durch seine Experimente für diese Fragen und nur für sie eine Entscheidung zu erzwingen." (Popper 2005, S. 84). Ein einschlägiges Beispiel für so ein theoriegeleitetes Experiment ist Arthur Eddingtons Expedition im Jahr 1919 auf die Vulkaninsel Principe vor der westafrikanischen Küste, um bei einer Sonnenfinsternis die von Einsteins allgemeiner Relativitätstheorie vorhergesagte Lichtablenkung im Schwerefeld zu überprüfen.

Eine treffende Beschreibung explorativer Experimente findet sich dagegen in Ernst Machs „Erkenntnis und Irrtum": „Alles, was wir durch ein Experiment erfahren können, ist durch die *Abhängigkeit* und *Unabhängigkeit* der Elemente (oder Umstände) einer Erscheinung voneinander gegeben und erschöpft. Indem wir eine gewisse Gruppe oder auch *ein* Element, willkürlich variieren, ändern sich hiermit auch andere Elemente oder bleiben unter Umständen unverändert. Die Grundmethode des Experimentes ist die Methode der *Variation*." (Mach 1905, S. 199-200; Hervorhebungen im Original) Dieser Ansatz der Variablenvariation entspricht dem klassischen Vorgehen bei den meisten Laborexperimenten. Als zum Beispiel Wilhelm Röntgen Ende des 19. Jahrhunderts eine neue Art von Strahlung entdeckte, stand ihm anders als Eddington keine ausgearbeitete Theorie zur Verfügung. Um das Phänomen theoretisch zu erschließen, veränderte er systematisch alle Variablen, die er als potentiell relevant erachtete, und überprüfte ihren Einfluss auf die Strahlung.

Ein ganz wichtiges Merkmal zur Unterscheidung dieser beiden Grundtypen von Experimenten betrifft also die theoretische Einbindung. Theoriegeleitete Experimente vergleichen Vorhersagen einer bereits detailliert ausgearbeiteten Theorie mit der Erfahrung, während explorative Experimente eine solche Theorie erst erschließen sollen oder zumindest ein System phänomenologischer Gesetzmäßigkeiten. Anders gesagt besteht im ersten Fall eine starke Theoriebeladenheit, im zweiten Fall hingegen fehlt diese, weil das untersuchte Phänomen noch kaum erschlossen ist. Das gilt ebenfalls für die sprachliche Ebene. Beim explorativen Experimentieren müssen die zentralen Begrifflichkeiten zusammen mit den kausalen Abhängigkeiten erst noch entwickelt werden, während sie beim theoriegeleiteten Experimentieren natürlicherweise durch die Theorie vorgegeben sind.

Daraus ergeben sich einige auffällige Gemeinsamkeiten zwischen explorativem Experimentieren und datenintensiver Wissenschaft. Zentral ist die genannte Theoriefreiheit, die beide wissenschaftliche Praktiken kennzeichnet.[6] Wir haben es zudem in beiden Fällen mit einer Logik der Variablenvariation zu tun, das heißt man untersucht den Einfluss sich ändernder Rahmenbedingungen. Ziel dieses Vorgehens ist es die kausale Struktur eines bisher noch nicht verstandenen Phänomens zu erschließen.

Es gibt allerdings auch einige Unterschiede. Aus erkenntnistheoretischer Sicht ist vielleicht am interessantesten, dass datenintensive Wissenschaft zumeist nicht auf experimentell gewonnene Daten, sondern auf Beobachtungsdaten zurückgreift. Ein weiterer wichtiger Punkt betrifft die Komplexität der untersuchten Phänomene. So lassen sich diese in der datenintensiven Wissenschaft meist nicht in eine kontrollierte Laborumgebung zwängen.

Während es allgemein akzeptiert ist, dass exploratives Experimentieren auf kausales Wissen abzielt, steht im Fall datenintensiver Wissenschaft die gerade behauptete zentrale Rolle für Kausalität im Gegensatz zur einschlägigen Literatur. So schreibt Chris Anderson in einem vielzitierten Aufsatz im Technologiemagazin WIRED, dass im Zuge von Big Data Kausalität durch Korrelation ersetzt wird (2008). Viktor Mayer-Schönberger und Kenneth Cukier verbinden mit Big Data „eine Abkehr von der uralten Suche nach Kausalität" (2013, S. 14, unsere Übersetzung; vgl. auch Kap. 4). Das Gegenteil ist der Fall:

Sechste These: Kausalität ist das zentrale Konzept um zu verstehen, warum datenintensive Ansätze wissenschaftlich relevant sein können, u.a. verlässliche Vorhersagen liefern oder effektive Eingriffe erlauben.

Das Argument für diese These ist bestechend einfach und lässt sich gut anhand des folgenden Zitats der Wissenschaftstheoretikern Nancy Cartwright illustrieren: „Auf kausale Gesetze kann man nicht verzichten, weil sie den Unterschied zwischen wirksamen und nutzlosen Strategien begründen." (Cartwright 1979, S. 420; unsere Übersetzung) Eine reine Korrelation, selbst wenn sie empirisch gut belegt ist, beispielsweise in manchen Regionen zwischen Storchenpopulation

---

6   Hacking hat mit Blick auf die relative Theoriefreiheit explorativer Experimente davon gesprochen, dass das Experiment ein Eigenleben hat (1983). Analog könnte man nun auch von einem Eigenleben datenintensiver Wissenschaft sprechen, dass sie also eine gewisse Unabhängigkeit von anderen wissenschaftlichen Praktiken aufweist.

und menschlicher Geburtenrate, lässt sich eben nicht dafür einsetzen Phänomene zu beeinflussen. Auch das Ansiedeln einer großen Zahl von Störchen wird kaum dazu führen, dass die Leute mehr Kinder bekommen. Im Vergleich dazu lässt sich mit Kausalzusammenhängen immer planvoll in die Welt eingreifen. Das gilt auch, wenn diese Zusammenhänge lediglich statistischer Natur sind, wobei dann entsprechende Strategien selbstverständlich nur im Mittel wirksam sind. Wenn man die Leute dazu ermutigt weniger zu rauchen, wird mit großer Wahrscheinlichkeit auch die Lungenkrebsrate in der Bevölkerung sinken.

Wenn es nun in der datenintensiven Wissenschaft darum geht Phänomene zu beeinflussen, dann genügen Korrelationen dafür offenbar nicht, sondern es muss ein Kausalzusammenhang bestehen – beispielsweise wenn Facebook seine Mitglieder dazu bewegen möchte, möglichst viel Zeit in dem sozialen Netzwerk zu verbringen, oder Google erreichen möchte, dass seine Nutzer bestimmte Werbelinks anklicken.

Man könnte an dieser Stelle einwenden, dass es in der datenintensiven Wissenschaft manchmal nicht darum geht Phänomene zu beeinflussen sondern nur um verlässliche Vorhersage und dafür wiederum sollten Korrelationen ausreichen. Man muss dann aber zwei unterschiedliche Typen von Korrelationen unterscheiden: zuerst solche, die sich auf eine gemeinsame Ursache zurückführen lassen, und dann solche, die rein zufällig entstanden sind. Nur im ersten Fall kann eine Korrelation eine verlässliche Vorhersage begründen. Mit dem aktuellen Barometerstand lässt sich relativ gut das Wetter vorhersagen, eben weil es die gemeinsame Ursache Luftdruck gibt. Mit dem Wasserstand in Venedig hingegen lassen sich nicht die Brotpreise in Großbritannien vorhersagen, selbst wenn diese Korrelation in der Vergangenheit empirisch gut belegt ist (Sober 2001). Zufällige Korrelationen treten immer dann auf, wenn eine große Anzahl unterschiedlicher Variablen betrachtet wird. Es zeigt sich also, dass in Einklang mit der sechsten These auch verlässliche Vorhersagen kausal begründet sein müssen, nur dass hierfür nicht unbedingt ein direkter Kausalzusammenhang bestehen muss, sondern ein indirekter Kausalzusammenhang über eine gemeinsame Ursache genügt.

Natürlich ist damit die Kausalitätsproblematik nicht ausreichend abgehandelt, ist doch bekanntermaßen Kausalität einer der umstrittensten Begriffe im Bereich wissenschaftliche Methodik, dem man in einer kurzen Einleitung kaum gerecht werden kann. Man findet in der Geschichte und auch noch in der gegenwärtigen Wissenschaft jedenfalls beide Extreme: prominente Stimmen, die Kausalität vollständig ablehnen, aber auch den entgegengesetzten Standpunkt, dass wissenschaftliches Wissen grundsätzlich kausaler Natur sein muss. Dieser Konflikt lässt sich offenbar nur klären, wenn präzisiert wird, was genau mit Kausalität gemeint ist. Das führt uns zur nächsten These.

Siebte These: Die Rolle von Kausalität in datenintensiver Wissenschaft besteht in ‚Difference-Making‘, dass also die Veränderung bestimmter Rahmenbedingungen eine Veränderung des untersuchten Phänomens hervorruft.

Tatsächlich scheinen viele der gängigen begrifflichen Analysen von Kausalität ungeeignet für den Kontext datenintensiver Wissenschaft. Interventionistische Ansätze (etwa Woodward 2003) setzen einen starken Interventionsbegriff voraus, der sich nur schwer damit in Einklang bringen lässt, dass in datenintensiver Wissenschaft Kausalbeziehungen durch Beobachtungsdaten, das heißt in Abwesenheit einer expliziten Intervention, begründet werden müssen.[7]

Mechanistische Ansätze oder die verwandten Prozesstheorien, eine weitere große Gruppe von Interpretationen, setzen voraus, dass der Zusammenhang zwischen Ursache und Wirkung aufgrund fundamentaler Gesetze oder Prozesse nachvollzogen werden kann (etwa Salmon 1984). Das aber ist in der datenintensiven Wissenschaft gerade nicht möglich. Bedeutet doch die mehrfach erwähnte Theoriefreiheit, dass tieferliegende Gründe für den Zusammenhang zwischen Ursache und Wirkung gemeinhin nicht bekannt sind.

Schließlich lässt sich mit Bezug auf eine dritte große Klasse von Interpretationen, sogenannte Regularitätstheorien der Kausalität, feststellen, dass diese nur schwer vereinbar mit der in datenintensiver Wissenschaft gebräuchlichen variationalen Evidenz sind. Wie zu Anfang des Abschnitts bereits erwähnt, geht es in datenintensiver Wissenschaft primär nicht um die Beobachtung von Regularitäten, bei denen auf Rahmenbedingung B immer Phänomen A folgt, sondern vor allem darum, wie sich Phänomen A unter möglichen Variationen der Rahmenbedingungen verändert.

Tatsächlich müsste man an dieser Stelle in viel größerem Detail auf die Kausalitätsdebatte eingehen, was aus Platzgründen aber nicht möglich ist. Stattdessen soll kurz argumentiert werden, dass Difference-Making entscheidend ist, um die Rolle von Kausalität für datenintensive Wissenschaft zu verstehen. Es wird also untersucht, welche Rahmenbedingungen für ein Phänomen einen Unterschied machen,

---

7   Selbst der relativ schwache Interventionsbegriff von Woodward setzt beispielsweise voraus, dass Interventionen alternative kausale Pfade zwischen Interventions- und Zielvariable eliminieren. Es bleibt damit bei einer ontologisch harten und für den Kontext datenintensiver Wissenschaft völlig ungeeigneten Unterscheidung zwischen experimentellen und Beobachtungsdaten.

ganz wie in John Stuart Mills berühmter Formulierung der Methode des Unterschieds gefordert: „Wenn eine Situation, in der das untersuchte Phänomen auftritt,
und eine andere Situation, in der das untersuchte Phänomen nicht auftritt, bis auf
eine einzige Bedingung völlig gleich sind, wobei diese nur im ersten Fall auftritt;
dann ist diese Bedingung, in der sich die Situationen unterscheiden, die Wirkung,
die Ursache oder ein unabdingbarer Teil der Ursache des Phänomens." (Mill 1886,
S. 256; unsere Übersetzung)

Sowohl die Methode des Unterschieds als auch verwandte kontrafaktische Ansätze zu Kausalität haben eine Vielzahl von Problemen, auf die wir hier nicht weiter
eingehen können.[8] Es lässt sich dennoch festhalten, dass die Methode des Unterschieds den zuvor genannten Kriterien gerecht wird. Sie setzt keinen starken Interventionsbegriff voraus, zumindest im Prinzip sollten sich Kausalzusammenhänge
demnach auch durch Beobachtungsdaten bestimmen lassen. Zudem wird kein theoretisches Verständnis der Zusammenhänge zwischen Ursache und Wirkung verlangt, im Idealfall können kausale Relationen ohne Wissen über zugrundeliegende
kausale Prozesse identifiziert werden. Und schließlich beruht die Methode des Unterschieds nicht auf Regularitäten, sondern auf variationaler Evidenz. Das heißt,
es wird untersucht, welchen Einfluss die Änderung von Rahmenbedingungen hat.

In der einschlägigen Literatur wird die These, dass Korrelationen Kausalität
ersetzen, häufig in einen engen Zusammenhang gestellt mit einer weiteren Charakteristik von datenintensiver Wissenschaft, dass diese die Phänomene angeblich
nicht mehr erklären kann. Ohne Kausalität keine Erklärungen, so lautet die übliche
Schlussweise. Dieser Zusammenhang gilt allerdings nur für bestimmte Kausalitäts-
begriffe. Beispielsweise ist es im Fall mechanistischer Kausalität plausibel, dass der
zugrundeliegende Mechanismus auch den entsprechenden Kausalzusammenhang
erklären kann. Wird Kausalität allerdings, wie hier vorgeschlagen, im Sinne von
Difference-Making verstanden, sind Kausalzusammenhänge nicht grundsätzlich in
einem starken Sinne erklärend. Das führt zur achten These:

*Achte These: Datenintensive Wissenschaft kann durch die Angabe von Ursachen
erklären, aber nicht durch vereinheitlichende Prinzipien.*

Diese These hat viel zu tun mit der bereits eingeführten Unterscheidung zwischen
phänomenologischer und theoretischer Wissenschaft. In diesem Kontext hat Nancy Cartwright zwei unterschiedliche Typen von Erklärung beschrieben: „Es gibt
zwei ganz unterschiedliche Arten von Dingen, die wir tun, wenn wir ein Phänomen

---

8 Siehe bspw. das Manuskript eines der Autoren zu Kausalität als Difference-Making:
   http://philsci-archive.pitt.edu/11913/.

in der Physik erklären. Erstens beschreiben wir seine Ursachen. Zweitens ordnen wir das Phänomen in einen theoretischen Rahmen ein." (Cartwright 1983, S. 16; unsere Übersetzung) Offenbar bleibt die zweite Art von Erklärung weitgehend den theoretischen Wissenschaften vorbehalten, während die erste Art auch in phänomenologischen Wissenschaften zu finden ist.

Wie bereits erwähnt, bewegt sich datenintensive Wissenschaft vornehmlich auf phänomenologischer Ebene, stellt kausale Abhängigkeiten fest, entwickelt aber kaum vereinheitlichende Prinzipien. Das liegt vor allem daran, dass sich datenintensive Wissenschaft meist mit irreduzibel komplexen Phänomenen auseinandersetzt, für die solche Prinzipien nicht existieren. Damit sind selbstverständlich auch keine Erklärungen durch Verweis auf vereinheitlichende Prinzipien möglich, wobei man hinzufügen muss, dass diese vereinheitlichende Art von Erklärung in vielen Wissenschaften als grundlegend und geradezu unverzichtbar angesehen wird. Das gilt insbesondere für die Physik. Eine zufriedenstellende Erklärung der Gezeiten zum Beispiel sollte letztlich auf allgemeine Gesetze der Mechanik und Gravitation verweisen.

In phänomenologischen Wissenschaften lässt sich nur eine viel schwächere Form von Erklärung realisieren, nämlich durch Verweis auf Ursachen im Sinne des beschriebenen Difference-Making. So könnte man das Auftreten der Gezeiten auch durch die Stellung von Mond und Sonne erklären. Solche Erklärungen können begründen, warum bestimmte Vorhersagen erfolgreich sind, ihnen fehlt aber weitgehend der vereinheitlichende Aspekt. Sie lassen sich daher kaum als Anhaltspunkt verwenden, wenn es um das Verständnis verwandter Phänomene geht. Ein möglicher Zusammenhang zwischen einem fallenden Stein auf der Erde und den Gezeiten ließe sich derart jedenfalls nicht etablieren.

Das Aufkommen datenintensiver Wissenschaft scheint also zur Folge zu haben, dass verschiedene anwendungsnahe Disziplinen wie die Medizin oder die Sozialwissenschaften, welche in Anlehnung an die fundamentalen Naturwissenschaften häufig als theoretische Wissenschaften aufgefasst wurden, stärker als phänomenologische Wissenschaften wahrgenommen werden. Gerade für irreduzibel komplexe Phänomene in diesen Bereichen scheinen wissenschaftliche Erklärungen im klassischen Sinne nicht möglich, kausale Beeinflussung und Vorhersage hingegen schon. Und da in Zukunft nur der Computer und weniger der Mensch die komplexen Zusammenhänge überblicken kann, gilt:

Neunte These: Nicht zuletzt durch die Automatisierung von Wissenschaft verändert sich grundlegend die Rolle von wissenschaftlichen Experten.

Wir müssen uns offenbar damit abfinden, dass aus prinzipiellen Gründen, die in den Gegebenheiten des menschlichen Erkenntnisapparats liegen, viele der wissenschaftlichen Prozesse, die im Computer ablaufen, nicht mehr nachvollziehbar sind. Ein gutes Beispiel ist der Beweis des Vier-Farben-Theorems, bei dem es nur noch einem Computer möglich ist, alle verbleibenden Varianten durchzuspielen. In manchen Bereichen ist der daraus resultierende Wandel in der Rolle menschlicher Experten bereits im Gange, wenn wie geschildert in Übersetzungsprogrammen explizites grammatikalisches Wissen keine Rolle mehr spielt oder wenn Amazons Buchempfehlungen nicht mehr von Kulturjournalisten verfasst, sondern algorithmisch generiert werden, was zumindest aus finanzieller Sicht extrem erfolgreich ist.

Diese Veränderungen und die Auswirkungen für den Arbeitsmarkt sind ein gutes Beispiel für die gesellschaftliche Bedeutung datenintensiver Wissenschaft. Auch die ethische Dimension der aktuellen Entwicklungen geht weit über die allgegenwärtige Debatte zu Privatheit hinaus. Letztere ist unbestrittenermaßen wichtig, aber durch die verbesserte Vorhersagefähigkeit unter anderem im sozialen und medizinischen Bereich sowie durch die Verwendung datengetriebener Vorhersage bei der Entwicklung neuer Produkte wie dem automatisierten Fahren entsteht eine Vielzahl völlig offener ethischer Fragen.

Es bleibt festzuhalten:

Zehnte These: Ethik und Erkenntnistheorie datenintensiver Wissenschaft sind schwer voneinander zu trennen.

Ethische Probleme im Zusammenhang mit datenintensiver Wissenschaft können nur dann angegangen werden, wenn die erkenntnis- und wissenschaftstheoretischen Zusammenhänge durchdrungen sind. Letztere loten den Rahmen des Machbaren aus, innerhalb dessen dann festgelegt werden kann, was ethisch gewollt ist.

1.3   Überblick über den Band

Den eben besprochenen Themen folgend, gliedert sich der Band in vier große Abschnitte, „Big Data und die Wissenschaften", „Berechenbarkeit", „Komplexität und Information" sowie „Ethische und politische Perspektiven".

Der erste Teil beginnt mit einem ursprünglich als Blog-Beitrag veröffentlichten Essay von Google Forschungschef Peter Norvig. Dieser Aufsatz hat für einen der Herausgeber (Wolfgang Pietsch) eine entscheidende Rolle dabei gespielt, wichtige erkenntnistheoretische Aspekte und Zusammenhänge datenintensiver Wissenschaft zu verstehen. Der nachfolgende Beitrag der Kopenhagener Technikphilosophen Gernot Rieder und Judith Simon untersucht ebenfalls zentrale epistemologische Fragestellungen im Zusammenhang mit Big Data, beispielsweise Theoriefreiheit oder Objektivität. Sie hinterfragen kritisch die These eines neuen Empirismus und beleuchten einige politische und gesellschaftliche Konsequenzen. Der dritte Aufsatz des Cottbuser Technikphilosophen Klaus Kornwachs nimmt auch eine eher erkenntniskritische Perspektive ein und argumentiert für die Tugenden und Vorteile traditioneller wissenschaftlicher Modellbildung.

Die folgenden Beiträge fokussieren auf ausgewählte erkenntnistheoretische Aspekte datenintensiver Wissenschaft. Der texanische Mathematiker und Bioinformatiker Edward Dougherty argumentiert, dass auch im Zeitalter von Big Data in vielen Bereichen eher ein Mangel an Daten das Problem ist, wenn es darum geht, komplexe stochastische und nichtlineare Phänomene zu modellieren, beispielsweise die Zellregulierung. Die Statistiker Anne-Laure Boulesteix, Roman Hornung und Willi Sauerbrei untersuchen in ihrem Beitrag, welcher Interpretationsspielraum bei der Analyse großer Datensätze bleibt. Insbesondere besprechen sie verschiedene Ansätze, wie sich ein Missbrauch durch das sogenannte „fishing for significance" vermeiden lässt. Im letzten Beitrag des ersten Teils untersuchen die Münchner Mathematiker Nadine Gissibl, Claudia Klüppelberg und Johanna Mager die Bedeutung großer Datensätze für die Analyse extremer Risiken, beispielsweise beim sicheren Landen von Flugzeugen.

Der zweite Teil des Bandes betrachtet Aspekte der Berechenbarkeit der Welt aus unterschiedlichen disziplinären Perspektiven. Er beginnt mit einem Essay, in welchem der russische Philosoph Andrei Rodin untersucht, inwieweit neben temporalen Aspekten auch die räumliche Verortung in Berechenbarkeitsmodellen berücksichtigt werden sollte. Es folgt ein Aufsatz des Biophysikers Leo van Hemmen über die Möglichkeit mathematischer Modellbildung in komplexen biologischen Systemen, insbesondere im Bereich Neurobiologie. Der Physiker Konrad Kleinknecht thematisiert die Frage, inwieweit in Anbetracht vielfältiger Nicht-Linearitäten Klimaberechnungen zuverlässig sein können. Der Wissenschaftsmoderator, Physiker und Naturphilosoph Harald Lesch stellt in seinem Aufsatz weitergehende Fragen nach dem Zusammenhang zwischen Physik und Metaphysik sowie den Grenzen physikalischer Erkenntnis, die sich unter anderem darin zeigen, dass der Anfang des Universums physikalisch nicht berechenbar ist.

Im Weiteren verschiebt sich der Fokus weg von den Naturwissenschaften hin zu den Sozialwissenschaften. Chadwick Wang, ein chinesischer Forscher im Bereich Science and Technology Studies, schildert eine kontroverse Debatte unter verschiedenen Fraktionen chinesischer Wissenschaftler, ob sich die Gesellschaft berechnen lässt oder nicht. Der Stuttgarter Physiker und Pionier der Soziophysik Wolfgang Weidlich gibt einen kurzen Überblick über seinen Ansatz und diskutiert einige wichtige methodologische Probleme. Schließlich skizziert der Augsburger Wissenschaftstheoretiker Theodor Leiber eine SWOT-Analyse neuer methodischer Ansätze im Bereich Computational Social Science und Big Data.

Der Salzburger Psychologe Günter Schiepek, der wie Wolfgang Weidlich aus dem Umfeld der Stuttgarter Schule der Synergetik stammt, untersucht in seinem Aufsatz Fragen der Komplexität, der Berechenbarkeit und des Einsatzes von großen Datenmengen in der Psychotherapie. Die Paderborner Philosophin Ruth Hagengruber diskutiert Unterschiede zwischen menschlichem und maschinellem Wissen und wirft dabei die Frage auf, inwieweit Maschinen Kreativität zugeschrieben werden kann. Der zweite Teil schließt dann mit einer historischen Perspektive des Münchner Wissenschaftstheoretikers Tobias Jung, der der Frage nachgeht, welchen Stellenwert Aspekte der Berechenbarkeit im Werk von Immanuel Kant einnehmen. Ähnlich wie Harald Lesch richtet Jung den Blick vor allem auf die Grenzen menschlicher Erkenntnisfähigkeit.

Der dritte Teil widmet sich genauer dem Begriff der Komplexität, dem großen Gegenspieler der Berechenbarkeit. Zuerst gibt die Moskauer Philosophieprofessorin und Komplexitätstheoretikerin Helena Knyazeva einen Überblick über grundlegende Prinzipien und Aspekte komplexer Systeme und unterstreicht die interdisziplinäre Natur jedweder Komplexitätswissenschaft. Der Darmstädter Physiker und Philosoph Jan Schmidt arbeitet bisher kaum beachtete Aspekte im wissenschaftstheoretischen Werk von Pierre Duhem heraus, die ihn als einen der Pioniere moderner Komplexitätsforschung erscheinen lassen. Der Technikphilosoph Alfred Nordmann, ebenfalls aus Darmstadt, vergleicht zwei sehr unterschiedliche Ansätze von Stuart Kauffman und Brian Goodwin zur Frage, wie sich die Biologie auf ihrem Weg hin zu einer Wissenschaft des Komplexen verändert. Daran anschließend untersucht der Oxforder Informatiker und Philosoph Hector Zenil, inwieweit die Theorie algorithmischer Information dazu beitragen kann, komplexe Systeme zu analysieren. Abschließend diskutiert der Physiker und Philosoph Holger Lyre die Frage, ob der Informationsbegriff eine fundamentale Rolle in den Naturwissenschaften spielen sollte.

Im vierten und letzten Teil weitet sich der Blick hin zu ethischen und politischen Perspektiven. Julian Nida-Rümelin, der Münchner Philosoph und ehemalige Staatsminister für Kultur und Medien analysiert, ob wir in Anbetracht autonomer Roboter den Verantwortungsbegriff überdenken müssen. Im nachfolgenden Beitrag untersucht der Konstanzer Philosoph Jürgen Mittelstrass den eng mit Komplexitätsfragen verwandten Begriff der Emergenz und zeigt insbesondere wie Emergenz im Zusammenspiel der Disziplinen zum Entstehen von Neuem und

Unvorhergesehenem beitragen kann. Der Münchner Wirtschaftsethiker Christoph Lütge skizziert den Forschungsansatz einer experimentellen Ethik und stößt die Frage an, inwieweit sich auch ethische Probleme empirisch untersuchen und berechnen lassen. Die Philosophin und Informatikerin Sabine Thürmel analysiert das Wechselspiel von Autonomie und Kontrolle in Big Data basierten Systemen. In einem auch methodisch interessanten Ansatz mit einem stark literaturwissenschaftlichen Bezug geht der Physiker und Energieforscher Thomas Hamacher der Frage nach, ob man aus einem Roman der Gegenwartsliteratur neue Ansätze zu Energie und Nachhaltigkeit herauslesen kann. Der Band schließt mit einer Untersuchung des japanischen Sozialphilosophen Naoshi Yamawaki, inwieweit eine kaum berechenbare Katastrophe wie der Atomunfall von Fukushima ein Umdenken bezüglich der Rolle der Ethik im Zusammenspiel der Wissenschaften erfordert.

Literatur

Anderson, Chris. 2008. The End of Theory: The Data Deluge Makes the Scientific Method Obsolete. *WIRED Magazine* 16/07. http://www.wired.com/science/discoveries/magazine/16-07/pb_theory.

Breiman, Leo. 2001. Statistical Modeling: The Two Cultures. *Statistical Science* 16 (3): 199-231.

Burian, Richard. 1997. Exploratory Experimentation and the Role of Histochemical Techniques in the Work of Jean Brachet, 1938-1952. *History and Philosophy of the Life Sciences* 19: 27-45.

Cartwright , Nancy. 1979. Causal Laws and Effective Strategies. Noûs 13 (4): 419-437.

Cartwright, Nancy. 1983. *How the Laws of Physics Lie*. Oxford: Oxford University Press.

Duhem, Pierre. 1954. *The Aim and Structure of Physical Theory*. Princeton, PA: Princeton University Press.

Hacking, Ian. 1983. *Representing and Intervening*. Cambridge: Cambridge University Press.

Hacking, Ian. 1990. *The Taming of Chance*. Cambridge: Cambridge University Press.

Halevy, Alon, Peter Norvig und Fernando Pereira. 2009. The Unreasonable Effectiveness of Data. *IEEE Intelligent Systems* 24(2): 8-12.

Kolmogorow, Andrei N. 1983. Combinatorial foundations of information theory and the calculus of probabilities. *Russian Mathematical Surveys* 38 (4): 29-40.

Kuhlmann, Meinard. 2011. Mechanisms in Dynamically Complex Systems. In *Causality in the Sciences*, hrsg. P. McKay Illari, F. Russo, and J. Williamson, 880-906. Oxford: Oxford University Press.

Laney, Doug. 2001. 3D Data Management: Controlling Data Volume, Velocity, and Variety. Research Report. http://blogs.gartner.com/doug-laney/files/2012/01/ad949-3D-Data-Management-Controlling-Data-Volume-Velocity-and-Variety.pdf.

Leonelli, Sabina. 2012. Introduction: Making sense of data-driven research in the biological and biomedical sciences. *Studies in History and Philosophy of Biological and Biomedical Sciences* 43 (1): 1-3.

Mach, Ernst. 1905. *Erkenntnis und Irrtum. Skizzen zur Psychologie der Forschung.* Leipzig: Johann Ambrosius Barth.

Mayer-Schönberger, Viktor & Kenneth Cukier. 2013. *Big Data.* London: John Murray.

Mill, John S. 1886. *System of Logic.* London: Longmans, Green & Co.

Mitchell, Sandra. 2008. *Komplexitäten. Warum wir erst anfangen, die Welt zu verstehen.* Frankfurt a.M.: Suhrkamp.

Pietsch, Wolfgang. 2015. Aspects of Theory-Ladenness in Data-Intensive Science. *Philosophy of Science* 82 (5): 905-916.

Pietsch, Wolfgang. 2016. The Causal Nature of Modeling with Big Data. *Philosophy & Technology* 29 (2): 137-171.

Popper, Karl. 2005. *Logik der Forschung.* Tübingen: Mohr Siebeck.

Salmon, Wesley. 1984. *Scientific Explanation and the Causal Structure of the World.* Princeton: Princeton University Press.

Sober, Elliott. 2001. Venetian Sea Levels, British Bread Prices, and the Principle of the Common Cause. *British Journal for the Philosophy of Science* 52: 331-346.

Steinle, Friedrich. 1997. Entering New Fields: Exploratory Uses of Experimentation. *Philosophy of Science* 64: S65-S74.

Woodward, James. 2003. *Making Things Happen: A Theory of Causal Explanation.* Oxford: Oxford University Press.

1. Introduction: Ten Theses on Big Data and Computability

Wolfgang Pietsch, Jörg Wernecke

1.1 An experiment in interdisciplinarity

This volume is an experiment. It addresses various aspects of computability in a thematic and methodological breadth as it was once demanded by Francis Bacon at the beginning of modern Western science, but as it is rarely practiced today in a time of ever-increasing scientific specialization. At the beginning of the third millennium, the question of the computability of the physical, the psychological, and the social world arises once again—mainly due to advances in information and communication technology and the concurrent data deluge not only in the sciences but also in many other areas, e.g. in the economy and even our personal lives.

The volume is *interdisciplinary*, it tries to gather a large number of perspectives —from the formal sciences like mathematics, statistics, or computer science to the natural and social sciences to the humanities including philosophy. Furthermore, the volume assumes a *foundational* viewpoint, in that many of the issues that are discussed concern fundamental concepts and methods. Finally, it is *application-oriented* by examining a current issue in the empirical sciences, namely to what extent the data deluge has an impact on scientific concepts and methodology.

These three aspects also characterize the work of the Munich philosopher of science Klaus Mainzer, to whom we dedicate this volume at the occasion of his retirement. The themes of the book reflect quite well some of the key research concerns of Klaus Mainzer, who has repeatedly demonstrated a keen sense for uncovering innovative questions at the interface between philosophy and science. For example, he was among the first philosophers of science to reflect critically on developments in complexity research or to recognize the epistemological significance of Big Data.

Mainzer never approaches such issues from a philosophical and epistemological viewpoint only, but in his multifaceted scientific career has always undertaken great efforts for an interdisciplinary exchange. This grounding in scientific practice is a crucial premise that abstract philosophical reflection can be relevant beyond the narrow confines of specialized debates. Relatedly, Mainzer has always been a great communicator not only with other disciplines, but he has in many lectures, interviews, and books also approached a wider public.

If one opens the science pages of major newspapers today, outsiders may get the impression that interdisciplinary research constitutes the norm rather than the exception. This, however, is not the case. The path to academic success is still almost exclusively organized in a disciplinary manner. Besides, the problems of interdisciplinarity can only be understood if acknowledging the existence of various scientific cultures that start from different knowledge bases and in particular have different epistemic interests and value systems.

Despite all these difficulties, there are many cross-sectoral issues that can be addressed only in an interdisciplinary manner, ultimately because relevant knowledge and understanding are spread among different academic fields. A novel view on computability in the age of Big Data is certainly an excellent example for such a cross-sectoral issue.

1.2   Ten theses on computability and data-intensive science

The question of the computability of nature is as old as science itself, as old as human efforts to predict the physical and social environment in order to survive. Nevertheless, the perspective has repeatedly changed over the centuries, often in parallel with significant developments regarding both scientific method and the availability of data. The inductivism of a Francis Bacon as well as the novel focus on experimentation and observation during the Renaissance fostered the breakthrough of modern empirical science liberating itself from the dogmas of scholasticism. The beginnings of modern statistics, to name another example, are closely tied with efforts to cope with the 'Big Data' of the 18th and 19th centuries, when many administrative institutions and statistical offices were founded resulting in a flood of printed numbers (Hacking 1990, Ch. 4). There are many indications that today we are once again at a turning point where novel information technologies in combination with huge data sets raise the question of the computability of the physical and social world.

As often in the history of science, Big Data as a slogan and leitmotif could prevail because the concept is so ambiguous and malleable that it can be interpreted differently in various fields from computer science to economics. While in the sciences conceptual inaccuracies can occasionally be fruitful, philosophy calls for greater precision. Therefore we must first clarify the meaning that is intended with the terms Big Data and data-intensive science.

In the literature, Big Data is usually characterized with reference to the so-called 3 Vs, i.e. the amount of data (Volume), the data rate (Velocity) and, finally, the heterogeneous structure of the data (Variety) (see Laney 2001). This approach has sometimes been extended to include 5 or even 7 Vs. While such a definition seems adequate when focusing on the technical challenges of large datasets, it proves largely unsuitable for addressing epistemological and methodological issues in connection with data-intensive science. The main problem lies in the exclusive focus on the data and its structure while neglecting the algorithms and methods of data processing. Quite plausibly, these latter are much more important to establish novel epistemological aspects in data-intensive science.

Therefore, let us highlight in the following two methodological aspects of data-intensive science. First, there is the often-invoked assertion that Big Data is supposedly able to map all individuals of a population, what is often summarized as the "N = all" quality of Big Data (e.g. Mayer-Schönberger & Cukier 2013, p. 26-31). If one were to accept this characteristic as part of a definition of Big Data, then the aforementioned objection is mitigated since "N = all" has clear methodological implications. For example, conventional statistical approaches typically start with samples, which should be as representative as possible of a population with respect to a given research question. On the basis of such samples, probabilities can be determined for other individuals in the studied population. Such a methodologically sophisticated process of extrapolation apparently no longer happens in the case of a data-intensive science defined by means of "N = all". Instead, the novel approach seems much simpler resembling a process of looking up in a table of all possibilities what is the case for a particular individual.

Of course, it is unrealistic to capture in a data set all the individuals of a population, for example future cases can hardly be taken into account. It is also implausible that there is no reduction of complexity at all, for example by eliminating completely identical data points. Another undesirable consequence of "N = all" is that for simple systems a small number of data already covers all possibilities, for example, the data points (switch on, light on), (switch off, light off) already contain the full variation of a system consisting of a switch and a light bulb. Obviously, it is not appropriate to speak of Big Data or data-intensive science in such cases. To work around this problem, one could explicitly restrict the definition to sufficiently complex phenomena, which then necessarily require large amounts of data.

A proposal in the spirit of "N = all", which takes into account the mentioned objections, is:

Data-intensive science refers to the systematic collection and analysis of data sets which represent a large part of all possible variations of a complex phenomenon[1], such that the relevant causal structure can be inferred algorithmically without further theoretical background assumptions.

Thus, data-intensive science is above all an inductivist research program. We will later return to some details of this definition, in particular the relative theory-independence and the focus on causality. Furthermore, substantial similarities between data-intensive science and exploratory experimentation will be pointed out.

The second methodological aspect concerns the automation of the entire scientific process (see for example Leonelli 2012). This has the immediate consequence that science takes place under epistemic premises that fundamentally differ from those of the human cognitive apparatus, particularly as regards storage capacity and processing speed of the data. Thus, scientific knowledge can be generated, which is no longer accessible to the human mind. In particular, a 'machine science' can examine phenomena, which from the perspective of a 'human science' are too complex to make reliable predictions or to establish effective interventions, because these phenomena involve too many variables or because the dependencies between the variables cannot be represented by simple functions. One might thus specify that the phenomena studied in data-intensive science commonly are so complex that automated methods yield more accurate predictions compared with traditional approaches of a 'human science'.

In the following, the topic of computability shall be considered from the perspective of data-intensive science in ten theses. Let us start with the big promise:

First thesis: Big Data leads to the increasing predictability of complex phenomena, especially to more reliable short-term predictions.

---

1   In this context, one often speaks of variational evidence.

This property follows from the two above-mentioned characteristics that, on the one hand, reliable predictions are feasible if the data reflect a sufficient amount of different states of a complex phenomenon, and that, on the other hand, the change in epistemic conditions, particularly the improved ways of storing and processing data, allows for new approaches to analyze complexity.

Applications in data-intensive science often exhibit the following structure: A data set is given linking a large number of predictor variables with a smaller number of response variables. Using a part of the data set, called the training set, a model of the data is algorithmically developed, which is then validated with the aid of the remaining data, called the test set. For example, the results of an Internet search may be determined using the search history or products may be recommended in an online store based on user profiles and past purchases. The crucial novelty is that due to the automation of data collection and analysis many more variables as well as data points can be taken into account compared with conventional statistical approaches. In this twofold sense, we are thus dealing with Big Data.

Let us now take a brief look at the concept of complexity. First, one might distinguish complexity in the descriptions *(descriptive* complexity) from complexity in the phenomena *(phenomenological* complexity). The best-known example of the first type is the so-called Kolmogorov complexity, according to which complexity is perceived as "the minimal length of a programme by which the object x can be obtained following the programming method S." (Kolmogorov 1983, p. 33) According to Kolmogorov, there is a close relationship between complexity and randomness, as random objects exhibit maximum complexity. By contrast, a small complexity must be attributed to objects that can be derived from a few basic assumptions. Furthermore, such descriptive complexity depends on the specific representation, in particular on the programming method, the assumptions made and the allowed derivation steps.

Herein lies the main difference compared with phenomenological complexity, which is located in the phenomena themselves, and thus should ideally be objective and independent of the chosen representation. In the case of phenomenological complexity it is plausible to introduce a further distinction between *emergent* and *irreducible* complexity.[2] In the first case the complex phenomenon results from a relatively simple system of laws or instructions, i.e. from simple basic phenomena that can be accounted for by only a few variables and well defined dependencies between these variables. The complex dynamics then is emergent, it usually results from nonlinearities, which often imply a strong sensitivity in the system behavior

---

2   A closely related distinction between compositional and dynamic complexity is developed by Meinard Kuhlmann (2011; see also Mitchell 2008).

with respect to small variations in the initial conditions. Typical systems with emergent complexity are treated in the physics of non-equilibrium and in chaos theory, for example, the double pendulum, simple predator-prey systems, or the Rayleigh-Bénard convection.

In contrast, irreducible complexity is characterized precisely by the fact that such phenomena cannot be reduced to a few equations with a small number of variables and clearly defined dependencies. Rather, one is confronted with a complicated arrangement of a large number of very diverse components interacting according to complicated causal interdependencies. Here, dynamic complexity is not emergent in the sense that it can be derived from an underlying simple model. Rather, the system itself is irreducibly complex. Of course, all the characteristics of emergent complexity may still occur such as nonlinearities, feedback loops, or the lack of additivity. But crucially, the representation of the phenomenon cannot be reduced to a few simple assumptions. Relatedly, in the case of irreducible complexity one can no longer introduce a strictly hierarchical structure of various levels of coarse-grained description, as it exists for example in physics: quarks, elementary particles, atoms, molecules, etc. Various interactions between these levels are possible, ultimately leading to a de facto dissolution of the description levels.

It is particularly this latter form of complexity for which data-intensive science constitutes a great advantage. After all, in order to properly represent irreducibly complex dependencies, which cannot be fully grasped by the human cognitive apparatus, one necessarily requires a large amount of data. From what we know today, such complex phenomena occur in almost all application-oriented sciences, especially the social sciences, medicine, or biology. By contrast, phenomena in the theoretical sciences, especially in physics, are usually only complex in the emergent sense, which in part already results from the deliberate focus on a small number of supposedly paradigmatic phenomena (see discussion of the second thesis).

With respect to the subject of the collected volume, a related and just as many-faceted term should also be briefly analyzed, namely computability. In computer science, computability mainly refers to the solvability of mathematical problems. If for a particular field, e.g. arithmetic, general statements or axioms are given, can one determine for each proposition in this field, whether it is true or false? This form of computability addresses the question what can be concluded from a small number of general statements.

Especially for phenomena of irreducible complexity the mentioned conception of computability is not appropriate, precisely because by definition there are no general axioms or fundamental laws. Computability in the context of data-intensive science then is not so much about deriving valid statements from a limited number of universal propositions, but rather about predicting whether something is the case in the world or not, based on a data set of individual events that is at least in principle arbitrarily extensible. One might again roughly classify these two viewpoints as descriptive computability on the one hand and phenomenological computability on the other. Obviously, in the case of phenomenological computability, the much discussed problem of undecidable propositions no longer occurs, for the simple reason that the interest shifts to propositions which are either realized in the world or not, i.e. which always have a determined truth value. Thus, different concepts of truth are employed in both cases.

Now, the fact that data-intensive science analyzes irreducibly complex phenomena is an important reason that changes in modeling occur (see also Norvig, this volume):

Second thesis: A change in modeling occurs from heavily theory-laden approaches with little data to simple models using a lot of data.

Bearing in mind the distinction between emergent and irreducible complexity, this development is hardly surprising. When dealing with emergent complexity, theoretical background knowledge can fix the basic structure of the dependencies between the individual variables. For example, the relevant equations of the double pendulum can be derived from the fundamental laws of mechanics. Then, a few sufficiently accurate and well-chosen data points will suffice to determine the additional parameters of the model. For example, an ordinary double pendulum has four parameters, the two masses and the lengths of the two arms.

However, in the case of irreducible complexity, more or less by definition, the structure of the equations cannot be determined using theoretical background assumptions, but must instead be inferred from the data, if a sufficient amount is available. This leads to the mentioned inductivist and largely theory-free approach of data-intensive science.

The classic example for the change in modeling as depicted in the second thesis stems from linguistics (Halevy et al. 2009). From a much simplified perspective, two distinct approaches exist in the field of machine translation, one rule-based, the other data-driven. The first proceeds by modeling the different languages in terms of all their grammatical rules. A translation program must then recognize by means of these rules the grammatical structure of a sentence in the source language, then using an ordinary dictionary translate each word of the sentence into the target language, and ultimately implement the sentence into the grammatical structure of the target language. However, this historically prior approach quickly reached its limits. The main reason for its failure presumably lies in the (irreducible) complexity of language. There are just too many grammatical rules and again too many exceptions to these rules as that it would be possible to manually implement all of them into a computer program.

Somewhat surprisingly, a data-driven approach turned out much more successful, which completely refrains from an explicit modeling of the grammatical structure. Instead, it relies on large text corpora, derived for example from the Google Books project or from the Internet. The predetermined model structure is now far less complex than in the rule-based approach. Instead of modeling a large number of grammatical rules, a simple Bayesian algorithm is used that determines the most probable translation based on the relative frequencies of word sequences in the different corpora.[3] Machine translation still exhibits serious weaknesses, but the data-driven approach makes it possible today to at least grasp the basic ideas of a text in a foreign language. Thus, in the example one can observe just as alleged in the second thesis that simple model structures with a lot of data are sometimes much more successful than complex modeling approaches with little data.

This change in modeling can be further illustrated by means of a distinction between phenomenological and theoretical science that can be found with a number of authors, e.g. Pierre Duhem (1954) or Nancy Cartwright (1983). Pertinent examples for a phenomenological approach are the engineering sciences, while physics, on the other hand, is a classic theoretical science.

Differences exist on almost all epistemological levels. For example, while the laws in phenomenological sciences are causal and contextual, theoretical sciences mainly aim at abstract and universal relationships (see also thesis six below). Thus, the engineering sciences, to use the aforementioned example, are interested in diverse and highly specialized causal knowledge to solve technical problems, while physics mainly wants to discover and explore a small number of fundamental laws of nature[4], from Newton's axioms to the fundamental laws of quantum mechanics. Relatedly, phenomenological sciences investigate a much greater breadth of phenomena compared with theoretical sciences. The engineering sciences for example create an enormous variety of artifacts that are to fulfill their respective purposes under an extremely diverse range of conditions. By contrast, physics is interested mostly in a very limited range of exemplary and paradigmatic phenomena such as

---

3   In essence, the model structure is as follows: $Pr(e|f) = \mathrm{argmax}_e\, Pr(e)\, Pr(f|e)$, where $Pr(e)$ denotes the relative frequency of a word sequence e in the target language and $Pr(f|e)$ the relative frequency of the word sequence f in the source language, given that the word sequence e can be found at the corresponding position in the target language (cf. 'The Unreasonable Effectiveness of Data', talk given by Peter Norvig at UBC, 23.9.2010, http://www.youtube.com/watch?v=yvDCzhbjYWs at 38:00).

4   These laws are no longer in a strong sense causal, a discussion of this point, however, would lead us too far astray.

the pendulum, the inclined plane or, in more recent times, particle accelerators. Apparently, an underlying assumption is that such paradigmatic phenomena demonstrate the relationships in nature in a pristine manner and thereby help uncover unifying principles of nature. The expectation is that the world can be understood in terms of these unifying principles. Thus, to a strong degree, theoretical sciences aim at explanation and unification, while phenomenological sciences mainly want to provide reliable predictions and justify successful interventions in the world (see thesis eight below). In other words, the epistemic aims differ substantially.

With respect to this distinction, data-intensive science chiefly remains on the phenomenological level, as can clearly be seen in the example of machine translation. After all, the data-driven approach in machine translation dispenses with the explicit modeling of grammatical rules, which could otherwise be used for explanations, and instead focuses on the 'prediction' of useful translations. In the employed text corpora, linguistic practice is recorded in its full breadth instead of focusing on examples that might illustrate particularly well specific grammatical rules. Finally, the aim of the data-driven approach is not to establish a small number of universally valid grammatical rules, but to find rules that in very specific contexts provide the correct translation.

To summarize, in the example of machine translation an inductive approach proved much more successful than a hypothetico-deductive approach based on theoretical background knowledge and complex modeling assumptions. This is insofar remarkable as inductivist approaches were mostly frowned upon in the 20th century—not least in statistics. To this day, many statisticians consider any statistical approach as necessarily involving sophisticated modeling, thereby excluding from the outset many inductive methods, for example from machine learning. This leads us to the third thesis:

Third thesis: Conventional statistics is only partly equipped to deal with the data deluge, novel inductive methodology is necessary.

This statement is well illustrated by a much-discussed essay of Leo Breiman "Statistical Modeling: The Two Cultures" (2001; cp. also Norvig, this volume), in which he distinguishes a culture of data modeling from a culture of algorithmic modeling. The first denotes the conventional approach in statistics, in which informed by background theory a parametric stochastic model is postulated usually depicting an explicit functional relationship between predictor and response variables. Subsequently, it is examined how well this model matches the data, e.g. whether the deviations from the postulated function satisfy a normal distribution. Linear regression methods are a typical example of such a data modeling approach.

By contrast, according to the algorithmic modeling approach, which was developed mainly outside of academic statistics for example by computer scientists or physicists, an algorithm is used to calculate the response variables from the predictor variables. There is no model anymore specifying a clearly defined functional relationship, the parameters of which are to be determined. Rather, the relationship between predictor and response variables remains largely a black box and the quality of the algorithm is measured just by how precise the predictions turn out. Examples of this second type of modeling are decision trees or neural networks. According to Breiman, in 2001 about 98% of all professional statisticians use data modeling, while only 2% employ algorithmic models. These figures have surely changed since then.

A closely related, somewhat older distinction in statistics is between parametric and non-parametric modeling.[5] Parametric models are fixed by a limited number of free parameters, while the structure of non-parametric models is undetermined in the sense that there is no restriction in the number of possible parameters, as is the case for neural networks or decision trees. Obviously, parametric models are closely related with Breiman's data models, both of which rely to a considerable extent on theoretical background knowledge, while non-parametric models are quite similar to algorithmic models, both proceeding in a strongly exploratory and inductive manner.

Several authors have pointed out that non-parametric modeling only became scientifically interesting and promising with the advent of the first computers. In fact, this type of modeling is almost always data-intensive, because the model structure itself is indeterminate to a considerable extent and has to be inferred from the data. Consequently, there are also different requirements regarding the data. For the determination of a few parameters in parametric modeling usually a relatively small number of precise data points suffice, while in non-parametric approaches the quantity of data becomes more important than the precision of the individual data points. In summary, then, if irreducibly complex phenomena are to be modeled on the basis of large data sets, one almost always deals with algorithmic and non-parametric modeling, as is the case for example in data-driven machine translation.

Apparently, such non-parametric models no longer can be represented in terms of coupled systems of equations, the mathematical paradigm that has dominated science for centuries. Particularly in physics, partial differential equations reflected

---

5   A query to the Google Books Ngram Viewer shows that terms such as "parametric statistics" and "parametric modeling" have been used only since the 50s and 60s of the last century, exactly since the time when the first computers entered scientific research.

the dynamics of the system variables, for example, Hamilton's equations in classical mechanics or the Schrödinger equation in quantum mechanics. Obviously, such coupled systems of equations are always parametric models. It follows:

Fourth thesis: New types of formal representation are required for modeling irreducibly complex phenomena.

In particular, one increasingly finds discrete, fully scalable modeling structures, especially networks (e.g. decision trees, neural networks, or Bayesian networks). These models are often so extensive that they can no longer be manually displayed. Thus, the medium for representation changes as well. The sheet of paper, the book or the human brain are increasingly being replaced by the computer, which alone can develop these models.

Central to the two previous theses is the distinction between theory-driven and exploratory approaches in statistics. Remarkably, there exists an analogous distinction with respect to experimentation, which leads us to the next thesis:

Fifth thesis: Strong analogies exist between exploratory experimentation and data-intensive science.

First, let us delineate the important distinction between exploratory and theory-driven experimentation, which, somewhat astonishingly, has been explicitly introduced into the philosophy of science literature only a few years ago independently by Richard Burian and Friedrich Steinle in 1997. This late moment is surprising especially because the distinction concerns the most basic categorization of experimental research that is conceivable.

Nevertheless, both viewpoints on the role of experiments can already be found in the works of earlier authors, while the juxtaposition is lacking. Karl Popper, for example, characterizes experiments exclusively as theory-driven: "the theoretician puts certain definite questions to the experimenter, and the latter, by his experiments, tries to elicit a decisive answer to these questions, and to no others." (Popper 2002, p. 89) A pertinent example of such a theory-driven experiment is Arthur Eddington's expedition in 1919 to the volcanic island Principe off the West African coast in order to verify during a solar eclipse the deflection of light rays in a gravitational field as predicted by Einstein's general theory of relativity.

An apt description of exploratory experiments, by contrast, can be found in Ernst Mach's "Knowledge and Error": "What we can learn from an experiment resides wholly and solely in the dependence or independence of the elements or conditions of a phenomenon. By arbitrarily varying a certain group of elements or a single one, other elements will vary too or perhaps remain unchanged. The basic method of experiment is the method of variation." (Mach 1976, p. 149) Such variable variation constitutes the classical procedure in most laboratory experiments. For example, when Wilhelm Röntgen in the late 19th century discovered a new type of radiation, for him, in contrast to Eddington, an elaborate theory was not available. In order to explore the phenomenon, he changed systematically all the variables that he considered potentially relevant, examining their respective impact on the radiation.

Thus, a crucial feature to distinguish these two types of experimentation concerns the theoretical integration. Theory-driven experiments compare the predictions of an elaborate and detailed theoretical framework with experience, while exploratory experiments aim at developing such a theory in the first place or at least discovering the most important phenomenological laws. In other words, in the first case there is a considerable theory-ladenness in experimental practice, in contrast to the second case, where a theoretical understanding is largely unavailable. Relatedly, in exploratory experimentation, the central concepts still have to be developed together with the causal dependencies while the scientific terminology is determined by the theoretical framework in theory-driven experimentation.

There are some striking similarities between exploratory experimentation and data-intensive science. Most importantly, a substantial theory-independence characterizes the two scientific practices.[6] Also, both share the logic of variable variation, that the impact of changing conditions or circumstances on the phenomenon is examined. Finally, the ultimate aim of this procedure is in both cases to determine the causal structure of the examined phenomena.

There are some differences as well. From an epistemological perspective, it is perhaps most remarkable that data-intensive science usually does not start from experimentally obtained data, but rather works with observational data. Another crucial issue concerns the complexity of the phenomena that are being studied. Those examined in data-intensive science usually are so complicated and context-dependent that they do not fit into a controlled laboratory environment.

While it is generally accepted that exploratory experimentation aims at causal knowledge, in the case of data-intensive science the alleged central role for causality stands in contradiction with the relevant literature. For example, Chris Anderson wrote in a much-quoted article in the technology magazine WIRED that in

---

6   Referring to the relative theory-independence of exploratory experimentation, Ian Hacking could claim that experiments have a life of their own (Hacking 1983). Similarly, one could say that data-intensive science also has a life of its own, in that it is largely independent if more theory-laden scientific practices.

the wake of Big Data causality is replaced by correlation (2008). Similarly, Viktor Mayer-Schönberger and Kenneth Cukier claim that Big Data implies "a move away from the age-old search for causality" (2013, p. 14. See also Ch. 4). The opposite is the case:

Sixth thesis: Causality is the central concept to understand why data-intensive approaches can be scientifically relevant, in particular establish reliable predictions or allow for effective interventions.

The argument for this thesis is quite simple and can be illustrated with the following quotation by the British philosopher of science Nancy Cartwright: "causal laws cannot be done away with, for they are needed to ground the distinction between effective strategies and ineffective ones." (Cartwright 1979, p. 420) A mere correlation, even if it is empirically well documented, e.g. between the stork population and the human birth rate in some regions, cannot be used to influence phenomena. Even the settlement of a large number of storks will hardly cause people to have more children. In comparison, causal relationships can always be used to systematically interact with the world. This holds even if the relationships are only of statistical nature, then the corresponding strategies are of course effective only on average. When people are encouraged to smoke less, the lung cancer rate will very probably decrease in the population.

Now if data-intensive science is about manipulating phenomena then correlations are obviously not sufficient, but there must be a causal connection—for example, if Facebook wants its members to spend as much time as possible on the social network or if Google wants its users to click on certain advertising links.

One might object that sometimes data-intensive science is not about influencing or manipulating phenomena but only concerned with reliable predictions and for this correlations should be sufficient. However, two different types of correlations must be further distinguished, first those that can be attributed to a common cause, and then those which have arisen purely by chance. Correlations can establish reliable predictions only in the former case. Given a barometer reading one can relatively well predict the weather, just because air pressure is a common cause of both variables. However, water levels in Venice cannot help to predict bread prices in the UK even if the corresponding correlation is empirically well established (Sober 2001). Such random correlations occur with high probability whenever a very large number of different variables is considered. It is clear then that in accordance with the sixth thesis reliable predictions must just as well be causally justified, only that there need not be a direct causal link, but an indirect causal relationship via a common cause is enough.

Of course, with these few remarks the problem of causality is not sufficiently dealt with. After all it is one of the most controversial concepts in scientific methodology, to which one can hardly do justice in a brief introduction. One finds in history and still in current science both extremes: prominent voices that completely reject causation as an incoherent concept, but also the opposite view that scientific knowledge must always be of causal nature. This conflict can obviously only be resolved if it is further specified what exactly one means by causality—leading us to the next thesis.

Seventh thesis: The conceptual core of causality in data-intensive science consists in difference-making that a change in circumstances produces a change in the examined phenomena.

In fact, most of the familiar conceptual analyses of causality seem inappropriate for the context of data-intensive science. For example, the currently popular interventionist approaches (e.g. Woodward 2003) rely on a strong notion of intervention which is difficult to reconcile with the fact that in data-intensive science causal relationships are often established based on observational data, i.e. data collected in the absence of explicit interventions.[7]

Mechanistic approaches or the related process theories (e.g. Salmon 1984), another large group of interpretations, presuppose that the connection between cause and effect can be reconstructed in terms of fundamental laws, for example by tracing the electric current moving from a light switch to the bulb according to Maxwell's equations. But that is just not possible in the case of data-intensive science. After all, the mentioned theory-independence of data-intensive science implies that deeper reasons for the connection between cause and effect are commonly unknown.

Finally, with respect to a third large class of interpretations, so-called regularity theories of causality, these are difficult to reconcile with the variational evidence commonly employed in data-intensive science. As pointed out at the beginning of the section, data-intensive science is not so much concerned with the observation of regularities, in which a condition B is always followed by phenomenon A, but above all with the observation of variation, how a phenomenon is influenced by the change of circumstances or conditions.

---

7   Even Woodward's relatively weak construal of interventions assumes for example that interventions eliminate alternative causal paths between the predictor and the response variables. Thus, an ontologically hard distinction remains between experimental and observational data that is totally unsuitable for the context of data-intensive science.

One should discuss here in much greater detail the ramified philosophical debate on causation, which for lack of space is not possible. Instead, I want to briefly argue that the notion of difference-making is crucial to understanding the role of causality in data-intensive science. After all, it is generally examined which conditions make a difference for a phenomenon, just as required in John Stuart Mill's famous formulation of the method of the difference: "If an instance in which the phenomenon under investigation occurs, and an instance in which it does not occur, have every circumstance save one in common, that one occurring only in the former; the circumstance in which alone the two instances differ is the effect, or the cause, or an indispensable part of the cause, of the phenomenon." (Mill 1886, p. 256)

Both the method of difference as well as related counterfactual approaches to causality exhibit a variety of problems, which we cannot discuss here due to lack of space.[8] We can nevertheless conclude that the method of difference meets the criteria that were discussed above. First of all, it does not presuppose a strong notion of intervention. At least in principle, then, causal relationships should be determinable on the basis of observational data. Also, a deeper theoretical understanding of the relationship between cause and effect is not required for the method of difference. Ideally, causal relations can be identified without knowledge of the underlying causal processes if as many conditions as possible can be held fix. Finally, the method of difference does not rely on evidence in terms of regularities but rather in terms of variation. After all, it examines the influence of a change in circumstances.

In the literature, the idea that correlation replaces causation is often connected with another characteristic of data-intensive science that it supposedly cannot explain the phenomena. Without causation no explanations, thus the common logic. However, this conclusion holds only under certain construals of causality. For example, in the case of mechanistic causality it is plausible that the underlying mechanism may also explain the corresponding causal relationship. By contrast, if causality is understood in terms of difference-making, as suggested here, then causal relationships are generally not in a strong sense explanatory. This leads to the eighth thesis:

Eighth thesis: Data-intensive science can explain by specifying causes, but not by referring to unifying principles.

---

8   See for example the manuscript on causation as difference making by one of the authors: http://philsci-archive.pitt.edu/11913/.

This thesis has a lot to do with the aforementioned distinction between phenomenological and theoretical science. In this context, Nancy Cartwright has introduced two types of explanation that one might term causal explanation on the one hand and unifying explanation on the other: "there are two quite different kinds of things we do when we explain a phenomenon in physics. First, we describe its causes. Second, we fit the phenomenon into a theoretical frame." (Cartwright 1983, p. 16) Obviously, the second type of explanation remains reserved largely for theoretical sciences, while the first type can also be found in phenomenological sciences.

As stated before, data-intensive science is confined primarily to the phenomenological level, it determines causal dependencies but refrains from searching for unifying principles. Again the main reason is that data-intensive science usually deals with irreducibly complex phenomena for which such principles do not exist. Obviously then, explanations that refer to unifying principles are impossible in data-intensive science, even though such unifying explanation is regarded by many as a hallmark of science. This applies particularly to physics. A satisfactory explanation of the tides, for example, should ultimately refer to general laws of mechanics and gravitation.

In phenomenological sciences, only a much weaker form of explanation can be realized that refers to causes in the sense of difference-making. Thus, one can explain the occurrence of the tides by referring to the positions of the moon and sun. Such explanations can establish why certain predictions are successful, but they almost entirely lack a unifying nature. In particular, they can hardly be used for guidance to understand related phenomena. A possible link between a stone falling to earth and the tides could hardly be established in this manner.

The advent of data-intensive science thus seems to have the effect that various disciplines such as medicine or the social sciences, which were often seen as theoretical sciences in analogy with the fundamental natural sciences, are now increasingly perceived as phenomenological sciences. In particular for irreducibly complex phenomena in these fields, scientific explanations in the conventional sense are not feasible, causal manipulation and prediction however remain possible. And since in the future, only the computer and less the human scientist can "understand" these complex phenomena, the following holds:

Ninth thesis: The increasing automation of science changes fundamentally the role of scientific experts.

We must accept that due to the limitations of the human cognitive apparatus many scientific processes that take place in the computer are no longer traceable for us humans. A good example is the proof of the four-color theorem, in which only a computer could check all the remaining variants. In some areas the resulting changes for the role of human experts are already under way, when, as mentioned, grammatical knowledge is no longer relevant for writing translation programs or when reading recommendations of online book shops are no longer written by cultural journalists, but rather are algorithmically generated, which at least from a financial perspective turns out extremely successful.

These changes and the consequences for the labor market exemplify the social impact of data-intensive science. In fact, the ethical dimension of the current developments reaches far beyond the ubiquitous debate on privacy. The latter is indisputably important, but the improved capabilities for predicting complex phenomena, for example in the social sciences or in medicine, as well as the use of data-driven predictions for developing new products such as automated driving lead to a large variety of mostly unanswered ethical questions.

It should however be noted:

Tenth thesis: The ethics and the epistemology of data-intensive science can hardly be separated.

Most ethical problems related to data-intensive science can only be addressed if the epistemological framework is already understood. In other words, the epistemology determines the scope of what is possible, within the confines of which one may deliberate what is ethically desired.

1.3   Overview of the volume

In line with the topics that were just discussed, the volume is divided into four sections: "Big Data and the Sciences", "Computability", "Complexity and Information" and "Ethical and Political Perspectives".

The first part begins with an essay by Google research director Peter Norvig that was originally published as a blog post on his website. For one of the editors (Wolfgang Pietsch), this essay played a crucial role in understanding the epistemological significance of data-intensive science. The subsequent contribution by the Copenhagen philosophers of technology Gernot Rieder and Judith Simon also examines several epistemological issues related to Big Data, for example, theory-independence or the supposed objectivity. They question the thesis of a new empiricism and highlight some political and social consequences of the use of big data.

The third article by the Cottbus philosopher of technology Klaus Kornwachs also takes on a critical perspective with respect to some standard epistemological claims in connection with big data, arguing for the virtues and advantages of traditional scientific modeling.

The following articles focus on selected epistemological aspects of data-intensive science. The Texan mathematician and computational biologist Edward Dougherty argues that even in the age of big data a pressing problem in many areas remains the lack of data when it comes to modeling complex stochastic and nonlinear phenomena, such as cell regulation. The statisticians Anne-Laure Boulesteix, Roman Hornung, and Willi Sauerbrei then discuss in their article how much room for interpretation remains in the analysis of large data sets. In particular, they compare different measures how so-called "fishing for significance" can be avoided. In the last contribution of the first part, the Munich mathematicians Nadine Gissibl, Claudia Klüppelberg, and Johanna Mager investigate the significance of large data sets for the analysis of extreme risks, such as plane crashes.

The second part of the volume considers aspects of computability from various disciplinary perspectives. It begins with an essay, in which the Russian philosopher Andrei Rodin discusses whether spatial aspects should also be taken into account in models of computation besides the temporal ordering. Next is an article by the biophysicist Leo van Hemmen on the feasibility of mathematical modeling in complex biological systems, especially in neurobiology. The physicist Konrad Kleinknecht then asks to what extent climate modeling can yield reliable predictions in view of abundant non-linearities. The science TV-host, physicist, and natural philosopher Harald Lesch considers the relationship between physics and metaphysics concentrating in particular on the limits of physical knowledge, for example that the ultimate beginning of the universe is not computable.

The focus then shifts from the natural sciences towards the social sciences. Chadwick Wang, a Chinese researcher in science and technology studies, portrays a controversial debate among various factions of Chinese scientists, whether society is computable or not. The Stuttgart physicist and pioneer of sociophysics Wolfgang Weidlich provides a brief overview of his approach and discusses some important methodological problems. Finally, the Augsburg philosopher of science Theodor Leiber outlines a SWOT analysis of new methodological approaches in computational social science and Big Data.

The Salzburg psychologist Günter Schiepek who like Wolfgang Weidlich belongs to the broader circle of the Stuttgart school of synergetics, examines in his essay issues of complexity, computability, and the use of large data sets in psychotherapy. The Paderborn philosopher Ruth Hagengruber then discusses differences between human and machine knowledge raising the question to what extent creativity can ever be attributed to machines. The second part concludes with a historical perspective by the Munich philosopher of science Tobias Jung, who pursues the question how important aspects of computability are in the work of Immanuel Kant. Similar to Harald Lesch, Jung focuses primarily on the limits of human cognition and knowledge.

The third part is devoted to the great antagonist of computability, the notion of complexity. First, the Moscow philosopher and complexity-theorist Helena Knyazeva gives an overview of basic principles and aspects of complex systems emphasizing the interdisciplinary nature of any complexity science. The Darmstadt physicist and philosopher Jan Schmidt presents largely ignored aspects in the epistemological work of Pierre Duhem that make him appear as one of the pioneers of modern complexity research. The philosopher of technology Alfred Nordmann, also from Darmstadt, compares two very different approaches by Stuart Kauffman and Brian Goodwin on how biology changes on its way to becoming a science of complexity. Subsequently, the Oxford computer scientist and philosopher Hector Zenil examines the extent to which the theory of algorithmic information can help to analyze complex systems. Finally, the physicist and philosopher Holger Lyre discusses whether the concept of information should play a fundamental role in the natural sciences at all.

In the fourth and last part, the view widens towards ethical and political perspectives. Julian Nida-Rümelin, the Munich philosopher and former German Minister of State for Culture and the Media analyzes whether we need to rethink the concept of responsibility in light of autonomous robots. In the following contribution, the Konstanz philosopher Jürgen Mittelstrass examines the concept of emergence, which of course is closely intertwined with complexity. He shows in particular how emergence can contribute to the creation of novelty at the interface of various disciplines. The business ethicist Christoph Lütge outlines the research agenda of the new field of experimental ethics and raises the question to what extent ethical problems can be empirically examined and calculated. The philosopher and computer scientist Sabine Thürmel then analyzes the interplay of autonomy and control in Big Data based systems. A contribution by the physicist and energy researcher Thomas Hamacher follows that is also methodologically innovative by taking recourse to literature studies. It asks, whether the analysis of a contemporary novel can foster insights regarding energy and sustainability. The volume concludes with a study by the Japanese social philosopher Naoshi Yamawaki how an unpredictable and uncalculable disaster like the nuclear accident of Fukushima may require rethinking the role of ethics for the sciences.

References

Anderson, Chris. 2008. The End of Theory: The Data Deluge Makes the Scientific Method Obsolete. *WIRED Magazine* 16/07. http://www.wired.com/science/discoveries/magazine/16-07/pb_theory.

Breiman, Leo. 2001. Statistical Modeling: The Two Cultures. *Statistical Science* 16 (3): 199-231.

Burian, Richard. 1997. Exploratory Experimentation and the Role of Histochemical Techniques in the Work of Jean Brachet, 1938-1952. *History and Philosophy of the Life Sciences* 19: 27-45.

Cartwright , Nancy. 1979. Causal Laws and Effective Strategies. Noûs 13 (4): 419-437.

Cartwright, Nancy. 1983. *How the Laws of Physics Lie.* Oxford: Oxford University Press.

Duhem, Pierre. 1954. *The Aim and Structure of Physical Theory.* Princeton, PA: Princeton University Press.

Hacking, Ian. 1983. *Representing and Intervening.* Cambridge: Cambridge University Press.

Hacking, Ian. 1990. *The Taming of Chance.* Cambridge: Cambridge University Press.

Halevy, Alon, Peter Norvig und Fernando Pereira. 2009. The Unreasonable Effectiveness of Data. *IEEE Intelligent Systems* 24(2): 8-12.

Kolmogorov, Andrey N. 1983. Combinatorial foundations of information theory and the calculus of probabilities. *Russian Mathematical Surveys* 38 (4): 29-40.

Kuhlmann, Meinard. 2011. Mechanisms in Dynamically Complex Systems. In *Causality in the Sciences,* hrsg. P. McKay Illari, F. Russo, and J. Williamson, 880-906. Oxford: Oxford University Press.

Laney, Doug. 2001. 3D Data Management: Controlling Data Volume, Velocity, and Variety. Research Report. http://blogs.gartner.com/doug-laney/files/2012/01/ad949-3D-Data-Management-Controlling-Data-Volume-Velocity-and-Variety.pdf.

Leonelli, Sabina. 2012. Introduction: Making sense of data-driven research in the biological and biomedical sciences. *Studies in History and Philosophy of Biological and Biomedical Sciences* 43 (1): 1-3.

Mach, Ernst. 1976. *Knowledge and Error: Sketches on the Psychology of Enquiry.* Dordrecht: D. Reidel.

Mayer-Schönberger, Viktor & Kenneth Cukier. 2013. *Big Data.* London: John Murray.

Mill, John S. 1886. *System of Logic.* London: Longmans, Green & Co.

Mitchell, Sandra. 2008. *Komplexitäten. Warum wir erst anfangen, die Welt zu verstehen.* Frankfurt a.M.: Suhrkamp.

Pietsch, Wolfgang. 2015. Aspects of Theory-Ladenness in Data-Intensive Science. *Philosophy of Science* 82 (5): 905-916.

Pietsch, Wolfgang. 2016. The Causal Nature of Modeling with Big Data. *Philosophy & Technology* 29 (2): 137-171.

Popper, Karl. 2002. *The Logic of Scientific Discovery.* London: Routledge Classics.

Salmon, Wesley. 1984. *Scientific Explanation and the Causal Structure of the World.* Princeton: Princeton University Press.

Sober, Elliott. 2001. Venetian Sea Levels, British Bread Prices, and the Principle of the Common Cause. *British Journal for the Philosophy of Science* 52: 331-346.

Steinle, Friedrich. 1997. Entering New Fields: Exploratory Uses of Experimentation. *Philosophy of Science* 64: S65-S74.

Woodward, James. 2003. *Making Things Happen: A Theory of Causal Explanation.* Oxford: Oxford University Press.

I
Big Data und die Wissenschaften

2. On Chomsky and the Two Cultures of Statistical Learning[1]

Peter Norvig[2]

Abstract

At the Brains, Minds, and Machines symposium held during MIT's 150th birthday party, Technology Review reports that Prof. Noam Chomsky derided researchers in machine learning who use purely statistical methods to produce behavior that mimics something in the world, but who don't try to understand the meaning of that behavior. This essay discusses what Chomsky said, speculates on what he might have meant, and tries to determine the truth and importance of his claims.

At the Brains, Minds, and Machines symposium held during MIT's 150th birthday party, Technology Review reports[3] that Prof. Noam Chomsky derided researchers in machine learning who use purely statistical methods to produce behavior that mimics something in the world, but who don't try to understand the meaning of that behavior.

1 This essay first appeared on Peter Norvig's blog: http://norvig.com/chomsky.html
2 Peter Norvig is a Director of Research at Google Inc; previously he directed Google's core search algorithms group. He is co-author of *Artificial Intelligence: A Modern Approach*, the leading textbook in the field, and co-teacher of an Artificial Intelligence class that signed up 160,000 students, helping to kick off the current round of massive open online classes. He is a fellow of the AAAI, ACM, California Academy of Science and American Academy of Arts & Sciences.
3 https://www.technologyreview.com/s/423917/unthinking-machines/

The transcript[4] is now available, so let's quote Chomsky himself:

> It's true there's been a lot of work on trying to apply statistical models to various linguistic problems. I think there have been some successes, but a lot of failures. There is a notion of success ... which I think is novel in the history of science. It interprets success as approximating unanalyzed data.

This essay discusses what Chomsky said, speculates on what he might have meant, and tries to determine the truth and importance of his claims.

Chomsky's remarks were in response to Steven Pinker's question about the success of probabilistic models trained with statistical methods.

2.1   What did Chomsky mean, and is he right?

I take Chomsky's points to be the following:

1. Statistical language models have had engineering success, but that is irrelevant to science.
2. Accurately modeling linguistic facts is just butterfly collecting; what matters in science (and specifically linguistics) is the underlying principles.
3. Statistical models are incomprehensible; they provide no insight.
4. Statistical models may provide an accurate simulation of some phenomena, but the simulation is done completely the wrong way; people don't decide what the third word of a sentence should be by consulting a probability table keyed on the previous two words, rather they map from an internal semantic form to a syntactic tree-structure, which is then linearized into words. This is done without any probability or statistics.
5. Statistical models have been proven incapable of learning language; therefore language must be innate, so why are these statistical modelers wasting their time on the wrong enterprise?

Is he right? That's a long-standing debate. These are my answers:

A. I agree that engineering success is not the goal or the measure of science. But I observe that science and engineering develop together, and that engineer-

---

4   http://languagelog.ldc.upenn.edu/myl/PinkerChomskyMIT.html

ing success shows that something is working right, and so is evidence (but not proof) of a scientifically successful model.

B. Science is a combination of gathering facts and making theories; neither can progress on its own. I think Chomsky is wrong to push the needle so far towards theory over facts; in the history of science, the laborious accumulation of facts is the dominant mode, not a novelty. The science of understanding language is no different than other sciences in this respect.

C. I agree that it can be difficult to make sense of a model containing billions of parameters. Certainly a human can't understand such a model by inspecting the values of each parameter individually. But one can gain insight by examining the *properties* of the model—where it succeeds and fails, how well it learns as a function of data, etc.

D. I agree that a Markov model of word probabilities cannot model all of language. It is equally true that a concise tree-structure model without probabilities cannot model all of language. What is needed is a probabilistic model that covers words, trees, semantics, context, discourse, etc. Chomsky dismisses all probabilistic models because of shortcomings of particular 50-year old models. I understand how Chomsky arrives at the conclusion that probabilistic models are unnecessary, from his study of the generation of language. But the vast majority of people who study *interpretation* tasks, such as speech recognition, quickly see that interpretation is an inherently probabilistic problem: given a stream of noisy input to my ears, what did the speaker most likely mean? Einstein said to make everything as simple as possible, but no simpler. Many phenomena in science are stochastic, and the simplest model of them is a probabilistic model; I believe language is such a phenomenon and therefore that probabilistic models are our best tool for representing facts about language, for algorithmically processing language, and for understanding how humans process language.

E. In 1967, Gold's Theorem showed some theoretical limitations of logical deduction on formal mathematical languages. But this result has nothing to do with the task faced by learners of natural language. In any event, by 1969 we knew that probabilistic inference (over probabilistic context-free grammars) is not subject to those limitations (Horning showed that learning of PCFGs[5] is possible). I agree with Chomsky that it is undeniable that humans have some innate capability to learn natural language, but we don't know enough about that capability to rule out probabilistic language representations, nor statistical learning. I think it is much more likely that human language learning involves something like probabilistic and statistical inference, but we just don't know yet.

Now let me back up my answers with a more detailed look at the remaining questions.

---

5   Probabilistic Context-Free Grammars

2.2   What is a statistical model?

A **statistical model** is a mathematical model which is modified or trained by the input of data points. Statistical models are often but not always probabilistic. Where the distinction is important we will be careful not to just say "statistical" but to use the following component terms:

- A **mathematical model** specifies a relation among variables, either in functional form that maps inputs to outputs (e.g. $y = m\,x + b$) or in relation form (e.g. the following (x, y) pairs are part of the relation).
- A **probabilistic model** specifies a probability distribution over possible values of random variables, e.g., $P(x, y)$, rather than a strict deterministic relationship, e.g., $y = f(x)$.
- A **trained model** uses some training/learning algorithm to take as input a collection of possible models and a collection of data points (e.g. (x, y) pairs) and select the best model. Often this is in the form of choosing the values of parameters (such as m and b above) through a process of statistical inference.

For example, a decade before Chomsky, Claude Shannon (1948) proposed probabilistic models of communication based on Markov chains of words. If you have a vocabulary of 100,000 words and a second-order Markov model in which the probability of a word depends on the previous two words, then you need a quadrillion (10^{15}) probability values to specify the model. The only feasible way to learn these 10^{15} values is to gather statistics from data and introduce some smoothing method for the many cases where there is no data. Therefore, most (but not all) probabilistic models are trained. Also, many (but not all) trained models are probabilistic.

As another example, consider the Newtonian model of gravitational attraction, which says that the force between two objects of mass m_1 and m_2 a distance r apart is given by

$$F = G\,m_1\,m_2\,/\,r^2$$

where G is the universal gravitational constant. This is a trained model because the gravitational constant G is determined by statistical inference over the results of a series of experiments that contain stochastic experimental error. It is also a deterministic (non-probabilistic) model because it states an exact functional relationship. I believe that Chomsky has no objection to this kind of statistical model. Rather, he seems to reserve his criticism for statistical models like Shannon's that have quadrillions of parameters, not just one or two.

(This example brings up another distinction: the gravitational model is **continuous** and **quantitative** whereas the linguistic tradition has favored models that are discrete, categorical, and qualitative: a word is or is not a verb, there is no question of its degree of verbiness. For more on these distinctions, see Chris Manning's article on Probabilistic Syntax (2002).)

A relevant probabilistic statistical model is the ideal gas law, which describes the pressure P of a gas in terms of the number of molecules, N, the temperature, T, and Boltzmann's constant, k:

$$P = N k T / V.$$

The equation can be derived from first principles using the tools of statistical mechanics. It is an uncertain, incorrect model; the *true* model would have to describe the motions of individual gas molecules. This model ignores that complexity and *summarizes* our uncertainty about the location of individual molecules. Thus, even though it is statistical and probabilistic, even though it does not completely model reality, it does provide both good predictions and insight—insight that is not available from trying to understand the *true* movements of individual molecules.

Now let's consider the non-statistical model of spelling expressed by the rule "*I before E except after C.*" Compare that to the probabilistic, trained statistical model:

```
P(IE)  = 0.0177     P(CIE)  = 0.0014     P(*IE)  = 0.0163
P(EI)  = 0.0046     P(CEI)  = 0.0005     P(*EI)  = 0.0041
```

This model comes from statistics on a corpus of a trillion words[6] of English text. The notation P(IE) is the probability that a word sampled from this corpus contains the consecutive letters "IE." P(CIE) is the probability that a word contains the consecutive letters "CIE," and P(*IE) is the probability of any letter other than C followed by IE. The statistical data confirms that IE is in fact more common than EI, and that the dominance of IE lessens when following a C, but contrary to the rule, CIE is still more common than CEI. Examples of "CIE" words include "science," "society," "ancient" and "species." The disadvantage of the "I before E except after C" model is that it is not very accurate. Consider:

```
Accuracy ("I before E") = 0.0177/(0.0177+0.0046) = 0.793


     Accuracy ("I before E except after C") =
(0.0005+0.0163)/(0.0005+0.0163+0.0014+0.0041) = 0.753
```

---

6   http://norvig.com/ngrams/

A more complex statistical model (say, one that gave the probability of all 4-letter sequences, and/or of all known words) could be ten times more accurate[7] at the task of spelling, but offers little **insight** into what is going on. (Insight would require a model that knows about phonemes, syllabification, and language of origin. Such a model could be trained (or not) and probabilistic (or not).)

As a final example (not of statistical models, but of insight), consider the Theory of Supreme Court Justice Hand-Shaking: when the supreme court convenes, all attending justices shake hands with every other justice. The number of attendees, n, must be an integer in the range 0 to 9; what is the total number of handshakes, h, for a given n? Here are three possible explanations:

A. Each of n justices shakes hands with the other n - 1 justices, but that counts Ali-to/Breyer and Breyer/Alito as two separate shakes, so we should cut the total in half, and we end up with $h = n \times (n - 1) / 2$.
B. To avoid double-counting, we will order the justices by seniority and only count a more-senior/more-junior handshake, not a more-junior/more-senior one. So we count, for each justice, the shakes with the more junior justices, and sum them up, giving $h = \Sigma_{i = 1 \, .. \, n} (i - 1)$.
C. Just look at this table:

n:	0	1	2	3	4	5	6	7	8	9
h:	0	0	1	3	6	10	15	21	28	36

Some people might prefer A, some might prefer B, and if you are slow at doing multiplication or addition you might prefer C. Why? All three explanations describe *exactly the same theory*—the same function from n to h, over the entire domain of possible values of n. Thus we could prefer A (or B) over C only for reasons other than the theory itself. We might find that A or B gave us a better understanding of the problem. A and B are certainly more useful than C for figuring out what happens if Congress exercises its power to add an additional associate justice. Theory A might be most helpful in developing a theory of handshakes at the end of a hockey game (when each player shakes hands with players on the opposing team) or in proving that the number of people who shook an odd number of hands at the MIT Symposium is even.

---

7   http://norvig.com/spell-correct.html

2.3 How successful are statistical language models?

Chomsky said words to the effect that statistical language models have had some limited success in some application areas. Let's look at computer systems that deal with language, and at the notion of "success" defined by "making accurate predictions about the world." First, the major application areas:

- **Search engines:** 100% of major players are trained and probabilistic. Their operation cannot be described by a simple function.
- **Speech recognition:** 100% of major systems are trained and probabilistic, mostly relying on probabilistic hidden Markov models or neural networks.
- **Machine translation:** 100% of top competitors in competitions such as NIST[8] use statistical methods. Some commercial systems use a hybrid of trained and rule-based approaches. Of the 4000 language pairs covered by machine translation systems, a statistical system is by far the best for every pair except Japanese-English, where the top statistical system is roughly equal to the top hybrid system. The most recent top-performing systems use deep neural networks.
- **Question answering:** this application is less well-developed, and many systems build heavily on the statistical and probabilistic approach used by search engines. The IBM Watson[9] system that recently won on Jeopardy is thoroughly probabilistic and trained, while Boris Katz's START[10] is a hybrid. All systems use at least some statistical techniques.

Now let's look at some components that are of interest only to the computational linguist, not to the end user:

- **Word sense disambiguation:** 100% of top competitors at the SemEval-2[11] competition used statistical techniques; most are probabilistic; some use a hybrid approach incorporating rules from sources such as Wordnet.
- **Coreference resolution:** The majority of current systems are statistical, although we should mention the system of Haghighi and Klein[12], which can be described as a hybrid system that is mostly rule-based rather than trained, and performs on par with top statistical systems. The highest-scoring system uses deep reinforcement learning.

---

8 http://www.itl.nist.gov/iad/mig/tests/mt/2009/ResultsRelease/currentArabic.html
9 http://www.ibm.com/smarterplanet/us/en/ibmwatson/
10 http://groups.csail.mit.edu/infolab/publications/Katz-etal-TREC2003.pdf
11 http://semeval2.fbk.eu/semeval2.php?location=SemEval2010-Program
12 http://www.aclweb.org/anthology/D/D09/D09-1120.pdf

- **Part of speech tagging:** Most current systems are statistical. The Brill tagger[13] stands out as a successful hybrid system: it learns a set of deterministic rules from statistical data.
- **Parsing:** There are many parsing systems, using multiple approaches. Almost all of the most successful are statistical, and the majority are probabilistic (with a substantial minority of deterministic parsers).

Clearly, it is inaccurate to say that statistical models (and probabilistic models) have achieved *limited* success; rather they have achieved a *dominant* (although not exclusive) position.

Another measure of success is the degree to which an idea captures a community of researchers. As Steve Abney wrote in 1996, "In the space of the last ten years, statistical methods have gone from being virtually unknown in computational linguistics to being a fundamental given. ... anyone who cannot at least use the terminology persuasively risks being mistaken for kitchen help at the ACL [Association for Computational Linguistics] banquet."

Now of course, the majority doesn't rule—just because everyone is jumping on some bandwagon, that doesn't make it right. But I made the switch: after about 14 years of trying to get language models to work using logical rules, I started to adopt probabilistic approaches (thanks to pioneers like Gene Charniak (and Judea Pearl for probability in general) and to my colleagues who were early adopters, like Dekai Wu). And I saw everyone around me making the same switch. (And I didn't see anyone going in the other direction.) We all saw the limitations of the old tools, and the benefits of the new.

And while it may seem crass and anti-intellectual to consider a financial measure of success, it is worth noting that the intellectual offspring of Shannon's theory create several trillion dollars of revenue each year, while the offspring of Chomsky's theories generate well under a billion.

This section has shown that one reason why the vast majority of researchers in computational linguistics use statistical models is an *engineering* reason: statistical models have state-of-the-art performance, and in most cases non-statistical models perform worst. For the remainder of this essay we will concentrate on *scientific* reasons: that probabilistic models better represent linguistic facts, and statistical techniques make it easier for us to make sense of those facts.

---

13 https://en.wikipedia.org/wiki/Brill_tagger

2.4   Is there anything like [the statistical model] notion of success in the history of science?

When Chomsky said *"That's a notion of [scientific] success that's very novel. I don't know of anything like it in the history of science"* he apparently meant that the notion of success of "accurately modeling the world" is novel, and that the only true measure of success in the history of science is "providing insight" – of answering *why* things are the way they are, not just describing *how* they are.

A dictionary definition[14] of science is "the systematic study of the structure and behavior of the physical and natural world through observation and experiment," which stresses accurate modeling over insight, but it seems to me that both notions have always coexisted as part of doing science. To test that, I consulted the epitome of doing science, namely *Science*. I looked at the current issue[15] and chose a title and abstract at random:

Chlorinated Indium Tin Oxide Electrodes with High Work Function for Organic Device Compatibility

In organic light-emitting diodes (OLEDs), a stack of multiple organic layers facilitates charge flow from the low work function [~4.7 electron volts (eV)] of the transparent electrode (tin-doped indium oxide, ITO) to the deep energy levels (~6 eV) of the active light-emitting organic materials. We demonstrate a chlorinated ITO transparent electrode with a work function of >6.1 eV that provides a direct match to the energy levels of the active light-emitting materials in state-of-the art OLEDs. A highly simplified green OLED with a maximum external quantum efficiency (EQE) of 54% and power efficiency of 230 lumens per watt using outcoupling enhancement was demonstrated, as were EQE of 50% and power efficiency of 110 lumens per watt at 10,000 candelas per square meter.

It certainly seems that this article is much more focused on "accurately modeling the world" than on "providing insight." The paper does indeed fit in to a body of theories, but it is mostly reporting on specific experiments and the results obtained from them (e.g. efficiency of 54%).

I then looked at all the titles and abstracts from the current issue of *Science*:

• Comparative Functional Genomics of the Fission Yeasts

---

14 *The New Oxford Dictionary of English*, 1998.
15 Vol. 332, Issue 6032 (20 May 2011)

- Dimensionality Control of Electronic Phase Transitions in Nickel-Oxide Superlattices
- Competition of Superconducting Phenomena and Kondo Screening at the Nanoscale
- Chlorinated Indium Tin Oxide Electrodes with High Work Function for Organic Device Compatibility
- Probing Asthenospheric Density, Temperature, and Elastic Moduli Below the Western United States
- Impact of Polar Ozone Depletion on Subtropical Precipitation
- Fossil Evidence on Origin of the Mammalian Brain
- Industrial Melanism in British Peppered Moths Has a Singular and Recent Mutational Origin
- The Selaginella Genome Identifies Genetic Changes Associated with the Evolution of Vascular Plants
- Chromatin "Prepattern" and Histone Modifiers in a Fate Choice for Liver and Pancreas
- Spatial Coupling of mTOR and Autophagy Augments Secretory Phenotypes
- Diet Drives Convergence in Gut Microbiome Functions Across Mammalian Phylogeny and Within Humans
- The Toll-Like Receptor 2 Pathway Establishes Colonization by a Commensal of the Human Microbiota
- A Packing Mechanism for Nucleosome Organization Reconstituted Across a Eukaryotic Genome
- Structures of the Bacterial Ribosome in Classical and Hybrid States of tRNA Binding

and did the same for the current issue[16] of *Cell*:

- Mapping the NPHP-JBTS-MKS Protein Network Reveals Ciliopathy Disease Genes and Pathways
- Double-Strand Break Repair-Independent Role for BRCA2 in Blocking Stalled Replication Fork Degradation by MRE11
- Establishment and Maintenance of Alternative Chromatin States at a Multicopy Gene Locus
- An Epigenetic Signature for Monoallelic Olfactory Receptor Expression
- Distinct p53 Transcriptional Programs Dictate Acute DNA-Damage Responses and Tumor Suppression

---

16 Vol. 154, Issue 4 (13 May 2011)

- An ADIOL-ERβ-CtBP Transrepression Pathway Negatively Regulates Microglia-Mediated Inflammation
- A Hormone-Dependent Module Regulating Energy Balance
- Class IIa Histone Deacetylases Are Hormone-Activated Regulators of FOXO and Mammalian Glucose Homeostasis

and for the 2010 Nobel Prizes in science:

- Physics: *for groundbreaking experiments regarding the two-dimensional material graphene*
- Chemistry: *for palladium-catalyzed cross couplings in organic synthesis*
- Physiology or Medicine: *for the development of in vitro fertilization*

My conclusion is that 100% of these articles and awards are more about "accurately modeling the world" than they are about "providing insight," although they all have some theoretical insight component as well. I recognize that judging one way or the other is a difficult ill-defined task, and that you shouldn't accept my judgements, because I have an inherent bias. (I was considering running an experiment on Mechanical Turk to get an unbiased answer, but those familiar with Mechanical Turk told me these questions are probably too hard. So you the reader can do your own experiment and see if you agree.)

2.5   What doesn't Chomsky like about statistical models?

I said that statistical models are sometimes confused with probabilistic models; let's first consider the extent to which Chomsky's objections are actually about probabilistic models. In 1969 he famously wrote:

> But it must be recognized that the notion of "probability of a sentence" is an entirely useless one, under any known interpretation of this term.

His main argument being that, under any interpretation known to him, the probability of a novel sentence must be zero, and since novel sentences are in fact generated all the time, there is a contradiction. The resolution of this contradiction is of course that it is not necessary to assign a probability of zero to a novel sentence; in fact, with current probabilistic models it is well-known how to assign a non-zero probability to novel occurrences, so this criticism is invalid, but was very influential for decades. Previously, in *Syntactic Structures* (1957) Chomsky wrote:

> I think we are forced to conclude that ... probabilistic models give no particular insight into some of the basic problems of syntactic structure.

In the footnote to this conclusion he considers the possibility of a useful probabilistic/statistical model, saying "I would certainly not care to argue that ... is unthinkable, but I know of no suggestion to this effect that does not have obvious flaws." The main "obvious flaw" is this: Consider:

1. I never, ever, ever, ever, ... **fiddle** around in any way with electrical equipment.
2. **She** never, ever, ever, ever, ... **fiddles** around in any way with electrical equipment.
3. * I never, ever, ever, ever, ... **fiddles** around in any way with electrical equipment.
4. * **She** never, ever, ever, ever, ... **fiddle** around in any way with electrical equipment.

No matter how many repetitions of "ever" you insert, sentences 1 and 2 are grammatical and 3 and 4 are ungrammatical. A probabilistic Markov-chain model with n states can never make the necessary distinction (between 1 or 2 versus 3 or 4) when there are more than n copies of "ever." Therefore, a probabilistic Markov-chain model cannot handle all of English.

This criticism is correct, but it is a criticism of Markov-chain models—it has nothing to do with probabilistic models (or trained models) at all. Moreover, since 1957 we have seen many types of probabilistic language models beyond the Markov-chain word models. Examples 1-4 above can in fact be distinguished with a finite-state model that is not a chain, but other examples require more sophisticated models. The best studied is probabilistic context-free grammar (PCFG), which operates over trees, categories of words, and individual lexical items, and has none of the restrictions of finite-state models. We find that PCFGs are state-of-the-art for parsing performance and are easier to learn from data than categorical context-free grammars. Other types of probabilistic models cover semantic and discourse structures. Every probabilistic model is a superset of a deterministic model (because the deterministic model could be seen as a probabilistic model where the probabilities are restricted to be 0 or 1), so any valid criticism of probabilistic models would have to be because they are too expressive, not because they are not expressive enough.

In *Syntactic Structures*, Chomsky introduces a now-famous example that is another criticism of finite-state probabilistic models:

Neither (a) 'colorless green ideas sleep furiously' nor (b) 'furiously sleep ideas green colorless', nor any of their parts, has ever occurred in the past linguistic experience of an English speaker. But (a) is grammatical, while (b) is not.

Chomsky appears to be correct that neither sentence appeared in the published literature before 1955. I'm not sure what he meant by "any of their parts," but certainly every two-word part had occurred, for example:

- "It is neutral green, **colorless green**, like the glaucous water lying in a cellar." The Paris we remember, Elisabeth Finley Thomas (1942).
- "To specify those **green ideas** is hardly necessary, but you may observe Mr. [D. H.] Lawrence in the role of the satiated aesthete." The New Republic: Volume 29 p. 184, William White (1922).
- "**Ideas sleep** in books." Current Opinion: Volume 52, (1912).

But regardless of what is meant by "part," a statistically-trained finite-state model *can* in fact distinguish between these two sentences. Pereira (2001) showed that such a model, augmented with word categories and trained by expectation maximization on newspaper text, computes that (a) is 200,000 times more probable than (b). To prove that this was not the result of Chomsky's sentence itself sneaking into newspaper text, I repeated the experiment, using a much cruder model with Laplacian smoothing and no categories, trained over the Google Book corpus from 1800 to 1954, and found that (a) is about 10,000 times more probable. If we had a probabilistic model over trees as well as word sequences, we could perhaps do an even better job of computing degree of grammaticality.

Furthermore, the statistical models are capable of delivering the judgment that both sentences are *extremely* improbable, when compared to, say, "Effective green products sell well." Chomsky's theory, being categorical, cannot make this distinction; all it can distinguish is grammatical/ungrammatical.

Another part of Chomsky's objection is "we cannot seriously propose that a child learns the values of 10^9 parameters in a childhood lasting only 10^8 seconds." (Note that modern models are much larger than the 10^9 parameters that were contemplated in the 1960s.) But of course nobody is proposing that these parameters are learned one-by-one; the right way to do learning is to set large swaths of near-zero parameters simultaneously with a smoothing or regularization procedure, and update the high-probability parameters continuously as observations comes in. And noone is suggesting that Markov models by themselves are a serious model of human language performance. But I (and others) suggest that probabilistic, trained models are a better model of human language performance than are categorical, untrained models. And yes, it seems clear that an adult speaker of English does know billions of language facts (for example, that one says "big game" rather than "large game" when talking about an important football game). These facts must somehow be encoded in the brain.

It seems clear that probabilistic models are better for judging the likelihood of a sentence, or its degree of sensibility. But even if you are not interested in these factors and are only interested in the grammaticality of sentences, it still seems that probabilistic models do a better job at describing the linguistic facts. The *mathematical* theory of formal languages[17] defines a language as a set of sentences. That is, every sentence is either grammatical or ungrammatical; there is no need for probability in this framework. But natural languages are not like that. A *scientific* theory of natural languages must account for the many phrases and sentences which leave a native speaker uncertain about their grammaticality (see Chris Manning's article (2002) and its discussion of the phrase "as least as"), and there are phrases which some speakers find perfectly grammatical, others perfectly ungrammatical, and still others will flip-flop from one occasion to the next. Finally, there are usages which are rare in a language, but cannot be dismissed if one is concerned with actual data. For example, the verb *quake* is listed as intransitive in dictionaries, meaning that (1) below is grammatical, and (2) is not, according to a categorical theory of grammar.

1. The earth quaked.
2. ? It quaked her bowels.

But (2) actually appears as a sentence of English.[18] This poses a dilemma for the categorical theory. When (2) is observed we must either arbitrarily dismiss it as an error that is outside the bounds of our model (without any theoretical grounds for doing so), or we must change the theory to allow (2), which often results in the acceptance of a flood of sentences that we would prefer to remain ungrammatical. As Edward Sapir said[19] in 1921, "All grammars leak." But in a probabilistic model there is no difficulty; we can say that *quake* has a high probability of being used intransitively, and a low probability of transitive use (and we can, if we care, further describe those uses through subcategorization).

---

17 https://en.wikipedia.org/wiki/Formal_language
18 Cp. McEnery, Tony and Andrew Wilson. 2001. *Corpus Linguistics. An Introduction.* Edinburgh: Edinburgh University Press, p. 107.
19 Sapir, Edward. 1921. *Language: An Introduction to the Study of Speech.* New York: Harcourt, Brace and Company

Steve Abney (1996) points out that probabilistic models are better suited for modeling language change. He cites the example of a 15th century Englishman who goes to the pub every day and orders "Ale!" Under a categorical model, you could reasonably expect that one day he would be served eel, because the great vowel shift[20] flipped a Boolean parameter in his mind a day before it flipped the parameter in the publican's. In a probabilistic framework, there will be multiple parameters, perhaps with continuous values, and it is easy to see how the shift can take place gradually over two centuries.

Thus it seems that grammaticality is not a categorical, deterministic judgment but rather an inherently probabilistic one. This becomes clear to anyone who spends time making *observations* of a corpus of actual sentences, but can remain unknown to those who think that the object of study is their own set of *intuitions* about grammaticality. Both observation and intuition have been used in the history of science, so neither is "novel," but it is observation, not intuition that is the dominant model for science.

Now let's consider what I think is Chomsky's main point of disagreement with statistical models: the tension between "accurate description" and "insight." This is an old distinction. Charles Darwin (biologist, 1809–1882) is best known for his insightful theories but he stressed the importance of accurate description, saying "False facts are highly injurious to the progress of science, for they often endure long; but false views, if supported by some evidence, do little harm, for every one takes a salutary pleasure in proving their falseness." More recently, Richard Feynman (physicist, 1918–1988) wrote "Physics can progress without the proofs, but we can't go on without the facts."

On the other side, Ernest Rutherford (physicist, 1871–1937) disdained mere description, saying "All science is either physics or stamp collecting." Chomsky stands with him: "You can also collect butterflies and make many observations. If you like butterflies, that's fine; but such work must not be confounded with research, which is concerned to discover explanatory principles."

Acknowledging both sides is Robert Millikan (physicist, 1868–1953) who said in his Nobel acceptance speech "Science walks forward on two feet, namely theory and experiment ... Sometimes it is one foot that is put forward first, sometimes the other, but continuous progress is only made by the use of both."

---

20 https://en.wikipedia.org/wiki/Great_Vowel_Shift

2.6   The two cultures

After all those distinguished scientists have weighed in, I think the most relevant contribution to the current discussion is the 2001 paper by Leo Breiman (statistician, 1928–2005), Statistical Modeling: The Two Cultures. In this paper Breiman, alluding to C.P. Snow, describes two cultures:

First the **data modeling culture** (to which, Breiman estimates, 98% of statisticians subscribe) holds that nature can be described as a black box that has a relatively simple underlying model which maps from input variables to output variables (with perhaps some random noise thrown in). It is the job of the statistician to wisely choose an underlying model that reflects the reality of nature, and then use statistical data to estimate the parameters of the model.

Second the **algorithmic modeling culture** (subscribed to by 2% of statisticians and many researchers in biology, artificial intelligence, and other fields that deal with complex phenomena), which holds that nature's black box cannot necessarily be described by a simple model. Complex algorithmic approaches (such as support vector machines or boosted decision trees or deep belief networks) are used to estimate the function that maps from input to output variables, but we have no expectation that the *form* of the function that emerges from this complex algorithm reflects the true underlying nature.

It seems that the algorithmic modeling culture is what Chomsky is objecting to most vigorously. It is not just that the models are statistical (or probabilistic), it is that they produce a form that, while accurately modeling reality, is not easily interpretable by humans, and makes no claim to correspond to the generative process used by nature. In other words, algorithmic modeling describes what *does* happen, but it doesn't answer the question of *why*.

Breiman's article explains his objections to the first culture, data modeling. Basically, the conclusions made by data modeling are about the model, not about nature. (Aside: I remember in 2000 hearing James Martin,[21] the leader of the Viking missions to Mars, saying that his job as a spacecraft engineer was not to land on Mars, but to land on the model of Mars provided by the geologists.) The problem is, if the model does not emulate nature well, then the conclusions may be wrong. For example, linear regression is one of the most powerful tools in the statistician's toolbox. Therefore, many analyses start out with "Assume the data are generated by a linear model..." and lack sufficient analysis of what happens if the data are not in fact generated that way. In addition, for complex problems there are usually many alternative good models, each with very similar measures of goodness of fit. How

---

21 http://www.nasa.gov/home/hqnews/2002/02-072.txt

is the data modeler to choose between them? Something has to give. Breiman is inviting us to give up on the idea that we can uniquely model the true underlying *form* of nature's function from inputs to outputs. Instead he asks us to be satisfied with a function that accounts for the observed data well, and generalizes to new, previously unseen data well, but may be expressed in a complex mathematical form that may bear no relation to the "true" function's form (if such a true function even exists). Chomsky takes the opposite approach: he prefers to keep a simple, elegant model, and give up on the idea that the model will represent the data well. Instead, he declares that what he calls *performance* data—what people actually do—is off limits to linguistics; what really matters is *competence*—what he imagines that they should do.

In January of 2011, television personality Bill O'Reilly weighed in on more than one culture war with his statement *"tide goes in, tide goes out. Never a miscommunication. You can't explain that,"* which he proposed as an argument for the existence of God. O'Reilly was ridiculed by his detractors for not knowing that tides can be readily explained by a system of partial differential equations describing the gravitational interaction of sun, earth, and moon (a fact that was first worked out by Laplace in 1776 and has been considerably refined since; when asked by Napoleon why the creator did not enter into his calculations, Laplace said "I had no need of that hypothesis."). (O'Reilly also seems not to know about Deimos and Phobos (two of my favorite moons in the entire solar system, along with Europa, Io, and Titan), nor that Mars and Venus orbit the sun, nor that the reason Venus has no moons is because it is so close to the sun that there is scant room for a stable lunar orbit.) But O'Reilly realizes that it doesn't matter what his detractors think of his astronomical ignorance, because his supporters think he has gotten exactly to the key issue: *why?* He doesn't care *how* the tides work, tell him *why* they work. *Why* is the moon at the right distance to provide a gentle tide, and exert a stabilizing effect on earth's axis of rotation, thus protecting life here? *Why* does gravity work the way it does? *Why* does anything at all exist rather than not exist? O'Reilly is correct that these questions can only be addressed by mythmaking, religion or philosophy, not by science.

Chomsky has a philosophy based on the idea that we should focus on the deep *whys* and that mere explanations of reality don't matter. In this, Chomsky is in complete agreement with O'Reilly. (I recognize that the previous sentence would have an extremely low probability in a probabilistic model trained on a newspaper or TV corpus.) Chomsky believes a theory of language should be simple and understandable, like a linear regression model where we know the underlying process is a straight line, and all we have to do is estimate the slope and intercept.

For example, consider the notion of a pro-drop language[22] from Chomsky's Lectures on Government and Binding (1981). In English we say, for example, "I'm hungry," expressing the pronoun "I". But in Spanish, one expresses the same thought with "Tengo hambre" (literally "have hunger"), dropping the pronoun "Yo". Chomsky's theory is that there is a "pro-drop parameter" which is "true" in Spanish and "false" in English, and that once we discover the small set of parameters that describe all languages, and the values of those parameters for each language, we will have achieved true understanding.

The problem is that reality is messier than this theory. Here are some dropped pronouns in English:

- "Not gonna do it. Wouldn't be prudent." (Dana Carvey, impersonating George H. W. Bush)
- "Thinks he can outsmart us, does he?" (Evelyn Waugh, The Loved One)
- "Likes to fight, does he?" (S.M. Stirling, The Sunrise Lands)
- "Thinks he's all that." (Kate Brian, Lucky T)
- "Go for a walk?" (countless dog owners)
- "Gotcha!" "Found it!" "Looks good to me!" (common expressions)

Linguists can argue over the interpretation of these facts for hours on end, but the diversity of language seems to be much more complex than a single Boolean value for a pro-drop parameter. We shouldn't accept a theoretical framework that places a priority on making the model simple over making it accurately reflect reality.

From the beginning, Chomsky has focused on the *generative* side of language. From this side, it is reasonable to tell a non-probabilistic story: I *know* definitively the idea I want to express—I'm starting from a single semantic form—thus all I have to do is choose the words to say it; why can't that be a deterministic, categorical process? If Chomsky had focused on the other side, *interpretation*, as Claude Shannon did, he may have changed his tune. In interpretation (such as speech recognition) the listener receives a noisy, ambiguous signal and needs to decide which of many possible intended messages is most likely. Thus, it is obvious that this is inherently a probabilistic problem, as was recognized early on by all researchers in speech recognition, and by scientists in other fields that do interpretation: the astronomer Laplace said in 1819 "Probability theory is nothing more than common sense reduced to calculation," and the physicist James Maxwell said in 1850 "The true logic for this world is the calculus of Probabilities, which takes account of the magnitude of the probability which is, or ought to be, in a reasonable man's mind."

---

22 https://en.wikipedia.org/wiki/Pro-drop_language

Finally, one more reason why Chomsky dislikes statistical models is that they tend to make linguistics an empirical science (a science about how people actually use language) rather than a mathematical science (an investigation of the mathematical properties of models of formal language). Chomsky prefers the later, as evidenced by his statement in *Aspects of the Theory of Syntax* (1965):

> Linguistic theory is mentalistic, since it is concerned with discovering a mental reality underlying actual behavior. Observed use of language ... may provide evidence ... but surely cannot constitute the subject-matter of linguistics, if this is to be a serious discipline.

I can't imagine Laplace saying that observations of the planets cannot constitute the subject-matter of orbital mechanics, or Maxwell saying that observations of electrical charge cannot constitute the subject-matter of electromagnetism. It is true that physics considers idealizations that are abstractions from the messy real world. For example, a class of mechanics problems ignores friction. But that doesn't mean that friction is not considered part of the subject-matter of physics.

So how could Chomsky say that observations of language cannot be the subject-matter of linguistics? It seems to come from his viewpoint as a Platonist and a Rationalist and perhaps a bit of a Mystic. As in Plato's allegory of the cave, Chomsky thinks we should focus on the ideal, abstract forms that underlie language, not on the superficial manifestations of language that happen to be perceivable in the real world. That is why he is not interested in language performance. But Chomsky, like Plato, has to answer where these ideal forms come from. Chomsky (1991) shows that he is happy with a Mystical answer, although he shifts vocabulary from "soul" to "biological endowment."

> Plato's answer was that the knowledge is 'remembered' from an earlier existence. The answer calls for a mechanism: perhaps the immortal soul ... rephrasing Plato's answer in terms more congenial to us today, we will say that the basic properties of cognitive systems are innate to the mind, part of human biological endowment.

It was reasonable for Plato to think that the ideal of, say, a horse, was more important than any individual horse we can perceive in the world. In 400BC, species were thought to be eternal and unchanging. We now know that is not true; that the horses on another cave wall—in Lascaux—are now extinct, and that current horses continue to evolve slowly over time. Thus there is no such thing as a single ideal eternal "horse" form.

We also now know that language is like that as well: languages are complex, random, contingent biological processes that are subject to the whims of evolution and cultural change. What constitutes a language is not an eternal ideal form, represented by the settings of a small number of parameters, but rather is the contingent outcome of complex processes. Since they are contingent, it seems they can only be analyzed with probabilistic models. Since people have to continually understand the uncertain, ambiguous, noisy speech of others, it seems they must be using something like probabilistic reasoning. Chomsky for some reason wants to avoid this, and therefore he must declare the actual facts of language use out of bounds and declare that true linguistics only exists in the mathematical realm, where he can impose the formalism he wants. Then, to get language from this abstract, eternal, mathematical realm into the heads of people, he must fabricate a mystical facility that is exactly tuned to the eternal realm. This may be very interesting from a mathematical point of view, but it misses the point about what language is, and how it works.

Thanks

Thanks to Ann Farmer, Fernando Pereira, Dan Jurafsky, Hal Varian, and others for comments and suggestions on this essay.

Annotated Bibliography

Abney, Steve. 1996. Statistical Methods and Linguistics. In *The Balancing Act: Combining Symbolic and Statistical Approaches to Language*, ed. Judith L. Klavans and Philip Resnik, 1-26. Cambridge, MA: MIT Press.

An excellent overall introduction to the statistical approach to language processing, and covers some ground that is not addressed often, such as language change and individual differences.

Breiman, Leo. 2001. Statistical Modeling: The Two Cultures. *Statistical Science* 16(3): 199-231.

Breiman does a great job of describing the two approaches, explaining the benefits of his approach, and defending his points in the very interesting commentary with eminent statisticians: Cox, Efron, Hoadley, and Parzen.

Chomsky, Noam. 1956. Three Models for the Description of Language. *IRE Transactions on Information Theory* 2(3): 113-124.

Compares finite state, phrase structure, and transformational grammars. Introduces "colorless green ideas sleep furiously."

Chomsky, Noam. 1957. *Syntactic Structures*. The Hague: Mouton.

A book-length exposition of Chomsky's theory that was the leading exposition of linguistics for a decade. Claims that probabilistic models give no insight into syntax.

Chomsky, Noam. 1969. Some Empirical Assumptions in Modern Philosophy of Language. In *Philosophy, Science and Method: Essays in Honor of Ernest Nagel*, ed. S. Morgenbesser, P. Suppes, and M. White, 260-285. New York: St. Martin's Press.

Claims that the notion "probability of a sentence" is an entirely useless notion.

Chomsky, Noam. 1981. *Lectures on Government and Binding*. Dordrecht: Foris Publications.

A revision of Chomsky's theory; this version introduces Universal Grammar. We cite it for the coverage of parameters such as pro-drop.

Chomsky, Noam. 1991. Linguistics and Adjacent Fields: A Personal View, in *The Chomskyan Turn*, ed. A. Kasher, 3-25. Oxford and Cambridge, MA: Basil Blackwell.

I found the Plato quotes in this[23] article, published by the Communist Party of Great Britain, and apparently published by someone with no linguistics training whatsoever, but with a political agenda.

Gold, E. M. 1967. Language Identification in the Limit. *Information and Control* 10(5): 447-474.

Gold proved a result in formal language theory that we can state (with some artistic license) as this: imagine a game between two players, guesser and chooser. Chooser says to guesser, "Here is an infinite number of languages. I'm going to choose one of them, and start reading sentences to you that come from that language. On your N-th birthday there will be a True-False quiz where I give you 100 sentences you haven't heard yet, and you have to say whether they come from the language or not." There are some limits on what the infinite set looks like and on how the chooser can pick sentences (he can be deliberately tricky, but he can't just repeat the same sentence over and over, for example). Gold's result is that if the infinite set of languages are all generated by context-free grammars then there is no strategy for guesser that guarantees she gets 100% correct every time, no matter what N you choose for the birthday. This result was taken by Chomsky and others to mean that it is impossible for children to learn human languages without having an innate "language organ." As Johnson (2004) and others show, this was an invalid conclusion; the task of getting 100% on the quiz (which

---

23 http://www.cpgb.org.uk/article.php?article_id=1004261#23

Gold called language identification) really has nothing in common with the task of language acquisition performed by children, so Gold's Theorem has no relevance.

Horning, J. J. 1969. *A study of grammatical inference.* Ph.D. thesis, Stanford University.

Where Gold found a negative result—that context-free languages were not identifiable from examples, Horning found a positive result—that probabilistic context-free languages are identifiable (to within an arbitrarily small level of error). Nobody doubts that humans have unique innate capabilities for understanding language (although it is unknown to what extent these capabilities are specific to language and to what extent they are general cognitive abilities related to sequencing and forming abstractions). But Horning proved in 1969 that Gold cannot be used as a convincing argument for an innate language organ that specifies all of language except for the setting of a few parameters.

Johnson, Kent. 2004. Gold's Theorem and cognitive science. *Philosophy of Science* 71: 571-592.

The best article I've seen on what Gold's Theorem actually says and what has been claimed about it (correctly and incorrectly). Concludes that Gold has something to say about formal languages, but nothing about child language acquisition.

Lappin, Shalom, and Shieber, Stuart M. 2007. Machine learning theory and practice as a source of insight into universal grammar. *Journal of Linguistics* 43(2): 393-427.

An excellent article discussing the poverty of the stimulus, the fact that all models have bias, the difference between supervised and unsupervised learning, and modern (PAC or VC) learning theory. It provides alternatives to the model of Universal Grammar consisting of a fixed set of binary parameters.

Manning, Christopher. 2002. Probabilistic Syntax. In *Probabilistic Linguistics*, ed. Rens Bod, Jennifer Hay, and Stefanie Jannedy, 289-341. Cambridge, MA: MIT Press.

A compelling introduction to probabilistic syntax, and how it is a better model for linguistic facts than categorical syntax. Covers "the joys and perils of corpus linguistics."

Norvig, Peter. 2007. How to Write a Spelling Corrector. http://norvig.com/spell-correct.html

Shows working code to implement a probabilistic, statistical spelling correction algorithm.

Norvig, Peter. 2009. Natural Language Corpus Data. In *Beautiful Data*, ed. Toby Segaran and Jeff Hammerbacher, 219-242. Sebastopol, CA: O'Reilly.

Expands on the essay above; shows how to implement three tasks: text segmentation, cryptographic decoding, and spelling correction (in a slightly more complete form than the previous essay).

Pereira, Fernando. 2002. Formal grammar and information theory: together again? In *The Legacy of Zellig Harris*, ed. Bruce E. Nevin and Stephen B. Johnson, 13-32. Amsterdam: Benjamins.

When I set out to write the page you are reading now, I was concentrating on the events that took place in Cambridge, Mass., 4800 km from home. After doing some research I was surprised to learn that the authors of two of the three best articles on this subject sit within a total of 10 meters from my desk: Fernando Pereira and Chris Manning. (The third, Steve Abney, sits 3700 km away.) But perhaps I shouldn't have been surprised. I remember giving a talk at ACL on the corpus-based language models used at Google, and having Fernando, then a professor at U. Penn., comment "I feel like I'm a particle physicist and you've got the only super-collider." A few years later he moved to Google. Fernando is also famous for his quote "The older I get, the further down the Chomsky Hierarchy I go." His article here covers some of the same ground as mine, but he goes farther in explaining the range of probabilistic models available and how they are useful.

Plato. c. 380BC. *The Republic*.

Cited here for the allegory of the cave.

Shannon, C.E. 1948. A Mathematical Theory of Communication. *The Bell System Technical Journal* 27: 379-423.

An enormously influential article that started the field of information theory and introduced the term "bit" and the noisy channel model, demonstrated successive n-gram approximations of English, described Markov models of language, defined entropy with respect to these models, and enabled the growth of the telecommunications industry.

3. Big Data: A New Empiricism and its Epistemic and Socio-Political Consequences

Gernot Rieder[1], Judith Simon[2]

Abstract

The paper investigates the rise of Big Data in contemporary society. It examines the most prominent epistemological claims made by Big Data proponents, calls attention to the potential socio-political consequences of blind data trust, and proposes a possible way forward. The paper's main focus is on the interplay between an emerging new empiricism and an increasingly opaque algorithmic environment that challenges democratic demands for transparency and accountability. It concludes that a responsible culture of quantification requires epistemic vigilance as well as a greater awareness of the potential dangers and pitfalls of an ever more data-driven society.

1   Gernot Rieder is a PhD fellow at the IT University of Copenhagen and member of the research project "Epistemic Trust in Socio-Technical Epistemic Systems" at the University of Vienna. His dissertation investigates the rise of Big Data in public policy and the social, ethical, and epistemological implications of data-driven decision making. Gernot serves as an assistant editor for the journal "Big Data & Society".

2   Judith Simon is Associate Professor for Philosophy of Science and Technology at the IT University of Copenhagen and PI of the research project "Epistemic Trust in Socio-Technical Epistemic Systems" at the University of Vienna. She is co-editor of the journals "Philosophy & Technology" and "Big Data & Society". She also serves on the executive boards of the International Society for Ethics and Information Technology (inseit.net) and the International Association for Computing and Philosophy (iacap.org).

3.1   Introduction

The "Age of Big Data" (Lohr 2012) is firmly upon us, and it promises to change not only how "we live, work, and think" (Mayer-Schönberger and Cukier 2013), but also, and perhaps most fundamentally, *how we know*.

People, sensors, and systems generate increasingly large amounts of data. The networking company Cisco (2015) estimates that global Internet traffic has increased fivefold over the past five years, and will have tripled again by 2019. In the same year, driven by new users, products, and the quickly expanding Internet of Things, the number of Internet-connected devices is expected to reach 24 billion, compared to about 14 billion in 2015 (Ericsson 2015).[3] Already there are "more data [...] being generated every week than in the last millennia" (OECD 2015), and at a rate that is likely to accelerate (UNECE 2014).

But Big Data as a complex *techno-scientific phenomenon* is not just about growth in data "volume, velocity, and variety" (Laney 2001), it is also seen as the ability to mine and manipulate data in ways that allow to "extract meaning and insight" and "reveal [hidden] trends and patterns" (IBM 2014). Consequently, and heavily lobbied by industry stakeholders, there have been considerable investments in analytical capabilities (i.e., hardware, software, and skills) across both public and private sectors. The European Commission, for instance, has only recently announced a €2.5 billion public-private partnership in an effort to put "Europe at the forefront of the global data race" and "master Big Data" (EC 2014a).

Enthusiasts and advocates from research and industry have argued that Big Data presents a new scientific paradigm (Hey et al. 2009), a data-intensive exploratory science with the "dream of establishing a 'sensors everywhere' data infrastructure" (Bell 2009), enabling us to "measure more, faster, than ever before" (Wilbanks 2009). From such a perspective, the impact of advanced data analytics is nothing short of revolutionary: In addition to transforming a wide array of areas such as health care (Science Europe 2014), education (Dede 2015), or law enforcement (Bachner 2013), Big Data is said to produce a new kind of knowledge, one that is more *comprehensive*, more *objective*, and more *predictive*.

Against this backdrop, scholars from the social sciences and humanities have warned of an emerging "new empiricism" (Kitchin 2014a), a certain belief that "the volume of data, accompanied by techniques that can reveal their inherent truth, enables data to speak for themselves free of theory" (ibid.). Taken to the extreme, such

---

3   Such forecasts should be regarded with caution. The numbers vary not only between research firms, but also between individual reports issued by the same company, and thus are quite volatile.

unbridled "trust in numbers" (Porter 1995) is said to lead to "data-ism" (Brooks 2013), a "deification of data" (Jenkins 2013) that promotes forms of "algorithmic governance" (Williamson 2014) and "digital age [...] Taylorism" (Lohr 2015) where a focus on performance and efficiency metrics replaces other norms and values.

Following a brief discussion of the rise of Big Data in contemporary society (3.2), this paper examines some of the most prominent epistemological claims made by Big Data proponents (3.3), calls attention to the potential socio-political ramifications of unrestrained 'datatrust' (3.4), and, in a concluding section, points to potential conceptual alternatives (3.5).

3.2   Big Data: Genesis, Definitions, Trust

While the exact origins of the term remain a matter of debate (Lohr 2013), biblio-metric studies have documented a growing number of Big Data articles since the early 2000s, with a sharp increase in publications since 2008 (Halevi and Moed 2012). At first, references to Big Data can mainly be found in the engineering and computer science literature; more recently, the use of the term has become wide-spread across a host of disciplines – from business and management to physics, biology, and medicine to the social sciences and the humanities. But the popular-ization of the notion has not been restricted to academia. A Google Trends search for "Big Data" indicates a strong surge in interest[4] since 2011, the tool's forecast fea-ture predicting a further increase from 2015 onwards. At present, there appears to be hardly any major news outlet that has not dealt with the phenomenon in one way or another, and its heavy use as an advertising and marketing term has solidified its reputation as the "buzzword of the decade" (Barocas and Selbst 2015).

Many definitions of Big Data have been given, but no consensus has been reached. As Schroeder (2014) notes, there are "no definitive, academic definitions of data and of Big Data". Probably best known, however, are the so-called "3Vs" sug-gested by former META Group (now Gartner) analyst Doug Laney (2012; 2001). According to this model, Big Data can be characterized as growth in data *volume* (i.e., a change in the depth and breadth of data available), *velocity* (i.e., an accelera-tion of data generation), and *variety* (i.e., a greater heterogeneity of data types and formats). But even when extended to four (IBM 2013), five (Marr 2014), or seven

---

4   Google's "interest over time" graph shows "total searches for a term relative to the total number of searches done on Google", see:
https://support.google.com/trends/answer/4355164?hl=en&rd=1 (accessed 30 May 2016).

Vs (Van Rijmenam 2013), frameworks of this kind mainly focus on *measures of magnitude* and related challenges, thus providing a very narrow view of what constitutes Big Data. Other approaches have shifted attention from data properties to new forms of analysis, conceptualizing Big Data as a "problem-solving philosophy" (Hartzog and Selinger 2013) that "link[s] seemingly disparate disciplines" (Berman 2013), enabling researchers to "discover relationships" (Schaeffer and Olson 2014) and "harness information in novel ways to produce useful insights" (Mayer-Schönberger and Cukier 2013). While such a perspective offers a richer understanding of Big Data as a complex techno-scientific phenomenon, it cannot account for the current *climate of hype* that surrounds modern data analytics. This is where a third type of definition provides further insight. As boyd and Crawford (2012) have argued, Big Data is not only about technological progress and advances in analytical techniques, but also about a "widespread belief that large data sets offer a higher form of intelligence and knowledge that can generate insights that were previously impossible". Leonelli (2014) makes a similar argument, stressing that the novelty of Big Data science does not lie in the sheer quantity of data involved, but in the "prominence and status acquired by data as commodity and recognized output." It is arguably this *trust* in the authority of data, this *faith* and *belief* in numerical evidence, that has greatly contributed to the buzz-laden narrative of Big Data discourse.

The rhetoric of hope and hype is not limited to industry and business circles, it also features prominently in political realms. Policymakers have referred to Big Data as, e.g., the "new oil of the digital age" (EC 2012a), a game-changing "key asset" (EC 2015a), or the next "industrial revolution" (EC 2014c). There appears to be agreement that "Big Data drives big benefits" (EOP 2014), and that one must "seize the opportunities afforded by this new, data-driven revolution" (NSF 2012). On the one hand, the conviction that data science can "change this [...] world for the better" (Obama 2015) is deeply rooted in the history of modern Western statecraft (Rieder and Simon 2016); on the other hand, it is sustained and nurtured by a very specific epistemic imaginary, a core set of knowledge claims that, although not entirely new, has (re)gained significant traction over the past few years. In the next section, we shall take a closer look at the composition of this imaginary, examining how it contributes to the current data hype.

3.3   The Rise of a New Empiricism

One of the most comprehensive critical overviews of Big Data epistemologies to date has been provided by Kitchin (2014a; b). In essence, Kitchin argues that in the context of Big Data one can observe the "articulation of a new empiricism", which "operates as a discursive rhetorical device designed to [...] convince vendors [and other stakeholders] of the utility and value of Big Data analytics" (ibid. 2014a). But what exactly are the promises and claims associated with this emerging new paradigm? If Big Data serves as a potent rhetorical device, what are the ascribed powers? While an extensive discussion of the Big Data imaginary[5] is beyond the scope of this paper, a brief outline of certain key assumptions may allow for a better understanding of the particular kind of knowledge Big Data practices are said to produce. It thus seems worthwhile to consider these assumptions in some more detail.

First, there is the notion that Big Data is exhaustive in scope, capable of capturing entire populations or domains (N=all) rather than being limited to sample-based surveys, allowing researchers to "get the complete picture" (Oracle 2012) and "make sense [...] without traditional reduction" (Strawn 2012). This belief in what Lagoze (2014) critically refers to as the "allness" of Big Data is, on the one hand, driven by the proliferation of digital data in today's increasingly computerized society and, on the other hand, the result of improved capacities to retrieve, store, and analyze those data. The expectations are high: At its best, Big Data is supposed to "give a view of life in all its complexity" (Pentland 2014), combining "millions, if not billions, of individual data points" to get "the full resolution on worldwide affairs" (Steadman 2013). In addition, this vision of completeness is assumed to be (a) more inclusive and representative than other forms of research, "encompass[ing] thousands of times more people than a Gallup or Pew study" (Rudder 2014); (b) analytically superior in the sense that "the more data available the better and more accurate the results" (EC-BIO 2013a); and (c) more direct and unmediated as it reveals "what people actually *do* rather than what they *say they do*" (Strong 2015). This last aspect points to another central claim.

A second key imaginary holds that with enough volume the data speak for themselves, replacing "the narrative with the empirical" (Brooks 2013) and eliminating any need for a priori theory. Instead, meaning is thought to emerge from the data "without human involvement" (Szal 2015), rendering established forms of scientific inquiry – i.e., hypothesize, model, test – obsolete (see Anderson 2008). Advanced algorithms are said to "find patterns where science cannot" (ibid.), shifting

---

5   For other mentions of such a distinct "Big Data imaginary", see Housley (2015) and Williamson (2015).

the focus from causal explanations to the discovery of statistical correlations that "inherently produce [...] insightful knowledge about social, political and economic processes" (Kitchin 2014b). What follows is a reduction in the perceived relevance of context since "knowing *what*, not *why*, is good enough" (Mayer-Schönberger and Cukier 2013) to make "human systems [...] run better and smarter" and "engineer a safer and healthier world" (Eagle & Greene 2014). Similarly, subject matter expertise and domain-specific knowledge is believed to "matter less" when "probability and correlation are paramount", suggesting that the pioneers and innovators of the Big Data era will "often come from fields outside the domain where they make their mark" (Mayer-Schönberger and Cukier 2013). Consequently, computer and data scientists rather than, e.g., physicians, biologists, or sociologists are considered the main protagonists of this new research paradigm (see Davenport and Patil 2012).

The third imaginary overlaps with the second, but extends the argument even further: Not only are data seen as speaking for themselves, free of human intervention, they are also, by their very nature, expected to be fair and objective. More specifically, "agnostic statistics" (Anderson 2008) are said to "eliminate human bias" (Richtel 2013) from decision-making processes, offering an impartial "view from nowhere" that reveals new truths and provides a "disinterested picture of reality" (Jurgenson 2014). Replacing "gut and intuition" with "numbers and metrics" (Gutierrez 2015), fully automated software is supposed to deliver fact-based recommendations, acting as a neutral corrective to people's "subjective judgments and hunches" (Clinton 2016). While confidence in the virtues of quantification is no new phenomenon (e.g., see Cohen 2005; Porter 1995; Hacking 1990), the promise of "algorithmic objectivity" (Gillespie 2014) has raised hopes that modern data analytics may serve as a "powerful weapon in the fight for equality" (Castro 2014), battling discrimination across a broad range of sectors – from employment and education to law enforcement and health care to housing and credit – "empower[ing] vulnerable groups" and "ensur[ing] equal opportunity for all" (FPF 2014). The idea is both simple and compelling: The more mechanized the process, the higher the chance that the results won't be tainted by researchers' interpretive subjectivity (see Venturini et al. 2014) – after all, "it's humans, not algorithms, that have a bias problem" (New 2015).

Fourth and finally, the application of sophisticated Big Data techniques is meant to provide "certainty in an uncertain world" (AppDynamics 2014), generating reliable knowledge for robust, evidence-informed decision making. As Hardy (2013) notes, the "promise of certainty has been a hallmark of the technology industry for decades", and the ability to reduce ambiguity, establish clarity, and determine risk is often touted as a major benefit of advanced data analytics. Thriving on rather than drowning in information overload, Big Data methods are expected to find the 'signal in the noise', the 'needle in the haystack', 'connecting the dots' with high precision and accuracy. What is particularly noteworthy is the scope of the claim: Governing principles – much like physical laws – are believed to "undergird virtually every interaction in society" (Silver 2012), and the arrival of dense, continuous data together with modern computation makes it possible to detect "statistical regularities that [...] are true of almost everyone almost all of the time", offering valuable "insights about human nature" (Pentland 2014). Once 'reality' has been mined[6] and patterns are found, "more and more aspects of our lives [...] become predictable" (EC-BIO 2013b), allowing us to "tame uncertainty" (Byrne 2012) and "sens[e] the future before it occurs" (Fitzgerald 2012). If to "know ahead [and] act before" (Quantacast 2013) is the industry's trademarked mantra, the "end of chance" (Müller et al. 2013) is its ultimate goal.

There are, of course, a number of other epistemological beliefs and assumptions that feed into the Big Data imaginary – such as the idea that "unless something can be measured, it cannot be improved" (Kelly 2007) or that "the law of large numbers [...] evens out the errors of any individual data point" (Phillips Mandaville 2014) – but these are arguably an extension of, or at least closely related to, the four central propositions outlined above: Namely, that Big Data represents nothing less than a computational means to know everything (I), of anything (II), free from bias (III), with a high degree of certainty (IV).

Scholars from various disciplines have challenged the bold claims of Big Data empiricism, arguing that modern analytical techniques neither provide a complete picture of entire populations (McFarland and McFarland 2015) nor eradicate the need for models and a priori theory (Pigliucci 2009), are neither neutral or free of bias (Hardt 2014) nor have the ability to predict with certainty (Silver 2012). Instead, Big Data collections can be "small" and "partial" (Leonelli 2014); algorithms may "perpetuate the prejudices of their creators" (CIHR 2015) or "learn bias from the data fed into them" (Kun 2015); and forecasts are never certain, but deal in probabilities, possibilities, and uncertainties that may be "narrow in reach, scope, and perspective" (Ekbia et al. 2013). Yet despite signs of growing awareness – e.g., the recent White House report *Big Data: Seizing Opportunities, Preserving Values* (EOP 2014) emphasizes the need for a "national conversation on big data, discrimination, and civil liberties" – the ideal of "impersonal rationality achieved through technical methods" (Porter 2011) continues to act as an important techno-political leitmotif. As Hildebrandt (2013) observes, "we have trouble to resist the seemingly clean, objective knowledge [Big Data] produces", and turn it, by making ourselves

---

6   For more on the concept of reality mining, see Eagle and Pentland (2006) and Eagle and Greene (2014).

dependent upon its oracles, "into a new pantheon, filled with novel gods." But what are the dangers of being enthralled by the power and possibilities of modern computing, of placing our faith in the veracity of ever more widespread predictive analytics? If society were to follow the path toward digital serfdom, what are the costs of becoming "slaves to Big Data" (ibid.)?

3.4   A Black Box Society

Discussions about the social and political ramifications of Big Data have predominantly focused on two interrelated sets of issues: questions of *privacy* (e.g., regarding surveillance, profiling, or data protection/security) on the one hand and instances of *discrimination* (e.g., through differential access or treatment) on the other. Journalists and academics have examined a variety of analytical tools and techniques – from Target's pregnancy-prediction model (Duhigg 2012) and the City of Boston's StreetBump app (Crawford 2013) to Google's Flu Trend service (Lazer et al. 2014) and Facebook's 'emotional contagion' study (Meyer 2014) – exposing both methodological biases and ethical transgressions.

While reports of this kind have contributed to public awareness and dialogue, the focus on a limited number of high-profile cases may give the impression that there is only cause for concern if things have either gone 'wrong' or 'too far', that is, if the analytical process is flawed or there is clear indication of unethical or illegal conduct. But this is not the case. In fact, even if done 'right', data mining may reflect "widespread biases that persist in society at large", and can thus "affect the fortunes of whole classes of people in consistently unfavorable ways" (Barocas and Selbst 2015). Similarly, the strong reactions to Facebook's emotional contagion study, in which researchers altered the News Feeds of almost 700,000 users (Kramer et al. 2014), fail to take into account that "manipulating the News Feed is Facebook's entire business" (Patel 2014). To be clear, the point here is not to defend any particular study or experiment; rather, it is to emphasize that the issues and concerns run much deeper, are systemic rather than case-specific, cultural rather than attributable to a few outliers. The main impact of the Big Data phenomenon is not that smart TVs may be listening to "everything you say" (Harris 2015), that social media posts "may damage your credit score" (McLannahan 2015), or that the police might pay you a visit because your name appeared on a software-generated "heat list" (Gorner 2013); much rather, it is that Big Data with its "aura of truth [and] objectivity" (boyd & Crawford 2012) contributes to a specific "algorithmic culture"

(Striphas 2015) that renders such practices not only "the new normal" (Andrejevic 2013), but also socially acceptable[7].

The consequences may be quite severe. Pasquale (2015), for instance, cautions of an emerging "black box society", i.e., a "system whose workings are mysterious", where the "distinction between state and market is fading", and where people submit to the "dictate of salient, measurable data", the "rule of scores and bets". Others have voiced similar concerns, arguing that unquestioned faith in the seemingly impartial workings of the machine may undermine civil liberties (Al-Rodhan 2014), threaten social and economic justice (Newman 2015), and prove "toxic to democratic governance and [...] democracy itself" (Howard 2014). But what exactly are the problems and challenges associated with this brave new data world? Though there are overlaps, it is possible to distinguish between three main issues and concerns: opacity, accountability, responsibility.

Opacity – While the use of algorithmic decision-making tools by governments and private entities has grown progressively (Zarsky 2016), analytical processes are often opaque, operating according to rules that are hidden, using input data that remain unknown. This lack of transparency has several reasons. For one, as Burrell (2016) highlights, opacity can be the result of intentional *corporate or state secrecy*, e.g., in order to secure a competitive advantage, to shield an algorithm from being 'gamed', or to avoid regulation and control. Second, opacity may be a consequence of high algorithmic *complexity*, especially in the case of machine learning applications. While it might be possible to "untangle the logic of the code within a complicated software system", being able to understand the algorithm in action as it operates on data may not be feasible as such machine optimizations "do not naturally accord with human semantic explanations" and hence "escape full understanding and interpretation" (ibid.). As Gillespie (2014) argues, there appears to be something "impenetrable about algorithms", even for well-trained programmers and computer scientists. Finally, there is the problem of *volatility*. As Facebook engineers explain, "code is not a fixed artifact but an evolving system, updated frequently and concurrently by many developers" (Calcagno et al. 2015). Google, for instance, is known to change its search algorithm around 500-600 times a year, including both minor and major updates (Moz 2016). In that sense, code may not only be secret and opaque, it may also be quite elusive.

---

7   Such social acceptance is actually quite contradictory: While a study of the Pew Research Center (2014) finds that 91% of the surveyed adults "agree" or "strongly agree" that consumers have lost control over how personal information is collected and used by companies, Internet and technology giants such as Amazon, Google, or Apple regularly rank not only amongst the most valuable (Ember 2015), but also amongst the most reputable corporations (Adams 2015).

Accountability – In a world where Big Data software takes command, "accountability is often murky" (Rosenblat et al. 2014). Internet companies, in particular, seek to avoid scrutiny and deflect concerns regarding their services and products. Whenever there is public outcry – as in the case of Facebook's emotional contagion study – the giants of the Web are quick to apologize, acknowledging that they "did a bad job", "really messed up", or "missed the mark" (Isaac 2014). But the tinkering and testing continues – without much oversight, meaningful public deliberation, or serious legal or financial consequences[8]. New legislative instruments, such as the EU's proposed General Data Protection Regulation (EC 2012b), are supposed to clarify industry obligations by introducing "binding corporate rules" (ibid.). Legal scholars, however, have warned that current reform efforts do not go far enough as Big Data "defeats traditional privacy law by undermining core principles and regulatory assumptions" (Rubinstein 2013), including, for instance, the informed choice model, the distinction between personal and non-personal data, or requirements of data minimization. Thus, the overall picture is somewhat bleak: While current laws seem unable to enforce greater corporate accountability, data-mining firms show little interest in disclosing details about their business practices, a sentiment not unique to the private sector as the Snowden revelations have made abundantly clear (Greenwald 2014).

Responsibility – Professional software systems are usually not created by a single person, but by groups of people with different skills and expertise working in institutional settings. If a system fails and causes harm, determining individual responsibility may prove difficult since "responsibility […] does not easily generalize to collective action" (Nissenbaum 1996). The question of who is to blame becomes even more challenging now that 'smart' technologies take over a growing number of knowledge and decision-making processes. Especially in machine learning, where "computers [are given] the ability to learn without being explicitly programmed" (Samuel 1959), the "influence of the creator over the machine decreases [while] the influence of the operating environment increases" (Matthias 2004), giving rise to so-called autonomous agents whose actions can be difficult to predict. Companies have sometimes emphasized this autonomy to shirk responsibility. Google, for example, maintains that its autocomplete suggestions, which have repeatedly been criticized for discriminating against protected classes (UN Women 2013) and

---

8   The monetary fines imposed on large technology companies have often been insignificant compared to the revenues their services generate (Pasquale 2015). More recently, however, EU regulators have increased pressure by filing antitrust charges against Google, which could lead to fines of more than €6 billion – about 10 percent of the company's 2014 revenue (Kanter and Scott 2015).

defaming individuals (Niggemeier 2012), are "generated by an algorithm without any human involvement" and merely "reflect what other people are searching for" (Google 2016). Courts around the globe have arrived at different verdicts – from imposing fines and ordering the removal of specific autocomplete suggestions (Valinsky 2013) to discharging the search company from liability (Bellezza and De Santis 2013) – demonstrating that in Big Data contexts, questions of responsibility are neither obvious nor easy.

Government and corporate secrecy paired with technical inscrutability; obsolescent legal safeguards that are no match for new forms of "digital feudalism" (Clark 2011); algorithmic scapegoating to avoid responsibility and curtail agency – these are the main ingredients of a thoroughly black-boxed data economy, in which "opaque technologies are spreading, unmonitored and unregulated" (Pasquale 2015). Such technologies do not just describe, but actively create social realities: They rank and recommend, classify and score, predict and prescribe, exercising social control through numerical judgment. What they offer are seemingly innocuous automated results; what they produce are specific systems of order. Algorithmic regulation is often described as fair and objective, and the epistemic claims of Big Data empiricism reinforce this image: By claiming to capture everything in great detail, Big Data diverts attention from what remains unseen; by claiming general applicability, it discourages critical debate about where data analytics should or should not be applied; by claiming unbiased neutrality, it discursively impedes public scrutiny; by claiming empirical certainty, it fends off doubts about analytical veracity. The epistemological promises of Big Data empiricism are both powerful and seductive, and if not challenged will contribute to an algorithmic culture that may make us abandon all checks and balances.

3.5   Conclusions

As outlined, the bold claims of Big Data empiricism can be contested both on epistemological grounds and for their socio-political consequences. Given the amount of criticism these claims have received not only from SSH scholars but also from computer and natural scientists (e.g., O'Neil 2016; Hardt 2014), their continued prominence in business and policy contexts may come as a surprise. But while the proliferation of Big Data excitement within the business sector may be attributed to vested commercial interests (IDC 2015), the uncritical reception in certain political domains can be understood as but the latest manifestation of a more generic "trust in numbers" (Porter 1995). Numbers provide authority and justify decision making, especially in times of crisis and uncertainty. The European Commission's recent push for "evidence-informed policy making" (EC 2014b) is a telling example. While we acknowledge the potential benefits of "data for policy" (EC 2015b), we hold that a responsible *culture of quantification* requires greater awareness of the possible dangers and pitfalls of an increasingly data-driven society. As "governance by algorithms" (Musiani 2013) is becoming ever more commonplace, a public conversation about the politics of algorithmic regulation and what it means to be a "citizen of a software society" (Manovich 2013) is crucial.

What is needed is "epistemic vigilance" (Sperber et al. 2010) instead of blind data trust. Such vigilance requires, on the one hand, access to both the data used and the algorithms applied and, on the other hand, competencies to understand the analytical process as an embedded socio-technical practice, a specific way of producing knowledge that is neither inherently objective nor unquestionably fair. Thus, while legally guaranteed access to data and algorithmic code is important, it is not sufficient; what must also be cultivated are capabilities to interpret and possibly contest algorithmic data practices. The more such automated processes shape our society, the greater the need to invest in the education of researchers and policymakers, but also of journalists and ordinary citizens, to facilitate a better understanding of the epistemic foundations and socio-political ramifications of pervasive data analytics. Without doubt, this will be a difficult task that may require not only educational measures, but also technological intervention. On the one hand, we are talking about tools and services that can support epistemic vigilance by making visible what is often thoroughly black-boxed and concealed. On the other hand, technology design can also be used as a means to implement ethically sound software solutions as a form of *governance by design*.

The importance of epistemic vigilance cannot be overstated: Any political, ethical, or legal assessment of Big Data practices hinges upon a proper understanding of the epistemological foundations – and limitations – of algorithmic knowledge production.

While transparency and education are aimed at tackling the problem of opacity, there is also a need to address issues of distributed agency and the challenges they pose for accountability and responsibility attribution. As noted, algorithmic systems are complex assemblages of human and non-human actors and identifying a single culprit – be it man or machine – can be difficult, if not impossible. Moreover, if disparate impact is not the result of intentional discrimination, but of widespread biases that exist in society at large (Barocas and Selbst 2015), who can ultimately be held responsible for the potentially severe real-life consequences of Big Data analytics? To deal with these distributed forms of agency, new concepts to understand

and enforce socio-technically distributed responsibility must be developed (Simon 2015; Floridi 2013).

In conclusion, if we wish to tackle the epistemological and socio-political challenges of a new empiricism, we must act on several fronts: Hard law, soft law, education and ethically-informed software design all need to be employed for a "good enough governance" (Pagallo 2015) of Big Data practices.

Acknowledgments

The authors wish to acknowledge the financial support of the Austrian Science Fund (P-23770).

References

Adams, Susan. 2015. The World's Most Reputable Companies in 2015. *Forbes*, April 21. http://www.forbes.com/sites/susanadams/2015/04/21/the-worlds-most-reputable-companies-in-2015/. Accessed 30 May 2016.

Al-Rodhan, Nayef. 2014. The Social Contract 2.0: Big Data and the Need to Guarantee Privacy and Civil Liberties. *Harvard International Review*, September 16. http://hir.harvard.edu/the-social-contract-2-0-big-data-and-the-need-to-guarantee-privacy-and-civil-liberties/. Accessed 30 May 2016.

Anderson, Chris. 2008. The End of Theory: The Data Deluge Makes the Scientific Method Obsolete. *Wired*, June 23. http://www.wired.com/2008/06/pb-theory/. Accessed 30 May 2016.

Andrejevic, Mark. 2013. *Infoglut: How Too Much Information Is Changing the Way We Think and Know.* New York: Routledge.

AppDynamics. 2014. AppDynamics Announces New Application Intelligence Platform. https://www.appdynamics.com/press-release/appdynamics-announces-new-application-intelligence-platform/. Accessed 30 May 2016.

Bachner, Jennifer. 2013. Predictive Policing: Preventing Crime with Data and Analytics. *IBM Center for the Business of Government. Improving Performance Series.* http://www.businessofgovernment.org/sites/default/files/Predictive%20Policing.pdf. Accessed 30 May 2016.

Barocas, Solon, and Andrew D. Selbst. 2015. Big Data's Disparate Impact. *California Law Review* 104.

Bell, Gordon. 2009. Foreword. In *The Fourth Paradigm. Data-Intensive Scientific Discovery*, ed. Tony Hey, Stewart Tansley, and Kristin Tolle, XI-XV. Redmond: Microsoft Research.

Bellezza, Marco, and Federica De Santis. 2013. Google Not Liable for Autocomplete and Re-
    lated Search Results, Italian Court Rules. *CGCS Media Wire*, April 22.
    http://www.global.asc.upenn.edu/google-not-liable-for-autocomplete-and-related-
    search-results-italian-court-rules/. Accessed 30 May 2016.
Berman, Jules J. 2013. *Principles of Big Data. Preparing, Sharing, and Analyzing Complex In-
    formation.* Amsterdam: Elsevier, Morgan Kaufmann.
boyd, danah, and Kate Crawford. 2012. Critical Question for Big Data. Provocations for a
    Cultural, Technological, and Scholarly Phenomenon. *Information, Communication & So-
    ciety* 15(5).
Brooks, David. 2013. The Philosophy of Data. *The New York Times*, February 4.
    http://www.nytimes.com/2013/02/05/opinion/brooks-the-philosophy-of-data.html.
    Accessed 30 May 2016.
Burrell, Jenna. 2016. How the Machine 'Thinks': Understanding Opacity in Machine Learn-
    ing Algorithms. *Big Data & Society*. doi: 10.1177/2053951715622512.
Byrne, Robert F. 2012. The Three Laws of Big Data. *SupplyBrainChain*, April 27.
    http://www.supplychainbrain.com/content/nc/technology-solutions/forecasting-demand-
    planning/single-article-page/article/the-three-laws-of-big-data/. Accessed 30 May 2016.
Calcagno, Cristiano, Dino Distefano, and Peter O'Hearn. 2015. Open-Sourcing Facebook In-
    fer: Identify Bugs Before You Ship. https://code.facebook.com/posts/1648953042007882/
    open-sourcing-facebook-infer-identify-bugs-before-you-ship/. Accessed 30 May 2016.
Castro, Daniel. 2014. Big Data Is a Powerful Weapon in the Fight for Equality. *The Hill*, Octo-
    ber 23. http://thehill.com/blogs/pundits-blog/technology/221583-big-data-is-a-powerful-
    weapon-in-the-fight-for-equality. Accessed 30 May 2016.
Centre for Internet and Human Rights (CIHR). 2015. Should Algorithms Decide Your Fu-
    ture? https://cihr.eu/wp-content/uploads/2015/07/EoA_web.pdf. Accessed 30 May 2016.
Cisco. 2015. Cisco Visual Networking Index: Forecast and Methodology, 2014–2019.
    http://www.cisco.com/c/en/us/solutions/collateral/service-provider/ip-ngn-ip-next-
    generation-network/white_paper_c11-481360.pdf. Accessed 30 May 2016.
Clark, Jack. 2011. Facebook, Google: Welcome to the New Feudalism. *ZDNet*, September
    10. http://www.zdnet.com/article/facebook-google-welcome-to-the-new-feudalism/. Ac-
    cessed 30 May 2016.
Clinton, Rachel. 2016. What's the Difference between Business Intelligence and Predictive
    Analytics? *Smart Vision Europe*, January 06. http://www.sv-europe.com/blog/whats-the-
    difference-between-business-intelligence-and-predictive-analytics/. Accessed 30 May
    2016.
Cohen, Bernard I. 2005. *Triumph of Numbers: How Counting Shaped Modern Life.* New York:
    W. W. Norton & Company.
Crawford, Kate. 2013. The Hidden Biases in Big Data. *Harvard Business Review*, April 1.
    https://hbr.org/2013/04/the-hidden-biases-in-big-data. Accessed 30 May 2016.
Davenport, Thomas H., and D.j. Patil. 2012. Data Scientist: The Sexiest Job of the 21st Cen-
    tury. *Harvard Business Review*, October 2012. https://hbr.org/2012/10/data-scientist-the-
    sexiest-job-of-the-21st-century/. Accessed 30 May 2016.
Dede, Chris. 2015. Data-Intensive Research in Education: Current Work and Next Steps.
    http://cra.org/wp-content/uploads/2015/10/CRAEducationReport2015.pdf. Accessed 30
    May 2016.

Duhigg, Charles. 2012. How Companies Learn Your Secrets. *The New York Times*, February 16. http://www.nytimes.com/2012/02/19/magazine/shopping-habits.html. Accessed 30 May 2016.

Eagle, Nathan, and Kate Greene. 2014. *Reality Mining: Using Big Data to Engineer a Better World*. Cambridge, MA: MIT Press.

Eagle, Nathan, and Alex Pentland. 2006. Reality Mining: Sensing Complex Social Systems. *Personal and Ubiquitous Computing* 10(4): 255-268.

Ekbia, Hamid, Michael Mattioli, Inna Kouper, G. Arave, Ali Ghazinejad, Timothy Bowman, Venkata Ratandeep Suri, Andrew Tsou, Scott Weingart, and Cassidy R. Sugimoto. 2015. Big Data, Bigger Dilemmas: A Critical Review. *Journal of the Association for Information and Technology* 66(8).

Ember, Sydney. 2015. Tech Giants Top Best Global Brands List. *The New York Times*, October 4. http://www.nytimes.com/2015/10/05/business/media/tech-giants-top-best-global-brands-list.html. Accessed 30 May 2016.

Ericsson. 2015. Ericsson Mobility Report. On the Pulse of the Networked Society. http://www.ericsson.com/res/docs/2015/ericsson-mobility-report-june-2015.pdf. Accessed 30 May 2016.

European Commission (EC). 2015a. Big Data. https://ec.europa.eu/digital-agenda/en/big-data. Accessed 30 May 2016.

European Commission (EC). 2015b. Data For Policy: When The Haystack Is Made of Needles. A Call for Contributions. https://ec.europa.eu/digital-single-market/news/data-policy-when-haystack-made-needles-call-contributions. Accessed 30 May 2016.

European Commission (EC). 2014a. European Commission and Data Industry Launch €2.5 Billion Partnership to Master Big Data. http://europa.eu/rapid/press-release_IP-14-1129_en.htm. Accessed 30 May 2016.

European Commission (EC). 2014b. Data Technologies for Evidence-Informed Policy Making (Including Big Data). https://ec.europa.eu/digital-single-market/news/data-technologies-evidence-informed-policy-making-including-big-data-smart-20140004. Accessed 30 May 2016.

European Commission (EC). 2014c. Towards a Thriving Data-Driven Economy. http://ec.europa.eu/information_society/newsroom/cf/dae/document.cfm?action=display&doc_id=6210. Accessed 30 May 2016.

European Commission (EC). 2012a. From Crisis of Trust to Open Governing. http://europa.eu/rapid/press-release_SPEECH-12-149_en.htm. Accessed 30 May 2016.

European Commission (EC). 2012b. Proposal for a Regulation of the European Parliament and of the Council on the Protection of Individuals with Regard to the Processing of Personal Data and on the Free Movement of such Data (General Data Protection Regulation). http://ec.europa.eu/justice/data-protection/document/review2012/com_2012_11_en.pdf. Accessed 30 May 2016.

European Commission – Business Innovation Observatory (EC-BIO). 2013a. Big Data: Analytics and Decision Making. http://ec.europa.eu/DocsRoom/documents/13411/attachments/1/translations/en/renditions/native. Accessed 30 May 2016.

European Commission – Business Innovation Observatory (EC-BIO). 2013b. Big Data: Artificial Intelligence. http://ec.europa.eu/DocsRoom/documents/13411/attachments/2/translations/en/renditions/native. Accessed 30 May 2016.

European Data Protection Supervisor (EDPS). 2015. Meeting the Challenges of Big Data. A Call for Transparency, User Control, Data Protection by Design and Accountability. https://secure.edps.europa.eu/EDPSWEB/webdav/site/mySite/shared/Documents/Consultation/Opinions/2015/15-11-19_Big_Data_EN.pdf. Accessed 30 May 2016.

Executive Office of the President (EOP). 2014. Big Data and Privacy: A Technological Perspective. https://www.whitehouse.gov/sites/default/files/microsites/ostp/PCAST/pcast_big_data_and_privacy_-_may_2014.pdf. Accessed 30 May 2016.

Fitzgerald, Michael. 2012. Sensing the Future Before it Occurs. *MIT Sloan Management Review*, December 20. http://sloanreview.mit.edu/article/sensing-the-future-before-it-occurs/. Accessed 30 May 2016.

Floridi, Luciano. 2013. Distributed Morality in an Information Society. *Science and Engineering Ethics* 19(3): 727-743.

Future of Privacy Forum (FPF). 2014. Big Data: A Tool for Fighting Discrimination and Empowering Groups. https://fpf.org/2014/09/11/big-data-a-tool-for-fighting-discrimination-and-empowering-groups/. Accessed 30 May 2016.

Gillespie, Tarleton. 2014. The Relevance of Algorithms. In *Media Technologies: Essays on Communication, Materiality, and Society*, ed. Tarleton Gillespie, Pablo J. Boczkowski, and Kirsten A. Foot, 167-194. Cambridge, MA: MIT Press.

Google. 2016. Search Help: Autocomplete. https://support.google.com/websearch/answer/106230?hl=en. Accessed 30 May 2016.

Gorner, Jeremy. 2013. Chicago Police Use 'Heat List' as Strategy to Prevent Violence. *Chicago Tribune*, August 21. http://articles.chicagotribune.com/2013-08-21/news/ct-met-heat-list-20130821_1_chicago-police-commander-andrew-papachristos-heat-list. Accessed 30 May 2016.

Greenwald, Glenn. 2014. *No Place to Hide: Edward Snowden, the NSA and the Surveillance State*. London: Penguin Books.

Gutierrez, Daniel. 2015. Will Big Data Kill the Art of Marketing? *Inside Big Data*, January 16. http://insidebigdata.com/2015/01/16/will-big-data-kill-art-marketing/. Accessed 30 May 2016.

Hacking, Ian. 1990. *The Taming of Chance*. Cambridge: Cambridge University Press.

Halevi, Gali, and Henk Moed. 2012. The Evolution of Big Data as a Research and Scientific Topic. *Research Trends* 30.

Hardt, Moritz. 2014. How Big Data Is Unfair: Understanding Sources of Unfairness in Data Driven Decision Making. *Medium*, September 26. https://medium.com/@mrtz/how-big-data-is-unfair-9aa544d739de#.8ix0w5v59. Accessed 30 May 2016.

Hardy, Quentin. 2013. Why Big Data is not Truth. *The New York Times*, June 1. http://bits.blogs.nytimes.com/2013/06/01/why-big-data-is-not-truth/. Accessed 30 May 2016.

Harris, Shane. 2015. Your Samsung SmartTV Is Spying on You, Basically. *The Daily Beast*, June 02. http://www.thedailybeast.com/articles/2015/02/05/your-samsung-smarttv-is-spying-on-you-basically.html. Accessed 30 May 2016.

Hartzog, Woodrow, and Evan Selinger. 2013. Big Data in Small Hands. *Stanford Law Review Online*. http://www.stanfordlawreview.org/online/privacy-and-big-data/big-data-small-hands. Accessed 30 May 2016.

Hey, Tony, Steward Tansley, and Kristin Tolle, ed. 2009. *The Fourth Paradigm. Data-Intensive Scientific Discovery*. Redmond: Microsoft Research.

Hildebrandt, Mireille. 2013. Slaves to Big Data. Or Are We? *IDP. Revista De Internet, Derecho Y Política*, October. http://journals.uoc.edu/index.php/idp/article/viewFile/n17-hildebrandt/n17-hildebrandt-en. Accessed 30 May 2016.

Housley, William. 2015. Focus: The Emerging Contours of Data Science. *Discovery Society*, August 3. http://discoversociety.org/2015/08/03/focus-the-emerging-contours-of-data-science/. Accessed 30 May 2016.

Howard, Alex. 2014. Data-Driven Policy and Commerce Requires Algorithmic Transparency. *TechRepublic*, January 31. http://www.techrepublic.com/article/data-driven-policy-and-commerce-requires-algorithmic-transparency/. Accessed 30 May 2016.

IBM. 2014. IBM Watson Explorer: Search, Analyze, and Interpret to Enable Cognitive Exploration. https://www.ibm.com/developerworks/community/files/basic/anonymous/api/library/1b22311f-c1ab-4a8c-8b27-ec9f39115adf/document/476814cc-636a-47fe-93c5-63c4b66c3b9b/media. Accessed 30 May 2016.

IBM. 2013. The Four V's of Big Data. *Big Data & Analytics Hub*. http://www.ibmbigdatahub.com/infographic/four-vs-big-data#comment-1011454161. Accessed 30 May 2016.

International Data Corporation (IDC). 2015. New IDC Forecast Sees Worldwide Big Data Technology and Services Market Growing to $48.6 Billion in 2019, Driven by Wide Adoption Across Industries. http://www.idc.com/getdoc.jsp?containerId=prUS40560115. Accessed 30 May 2016.

Isaac, Mike. 2014. Facebook Says It's Sorry. We've Heard That Before. *The New York Times*, June 30. http://nyti.ms/1mBqRKp. Accessed 30 May 2016.

Jenkins, Tiffany. 2013. Tiffany Jenkins: Don't Count on Big Data for Answers. *The Scotsman*, February 12. http://www.scotsman.com/news/tiffany-jenkins-don-t-count-on-big-data-for-answers-1-2785890. Accessed 30 May 2016.

Jurgenson, Nathan. 2014. View From Nowhere. *The New Inquiry*, October 9. http://thenewinquiry.com/essays/view-from-nowhere/. Accessed 30 May 2016.

Kanter, James, and Mark Scott. 2015. Europe Challenges Google, Seeing Violations of Its Antitrust Law. *The New York Times*, April 15. http://nyti.ms/1D0lnRX. Accessed 30 May 2016.

Kelly, Kevin. 2007. What Is the Quantified Self? *Quantified Self. Self Knowledge Through Numbers*, October 5. http://quantifiedself.com/2007/10/what-is-the-quantifiable-self/. Access via the Internet Archive's Wayback Machine.

Kitchin, Rob. 2014a. Big Data, New Epistemologies and Paradigm Shifts. *Big Data & Society*. doi: 10.1177/2053951714528481.

Kitchin, Rob. 2014b. *The Data Revolution. Big Data, Open Data, Data Infrastructures & Their Limitations*. London: Sage.

Kramer, Adam D. I., Jamie E. Guillory, and Jeffrey T. Hancock. 2014. Experimental Evidence of Massive-Scale Emotional Contagion through Social Networks. *Proceedings of the National Academy of Science of the United States of America* 111(24).

Kun, Jeremy. 2015. Big Data Algorithms Can Discriminate, and It's Not Clear What to Do About It. *The Conversation*, August 13. https://theconversation.com/big-data-algorithms-can-discriminate-and-its-not-clear-what-to-do-about-it-45849. Accessed 30 May 2016.

Lagoze, Carl. 2014. Big Data, Data Integrity, and the Fracturing of the Control Zone. *Big Data & Society*. doi: 10.1177/2053951714558281.

Laney, Doug. 2012. Deja VVVu: Others Claiming Gartner's Construct for Big Data. *Gartner Blog Network*. http://blogs.gartner.com/doug-laney/deja-vvvue-others-claiming-gartners-volume-velocity-variety-construct-for-big-data/. Accessed 30 May 2016.

Laney, Doug. 2001. 3D Data Management: Controlling Data Volume, Velocity, and Variety. *META Group. Application Delivery Strategies*. https://blogs.gartner.com/doug-laney/files/2012/01/ad949-3D-Data-Management-Controlling-Data-Volume-Velocity-and-Variety.pdf. Accessed 30 May 2016.

Lazer, David; Ryan Kennedy, Gary King, and Alessandro Vespignani. 2014. The Parable of Google Flu: Traps in Big Data Analysis. *Science* 343, March 14.

Leonelli, Sabina. 2014. What Difference Does Quantity Make? On the Epistemology of Big Data in Biology. *Big Data & Society*. doi: 10.1177/2053951714534395.

Lohr, Steve. 2015. *Data-ism. The Revolution Transforming Decision Making, Consumer Behavior, and Almost Everything Else*. New York: Harper Collins.

Lohr, Steve. 2013. The Origins of 'Big Data': An Etymological Detective Story. *The New York Times*, February 1. http://bits.blogs.nytimes.com/2013/02/01/the-origins-of-big-data-an-etymological-detective-story/. Accessed 30 May 2016.

Lohr, Steve. 2012. The Age of Big Data. *The New York Times*, February 11. http://www.nytimes.com/2012/02/12/sunday-review/big-datas-impact-in-the-world.html. Accessed 30 May 2016.

Manovich, Lev. 2013. The Algorithms of Our Lives. *The Chronicle Review*, December 16. http://chronicle.com/article/The-Algorithms-of-Our-Lives-/143557/. Accessed 30 May 2016.

Marr, Bernard. 2014. Big Data: The 5 Vs Everyone Must Know. https://www.linkedin.com/pulse/20140306073407-64875646-big-data-the-5-vs-everyone-must-know. Accessed 30 May 2016.

Matthias, Andreas. 2004. The Responsibility Gap: Ascribing Responsibility for the Actions of Learning Automata. *Ethics and Information Technology* 6: 175-183.

Mayer-Schönberger, Viktor, and Kenneth Cukier. 2013. *Big Data. A Revolution That Will Transform How We Live, Work, And Think*. New York: Houghton Mifflin Harcourt.

McFarland, Daniel A., and Richard H. McFarland. 2015. Big Data and the Danger of Being Precisely Inaccurate. *Big Data & Society*. doi: 10.1177/2053951715602495.

McLannahan, Ben. 2015. Being 'Wasted' on Facebook May Damage Your Credit Score. *The Financial Times*, October 15. http://www.ft.com/cms/s/0/d6daedee-706a-11e5-9b9e-690fdae72044.html. Accessed 30 May 2016.

Meyer, Robinson. 2014. Everything We Know About Facebook's Secret Mood Manipulation Experiment. *The Atlantic*, June 28. http://www.theatlantic.com/technology/archive/2014/06/everything-we-know-about-facebooks-secret-mood-manipulation-experiment/373648/. Accessed 30 May 2016.

Moz. 2016. Google Algorithm Change History. https://moz.com/google-algorithm-change. Accessed 30 May 2016.

Müller, Martin U., Marcel Rosenbach, and Thomas Schulz. 2013. Living by the Numbers: Big Data Knows What Your Future Holds. *Spiegel Online International*, May 17. http://www.spiegel.de/international/business/big-data-enables-companies-and-researchers-to-look-into-the-future-a-899964.html. Accessed 30 May 2016.

Musiani, Francesca. 2013. Governance by Algorithms. *Internet Policy Review* 2(3). http://policyreview.info/articles/analysis/governance-algorithms. Accessed 30 May 2016.

National Science Foundation (NSF). 2012. NSF Leads Federal Efforts in Big Data. http://www.nsf.gov/news/news_summ.jsp?cntn_id=123607&org=NSF&from=news. Accessed 30 May 2016.

New, Joshua. 2015. It's Humans, Not Algorithms, That Have a Bias Problem. *Center for Data Innovation*, November 16. https://www.datainnovation.org/2015/11/its-humans-not-algorithms-that-have-a-bias-problem/. Accessed 30 May 2016.

Newman, Nathan. 2015. Data Justice: Taking on Big Data as an Economic Justice Issue. http://www.datajustice.org/sites/default/files/Data%20Justice-%20Taking%20on%20 Big%20Data%20as%20an%20Economic%20Justice%20Issue.pdf. Accessed 30 May 2016.

Niggemeier, Stefan. 2012. Autocompleting Bettina Wulff: Can a Google Function Be Libelous? *Spiegel Online*, September 20. http://www.spiegel.de/international/zeitgeist/google-autocomplete-former-german-first-lady-defamation-case-a-856820.html. Accessed 30 May 2016.

Nissenbaum, Helen. 1996. Accountability in a Computerized Society. *Science and Engineering Ethics* 2(1): 25-42.

Obama, Barack. 2015. President Barack Obama's Big Data Keynote. https://www.youtube.com/watch?v=vbb-AjiXyh0. Accessed 30 May 2016.

O'Neil, Cathy. 2016. The Ethical Data Scientist. *Slate*, February 4. http://www.slate.com/articles/technology/future_tense/2016/02/how_to_bring_better_ethics_to_data_science.html. Accessed 30 May 2016.

Oracle. 2012. Integrate for Insight: Combining Big Data Tools with Traditional Data Management. http://www.oracle.com/us/technologies/big-data/big-data-strategy-guide-1536569.pdf. Accessed 30 May 2016.

Organisation for Economic Co-Operation and Development (OECD). 2015. Data-Driven Innovation for Growth and Well-Being. http://www.oecd.org/sti/ieconomy/Policy-Note-DDI.pdf. Accessed 30 May 2016.

Pagallo, Ugo. 2015. Good Onlife Governance: On Law, Spontaneuous Orders, and Design. In *The Onlife Manifesto: Being Human in a Hyperconnected Era*, ed. Luciano Floridi, 161-177. Cham: Springer.

Pasquale, Frank. 2015. *The Black Box Society: The Secret Algorithms That Control Money and Information*. Cambridge, MA: Harvard University Press.

Patel, Nilay. 2014. Screwing With Your Emotions is Facebook's Entire Business. *Vox*, July 3. http://www.vox.com/2014/7/3/5865279/the-truth-about-facebook. Accessed 30 May 2016.

Pentland, Alex. 2014. *Social Physics: How Good Ideas Spread – The Lessons from a New Science*. New York: Penguin Press.

Pew Research Center. 2014. Public Perceptions of Privacy and Security in the Post-Snowden Era. http://www.pewinternet.org/files/2014/11/PI_PublicPerceptionsofPrivacy_111214.pdf. Accessed 30 May 2016.

Phillips Mandaville, Alicia. 2014. The Revolution Is Not Everywhere Yet – and That's a Challenge for Global Development. *Millenium Challenge Corporation*, January 15. https://www.mcc.gov/blog/entry/blog-011514-the-revolution-is. Accessed 30 May 2016.

Pigliucci, Massimo. 2009. The End of Theory in Science? *EMBO Reports* 10(6). http://www.ncbi.nlm.nih.gov/pmc/articles/PMC2711825/. Accessed 30 May 2016.

Porter, Theodore M. 2011. Statistics and the Career of Public Reason: Engagement and Detachment in a Quantified World. In *Statistics and the Public Sphere: Numbers and Peo-*

ple in Modern Britain c. 1800-2000, ed. Tom Crook, and Glen O'Hara, 32-48. New York: Routledge.

Porter, Theodore M. 1995. *Trust in Numbers: The Pursuit of Objectivity in Science and Public Life*. Princeton, NJ: Princeton University Press.

Quantcast. 2013. Know Ahead. Act Before. https://www.quantcast.com/wp-content/uploads/2013/11/Quantcast-Advertise-Mediakit_2013.pdf. Accessed 30 May 2016.

Richtel, Matt. 2013. How Big Data Is Playing Recruiter for Specialized Workers. *The New York Times*, April 27. http://www.nytimes.com/2013/04/28/technology/how-big-data-is-playing-recruiter-for-specialized-workers.html. Accessed 30 May 2016.

Rieder, Gernot, and Judith Simon. 2016. Datatrust: Or, The Political Quest for Numerical Evidence and the Epistemologies of Big Data. *Big Data & Society*, June.

Rosenblat, Alex, Tamara Kneese, danah boyd. 2014. Algorithmic Accountability. http://papers.ssrn.com/sol3/papers.cfm?abstract_id=2535540. Accessed 30 May 2016.

Rubinstein, Ira S. 2013. Big Data: The End of Privacy or a New Beginning? *International Data Privacy Law*. doi: 10.1093/idpl/ips036.

Rudder, Christian. 2014. *Dataclysm: Who We Are (When We Think No One's Looking)*. New York: Crown Publishers.

Samuel, Arthur L. 1959. Some Studies in Machine Learning Using the Game of Checkers. *IBM Journal and Development* 3(3): 210-229.

Schaeffer, Donna M., and Patrick Olson. 2014. Big Data Options for Small and Medium Enterprises. *Review of Business Information Systems* 18(1).

Schneier, Bruce. 2015. *Data and Goliath. The Hidden Battles to Collect Your Data and Control the World*. New York: W. W. Norton & Company, Inc.

Schroeder, Ralph. 2014. Big Data and the Brave New World of Social Media Research. *Big Data & Society*. doi: 10.1177/2053951714563194.

Science Europe. 2014. How to Transform Big Data into Better Health: Envisioning a Health Big Data Exosystem for Advancing Biomedical Research and Improving Health Outcomes in Europe. http://www.scienceeurope.org/urls/bigdata1. Accessed 30 May 2016.

Silver, Nate. 2012. *The Signal and the Noise. Why So Many Predictions Fail and Some Don't*. New York: The Penguin Press.

Simon, Judith. 2015. Distributed Epistemic Responsibility in a Hyperconnected Era. In *The Onlife Manifesto: Being Human in a Hyperconnected Era*, ed. Luciano Floridi, 145-159. Cham: Springer.

Sperber, Dan, Fabrice Clément, Christophe Heintz, Olivier Mascaro, Hugo Mercier, Gloria Origgi, and Deirdre Wilson. 2010. Epistemic Vigilance. *Mind & Language* 25(4): 359-393.

Steadman, Ian. 2013. Big Data and the Death of the Terrorist. *Wired*, January 25. http://www.wired.co.uk/news/archive/2013-01/25/big-data-end-of-theory. Accessed 30 May 2016.

Strawn, George O. 2012. Scientific Research: How Many Paradigms? *Educause Review*. http://er.educause.edu/articles/2012/6/scientific-research-how-many-paradigms. Accessed 30 May 2016.

Striphas, Ted. 2015. Algorithmic Culture. *European Journal of Cultural Studies* 18(4-5): 395-412.

Strong, Colin. 2015. *Humanizing Big Data: Marketing at the Meeting of Data, Social Science and Consumer Insight*. London: Kogan Page.

Szal, Andy. 2015. MIT System Successfully Processes Big Data Without Human Involvement. *Manufacturing.net*, October 19. http://www.manufacturing.net/news/2015/10/mit-system-successfully-processes-big-data-without-human-involvement. Accessed 30 May 2016.

United Nations Economic Commission for Europe (UNECE). 2014. MSIS Wiki: Big Data. http://www1.unece.org/stat/platform/display/msis/Big+Data. Accessed 30 May 2016.

UN Women. 2013. UN Women Ad Series Reveals Widespread Sexism. http://www.unwomen.org/en/news/stories/2013/10/women-should-ads. Accessed 30 May 2016.

Valinsky, Jordan. 2013. Japanese Court Orders Google to Pay Fine for Embarrassing Autocomplete Results. *Observer*, April 16. http://observer.com/2013/04/japanese-court-search-ruling-google/. Accessed 30 May 2016.

Van Rijmenam. 2013. Why the 3V's Are Not Sufficient to Describe Big Data. *Datafloq. Connectiong Data and People.* https://datafloq.com/read/3vs-sufficient-describe-big-data/166. Accessed 30 May 2016.

Venturini, Tommaso, Nicolas Baya Laffite, Jean-Philippe Cointet, Ian Gray, Vinciane Zabban, and Kari de Pryck. 2014. Three Maps and Three Misunderstandings: A Digital Mapping of Climate Diplomacy. *Big Data & Society.* doi: 10.1177/2053951714543804.

Wilbanks, John. 2009. I Have Seen the Paradigm Shift and It Is US. In *The Fourth Paradigm. Data-Intensive Scientific Discovery*, ed. Tony Hey, Stewart Tansley, and Kristin Tolle, 209-214. Redmond: Microsoft Research.

Williamson, Ben. 2015. Smarter Learning Software: Education and the Big Data Imaginary. http://dspace.stir.ac.uk/handle/1893/22743. Accessed 30 May 2016.

Williamson, Ben. 2014. Knowing Public Services: Cross-Sector Intermediaries and Algorithmic Governance in Public Sector Reform. *Public Policy and Administration* 29(4): 292-312.

Zarsky, Tal. 2016. The Trouble with Algorithmic Decisions: An Analytical Road Map to Examine Efficiency and Fairness in Automated and Opaque Decision Making. *Science, Technology, & Human Values* 41(1): 118-132.

4. Our Thinking – Must it be Aligned only to the Given Data?

Klaus Kornwachs[1]

Abstract

New technological possibilities in Big Data allow finding unexpected structures and relations in datasets, provided by different realms and areas. This article distinguishes between signals, data, information and knowledge, and discusses ownership of data and information. Knowledge will be considered as the result of understanding information. The results of big data analyses cannot be adequately interpreted if the research question, i.e. the question of what to look for, has not been asked beforehand. Thus, a model is required to perform a satisfactory data analysis. A model, which allows a causal explanation, is better than a model, which delivers only extrapolations. The potential tendency to replace scientific models with merely numerical procedures will be discussed critically.

1   Prof. Dr. Klaus Kornwachs, former Chair for Philosophy of Technology at Brandenburg Technical University of Cottbus (1992-2011), since 1990 Honorary Professor for Philosophy at University of Ulm, studied Physics, Mathematics and Philosophy. He was Senior Research Fellow at the Fraunhofer-Institute for Industrial Engineering, Stuttgart (1979-1992). Guest Professor at Technical Universities in Budapest, Vienna, and Dalian (China), Alcatel-Lucent Fellowship at IZKT, University Stuttgart (2012). Since 2004 Full Member of the National Academy for Science and Engineering (acatech), since 2013 Honorary Professor at College for Architecture and Urban Planning, Tongji-University, Shanghai. His main fields in research are: Analytical Philosophy, General System Theory, Ethics. Numerous publications; for more see: www.kornwachs.de (in German).

"True wisdom, as the fruit of self-examination, dialogue and generous encounter between persons, is not acquired by a mere accumulation of data which eventually leads to overload and confusion, a sort of mental pollution." (Pope Francis (2015), IV, Sec. 47, p. 33)

4.1   Introduction

The answer to the question[2] in the title is a clear "No!" The task of this contribution is to provide reasons for this "No!" and to look for another, more positive answer: Our thinking should decide, which data and which data processing results can be meaningful and which not.

We are confronted with enormous technological possibilities: The so-called Big-Data Technology allows us to process a huge amount of data in a very short period of time. Thus, the data processing results such as correlations, patterns, regularities and unexpected relations, etc., which can be found by data mining methods, can be used not only for long-term decisions but also for commands and control functions in real time.

It is necessary to distinguish between data mining and Big Data. Data mining comprises the known methods of analyzing datasets to find relevant information within the dataset.[3] We talk about Big Data if the data volume is too large to be handled by an Excel spreadsheet or by regular database technologies. The conditions and questions, under which the analyzing procedures are performed, can vary considerably due to the achievable velocity of data processing (Techopedia, 2015).

New software solutions like the so-called Hadoop technology make it possible to use a large number of connected small computers in a company and in clouds of providers to simultaneously process data and produce results faster than ever. This allows quick changes in the way that data is analyzed. The mask, that is, the space-time structure, with which the data are ordered and regarded, can be repeatedly altered to such an extent that the discovery of unknown structures is highly likely.

---

2   This is an inverse statement taken from Gumbrecht's (2014) title.

3   Data Mining is a container term for a lot of different data processing procedures such as anomaly detection (identification of outliers worth to be investigated further), dependency detection or factor analysis (looking for correlations and relations between variables), clustering and classification, pattern recognition, regression analysis, aggregation, visualization, etc; see Palace (1996); for textbooks see Witten *et al.* (2011).

The so-called *Reconstructability Analysis* in the '80s was one of the forerunners for analyzing data[4] without reference to an established theory about the generating processes of the data. It was part of a program to develop a *General Problem Solver* in the context of Mathematical System Theory. The basic idea was to vary and optimize a mask over a set of data by applying the minimum entropy principle, and then to test the structure hypotheses according to the question whether and how a set of variables may determine another set of variables. The structure found was used to reproduce the frequency or relation distributions of the data outcome in order to test it. This instrument tried to find structures of the data generating processes, if a causal theory could not be developed.[5]

Another early example of how to handle data in an unforeseen way—that is, without any theoretical apriori-knowledge—is so-called Geomarketing. Usually, marketing and advertising strategists are strongly interested in consumer habits such as beer consumption, osteoarthritis risk, or payment behavior. Geomarketing is using these data and tries to link them with other data from digital city maps. Thus, it is easy to find out, who is buying what in which regions or city districts. Various relations can be recognized between different districts on the one hand and pattern of consumption behavior as well as economic performance of the inhabitants on the other. The conclusions from this "discovery" were straight forward, but eventually in conflict with the ideal of democracy, equality, and fairness. Not only the personal reputation, but also the region where the home is located, gives the bank a certain probability of an individual's credit-worthiness: "*Living in the wrong place, ... can lead to worse conditions for obtaining a loan*" (Peter Schaar, cited in Rauner, 2006, p. 38).

Even meaningless data collected for the sake of completeness within the context of some processes or observable events in technology, society and economy can be put together, and sometimes it is possible to find a lot of previously unknown correlations. Very often we are confronted with incautiously-stated alleged causalities.

It seems natural to assume that, in light of an exponential growth of available data volume, a lot of meaningful applications need to be handled and data to be

---

4   Because of space limitations we do not discuss all analyzing methods like Monte Carlo Method, Exponential Smoothing and others. In Mathematics, they have been known long before computer capacities allowed their extensive applications.

5   Reconstructability Analysis applies set-theoretic modeling of relations as well as information-theoretic modeling of frequency distributions. It is possible to find structures between variables of different categories, dimensions and resolutions. For an overview see Zwick (2004), the first paper was written by Klir (1976); see also Klir (1985) with respect to applications in many areas, e.g., Klir *et al.* (1988).

processed, even if the volume, variety and velocity of change (the so-called 'Big Three V's') are still increasing. Meanwhile, there are known applications in:

- Marketing: centralized data management for arbitrarily structured data, generated by different sources like Web analytics, Social Media, advertising, campaigning;
- Organization: real-time systems for inventory management, warehousing, dispatch handling;
- Process optimization: optimization of individual processes within complex production chains;
- Trade: sales forecast, trend analysis, retargeting, predictive buying;
- Internet of Things: connected cars, smart traffic control, smart city;
- Media analysis: diffusion of contents, Web analytics and others;
- E-Commerce, on-line Shops, frequencies of use;
- "Industry 4.0": data analysis of real-time production processes and applications for control purposes (*um & HP, 2014).

The abundance of possible applications of Big Data, of the relevant mathematical procedures,[6] and of org-ware and hardware technologies has produced many hopes and fears. The hopes run from the expectation of new discoveries in natural sciences, sociology, and economics. With respect to the latter, there is an extensive hope for new business models. The fears range from violation of privacy, unjustified social and economic control up to threatening democracy and individual liberty made possible by total transparency. Moreover, wrongly drawn conclusions from Big Data analysis may cause wrong or abusive control of economic and social processes, and up to the prosecution of innocent individuals.

In considering this situation, it makes sense to take a closer look at some epistemological and ethical aspects of Big Data.

4.2   Whose data is it anyway?

There are further potential questions: Who is allowed to use the "insight" that emerge from a completely new combination of n records, if the ownership of these records is distributed among more than one and up to the maximum of n authorized individuals?

---

6   See footnote 3.

We are always interested in data. They provide us with information under certain conditions. One of these conditions is normally that we know the realm of the inquiries or the experimental setup of the measurement procedures. Thus, we know the questions, and we hope that the data may be able to provide an answer due to a theoretical framework. In other words: who is able and who is allowed to ask the questions and to carry out data collections, analysis or inquiries? If "a certain individual or a well-known institution" is named, who is allowed to do this, the individual or the institution is considered to be the justified owner of the collected data.[7] Thus, the owner of the data can primarily use the information which is generated by the data analysis exclusively.

Different users of the data may produce different information due to different models they have. This reflects, of course, their different interests. The purpose of using information is to produce knowledge. Only information that can be understood by individuals can generate knowledge. In other words: Knowledge is the personalized result of understanding information.

Thus, we have to distinguish between signals, data, information and knowledge (Kornwachs, 1999, 2001). In general, an observed signal, i.e. a time series of changing properties, can be transformed into data by a measurement device, which has been designed according to a measurement hypothesis.[8] This measurement hypothesis may be called here model I. This model defines the category of variables, the range of values thereof, the resolution level, the time steps, etc. It is possible to proceed in a similar manner when series of answers from inquiries or interview sheets are transformed into data.[9]

Data processing, by means of algorithms, leads to information, if the algorithm computes the data with respect to a certain question (see chapter 3). The research question determines not only the mask, with which the data can be processed, but also the choice of an adequate algorithm. This algorithm represents model II. The result can be a piece of information, mostly in printed (written or in charts) or electronic form. If information on a carrier (paper, screen etc.) is data-driven,[10] these

---

7   In the case of inquiries, German Law stipulates that the approval of the individual in question is required. Data collected by measurements (in experiments or tests) are mostly the property of the respective research institute.

8   This presupposes recognition of patterns within the signal by the measurement device. This can be done for example by pre-defined thresholds and intervals. Regarding measurement theory in natural sciences, see for instance Tal (2015); in sociology, see the standards of empirical research, e.g., Neumann (2005).

9   Regarding qualitative methods see Yin (2003) or Keller & Kluge (1999).

10 For instance, formatted according to a mask and coded by using a finite repertoire of characters or signs.

data, for example, can be visually processed very quickly by individuals. Knowledge is generated by the cognitive understanding process of an individual. This process takes time, it requires a large amount of already existing prior knowledge, and consists of integrating new insights into an already existing knowledge base. Knowledge and knowledge generation are restricted to individuals. Computers, the Internet, etc. provide information, not knowledge. Thus, to understand information we need prior knowledge, and—according to Immanuel Kant—some categories and forms of intuition (German: "*Anschauungsformen*") like time and space.[11]

By taking this distinction seriously, it is possible to say that signals have no owners, whereas data and information may be possessed by somebody.

By using ICT devices, we permanently produce signals, and these signals can be observed and transformed into collectable data. The signals come from our own actions, and the resulting data generation produces something which we call a data-shadow, i.e. a set of data that can be assigned to a single person. Exactly here the basic problems of data protection, the problem of privacy, and of "informational self-determination" arise.[12]

To clarify the concept of data ownership, it is necessary to point out that we communicate via coded signals, which in turn can be considered as data. These data can be processed by our sensory perception into information, and we can understand this information provided that the well-known epistemological conditions are met.[13] We can communicate subjective knowledge by expressing it in form of information, which can be in turn coded by means of signals (e.g., written, acoustic or electronic signals). Thus, for example, Edward Snowden submitted information on the basis of data to which he had access, but of which he was not and is not the owner. Moreover, he has handed on his own knowledge about his experiences by expressing them publicly in the form of oral or written information (e.g., interviews in media).

Here, a wicked problem may arise. We share knowledge by communicating it as expressed information, coded in signals. If we are able to decode these signals, to order the decoded signs into data, and to interpret this information after data processing (e.g., the simplest case is reading), an individual can understand it. Thus, classified information can lead to illegal knowledge, which is not allowed to be communicated further to other individuals. The unauthorized reception of classi-

---

11 cf. Kant, Immanuel: Critique of Pure Reason, A50/B74, ff. In: Kant (1956), p. 94ff.

12 This has been introduced by the Court decision on December 15, 1983 of the German Federal Constitution Court (Bundesverfassungsgericht): file no. 1 BvR 209, 269, 362, 420, 440, 484/83, so-called "*Volkszählungsurteil*" (census decision).

13 See textbooks about communication theory, e.g., Mortensen (2008).

fied information produces an unauthorized ownership of knowledge. Nonetheless, if the "owner" of secret knowledge remains silent, nobody can directly prove ownership.

If a government collects data from its population and if the individuals are neither informed about this fact nor about the algorithms applied to these data, the government acquires illegitimate information, which should not be used. Nevertheless, any information that can be transformed into knowledge can be used, since relevant knowledge is trivially a decisive factor for decisions and actions.

Who has the right to collect data? Who has the right to ask questions? Who has the right to analyze the data in order to produce information? And who has the right to produce knowledge from this information?

The American way of approaching this issue is clearly expressed in US-President Barack Obama's remarks on the Administration's Review of Signals Intelligence:[14]

"This report largely leaves issues raised by the use of big data in signals intelligence to be addressed through the policy guidance that the President announced in January. However, this report considers many of the other ways government collects and uses large datasets for the public good. Public trust is required for the proper functioning of government, and governments must be held to a higher standard for the collection and use of personal data than private actors. As President Obama [2014] has unequivocally stated, 'It is not enough for leaders to say: trust us, we won't abuse the data we collect.'"

Here, we can explicitly find the difference to this position mentioned above: In the USA, the use of Big Data is not restricted, but if prospective violations are taken into account, there are efforts to avoid and to punish any misuse. In contrast, the trend in Germany is rather to rely on protectionist reactions and bans according to precautionary principle.

On the one hand the range of the respective regulatory requirements are controversial, on the other hand there are tools used to ensure these goals.

US authorities tend to limit the extent of state regulations and to rely more on individual responsibility and competition between vendors, whereas the Europeans, especially Germany, strongly emphasize the paternalistic function of the state. They prefer more restrictive conditions in favor of the protection of consumers and users and their real or presumed interests. With respect to the legal instruments,

---

14 cf. Podesta et al. (2014), p. 15-16. The quote of Obama's response to this report of his office; see Obama (2014).

Americans are more prone to ex post compensations. In contrast, Europeans prefer to prevent technical abuses or legal violations in advance.[15]

Nevertheless, the differences are much stronger between the European, or specifically, German approach and the Chinese approach regarding the use of Big Data. At a recent conference about the so-called iCity,[16] there was a controversial discussion about the issue of whether Big Data should be used to control the people's behavior, or to enhance living conditions in cities. It was further controversially discussed whether the improvement towards better living conditions in big cities requires better information about all possible measureable states of a city in order to provide governability and to prevent social instabilities. In any case, the German position was that the values, norms and criteria, on the basis of which the quality of an iCity will be judged, should be subject to a public discourse and not defined only by the government.

In their presentation, the Chinese approach tried to develop Big Data concepts for a smart city and/or for an iCity as an instrument, with which it is possible to forecast upcoming civil unrests and social destabilization in cities and to control the inhabitants' behavior at a very comprehensive level (i.e. not to control in each single case a possible deviant behavior of single individuals) and to learn about city dynamics and complexity by using Big Data. They hope that a suitable model might be available in the near future to interpret this gigantic amount of data. The German approach tried to find models to describe an iCity and its possible functions by starting with a kind of "wishful thinking": Firstly there is a concept of an ideal city, which may meet the general intuitions of a good city, good life, and a culturally defined urban concept elucidated by public discussions as it takes place in Germany. Then, companies start to develop concepts, software, products, and services according to business models, which are partly based on the results of these discussions. Data protection, intellectual property, privacy, and the protection against

---

15 According to Otthein Herzog's assessment, presented at the National Academy of Science and Engineering (acatech), Topic Network Information and Communication Technology, Berlin, November 11, 2014.

16 The 3rd Sino-German iCity Workshop "Smart Cities and Big Data", Oct. 29-30, 2014, Wuhan, China. The topics were: The information environment construction of smart city and big data, including the perception of cities, information networks, knowledge center; the industrial development of smart city, including intelligent manufacturing; information services; the construction of smart city with big data, including urban planning, transportation and logistics, urban construction, the management services construction of smart city and big data, including the urban environment; the evaluation standard of smart city construction. This list shows the wide range of prospective big data applications.

the threat of computers, nets or platforms are important issues in Germany. Thus, general principles (e.g., how to shape nets, data models, clouds, etc.) are strongly influenced by this discussion. Nevertheless, these issues seem not to be appreciated as much by the Chinese representatives as by the German representatives (Kornwachs, 2015).

Actually, it became obvious that in China there are tendencies to install databases, in which data about each citizen's social and economic behavior are collected. This includes purchase habits, criminal records, political activities and a survey of comments, blogs and postings on the Web. This system is declared an "important component of social governance" by the Chinese Council of State. This approach has been interpreted by opposing parties as an attempt to create a new citizen. There are concerns that such a collection of data needs support of ICT companies operating in China. It is presumed that these companies will not refuse collaboration with the authorities to avoid drawbacks in business (Grzanna, 2015).

4.3   What is a question?

"Since Aristotle, we have fought to understand the causes behind everything. But this ideology is fading. In the age of big data, we can crunch an incomprehensible amount of information, providing us with invaluable insights about the what rather than the why." (Mayer-Schönberger & Cukier, 2013).

It is well known that the interpretation of data runs into a completely different semantic universe or ontology,[17] if embedded in altered contexts or in different contexts compared to the source. As an example, data collections carried out in the context of psychological surveys in schools are linked to marketing data to determine consumer habits of young people. This represents an embedding of a well-defined ontology into another one. This embedding happens also when data are used in coupled systems, i.e. when everything is connected with everything else like in the propagated "Internet of Things". In certain circumstances the questions, which have given rise to the data collection, are not known anymore. New questions now define the interpretation, coming from new directions and new issues. Without knowing the question, the answer of the computer, i.e. the result produced by an

---

17 In computer science the term "ontology" is used to describe all objects and relations in database systems. This should not be confused with the term "ontology" in philosophy, where the term describes statements about being, things and modalities of existence.

algorithm when processing the data cannot be adequately interpreted. We call interpretable results information.

We will use the term "question" in a very general and operative fashion: Data may be collected in an experiment or test. This is implemented in the context of the respective experimental setup or the test device. All this presupposes more or less detailed knowledge of the boundaries and initial conditions and the possibility to prepare them. Furthermore, it can be assumed that the experiment was performed due to a nomothetic hypothesis that can be mathematically formulated.[18] In this realm, it is necessary to define which dependent and independent variables are observables and which of them are only theoretical state variables. Moreover, it has to be defined with which measurement specifications and devices the observables can be quantitatively detected. In such defined cases the questions are: What are the dynamics of the measured value of an observable in the concrete experiment or test with respect to the hypothesis which was the basis of the experiment, and with respect to the functional presumption (German: *"Funktionsvermutung"*), which in turn was the basis of the test?[19] In other words: Do the resulting data correspond to the theoretical expectations? Thus, the interpretation of the measured value has no meaning without questions like these.

To abandon the "ideology" of the cause-effect category means to refrain from the possibility to explain the nature of the processes. Hence, the generative structure of the mechanism that produces a measurement signal transferred to data by measurement devices, may give a theory-driven answer to the whereby-question. Nonetheless, this presupposes that the what-question has already been answered. This answer generates hypotheses and models. Thus, data processing (including statistics) can only provide hints—it is not a method to prove something or to answer a whereby-question, and leaves the why-question alone.

Now, if data are available and recorded for reasons to which the related questions are either no longer or only incompletely known, it is certainly possible to create new questions in order to "see", i.e. to interpret, data in a new context. Whether it is possible to gain completely new knowledge about the subject area, in which the data were collected without any prior theory is a problem, which has not been sufficiently discussed with respect to epistemology.

---

18 This is done in Natural Sciences either by tables and graphs, difference/differential equations, groups or symmetries. Additionally, recursive equations have opened up the broad field of recursively definable algorithms.

19 The difference between a test in technology and experiment in natural sciences is discussed by Kornwachs (2012, p. 123-132) and Kornwachs *et al.* (2013).

In any case, the premature proclamation of completely new sociological or criminological relations (correlations or even causal relations) from "analyzing" Big Data seems to be a little bit risky.[20]

4.4   Remarks about Models and Data

Not every new buzzword in computer and data sciences gives rise to a new epistemological revolution. Some may remember the hype regarding the so-called expert systems: The knowledge transfer from an expert to a "knowledge base" was thought to substitute experience of experts for a semi-smart database. Nothing happened; nobody is talking about expert systems anymore as a revolution; the pretensions have been considerably reduced (Bullinger and Kornwachs 1990).

These pretensions still have an active source; a conviction that our thinking should be aligned towards data (Gumbrecht 2014). This is a misunderstanding of an extreme positivistic position in the Philosophy of Natural Sciences according to which only experiments and observations (e.g., data provided) should deliver crucial experiences. These experiences can be used to decide, which of the competitive theories should be preferred.

It was David Hume, who discovered the problem of induction, who stated that there is no absolute reliability of our findings through observing and counting repetitive events alone. Further supporting knowledge is always necessary, e.g., a theory or a well-grounded hypothesis. Otherwise we are confronted with a purely inductive conclusion. It must be completed by deductive conclusions, that is, if the possible consequences of a hypothesis in form of a prognosis can be deductively derived, this prognosis helps to interpret the observations.

To observe natural processes -maybe the orbits of planets, the chemical properties of a biotope or, in technology, the temperature within a combustion engine—a measurement device is necessary. This device has to be built according to a model of the process that needs to be observed and measured. Even a simple hypothesis is already based on a (simple) model or a pre-theory. The measured entity can only be expressed as a quantitative value of a variable, if there is a quantitative concept, including a scale, a zero-point and a calibration rule. Recorded signals in natural sciences and technology can only be converted into data, if a model and a measurement rule are available. Such a model is an operative expression of a prior theory.

---

20 A more ironic collection of statistical analysis can be found in Stolz & Block (2012).

Normally, the natural laws of the process are independent of the measurement process.[21]

For instance, in the case of a sociological questionnaire survey or of behavioral observations, we are looking for institutional facts in society and economy. Market prices and quarterly figures represent data, but these data do not refer to processes in nature as found in biology, physics, etc., but rather to events and states in human interactions and communications, which are only possible due to agreements among people.

By examining such institutional facts, the underlying hypothesis to be tested can already be found when examining the questions and their arrangements within the questionnaires or interview guidelines, that is, the realm of possible answers is also based on a model. However, our models for institutional facts are far weaker than those for natural facts. There are two reasons for this: The use of physical analogies in economics, often praised as the solution of the prediction problems, has begun to show its weakness in the last years. Economic entities do not behave like physical entities, because the hypostatized economic "laws" are not independent from the way they are obeyed by different economies. The second reason is that the discussion about prognostic results changes the very process that has generated these results (see prognosis of polls) (Grunwald, Kornwachs et al., 2012).

Measurement and collection of data about consumer behavior with respect to internet, supermarkets, use of communication services, travel activities, etc. is even model-based, with a clear definition of the type of variables and the resolution and range of their values. It is not sufficient to own the data and to be able to process them—data only deliver useful information if data processing is based on a model in which the data were collected. A formal or symbolic model is a mathematical description (graphs or equations, etc.) of functional relations with a clear separation of independent and dependent variables.

The arrangement of data, often called a "mask", is pivotal for the computation. For example, by using the numbers of a dice, thrown several times with the outcome in time as values of an independent variable, we may get:

1,5,2,3,3,4,5,6,2,4,1,6,2,5,3,2,1,6,3,5,2,4.

---

21 The case of quantum mechanics does not contradict to this statement: The result is dependent on the measurement device, but the quantum mechanical process is ruled by laws which entangle the natural process and the measurement process as well to one process.

The re-arrangement of the numbers as

111222223333344445555666

is different from the arrangement above, since it is based on the idea of the model to look for the frequency of occurrence of individual numbers. Counting the numbers generates the values of an output or dependent variable. Another model could be, for example, to check the number according to aperiodicity or to find out the occurrence of certain combinations.

The visualization of data is another kind of re-arrangement. Let us have in our example a common bar chart. It is well known that the type of visual presentation strongly influences the interpretation of data (e.g., stretching the scale, cutting the zero line, etc.). It is an essential difference whether the data are used to produce a time series or a frequency diagram.

The next step is to understand the information, which presupposes knowledge as well as a model. If we want to know whether the dice has been fraudulently manipulated, we need a mathematical model that tells us, which deviations from a uniform distribution can be expected with respect to the number of trials. The deviation indicates that the cube is poorly constructed or manipulated. Only by applying this model, the chart or the series of numbers delivers us information whether the dice is satisfactory or not. Only when we have understood this information do we know the probability that the dice could have been manipulated or poorly manufactured.

However, the magnitude of the expected effect and the model requires a certain size of the sample, if acceptable error probabilities are given. It must be reflected in advance, which size of the test sample may be meaningful. Which error probabilities are tolerable is in turn dependent on the theory.

Let us take the following case: During the Ebola epidemic all infected individuals must be discovered. Let us assume that we have a test that detects with a certain probability an existing infection in an actually infected individual. Then, the first requirement is a high degree of sensitivity of the test –this means a high probability that the test detects an existing infection. There is always the risk of a false diagnosis that a non-infected person will be classified as infected. In comparison, a test must have high specificity if, for instance, a specific disease should be recognized with a high certainty and other diseases should be excluded. In particular, this is necessary if we want to take irreversible interventions or measures due to such a test. Therefore, such an error probability must be minimized.

Taking into consideration increasing amounts of data, it is necessary to readjust the error probability from the usual 5% down to 1% or even less. If we proceed in statistics in a rather exploratory way, from question to question, the error probability will be accumulative. If this methodological condition is not considered, it is always possible to "find some effects". This is the reason why some professional journals in Statistical Psychology require presentation of the hypothesis before the researcher starts with the exploration and the collection of data. To proceed from data analysis to the elaboration of questions afterwards is against scientific standards in statistics and is regarded as improper conduct.

The path from data to knowledge is paved with complexity; it needs an understanding of the chosen model and the strategy and how to minimize possible errors. This in turn takes time and this fact is often forgotten. As the global data volume currently doubles every two years, the required time increases to generate model-based information from these data. Only such information is capable of producing knowledge. Although computers may become exponentially faster and faster, this fact cannot ignore the requirement that human beings must develop an understanding of which decisions they have to make themselves and which decisions can be delegated to a computer program.

Computers are used to transform data to information to an incredible extent and at an incredible speed. Nevertheless, the time people need to "read" and understand the information produced has not been significantly shortened by computers.

The discussion about the technical feasibilities of Big Data should not shy away from the following point: Even if we had found an elegant technical solution of finding a needle in a haystack, we would need to have a model presentation of the needle. With appropriate filters and new combinations and mask arrangements it is always possible to find something unplanned. However, the question is what these structures and patterns tell us, or whether we believe that they should tell us something at all. Without theory it is possible to provide multiple data interpretations, including everything and nothing.

Let us illustrate an example in order to clarify the difference between fitting and explaining. We observe the motion of a pendulum and measure the position of the weight in short time intervals. The observed signal can be transformed into a time series, and a curve shows an oscillation around an axis as a zero line.

Without any knowledge about the physical nature of a pendulum, its length and mass, etc., we can fit the measured points of the time series with a complete orthogonal function system and calculate the coefficients according to the data by minimizing the difference between measured and theoretical values of the time series. To do this, we can use different polynomial systems, the Bessel functions or trigonometric functions (Fourier analysis). Along with the help of calculated coefficients we have a theoretical curve, with which we can make predictions of the future development of the observed variable by assuming that the nature of the generating process and thus the values of the variable can be considered to be sufficiently continuous and smooth.[22] The more data points we have and the higher the degree of the function,[23] the more reliable the prediction will be, but with greater distance in time from the previous measurement, the error probability increases. The possibility to generate predictions by this method does not deliver any explanation of the physical (i.e. causal) nature of the pendulum motion. Moreover, as we go farther into the future, the uncertainty increases.[24] In other words: A fit is not yet an explanation, and a set of fitted coefficients is not yet a model.

What we need here would be a model of the physical process. For model building, the crucial idea is that a restoring force proportional to the distance from the equilibrium point produces harmonic, i.e. periodic, vibrations. This leads to a differential equation, the solution to which is a periodic function. This periodic function can "explain" the observation due to its generation of a differential equation, based on a model, and it can also be used to predict if it can be calibrated within the boundary conditions.

The same holds more or less for statistical signals. The "Mathematical Theory of Communication", as C. Shannon and W. Weaver (1949) headlined their essay, shows this dilemma: by using the statistical analysis of time series, i.e. a variable signal in time, we can derive whether the signal fluctuates only in a random way, or whether the signal could include information and not only noise. It is possible to measure the amount of information (in bits). Knowing the coefficients of a fitted signal, as described above, provides us with the information that there is some regularity but not about the generating process. Nonetheless, often it has not been considered that it is not possible to conclude from such an amount of information what this information could tell us, i.e. what the meaning of the information given by a measurement is. We do not yet have any signs, but only an indication that a signal (usually a time series of variable states) can include information. With respect to a purely random distribution of signal states (noise), we can only conclude from the deviation that this signal results from a process which itself must exhibit certain regularities.

Whether there is information within in a signal "for us" cannot be decided by Shannon's Information Theory, which is based on a purely statistical treatment.

---

22 There are also complete function systems for state-discrete variables (e.g. Walsh functions).

23 In a polynomial approximation, this is the number of powers applied. The approximation by least square fit has been developed by Carl Friedrich Gauß (1777-1855), when he was 18 years old.

24 This uncertainty can be easily calculated (see for example Strutz, 2011).

Whether an intentionally written message is hidden beneath the data cannot be figured out only by statistical analysis.

Without any doubt, it is possible to find patterns in huge datasets by applying suitable arrangements and combinations of parts thereof and refined algorithms, for which it can be assumed that these are caused by processes that may be of interest for users of big data. Nevertheless, the criteria are again based on a model, i.e. they are already based on certain prior assumptions, and these assumptions are based on interests. We can be sure that the NSA processes and analyzes their available "world dataset" according to other models and viewpoints than Google, Amazon, Facebook and others. The corresponding algorithms represent these viewpoints and interests, that is, they do not provide an objective picture of the world, but they tell us only in a condensed fashion whether the data confirm or reject the hypotheses for which they were created.

4.5   The Renunciation of Rationality in Favor of Speed, Power and Greed

As already mentioned, China plans to use Big Data Technology in order to control and educate its citizens (Grzanna 2015). This is a reaction to the problem of Large Cities and the governance in a socially and economically fragmented society with economic liberty and political restrictions. Nevertheless, we should not underestimate the problem of governing such cities, and there is some hope to find solutions by collecting as much data as possible. But within the iCity project, there is—so far as known—no model to interpret the data with sufficiently explanatory power.

The belief that we do not need models anymore due to the abundant availability of data seems to be a contemporary movement.[25] Some authors conjure up the Petabyte-Age. Models are always imperfect, that is trivial, but do we really have something better at the moment?

The search engines' (Google is only the most prominent of such companies) access to available data has been characterized by Anderson (2008) as a *"massive corpus as a laboratory of the human condition"*. The storage of petabytes of data in clouds *"forces us to view data mathematically first and establish a context for it later"* (Anderson 2008). Thus, it is proclaimed, that the content of the data or a semantic

---

25 The title of an article in WIRED seems to be programmatic: *"The End of Theory: The Data Deluge Makes the Scientific Method Obsolete"* (see Anderson, 2008). The following remarks comment critically on this quite provocative article.

mapping does not matter in order to be able to "see" statistical surprise, unexpected relations or to be able to translate automatically from one language into another. In order to judge which advertisement will be more successful, it has been stated that only counting the clicks and links would be sufficient. Thus, man-made models are seemingly abandoned: *"The point I was making -- and I don't remember the exact words -- was that if the model is going to be wrong anyway, why not see if you can get the computer to quickly learn a model from the data, rather than have a human laboriously derive a model from a lot of thought."*[26]

There is a conviction that we do not need any theories about human behavior like sociology, psychology, neurosciences or linguistics, but that only measured facts about what people do can help to predict what they will do in the future, provided that the quantity of the data is large enough:

> "The new availability of huge amounts of data, along with the statistical tools to crunch these numbers, offers a whole new way of understanding the world. Correlation supersedes causation, and science can advance even without coherent models, unified theories, or really any mechanistic explanation at all." (Anderson 2008)

This very provocative statement aims directly at the scientific method described above when discussing the difference between fitting and explaining. A scientist knows that data without a model is only a string of numbers. The proposition is now, that it is no longer necessary to know some causal hypotheses about the connection of independent and dependent variables.

It is argued that, even in physics, the models are too simple—Newtonian Mechanics do not work in processes in which the velocity of light becomes essential or in submicroscopic atomic areas. This may be true. Nevertheless, natural sciences have an essential characteristic: The old models (such as Newtonian Mechanics) turned out to be a special case of Quantum Mechanics for $h \rightarrow 0$. In Relativity Theory one gets Newtonian Mechanics when putting $c \rightarrow \infty$. This correspondence principle has nothing to do with the mass of data in elementary particle physics: Even the standard model obeys the correspondence principle (either to electrodynamics or quantum mechanics). The detection of the Higgs Particle is also a triumph of model-based data analysis.

The discoveries of gene-protein interactions and other aspects of epigenetics seem to sweep away the Mendelian model and the hypothesis that acquired, or en-

---

26 Peter Norvig (2008). Google's research director answered to the article of Anderson (2008).

vironmentally modified, properties can never be inherited. But this does not mean that another new model for such a phenomena cannot be established, which is able to explain more than the old one.

It is stated that correlation with petabytes of data may be sufficient and we could stop looking for models:

> "We can analyze the data without hypotheses about what it might show. We can throw the numbers into the biggest computing clusters the world has ever seen and let statistical algorithms find patterns where science cannot." (Anderson 2008)

Nevertheless, the discovery of such phenomena is due to observation, measurement, experiments, tests, and the collected data. Without any expectation we cannot see anything: *"Concepts without perceptions are empty, intuitions without concepts are blind"*.[27] The surprise lies in the deviation from the expected data. This gives us a reason to look for new models, which can take a lot of time. But it is no argument to refrain from modeling or to work without concepts. All in all, the fact that there are more data available for analyzing purposes will not generate a new epistemological program. Only the application of the inductive method is perhaps a little bit more attractive in this context.

A reason for this renunciation of rationality can be found in the role of business models:

> "The idea is that businesses collect massive sets of data that may be homogeneous or automatically collected. Decision-makers need access to smaller, more specific pieces of data from those large sets. They use data mining to uncover the pieces of information that will inform leadership and help chart the course for a business." (Techopedia 2015)

Very often, risks and side effects are ignored when new technological possibilities are explored due to their promise to optimize already existing business models or to create new ones. Take the following two examples:

---

27 Cf. Kant, Immanuel: Critique of Pure Reason, KdRV B75, A 51. German Original: *„Gedanken ohne Inhalt sind leer, Anschauungen ohne Begriffe sind blind.* "Cf. Kant (1956), p. 95. Cited by the Gutenberg Project, in: http://www.gutenberg.org/files/4280/4280-h/4280-h.htm, Chapter: Introduction: Idea of a Transcendental Logic, Sec. I: Of Logic in General.

Dynamic pricing is one such development. It plays a role in smart grids, where the price of electrical energy dependents on time, when the supply of renewable energy is changing due to wind and sun, and if it is possible to shift energy consumption for less significant purposes to off-peak times. According to the size of the smart grid and the time interval of pricing, we have to handle big data to do this. This seems reasonable. Nevertheless, a dynamic pricing will be also applied to implement a variable price depending on whether the system supplier has estimated in this case that the costumer is willing to pay the costs. The basis for this estimation, which predicts behavior, consists in all traces of activity the customer has left on the Internet: visited web pages, residential data such as street or district where they live, preferred shops, travel, grocery stores and restaurants, age, estimated income, and last but not least, the same data about people, with whom they maintains relation (Lütge, 2014, Boeing, 2015).

Vehicles are already mobile computer centers, which are more and more connected with the ICT-environment. Integrated mobile phones, consumer electronics, immobilizing keys, navigation, devices for an increasing automation and driver support, customer, travel, and machine data for service, assurance, and maintenance, etc. are linked, resulting in a considerable amount of data. These data can be also used to dynamize the assurance premium in short intervals or as evidence in a criminal case, but also to produce a personal user profile. Moreover, the possibility of remote manipulation of a vehicle's vital function (e.g., the brakes) has been already reported. Thus, the data security is at the moment not clear, whereas the feasibility is discussed with highest priority.[28]

Buzzwords like Industry 4.0, Cyber-Physical Systems, and Internet of Things, etc. provoke hopes and fears. The protagonists promise abundant possibilities with a cornucopia of services and facilities, but they use the consumer's data not only for improving their business models and the performance thereof, but also in a way described above: Looking for a needle in the haystack without knowing whether there is a needle in it at all. Furthermore, if the needle is found, it may not necessarily be used to the benefit of the customer.[29]

It is therefore a good advice to think before we calculate, to make models before we act, to ask before we collect data, and to reflect again before we use results as solutions to questions not posed before.

---

28 German automobile firms have not yet joined the Trusted Computing Group (TCG), a union of all leading ICT and vehicle producers. This TCG tries to develop safety standards for computerized and automated vehicles (see Schulzki-Hadouti, 2015).

29 More examples can be found in Mainzer (2014) or Hofstetter (2014).

References

*um & HP - The unbelievable Machine Company & Hewlett Packard. 2015. *Hadoop2 – Big Data Projekte erfolgreich realisieren.* White Paper, Berlin 2014. https://www.unbelievable-machine.com/hadoop2/downloads/um-hp-whitepaper2014.pdf.

Anderson, Ch. 2008. The End of Theory. The Data Deluge Makes the Scientific Method Obsolete. *WIRED Magazine* June 23rd 2008, http://www.wired.com/2008/06/pb-theory/.

Boeing, N. 2015. Wie berechenbar ist der Mensch? *Technology Review* 03/2015. http://www.heise.de/tr/artikel/Wie-berechenbar-ist-der-Mensch-2599601.html.

Bullinger, H.-J., and K. Kornwachs. 1990. *Expertensysteme – Anwendungen und Auswirkungen im Produktionsbetrieb.* München: C.H. Beck.

Grunwald, A., K. Kornwachs et al. 2012. Technikzukünfte. Vorausdenken – Erstellen – Bewerten. In *acatech IMPULS*, ed. Acatech. Berlin, Heidelberg: Springer 2012. http://www.acatech.de/de/publikationen/impuls/acatech-impuls/detail/artikel/technikzukuenfte-vorausdenken-erstellen-bewerten.html.

Grzanna, M. 2015. Der Orwell'sche Bürger. *Technology Review* 2.7. 2015. http://www.heise.de/tr/artikel/Der-Orwell-sche-Buerger-2663947.html.

Gumbrecht, H.-U. 2014. Das Denken muss nun auch den Daten folgen. *Frankfurter Allgemeine Zeitung,* Feuilleton 12.3.2014. http://www.faz.net/aktuell/feuilleton/geisteswissenschaften/neue-serie-das-digitale-denken-das-denken-muss-nun-auch-den-daten-folgen-12840532.html

Hofstetter, Y. 2014. *Sie wissen alles – wie intelligente Maschinen in unser Leben eindringen und warum wir für unsere Freiheit kämpfen müssen.* Gütersloh: Bertelsmann.

Kant, I. 1956. *Kritik der Reinen Vernunft.* Hamburg: Meiner.

Keller, U., and S. Kluge. 1999. *Vom Einzelfall zum Typus – Fallvergleich und Fallkontrastierung in der qualitativen Sozialforschung.* Opladen: Leske+Budrich.

Klir, G.J. 1976. Identification of Generative Structures in Empirical Data. *International Journal of General Systems* 3, 89-104.

Klir, G.J. 1985. *The Architecture of Systems Problem Solving.* New York: Plenum Press.

Klir, G.J., M. Pitarelli, M. Mariano, and K. Kornwachs. 1988. The Potentiality of Reconstructability Analysis for Production Research. *International Journal for Production Research* 26, 629645.

Kornwachs, K. 1999. Von der Information zum Wissen? In *Gene, Neurone, Qubits & Co. Unsere Welten der Information/Forschung-Technik-Mensch. Verhandlungen der Gesellschaft Deutscher Naturforscher und Ärzte, 120. Versammlung, Berlin, 19.-22. September 1998,* ed. D. Ganten et al., 35-44. Stuttgart: Hirzel.

Kornwachs, K. 2001. Data - Information – Knowledge – A Trial for Technological Enlightenment. In *Toward the Information Society – The Case of Central and Eastern European Countries. Wissenschaftsethik und Technikfolgenbeurteilung, Bd. 9.,* ed. G. Banse, C.J. Langenbach, P. Machleidt, 109-123. Berlin, Heidelberg: Springer.

Kornwachs, K. 2012. *Strukturen technologischen Wissens. Analytische Studien zu einer Wissenschaftstheorie der Technik.* Berlin: Edition Sigma 2012.

Kornwachs, K. 2015. Short Reports and Impressions on the 3rd Sino-German i-City Workshop, Oct 29-30, 2014, Wuhan, China; submitted to the *National Academy of Science and Engineering (acatech),* Berlin, March 2015.

Kornwachs, K. et al. 2013. Technikwissenschaften. Erkennen – Gestalten – Verantworten. In *acatech IMPULS,* ed. acatech. Berlin, Heidelberg: Springer. http://www.acatech.de/de/publikationen/impuls.html. English: *Technological Sciences. Discovery – Design – Responsibility* (2014).http://www.acatech.de/de/publikationen/impuls/acatech-impuls/detail/artikel/ technological-science-discovery-design-responsibility.html.

Lütge, G. 2014. Schnüffeln verboten – Facebook, Google, Apple: Ignorieren sie den Datenschutz? *DIE ZEIT* 19, S. 22.

Mainzer, K. 2014. *Die Berechnung der Welt – Von der Weltformel zu Big Data*. München: Beck.

Mayer-Schönberger, V. und K. Cukier. 2013. *Big Data: A Revolution That Will Transform How We Live, Work and Think*. London: John Murray.

Mortensen C. D. (ed.). 2008. *Communication theory*. 2nded. New Brunswick, NJ: Transaction Publ.

Neumann, W. L. 2005. *Social Research Methods: Qualitative and Quantitative Approaches*. 6thed. Boston: Allyn & Bacon.

Norvig, P. 2008. *All we want are the facts, ma'm*. http://norvig.com/fact-check.html.

Obama, B. 2014. Remarks on the Administration's Review of Signals Intelligence from January 17, 2014. http://www.whitehouse.gov/the-pressoffice/2014/01/17/remarks-president-review-signals-intelligence. Accessed November 2015.

Palace, B. 1996. *Data Mining. Technology Note prepared for Management 274A*. Anderson Graduate School of Management at University of California, Los Angeles (UCLA) Spring 1996. http:// www.anderson.ucla.edu/faculty/jason.frand/teacher/technologies/palace/datamining.htm.

Podesta, J. et al. 2014. Big Data: Seizing Opportunities, preserving Values. Executive Office of the President. White House: Washington. http://www.whitehouse.gov/sites/default/files/ docs/big_data_privacy_report_may_1_2014.pdf.

Pope Francis. 2015. Encyclical Letter '*Laudato Si*' of the Holy Father Francis on Care for our Common Home. Vatican Press, Vatican, Rome, May, 24th, 2015. *http://w2.vatican.va/content/dam/francesco/pdf/encyclicals/documents/papa-francesco_ 20150524_enciclica-laudato-si_en.pdf.*

Rauner, M. 2006. Die Merkels von nebenan. *ZEIT Wissen* 4, 36-41.

Schulzki-Hadouti, Chr. 2015. Behauptete Sicherheiten. *VDI Nachrichten* 29/30, 11.

Shannon, C.E., and W. Weaver. 1949. *The Mathematical Theory of Communication*. 2nd ed. 1969. Chicago, London: Urbana. Deutsch: *Mathematische Grundlagen der Informationstheorie*. München: R. Oldenbourg.

Stolz, M., and J. Block. 2012. *Deutschlandkarte – 102 neue Wahrheiten*. München: Knaur.

Strutz T. 2011. *Data Fitting and Uncertainty – A practical introduction to weighted least squares and beyond*. Wiesbaden: Vieweg+Teubner.

Tal, E. 2015. Measurement Science. In *The Stanford Encyclopedia of Philosophy* (Summer Edition), ed. Edward N. Zalta. http://plato.stanford.edu/archives/sum2015/entries/ measurement-science/.

Techopedia. 2015. What is the difference between big data and data mining? http://www. techopedia.com/7/29678/technology-trends/what-is-the-difference-between-big-data-and-data-mining.

Witten, I. H., E. Frank, and M. A. Hall. 2011. *Data mining: practical machine learning tools and techniques*. 3rd ed. Burlington, MA: Morgan Kaufmann.

Yin, R. K. 2003. *Case Study Research – Design and Methods*. Thousand Oaks: Sage Publications.

Zwick, M. 2004. An Overview of Reconstructability Analysis. *Kybernetes* 33, 877-905.

All web-page accesses checked on January, 10th 2016.

5. Scientific Epistemology in the Context of Uncertainty

Edward R. Dougherty[1]

Abstract

Contemporary science confronts a huge wall: lack of data. Scientists desire to model large-scale systems that are both stochastic and nonlinear, for instance cell regulation; however, relative to the complexity of such systems there is a paucity of data that impedes our ability to both construct and validate models. With regard to large-scale systems and lack of data, this chapter has several aims: (1) discuss the relationship between scientific epistemology grounded in prediction and data insufficiency; (2) discuss the dichotomy between pure science, whose aim is to provide a mathematical representation of Nature, and translational science (engineering), whose aim is the pragmatic application of mathematical modeling to beneficially intervene in Nature, and the manner in which translational science mitigates the data requirement; (3) provide a detailed analysis of the effect of limited data as it applies to pattern recognition; and (4) provide a framework for translational science in the context of model uncertainty, including an epistemologically consequent concept of uncertainty quantification.

1 Edward R. Dougherty is the Robert M. Kennedy '26 Chair and Distinguished Professor in the Department of Electrical and Computer Engineering at Texas A&M University, and is Scientific Director of the Center for Bioinformatics and Genomic Systems Engineering. He holds a Ph.D. in mathematics from Rutgers University and the Doctor Honoris Causa from the Tampere University of Technology.

5.1   Requirements for a Scientific Theory

The epistemology that has evolved since the 17th century requires that a scientific theory contain two parts: (1) a *mathematical model* composed of symbols (variables and relations between the variables), and (2) a set of *operational definitions* relating the symbols to empirical observations. The latter requirement is necessary because a model must lead to predictions that are concordant with experimental observations. Specifically, model validation requires that the symbols be tied to observations by some semantic rules (operational definitions) that relate not necessarily to the general principles of the mathematical model but to conclusions drawn from the principles. In other words, the theory is tested by checking measurable consequences of the theory. The operational definitions are intrinsic to the theory, for without them there would be no connection between principles and observation. The mathematical equations relate abstract symbols, but there must be a well-defined procedure for relating the equations to quantifiable observations.

While the general epistemology requires the mathematical system to be tied to physical operations, it leaves open the manner and the extent to which the model must be related to experimental outcomes. According to Albert Einstein,

"In order that thinking might not degenerate into 'metaphysics', or into empty talk, it is only necessary that enough propositions of the conceptual system be firmly enough connected with sensory experiences and that the conceptual system, in view of its task of ordering and surveying sense experience, should show as much unity and parsimony as possible. Beyond that, however, the system is (as regards logic) a free play with symbols according to (logically) arbitrarily given rules of the game." (1944, p. 289)

An obvious question is what is meant by "enough propositions" being "firmly enough connected with sensory experiences." At once we see there is a subjective character to validation: the criteria must be posited by human beings.

Beyond the general statement of the criteria, how is concordance between measurements predicted by the model and corresponding empirical measurements to be quantified? The quantification of concordance can be a subtle issue, especially when dealing with stochastic models. Empirical observations of the phenomena must somehow be consonant with the manner in which predictions are generated by the model and concordance must take into account the randomness of both predications and measurements. The situation becomes more difficult when dealing with large-scale (complex) systems because there are so many possible predictions relating to a multitude of relationships within the model.

Operational definitions are required, but their exact formulation in a given circumstance is left open. Hence, their specification constitutes an epistemological requirement that must be addressed in mathematical (or logical) statements. Thus, whenever a theory is proposed, it should contain basic validation information: (1) the validation criteria and the manner of comparison; (2) the statistical theory supporting concordance of predictions with measurements; and (3) the theory justifying application of the statistical methods in the current circumstances.

While not part of the formal theory, experimentation constitutes a third requirement, in addition to the mathematical model and operational definitions. Because a model can only be validated to the extent that its symbols can be tied to observations in a predictive scheme, the ability to design and perform suitable experiments, including the availability of technology to perform the desired measurements, is mandatory. Limitations on experimentation can result in limitations on the complexity of a theory or a restriction on the symbols and operations constituting the theory. Hans Reichenbach states, "The reference to verifiability is a necessary constituent of the theory of meaning. A sentence the truth of which cannot be determined from possible observations is meaningless." (1971, p. 256)

Thus far we have focused on validation: what about model construction? According to Einstein, the model (conceptual system) is a creation of the "imagination." The manner of this creation is not part of the scientific theory and is free except that it must conform to the rules of the mathematical game. As part of the creative process, the scientist identifies a mathematical structure (system of differential equations, random process, directed graph, etc.) containing the variables of interest and then proceeds to infer from data various parameters and operational relationships in the model. The efficacy of the model will depend on both the chosen mathematical system and the accuracy of the parameter estimation. Both are problematic for large-scale systems. It is one thing to postulate a proportionality relationship between two variables and quite another to attempt to provide an abstract structure for interaction among hundreds of variables operating over different time scales and interacting nonlinearly. It is also quite a different matter to estimate two parameters from observations than to estimate hundreds (or thousands) of parameters of various types; indeed, owing to dependencies and physical limitations it may be virtually impossible to construct appropriate experiments.

Biological systems are subject to all of these impediments: model conception, a multitude of parameters to be estimated, and stringent experimental limitations (Dougherty and Shmulevich 2012). Activity within a cell is similar to that within a factory. In the latter, machines manufacture products, energy is consumed, information is stored, information is processed, decisions are made, and signals are sent to maintain proper factory organization and operation. All of these functions also take place within a cell and it is via analogy with a factory that one can approach the epistemology of the cell (Dougherty and Bittner 2011). To deal with such vast interactive complexity, the theorist and experimentalist must proceed in close connection.

With respect to both validation and model construction, the theory must not exceed the experimentalist's ability to conceive and perform appropriate experiments, and the experimentalist cannot produce directly meaningful experiments unless they are designed with a symbolic structure in mind. Gene and protein expressions must be measured dynamically in live cells because the cell is a dynamical system and absent dynamical measurements its regulation cannot be modeled accurately. Gene sequencing dead cells does not provide the needed measurements. Unfortunately, rather than designing experiments to address modeling demands, many biologists are collecting whatever data the latest technology produces, hoping that data mining algorithms will produce connections among the measurements. They may, but will they be meaningful? Norbert Wiener and Arturo Roesenblueth state the matter succinctly: "An experiment is a question. A precise answer is seldom obtained if the question is not precise; indeed, foolish answers—i.e., inconsistent, discrepant or irrelevant experimental results—are usually indicative of a foolish question." (Rosenblueth and Wiener 1945, p. 316) Not only does data mining not involve a precise question, it involves no question at all.

Shrewd experimental design is not a peripheral aspect of scientific inquiry, nor is it a recent necessity; rather, it is a fundamental differentiating factor between modern and ancient-medieval science. In his *Novum Organum* Francis Bacon makes the following fundamental point:

"There remains simple experience which, if taken as it comes, is called accident; if sought for, experiment. But this kind of experience is no better than a broom without its band, as the saying is—a mere groping, as of men in the dark, that feel all round them for the chance of finding their way, when they had much better wait for daylight, or light a candle, and then go. But the true method of experience, on the contrary, first lights the candle, and then by means of the candle shows the way; commencing as it does with experience duly ordered and digested, not bungling or erratic, and from it educing axioms, and from established axioms again new experiments." (Bacon 1620/1952)

A century and a half later, when reflecting on the monumental developments of the 17th and 18th centuries, Immanuel Kant, in the preface of the second edition of the *Critique of Pure Reason*, reiterates Bacon's view regarding experimental design: "To this single idea must the revolution be ascribed, by which, after groping in the dark for so many centuries, natural science was at length conducted into the path of certain progress." (Kant 1787/1952)

While this chapter focuses on theoretical issues regarding modeling, it is hoped that these few comments on experimentation support the view that modeling cannot be divorced from experiment. This is particularly true when we are starved for data, as we are today. Nature must be judiciously probed to extract the information needed.

5.2   Dilemmas of Large-scale Systems

Validation applies fairly directly to classical deterministic systems possessing a few degrees of freedom; however, it runs into increasing trouble as science attempts to handle larger scale systems and the concomitant issues of increasing complexity and inadequate data.

In the classical deterministic scenario, the model consists of a few variables and physical constants. The relational structure of the model is conceptualized by the scientist via intuition gained from thinking about the physical world. Intuition means that the scientist has some mental construct regarding the interactions beyond simply positing a skeletal mathematical system he believes is sufficiently rich to capture the interactions. There are few parameters to estimate and they are estimated from a handful of experiments. Owing to the deterministic character of the model, it can be tested with a few numerical predictions whose disagreement with future corresponding observations is due to either experimental error or model failure, with the former being mitigated by careful experimentation. As always, the theory is not positively validated but instead contingently accepted if predictions are deemed to be concordant with observations.

If behavior is represented stochastically, then predictions involve probability distributions. The model is stochastic in the sense that given initial conditions, the state at a subsequent time is undetermined, but the model itself is not uncertain. Many observations are required to test whether data distributions sufficiently match the predicted distributions. Statistical tests are required. For instance, with hypothesis testing one decides between two hypotheses—the distributions match or they do not match—and the decision invariably involves type I and type II errors. A decision to accept the theory depends on the acceptance threshold. The theory and test are intersubjective, but the decision to accept depends on subjective considerations.

The classic example of stochasticity is the quantum theory, with regard to which one can pose the following question: Is stochasticity inherent in human understanding of Nature or is it due to a temporary lack of understanding that will be overcome in the future and thereby yield a deterministic theory? Einstein argued for the latter interpretation but the general opinion appears to be that the uncertainty is inherent.

With large-scale systems containing hundreds or thousands of variables and parameters, this question does not arise. Even if one writes down a large-scale system, it must be reduced to make it mathematically and computationally tractable. Even if the full system is deterministic, the reduced model is stochastic because its behavior is affected by latent variables outside the reduced model whose changes make model behavior stochastic. A reduced perspective introduces uncertainty within the model. Werner Heisenberg's comment, meant for the quantum theory, applies equally here: "We decide, by our selection of the type of observation employed, which aspects of nature are to be determined and which are to be blurred." (1952, p. 82) Uncertainty can be diminished in one perspective at the cost of increasing it in another perspective: there are simply too many variables for all to be focused on equally.

For validation of large-scale systems there are too many relations for the full model to be tested and one must judiciously choose some predictions to check. For instance, a dynamical network can have a very complicated dependency structure so one might decide to validate the steady-state distribution by deriving the distribution and then comparing it to a data histogram obtained from observing the system in the steady state. This is a practical choice because many important properties of a dynamical system are based on the steady-state distribution. But even with this reduction, it may be necessary to obtain a large amount of data to get a decent representation of the empirical steady-state distribution and this data must be collected over a sufficiently long time frame as to allow the real system to reveal the steady-state distribution—keeping in mind that the stochasticity induced by the reduction requires that observations would have to sufficiently span the space of the unreduced model to capture the induced stochasticity. This may not be possible, even when we limit our focus to a single distribution induced by the network.

Validation is not the only obstacle with large-scale systems; model construction is also problematic because model parameters often cannot be satisfactorily estimated owing to lack of data. In many biological applications, even with a greatly reduced model, the lack of appropriate data makes pointwise parameter inference impossible. Thus, the analytic construct together with the data can at best produce an *uncertainty class* of models. The equations or graphical network constituting the structure itself gives rise to the class of models composed from all possible parameter values and the data effectively constrain this class via some non-pointwise inference procedure. In a standard stochastic setting, the model is stochastic in the sense that given initial conditions, the state at a subsequent time is undetermined, but the model itself is not uncertain; with an uncertainty class, in addition to randomness within the model, we can only identify a class of possible models.

With uncertainty classes of stochastic models we are a long way from the classical deterministic setting. There is no model discovery in the original sense because we can only obtain a class of models consistent with the basic mathematical structure and the data. Predictive validation in the classical sense does not even make sense.

5.3   Inherent Reduction in Translational Science

Scientific modeling aims to represent relations among natural phenomena and these relations can exhibit daunting complexity. Suppose, however, that we have a more focused objective, such as making a decision based on observations, filtering a signal or image in order to reduce noise or accentuate particular frequencies, or intervening in a natural system to force its behavior into one more in tune with our desires. The situation has changed from modeling behavior to affecting behavior. Translational science transforms a scientific mathematical model, whose purpose is to provide a conceptualization of some portion of the physical world, into a model characterizing human action in the physical world. Scientific knowledge is translated into practical knowledge by expanding a scientific system to include inputs that can be adjusted to affect the behavior of the system and outputs that can be used to monitor the effect of the external inputs and feed back information on how to adjust the inputs (Dougherty 2009). Rosenbluth and Wiener see science and translational science inextricably linked when they write, "The intention and the result of a scientific inquiry is to obtain an understanding and a control of some part of the universe." (1945, p. 316) A similar perspective can be taken regarding biomedical science, where models are constructed with the intention of using them for diagnosis, prognosis, and therapy.

Translational science involves designing an operator based on a model to determine some action in regard to Nature. It may be that one cannot obtain a model that can be validated via prediction—that is, a model that has scientific validity in the classical sense—but one may nevertheless find a model that can be used to determine a beneficial operator. The output of pure science is a validated model, whereas the output of applied science is a useful operator. For instance, when modeling the cell, the endpoint for pure science is a representation of the dynamical interaction between its macromolecules, whereas for translational science the endpoint might be determination of a drug that will block a signal involved with unwanted proliferation. In the latter case, the scientific model is a pragmatic construct used to facilitate control of Nature; its direct descriptive power is of secondary interest. For translational science, the epistemological requirements for accepting the model as scientifically valid are replaced by epistemological requirements regarding the performance of the operator relative to the purpose for which it has been constructed. The epistemology of pure science is replaced by the epistemology of practical science.

There are two basic operator problems concerning systems. One is *analysis*: given an input system and an operator, characterize the properties of the output system in terms of the properties of the input system. Often it is not mathematically feasible to characterize completely the output system given the complete input system or we may only know certain properties of the input system, so that the best we can hope for is to characterize related properties of the output system. But this is fine so long as we can characterize those properties of interest to our application.

The second basic operator problem is *synthesis*: given a system, design an operator to transform the system in some desirable manner. Synthesis represents the critical act for human intervention and forms the existential basis of modern engineering. One could proceed in a trial-and-error manner, trying one operation after another and observing the result; however, since this groping in the dark does not utilize scientific knowledge, we do not consider it to be translational science. In the context of translational science, synthesis begins with a mathematical theory constituting the relevant scientific knowledge and the theory is utilized to arrive at an optimal (or close to optimal) operator for accomplishing the desired transformation under the constraints imposed by the circumstances. A criterion, called a *cost function*, exists by which to judge the goodness of the response and the goal is to find an optimal way of manipulating the system. For a classifier the cost is typically its error rate and for an intervention it may be the probability of being in an undesirable state following intervention – the lower the cost, the better the operator.

Certainly one would prefer a good model when designing an operator because operator design would then be based on a system that accurately reflects real-world conditions and therefore would portend better performance for a synthesized operator; however, there is no demand that the model provides a predictive representation of Nature when application is the purpose. Taking Wiener's (and Bacon's) view with large-scale systems, we approach Nature in order to achieve some pragmatic benefit and it is always contextual—relative to the cost function and the conditions of application. Experimentally, mathematically, and computationally, a translational perspective may be the only option because only a targeted objective can reduce the scale of the problem so that it is tractable. The predictive capacity of the scientific model is of little concern because it is merely a tool and the relevant epistemology applies to the objective, not to the tool.

5.4 Epistemology of Classification

To illustrate the relationship between science and translational science, how the latter mitigates the predictive requirements of the underlying scientific model, and how translation entails a less restrictive epistemology, we consider pattern classification, where features are produced by measuring several aspects of an object and the object is classified based on the *feature vector*. At first glance this procedure may not look scientific, in the sense that the choice of measurements is based on the purpose of a specific classification problem, not the physical essence of the object, but in fact a feature vector does provide a scientific description. Niels Bohr makes the matter quite clear when he writes, "In our description of nature the purpose is not to disclose the real essence of the phenomena but only to track down, as far as possible, relations between the manifold aspects of our experience." (1958) The probability distribution of a feature vector characterizes the relations among the features (measurements), which are clearly aspects of our experience.

5.4.1 Model Construction

Suppose we wish to discriminate between phenotypes A_0 and A_1, and believe that the different phenotypes result from production of a single protein P controlled by transcription factors X_1 and X_2. When X_1 and X_2 bind to the promoter region for gene G the corresponding mRNA is produced and this translates into the production of protein P, thereby resulting in phenotype A_1. In the absence of either X_1 or X_2 binding, there is no transcription, which results in phenotype A_0. Letting phenotype be a *label* (binary variable) Y with $Y = 0$ for A_0 and $Y = 1$ for A_1, it is straightforward to provide a quantitative interpretation of this scenario: there exist expression levels κ_1 and κ_2 such that phenotype A_1 is obtained if $X_1 > \kappa_1$ and

$X_2 > \kappa_2$, whereas A_0 is obtained if either $X_1 \leq \kappa_1$ or $X_2 \leq \kappa_2$ (Dougherty, Hua, and Bittner 2007). These conditions characterize a classifier ψ defined by $\psi(X_1, X_2) = 1$ if $X_1 > \kappa_1$ and $X_2 > \kappa_2$, and $\psi(X_1, X_2) = 0$ if $X_1 \leq \kappa_1$ or $X_2 \leq \kappa_2$. Under the specified conditions, ψ is perfect. It never makes an error. Thus, its error, which is defined to be the probability of misclassification, is $\varepsilon[\psi] = 0$. No other classifier can do better, although there might be others that do as well.

In a second (more realistic) scenario, owing to concentration fluctuations, time delays, the effects of other genes/proteins, and inter-cellular communication, the preceding conditions are not likely to perfectly hold and a perfect classifier is unlikely. Hence, the actual error of ψ would be given by the probability

$$\varepsilon[\psi] = P(Y = 0 | X_1 > \kappa_1 \text{ and } X_2 > \kappa_1) + P(Y = 1 | X_1 \leq \kappa_1 \text{ or } X_2 \leq \kappa_2). \qquad (1)$$

Were the joint distribution for the transcription factors and phenotype known, this error could be directly computed. The result would be a *classifier model* consisting of the classifier ψ and its error $\varepsilon[\psi]$. However, ψ might not be optimal. It was deduced from assumptions on the two genes involved and these assumptions are no longer assumed to hold.

In general, a classifier possessing minimal error is known as a *Bayes classifier* and its error, and the error of any other Bayes classifier for the setting, is called the *Bayes error*. Given a feature vector $X = (X_1, X_2, \ldots, X_k)$ for discriminating between classes (labels) $Y = 0$ and $Y = 1$, the collection of Bayes classifiers and the Bayes error are intrinsic to the joint distribution of X and Y, which is known as the *feature-label distribution*. Considering the features and labels as physical measurements, the joint distribution represents our knowledge of the variables $X_1, X_2, \ldots, X_k, Y$. Since both the collection of Bayes classifiers and the Bayes error are intrinsic to the model, they are "physical" characteristics. In particular, the Bayes error quantifies the separability of the classes relative to $X_1, X_2, \ldots, X_k$. Given the feature-label distribution, one can in principle find a Bayes classifier and the Bayes error—only in rare cases have these been analytically derived from the feature-label distribution, but they can be accurately approximated by numerical methods.

In the first scenario above, we obtained a Bayes classifier from scientific knowledge without knowing the feature-label distribution or either of the *class-conditional distributions*, these being the distributions of X given class 0 or class 1. This shows that, in a sense, classification is easier than finding the feature-label distribution, because given the latter we can always obtain a Bayes classifier. In the second scenario, we again obtain a classifier without knowing the feature-label distribution. How good it is depends on the mixing of the classes, but if the stochastic contextual conditions are not too great, then we can expect a low-error classifier, albeit, perhaps not the best classifier.

From a scientific perspective, we would prefer to have the feature-label distribution because it gives a complete probabilistic description of the interacting random variables. In theory, under very general conditions one can approximate the feature-label distribution to any desirable accuracy by obtaining a sufficiently large random sample of pairs (X, Y) and forming the frequency histogram. This would, in principle, give us the feature-label distribution, a Bayes classifier, and the Bayes error with excellent accuracy; however, it would require an extremely large amount of data, more than is typically feasible, and the theory does not give us the size of the sample needed.

The inference problem can be eased if we possess some knowledge of the features and labels. For instance, we might have knowledge that the class-conditional distributions are Gaussian (or very closely so). Then all we need do is estimate the means and covariance matrices of the class-conditional densities and the *prior class probabilities*, c_0 and c_1, of a feature vector being selected from class 0 and class 1, respectively ($c_0 + c_1 = 1$), to obtain (a good approximation of) the feature-label distribution, a much easier inference problem than estimating the feature-label distribution directly.

To take a different tact, let us return to the original phenotype classification problem and suppose that we do not know the thresholds κ_1 and κ_2, only that phenotype A_1 occurs if and only if the transcription factors are both sufficiently expressed. Then we could proceed by developing a procedure, called a *classification rule*, that upon being applied to sample data yields estimates, $\hat{\kappa}_1$ and $\hat{\kappa}_2$, of κ_1 and κ_2, respectively. This would provide us with a classifier ψ whose performance is hopefully close to that of a Bayes classifier. Going further, we might not have any biological knowledge that gives us confidence that the classifier should be of the form $\psi(X_1, X_2) = 1$ if and only if $X_1 > \kappa_1$ and $X_2 > \kappa_2$. In this common situation, we need to use a classification rule that assumes some "reasonable" form for the classifier and then estimates the particulars of the classifier from the sample data. For instance, we might assume that the classifier is determined by a hyperplane in k-dimensional space and then a classification rule would estimate the parameters of the hyperplane.

Whenever a classification rule is used, we need to obtain an estimate $\hat{\varepsilon}[\psi]$ of the error $\varepsilon[\psi]$ of the designed classifier ψ. If we have access to a lot of data, then we can split the data into *training data* and *test data*, design a classifier on the training data, and let $\hat{\varepsilon}[\psi]$ be the proportion of incorrect classifications on the test data. This proportionality estimate, which is accurate for modest sized test-data samples, is an example of an *error estimation rule*. If data are limited, then we cannot split the data between training and testing without negatively impacting classifier design, error estimation, or both, but must form an error estimation rule using the training data. Error estimation via the training data is rarely accurate. We will have more to say on this shortly.

To summarize model discovery, when using a classification rule and an error estimation rule, we obtain sample data and determine a classifier ψ and an error estimate $\hat{\varepsilon}[\psi]$, which together form a classifier model.

5.4.2   Model Validation

Model validation in classification serves to illustrate the difference between model validation relative to description of Nature and model validation relative to a pragmatic objective.

Suppose we possess sufficient data to estimate the feature-label distribution to a high degree of accuracy without making any assumptions. Given virtually unlimited data, validation via prediction is straightforward: collect independent data and compare the model feature-label distribution to the data histogram. Once the feature-label distribution is validated, the derived Bayes classifier and its error, having been mathematically deduced from the distribution, are *ipso facto* valid. This whole procedure is artificial because if we utilize sufficient data for both model construction and validation, since in both cases the data histogram converges to the feature-label distribution, the two will be arbitrarily close. If they are not, it means that we have not used enough data.

The situation changes if we use modeling assumptions to assist model inference. For instance, with a Gaussian assumption (both class-conditional distributions are Gaussian), irrespective of how much data there is for inference, we are forming the feature-label distribution from the covariance matrices, means, and prior class probabilities. With a large independent test-data set, validation by comparing the constructed feature-label distribution to the data histogram (absent the Gaussian assumption) will show the extent to which our modeling assumptions are valid. If the constructed feature-label distribution is validated, once again we can derive a Bayes classifier and its error. However, suppose the model feature-label distribution is not validated. It still might be that we have derived a classifier whose error is very close to the Bayes error. To validate this intrinsic property of the feature-label distribution via the feature-label distribution, we would have to generate sufficient independent data to arrive at a new model feature-label distribution directly from the new data absent the modeling assumptions, generate another set of independent data to validate the new model feature-label distribution, derive a Bayes classifier and its error from the new feature-label distribution, and then compare the error of the original Bayes classifier to the new Bayes error. This process would be absurd because it would render using the original assumptions superfluous.

Rather than stop here, let us change our perspective to one of application. A classifier is found from the inferred feature-label model. Regardless of whether the model can be validated, is the classifier useful? That depends on its error. With a modest test sample, we can obtain an accurate estimate of its error and determine whether it is good enough for the application of interest. In fact, this method of classifier design has a long history. Under the Gaussian assumption, since the model feature-label distribution depends on the estimates of the covariance matrices, means, and prior probabilities, so too does the designed classifier and it can be expressed in terms of these. Because the decision boundary is hyperquadratic, this classification rule is known as *quadratic discriminant analysis* (QDA). If we make a further assumption that the covariance matrices are equal, then the decision boundary is a hyperplane and the method is known as *linear discriminant analysis* (LDA). These rules can produce good classifiers if the true feature-label distribution does not vary too far from the Gaussian assumptions and there are sufficient training data. Indeed, since LDA has fewer parameters to estimate, when data are scarce it can perform better than QDA when the covariance matrices are not identical. Overall, the basic approach is called the *plug-in* method: the form of the Bayes classifier is derived from an incomplete feature-label distribution model and estimates of the unknown parameters are plugged into the form.

Figure 5.1 illustrates LDA and QDA applied to a Gaussian model with different covariance matrices. The black curve shows the decision boundary for the Bayes classifier and the shaded and white regions show the decision regions obtained by the classification rule applied to the 90 data points. The samples size is reasonable for QDA and we see that it does better than LDA, whose form has been derived under the assumption of equal covariance matrices. For the given sample, the errors (found from the true feature-label distribution) are 0.0876 and 0.1036 for QDA and LDA, respectively.

The preceding analysis extends to the most general methodology of classifier design: use a classification rule (QDA, neural net, support vector machine, or any one of a host of rules) to obtain a classifier and then estimate its error. The issue of a Bayes classifier never arises. What matters is classifier error because this quantifies the designed classifier's predictive capacity when it is applied. Hence, the salient epistemological issue is the accuracy of the error estimator.

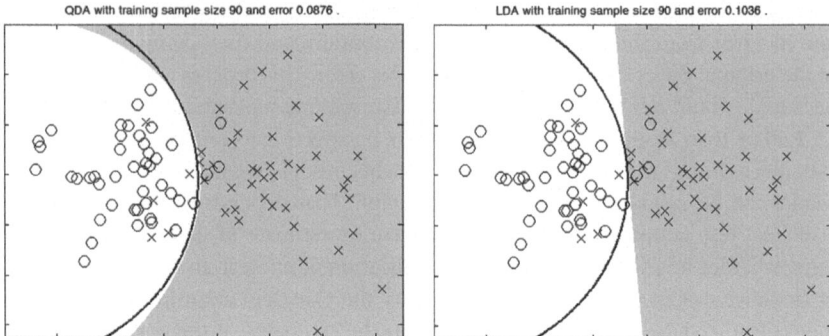

Fig. 5.1: QDA (left) and LDA (right) applied to a Gaussian model with unequal covariance matrices, where the black curve shows the Bayes classifier.

5.4.3 Error Estimation

The most commonly employed measure of error estimation accuracy is the *root-mean-square* (RMS) deviation, defined by $E^{1/2}[|\hat{\varepsilon}[\psi] - \varepsilon[\psi]|^2]$, where E denotes the expected value over all samples. Alternatively, the *mean-square error* (MSE) of the estimate is the expectation without the square root. If error estimation is done on test data, then it can be shown that RMS $\leq 1/2\sqrt{m}$, where m is the size of the test set, and this bound is independent of both the feature-label distribution and the classification rule (Devroye, Gyorfi, and Lugosi 1996). Error estimation on test data is quite accurate for modest size test sets. For instance, with $m = 100$, RMS ≤ 0.05.

Matters are much worse when estimating error on the training data. Only in rare instances do we possess a bound on the RMS and, when we do, the bound may depend on the classification rule and feature-label distribution (which we do not know), and it is usually so large as to be useless for even modest size samples, when in fact we are estimating the error on the training data because we only have a small sample. For instance, consider the *k-nearest-neighbor rule*, which assigns to a point the majority label among its nearest k neighbors in the sample, and *leave-one-out error estimation*, which sequentially leaves out a point from the sample, designs a classifier on the other points, applies the classifier on the left-out point, and estimates the error as the proportion of errors on the left-out points. For this classification rule and training-data error estimator, there is the following distribution-free upper bound on RMS (Devroye and Wagner 1979):

$$\text{RMS} \le \sqrt{\frac{1}{n} + \frac{24\sqrt{k}}{n\sqrt{2\pi}}} \; . \tag{2}$$

For the popular choice $k = 3$, at sample size $n = 100$, the bound is 0.419, quite useless.

Leave-one-out is a special case of cross-validation estimation where q points are left out at each stage. Owing to the computational burden, when $q > 1$, the left-out points are randomly selected and not all q-point combinations are left out. Other variations of cross-validation have been proposed. More generally, cross-validation belongs to the family of re-sampling training-data error estimators, with bootstrap estimation being another example. These estimators tend to perform poorly (large RMS) for small samples, but that is precisely the place where they need to be used. They tend to have large variance, very little (or even negative) correlation with the true error, and there tends to be little (or even negative) regression of the true error on the re-sampling estimator. All of this contributes to large RMS, the result being that they provide negligible knowledge of the true error and are therefore epistemologically vacuous with small samples.

A typical situation for leave-one-out is shown in Figure 5.2, which has resulted from applying LDA to a Gaussian model with sample size 60. The figure shows the scatter plot of estimate-true error pairs with the leave-one-out estimate on the horizontal axis and the true error on the vertical axis, resulting from many random samples. It also shows the linear regression between the true and estimated errors. There is large variance, little correlation, and flat regression. The variance is actually worse than shown in the figure because the plot stops at leave-one-out error equal to 0.5 and the actual scatter plot continues well to right of 0.5.

There are numerous other training-data error estimators, some of which provide somewhat better performance for certain feature-label distributions and classification rules; however, outside of LDA in the Gaussian model and multinomial discrimination (discrete classification), there are no known theoretical representations of the RMS. For the Gaussian model, closed-form RMS representations exist for the univariate model (Zollanvari, Braga-Neto, and Dougherty 2012) and asymptotic expressions exist for the multivariate model (Zollanvari, Braga-Neto, and Dougherty 2011) for leave-one-out and resubstitution (the error on the training data). Closed-form expressions are also known for multinomial discrimination (Braga-Neto and Dougherty 2005). Even though pattern recognition is among the most important applications of statistics, until recently very little attention has been paid to its epistemology. The situation has begun to improve but much greater effort is needed (Braga-Neto and Dougherty 2015).

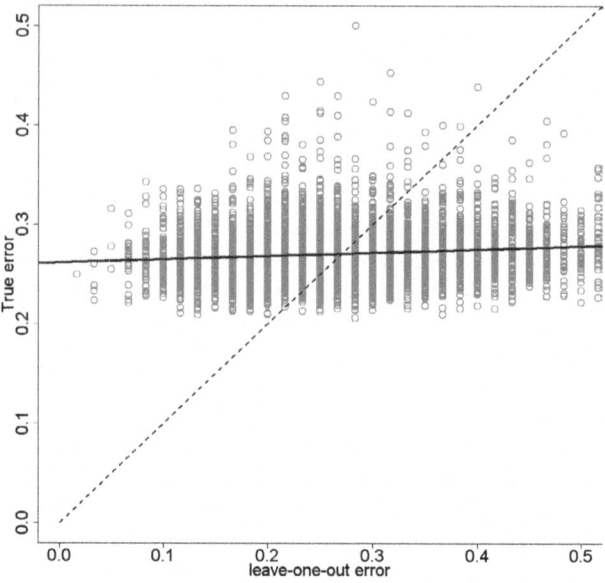

Fig. 5.2: Scatter plot and linear regression for the true error and the leave-one-out estimate.

5.4.4   Classifier Goodness

In practice, one typically chooses a classification rule that, due to experience, he hopes will produce a good classifier. Of course, a designed classifier can do no better than a Bayes classifier. The collection of Bayes classifiers (there may be more than one) for a feature-label distribution is intrinsic to the distribution and therefore a property of the features and label—a physical property if they correspond to physical measurements. Hence, one might raise the following epistemological question: How does the error of a classifier designed by a particular classification rule compare to the Bayes error? If we expect the errors to be close, then we can take the error of the designed classifier to be an approximation of the Bayes error, so that an accurate estimate of the classifier error provides an estimate of the Bayes error. To achieve this we need an accurate error estimator and some theoretical statement regarding the expected difference between the Bayes error and the error of the designed classifier. Having considered error estimation, we now take a brief look at the difference of the errors.

For any sample, the *design cost* is defined by $\Delta_{des} = \varepsilon_{des} - \varepsilon_{Bayes}$, where ε_{des} and ε_{Bayes} are the error of the designed classifier and the Bayes error, respectively. Δ_{des} and ε_{des} are sample-dependent random variables, whereas ε_{Bayes} is an intrinsic value of the feature-label distribution. The salient quantity for a classification rule is the expected design cost $E[\Delta_{des}]$, expectation being relative to the random sample. The expected error of a designed classifier is decomposed as

$$E[\varepsilon_{des}] = \varepsilon_{Bayes} + E[\Delta_{des}]. \tag{3}$$

A classification rule provides a good estimate of the Bayes error if $E[\Delta_{des}]$ is small (and the error estimator is accurate). The problem with small-sample design is that $E[\Delta_{des}]$ tends to be large.

A classification rule may yield a classifier that performs well on the sample data but has a large error. This phenomenon, known as *overfitting* the sample data, is especially prevalent with small samples. One tries to develop classification rules that are constrained so as to reduce overfitting, for instance, requiring the classifier to be linear (a hyperplane). The problem is that constraining the set of possible classifiers to ones that are less overfitting "bakes in" additional cost because the constrained class may not include a Bayes classifier. Letting ε_{const} be the error of the best classifier in the constrained class, Eq. 3 takes the form

$$E[\varepsilon_{des}] = \varepsilon_{Bayes} + \Delta_{const} + E[\Delta_{des}^{const}], \tag{4}$$

where $\Delta_{const} = \varepsilon_{const} - \varepsilon_{Bayes}$ is the *cost of constraint* and $\Delta_{des}^{const} = \varepsilon_{des} - \varepsilon_{const}$ is the design cost under constraint. Δ_{const} represents an added error guaranteed by the constraint. The constraint is beneficial if and only if

$$\Delta_{const} < E[\Delta_{des}] - E[\Delta_{des}^{const}], \tag{5}$$

the right-hand side being the decrease in design cost resulting from the constraint. The dilemma is that a strong constraint reduces $E[\Delta_{des}^{const}]$ at the cost of increasing Δ_{const}.

Basically, the more complex a class C of classifiers, the smaller the constraint cost and the greater the design cost. As written, this statement is far too vague to be meaningful. Much effort has gone into making it mathematically meaningful but this is outside our current scope. We will simply note a fundamental theorem that gives an upper bound on $E[\Delta_{des}^{const}]$ for a particular classification rule. For the *empirical-error classification rule*, which chooses a classifier in C that makes the least number of errors on the sample data,

$$E[\Delta_{\text{des}}^{\text{const}}] \leq 4\sqrt{\frac{V_C \log n + 4}{2n}}, \tag{6}$$

where V_C is the *VC* (*Vapnik-Chervonenkis*) *dimension* of C (Vapnik and Chervonenski 1971). Because its definition is somewhat involved, we shall not define the VC dimension, noting only that it provides a measure of classifier complexity. It is clear from Eq. 6 that the sample size n must greatly exceed V_C for the bound to be small.

To illustrate the relation of the VC dimension to classifier complexity, we note that, with k features, $V_C = k + 1$ for a linear classifier. For a neural net with an even number r of neurons, $V_C \geq kr$; with an odd number of neurons, $V_C \geq k(r - 1)$. As the complexity (number of neurons) of the neural net grows, its VC dimension becomes much larger than the VC dimension of a linear classifier. For an even number of neurons, the bound exceeds $4\sqrt{kr \log n / 2n}$, which makes it useless unless the sample size n is very large.

The bound in Eq. 6 is only meaningful if the VC dimension is finite, which means that C cannot be too complex. Since the bound assumes no feature-label distribution, it is "loose." It can be tighter under assumptions on the feature-label distribution. In practice, if no distributional assumptions are made, then one is stuck with this bound.

Combining Eqs. 5 and 6 yields the inequality

$$E[\varepsilon_{\text{des}}] - \varepsilon_{\text{Bayes}} \leq \Delta_{\text{const}} + \sqrt{\frac{V_C \log n + 4}{2n}}. \tag{7}$$

There is an epistemological quality to this inequality because it bounds the difference between the expected error of a designed classifier and the distribution-intrinsic Bayes error. Specifically, it bounds the error of the approximation $\varepsilon_{\text{Bayes}} \approx E[\varepsilon_{\text{des}}]$. There are two weaknesses to the bound. First, unless n is very large, the bound is useless. Nevertheless, one could let $n \to \infty$ and obtain the limiting bound $E[\varepsilon_{\text{des}}] - \varepsilon_{\text{Bayes}} \leq \Delta_{\text{const}}$. From the definition of the cost of constraint, $E[\varepsilon_{\text{des}}] - \varepsilon_{\text{Bayes}} \geq \Delta_{\text{const}}$. Taken together, these two inequalities say that, as $n \to \infty$, $E[\varepsilon_{\text{des}}] - \varepsilon_{\text{Bayes}} \to \Delta_{\text{const}}$. Intuitively, this says that, for very large sample size, $\varepsilon_{\text{Bayes}} \approx E[\varepsilon_{\text{des}}] + \Delta_{\text{const}}$. Hence, the error of a designed classifier, for the particular classification rule used to obtain the VC bound, gives us an idea of the Bayes error if we have some idea of the cost of constraint and this cost is small, the sample is very large, and the VC dimension of the class is finite. All of these provisos are problematic.

5.5   Modeling with Uncertainty

We began our discussion of classification using scientific knowledge to build the classifier and from there weakening our assumptions until we arrived at purely data-driven classifier design and error estimation, the result being that absent a large sample the resulting classifier model is virtually epistemologically vacuous. It takes a great deal of data, and data of the proper kind, to make scientific inferences—and with contemporary scientific problems *relevant* data is often in short supply. The situation is improved by abandoning pure science in favor of translational science because the data requirements for the latter are less demanding because applications utilize more focused (hence, less) knowledge than a full description of the underlying physical system. But even applications tend to demand more data than are available, even in the case of classification, which is among the least complex applications.

The solution is to abandon radical empiricism and return to basic scientific methodology by incorporating both existing knowledge and data in model formation. Rather than constructing a gene network by selecting a set of genes and then using some procedure to infer the relations among them, one can partially construct the network from existing genetic regulatory knowledge obtained via biochemical experimental methods. The result is an *uncertainty class* of networks which, if the prior knowledge is accurate, contains the desired network. From there, new data can be used to shrink the uncertainty class. We may not arrive at a single network by this procedure, but even if we have very limited data, it will be incorporated with existing knowledge and the resulting uncertainty class will represent our full knowledge of the system. Application can be based on an uncertainty class by finding an operator that is optimal with respect to the uncertainty class.

In the case of classification, if the feature-label distribution is known, then a Bayes classifier and the Bayes error can be found directly. Not knowing the feature-label distribution, the standard approach is to use a classification rule and error estimation rule in a purely data-driven methodology. This purely empirical approach is mitigated in the case of classification rules such as QDA, where a model form is postulated for the feature-label distribution and the classification rule derived from the form; however, it turns out that plug-in methods do not work well for small samples because these do not yield good estimation of the model parameters, such as the covariance matrices in the case of QDA. Moreover, plug-in error estimation, where the error of the designed classifier is estimated by finding its error on the model obtained from estimating the parameters, tends to perform poorly because it measures the error relative to a poor estimate of the true feature-label distribution and it is relative to this poor estimate that the classifier has been designed to begin with. The essential problem is that the data have been used relative to the assumed form of the feature-label distribution in an *ad hoc*-fashion. Moreover, no knowledge outside the form of the feature-label distribution has been assumed.

Suppose we wish to construct a classifier based on gene expression to decide between two treatment strategies and we have strong evidence that the class-conditional densities are both Gaussian. Then part of QDA classifier design involves estimating the covariances between the various genes constituting the features for each class-conditional density. Now suppose we have regulatory knowledge regarding the genes, for instance, when two of the genes are highly expressed, then a third will be highly expressed. This knowledge constrains the covariance values among the three genes, thereby providing information that can be used in conjunction with the data to reduce the uncertainty class. It is this idea that needs to be generalized into a rigorous mathematical framework.

For simplicity of exposition, we confine ourselves to a single unknown parameter θ; that is, the feature-label distribution is known except for the single parameter θ. But we have prior knowledge: we know that θ lies in the interval $[a, b]$. Moreover, there is a *prior distribution* on θ, meaning that there is a probability density, denoted $\pi(\theta)$, over $[a, b]$ characterizing the likelihood of θ's position in the interval. Should we lack such additional knowledge, there is no loss in making the assumption because we could simply make the prior distribution uniform over $[a, b]$. If we gather new data, then we can update the prior distribution by considering the distribution of θ, given the new observations, to arrive at a *posterior distribution*, denoted $\pi^*(\theta)$, which is mathematically derived by using Bayesian statistical analysis. The posterior distribution characterizes our full knowledge: the original prior (existing) knowledge together with the new knowledge supplied by the data.

If we design a classifier from the newly collected data, then we can find an estimate of its error that is optimal relative to the posterior distribution. The method is to use standard optimization theory: find the error estimate for which the mean-square error between the true error and the estimate is minimum with respect to the posterior distribution; that is, minimize the MSE between the estimated and true errors, relative to $\pi^*(\theta)$ (Dalton and Dougherty 2011a). In optimization terminology, this is called the *minimum-mean-square-error* (MMSE) error estimate. Owing to the Bayesian framework (prior distribution), it is also called the *optimal Bayesian error estimate* (BEE). We are guaranteed that the BEE is best, on average, over the uncertainty class, that is, best on average relative to the state of our knowledge. The mathematics can be messy but closed-form solutions are known in certain Gaussian models and for discrete classification (Dalton and Dougherty 2011b).

For other modeling assumptions, Markov-chain-Monte-Carlo (MCMC) methods can be used (Knight, Ivanov, and Dougherty 2014).

Why stop at error estimation? Given the posterior distribution, find an optimal classifier relative to $\pi^*(\theta)$ by finding the classifier that minimizes the average error over the posterior distribution; that is, find the classifier that minimizes the BEE (Dalton and Dougherty 2013). As in the case of error estimation, closed-form optimal Bayesian classifiers have been found in certain Gaussian models and for discrete classification. It is important to distinguish between a Bayes classifier and an optimal Bayesian classifier (OBC). The former is for a single feature-label distribution and the latter is for an uncertainty class of feature-label distributions. Consider the following: in a known common-covariance Gaussian model we obtain a linear Bayes classifier decision boundary; for an uncertainty class of common-covariance Gaussian models, each Bayes classifier is linear but an optimal Bayesian classifier is only linear in rare cases.

Optimal Bayesian error estimation weighs both prior knowledge and data. If there is perfect knowledge, then no data are needed because we have the feature-label-distribution. If there is unlimited data, then no prior knowledge is needed. In practice, we need to have enough of each to get an acceptable MSE between the true and estimated errors.

An advantage of the BEE is in regard to finding the MSE. In traditional data-driven classification, the MSE of an estimator is found relative to an assumed feature-label distribution. In the Bayesian setting, this would correspond to finding the MSE relative to the prior distribution; however, owing to the fact that the posterior distribution incorporates new data with the prior, for optimal Bayesian error estimation we can also find the MSE conditioned on the sample data (Dalton and Dougherty 2012). Thus, the MSE of the estimate can be updated as more data are acquired.

It is on account of the updating of the prior to the posterior distribution that the method is said to be Bayesian because Bayes rule is used to derive the posterior distribution. One could proceed with no data, just using prior knowledge, in which case the method has been referred to as *robust classification* (Dougherty et al. 2005). One could also proceed via iterative updating, $\pi \to \pi^* \to \pi^{**} \to \dots$, finding a new OBC and BEE at each stage until one is satisfied with the MSE (accuracy) of the BEE, which can be computed at each stage conditioned on the data.

The fundamental point is that *ad hoc* classifier design and error estimation have been replaced by optimization in the context of a prior distribution over an uncertainty class. In this vein, classification is a special case of the more general problem of deriving optimal filters on uncertainty classes of stochastic processes. This problem goes back to the 1970s, where minimax optimization was employed for linear operators on stochastic processes (Kassam and Lim 1977); it was placed into a Bayesian context for both linear and some nonlinear operators with a suboptimal operator being found (Grygorian and Dougherty 1999); and was recently solved completely for linear and certain classes of nonlinear operators (Dalton and Dougherty 2014). As for control, there is a keen interest in optimal control of gene regulatory networks because such control corresponds to optimal therapy (Shmulevich and Dougherty 2010). As opposed to classification, control under uncertainty has a long history (Silver 1963), but its application is obstructed by enormous computational demands.

All of this optimization relative to uncertainty depends on the accuracy of the prior distribution since everything is conditioned on it, the effect of the prior being stronger when the number of data points is small. In the case of MMSE classifier error estimation, robustness to incorrect modeling assumptions has been examined via simulation studies, with performance being only modestly affected for various kinds of incorrect modeling assumptions (Dalton and Dougherty 2011a). Robustness of optimal Bayesian classifiers to incorrect modeling assumptions has also been studied, with good robustness observed (Dalton and Dougherty 2013b). In general, if one is confident in his knowledge, then a tight prior is preferable because tighter priors require less data for good performance; however, when one lacks confidence, prudence calls for a less informative prior. A prior whose mass is concentrated far away from the true parameters will perform worse than one that is non-informative (flat). These issues have long been discussed in the Bayesian literature (Jaynes 1968).

The prior distribution has to be constructed from existing scientific knowledge and this requires a methodology. Put another way, existing scientific knowledge, which may be in the form of regulatory relationships, physical constraints, known or partially known stochastic differential equations, etc., must be mathematically transformed into distributional knowledge relevant to the classification (or other) translational problem. This transformation is unlikely to be direct, except in simple cases where we just assume the unknown parameters lie within derived intervals. More generally, the transformation methodology may include optimization relative to some cost function incorporating some measure of uncertainty (Esfahani and Dougherty 2014).

The choice of cost function and uncertainty measure introduces subjectivity into prior construction, but this is always the case with optimization because optimization must be with respect to some criteria. For pure science, this might be troublesome, but for translational science it is the paradigmatic approach. If one is content with operator derivation based on a cost function relative to the translational objective, why should he not be content with transforming existing physical knowledge into a prior distribution based on a cost function relative to the purpose to which it is being put?

From the perspective of epistemology, were we able to independently validate an optimal estimate, then there would be no downside to this overall approach; however, we have proceeded via uncertainty classes because data are hard to come by. The purist might be dissatisfied. His charge is that we have not only reduced the scientific enterprise to object-oriented application, but we have also removed standard validation based on concordant prediction. This is so. But as we have demonstrated, the alternative is to do nothing. The optimization approach is contextual relative to its cost functions. Put another way, it is conditioned on the prior distribution, which is dependent on its manner of derivation. Given this conditioning and the objective, the theory follows and provides one the possibility of obtaining useful results with rigorous error analysis. No doubt, there is grist here for the epistemologist's mill.

5.6 Objective Cost of Uncertainty

The basic principle we have been following is to derive an operator that is optimal over an uncertainty class of models governed by a prior (or posterior) distribution. Optimization is relative to a cost function—error rate for classification, some probabilistic error for filtering stochastic processes, or the probability mass of undesirable states when controlling a dynamical system. The operator suffers from our uncertainty, since the best procedure would be to derive an optimal operator relative to the true model.

To quantify the cost uncertainty, let M_θ denote the model with unknown parameter θ (which could be a vector of parameters) and let ξ_θ be the optimal operator for M_θ relative to the cost function C_θ on M_θ, so that $C_\theta(\xi_\theta) \leq C_\theta(\xi)$ for any operator ξ. Let Θ be the parameter space and let ξ_Θ have lowest average cost, $E[C_\theta(\xi_\Theta)]$, across the uncertainty class, where E denotes expectation relative to the distribution of θ. This means that $E[C_\theta(\xi_\Theta)] \leq E[C_\theta(\xi)]$ for any operator ξ. In the face of uncertainty, the best course of action is to apply operator ξ_Θ, but there is a cost to this choice relative to applying the optimal operator for the true network because $C_\theta(\xi_\theta) \leq C_\theta(\xi_\Theta)$ for any $\theta \in \Theta$.

For any $\theta \in \Theta$, we define the *objective cost of uncertainty* (OCU) as the difference in cost on M_θ between the optimal average performing operator and the optimal operator on M_θ:

$$\text{OCU}(\theta) = C_\theta(\xi_\Theta) - C_\theta(\xi_\theta).\tag{8}$$

Since the true value of θ is unknown, we take the *mean objective cost of uncertainty* (MOCU) as a measure of uncertainty (Yoon, Qian, and Dougherty 2013):

$$\text{MOCU}(\Theta) = E[\text{OCU}(\theta)] = E[C_\theta(\xi_\Theta) - C_\theta(\xi_\theta)].\tag{9}$$

If there is no uncertainty, then the uncertainty class contains only one model and $\text{MOCU}(\Theta) = 0$.

From a scientific perspective on uncertainty, one might prefer to use the entropy of the prior distribution, but distributional entropy does not focus attention on the translational objective. It may be that there is large entropy but that most of the uncertainty is irrelevant to the objective. For instance, in controlling a network it may be that there is much uncertainty in the network but there is a high degree of certainty regarding the mechanisms involved in the control. In this case, the entropy might be large but the MOCU be very small, and this is all we care about. The MOCU quantifies the uncertainty in our knowledge with respect to our objective and therefore is a fundamental epistemological parameter. It is intrinsic to the physical system, given our knowledge and objective.

Since the MOCU measures the uncertainty relevant to the objective, should we wish to update the prior with new data, a natural course is to find a posterior distribution with minimal MOCU. Hence, we would choose the experiment from the space of possible experiments that yields the minimum expected MOCU given the experiment (Dehghannasiri, Yoon, and Dougherty 2015). For network intervention, this choice would yield an experiment expected to maximally reduce the uncertainty regarding design of an optimal intervention policy.

References

Bacon, Francis. 1952 (originally published 1620). Novum Organum. Great Books of the Western World, 35, eds. R. M. Hutchins and M. J. Adler. Chicago: Encyclopedia Britannica.

Bohr, Niels. 1958. *Atomic Physics and Human Knowledge*. New York: John Wiley.

Braga-Neto, Ulisses M., and Edward R. Dougherty. 2005. Exact Performance of Error Estimators for Discrete Classifiers. *Pattern Recognition*. 38: 1799-1814.

Braga-Neto, Ulisses M., and Edward R. Dougherty. 2015. *Error Estimation for Pattern Recognition*. New York: Wiley-IEEE Press.

Dalton, Lori A., and Edward R. Dougherty. 2011a. Bayesian Minimum Mean-Square Error Estimation for Classification Error – Part I: Definition and the Bayesian MMSE Error Estimator for Discrete Classification. *IEEE Transactions on Signal Processing.* 59: 115-129.

Dalton, Lori A., and Edwrd R. Dougherty. 2011b. Bayesian Minimum Mean-Square Error Estimation for Classification Error – Part II: Linear Classification of Gaussian Models. *IEEE Transactions on Signal Processing.* 59: 130-144.

Dalton, Lori A., and Edward R. Dougherty. 2012. Exact MSE Performance of the Bayesian MMSE Estimator for Classification Error – Part I: Representation. *IEEE Transactions on Signal Processing.* 60: 2575-2587.

Dalton, Lori A., and Edward R. Dougherty. 2013a. Optimal Classifiers with Minimum Expected Error within a Bayesian Framework – Part I: Discrete and Gaussian Models. *Pattern Recognition.* 46: 1288-1300.

Dalton, Lori A., and Edward R. Dougherty. 2013b. Optimal Classifiers with Minimum Expected Error within a Bayesian Framework – Part II: Properties and Performance Analysis. *Pattern Recognition.* 46: 1301-1314.

Dalton, Lori A., Edward R. Dougherty. 2014. Intrinsically Optimal Bayesian Robust Filtering. *IEEE Transactions on Signal Processing.* 62: 657-670.

Dehghannasiri, Roozbeh, Byung-JunYoon, and Edward R. Dougherty. 2015. Optimal Experimental Design for Gene Regulatory Networks in the Presence of Uncertainty. *IEEE/ACM Transactions on Computational Biology and Bioinformatics.* 14: 938-950.

Devroye, Luc, and T. Wagner. 1979. Distribution-free Inequalities for the Deleted and Holdout Error Estimates. *IEEE Transactions on Information Theory.* 25: 202-207.

Devroye, Luc, Laszlo Gyorfi, and Gabor Lugosi. 1996. *A Probabilistic Theory of Pattern Recognition.* New York: Springer-Verlag.

Dougherty, Edward R. 2009. Translational Science: Epistemology and the Investigative Process. *Current Genomics.* 10: 102-109.

Dougherty, Edward R., and Michael L. Bittner. 2011. *Epistemology of the Cell: A Systems Perspective on Biological Knowledge.* New York: Wiley-IEEE Press.

Dougherty, Edward R., and Ilya Shmulevich. 2012. On the Limitations of Biological Knowledge. *Current Genomics.* 13: 574-587.

Dougherty, Edward R., Jianping Hua, and Michael L. Bittner. 2007. Validation of Computational Methods in Genomics. *Current Genomics.* 8: 1-19.

Dougherty, Edward R., Jianping Hua, Zixiang Xiong, and Yidong Chen. 2005. Optimal Robust Classifiers. *Pattern Recognition.* 38: 1520-1532.

Einstein, Albert. 1944. Remarks on Bertrand Russell's Theory of Knowledge. In *The Philosophy of Bertrand Russell,* ed. Paul A. Schlipp. Greensboro: The Library of Living Philosophers, 5.

Esfahani, Mohammad S., and Edward R. Dougherty. 2014. Incorporation of Biological Pathway Knowledge in the Construction of Priors for Optimal Bayesian Classification. *IEEE/ACM Transactions on Computational Biology and Bioinformatics.* 11: 202-218.

Grygorian, Artyom M., and Edward R. Dougherty. 1999. Design and Analysis of Robust Optimal Binary Filters in the Context of a Prior Distribution for the States of Nature. *Mathematical Imaging and Vision.* 11: 239-254.

Heisenberg, Werner. 1952. *Philosophic Problems of Nuclear Science.* New York: Pantheon.

Edwin T. Jaynes. 1968. Prior Probabilities. *IEEE Transactions on Systems Science and Cybernetics.* 4: 227-241.

Kant, Immanuel. 1952 (originally published 1787). *Critique of Pure Reason*, second edition. Great Books of the Western World, 42, eds. R. M. Hutchins and M. J. Adler. Chicago: Encyclopedia Britannica.

Kassam, Saleem, and T. I. Lim. 1977. Robust Wiener Filters. *J. Franklin Institute*. 304: 171–185.

Knight, Jason, Ivan Ivanov, and Edward R. Dougherty. 2014. MCMC Implementation of the Optimal Bayesian Classifier for Non-Gaussian Models: Model-based RNA-Seq Classification. *BMC Bioinformatics*. 15: 401.

Reichenbach, Hans. 1971. *The Rise of Scientific Philosophy*. Berkeley: University of California Press.

Rosenblueth, Arturo, and Norbert Wiener. 1945. The Role of Models in Science. *Philosophy of Science*. 12: 316-321.

Shmulevich, Ilya, and Edward R. Dougherty. 2010. *Probabilistic Boolean Networks: The Modeling and Control of Gene Regulatory Networks*. New York: SIAM Press.

Silver, Edward A. 1963. Markovian Decision Processes with Uncertain Transition Probabilities or Rewards. Technical report. Massachusetts Institute of Technology.

Vapnik, Vladimir N., and Alexey Y. Chervonenkis. 1971. On the Uniform Convergence of Relative Frequencies of Events to Their Probabilities. *Theory of Probability and its Applications*. 16: 264-280.

Yoon, Byung-Jun, Xiaoning Qian, and Edward R. Dougherty. 2013. Quantifying the Objective Cost of Uncertainty in Complex Dynamical Systems. *IEEE Transactions on Signal Processing*. 61: 2256-2266.

Zollanvari, Amin, Ulisses M. Braga-Neto, and Edward R. Dougherty. 2011. Analytic Study of Performance of Error Estimators for Linear Discriminant Analysis. *IEEE Transactions on Signal Processing*. 59: 4238-4255.

Zollanvari, Amin, Ulisses M. Braga-Neto, and Edward R. Dougherty. 2012. Exact Representation of the Second-order Moments for Resubstitution and Leave-one-out Error Estimation for Linear Discriminant Analysis in the Univariate Heteroskedastic Gaussian Model. *Pattern Recognition*. 45: 908-917.

6. On Fishing for Significance and Statistician's Degree of Freedom in the Era of Big Molecular Data

Anne-Laure Boulesteix[1], Roman Hornung[2], Willi Sauerbrei[3]

Abstract

There are usually plenty of conceivable approaches to statistically analyze data that both make sense from a substantive point of view and are defensible from a theoretical perspective. The data analyst has to make a lot of choices, a problem sometimes referred to as "researcher's degree of freedom". This leaves much room for (conscious or subconscious) fishing for significance: the researcher (data analyst) sometimes applies several analysis approaches successively and reports only the results that seem in some sense more satisfactory, for example in terms of statistical significance. This may lead to apparently interesting but false research findings that fail to get validated in independent studies. In this

---

1   Anne-Laure Boulesteix is associate professor of biostatistics with focus on computational molecular medicine at the University of Munich. She obtained her PhD in statistics in 2005 from the same university. Her research focuses on the statistical analysis of biomedical data with a special emphasis on prediction modeling and issues related to scientific practice from a statistical point of view.
2   Roman Hornung is a PhD student at the University of Munich, where he obtained his master's degree in statistics in 2011. He is working on statistical methodology for the analysis of high-dimensional molecular data with focus on prediction modeling and is currently involved in leukemia research based on molecular data.
3   Willi Sauerbrei is professor of medical biometry at the Medical Center – University of Freiburg. His areas of expertise include multivariable regression modeling and issues related to good statistical practice such as reporting. He is the chair of the STRATOS (STRengthening Analytical Thinking for Observational Studies) initiative which aims at providing guidance on the design and analysis of observational studies.

essay we describe and illustrate these problems and discuss possible strategies to (partially) address them such as validation, increased development of guidance documents, and publication of negative research findings, analysis plans, data and code.

6.1   Background

Suppose that a researcher – called consulting client in the rest of this paper, for example a medical doctor, asks a statistician to analyze his dataset in order to answer one or several questions of interest. A simple question may be whether a specific blood value is, on average, different in two groups of patients, say, patients who received a placebo and patients who received a new drug in a randomized clinical trial. A more complex question is whether exposure to a certain risk factor increases the risk of developing a disease or having a bad outcome, assuming that all other considered factors are fixed. A related but different question is how this risk can be predicted based on all these factors. Finally, another complex question is whether the patients under study can be clustered into distinct groups defining typical profiles. Medical studies often address several such questions simultaneously.

As a consequence of the advance of high-throughput molecular biology techniques such as DNA-chips or next-generation sequencing in the last 15 years, many of these questions have been increasingly often addressed in "big data" settings. By big data settings we mean, roughly speaking, that the dataset at hand does not only contain a handful of measurements and clinical markers for each patient as usual in classical settings, but thousands of measurements. For example, big molecular datasets may include the expression levels of thousands of genes or hundreds of metabolites or proteins in a sample. Our understanding of the term "big data" in the context of molecular research with medical applications refers to situations with many variables for each observation (patient) and a moderate number of observations. In other contexts the term "big data" may refer to a high number of observations (e.g., insurance data, web data) or to a high frequency of measurement in time (e.g., monitoring data, finance data), but we will not discuss these aspects here.

No matter whether they analyze big data or not, during the data analysis process analysts have many choices and have to make many decisions. These choices are referred as the "researcher's degree of freedom" by Simmons et al. (2011). Choices to be met may concern different aspects of data management, initial data analysis and main analysis. For example, one may have to choose to collapse two or more categories because the number of patients is very small or to perform a data transformation: a measurement may for example be transformed using the logarithmic function, or transformed into a binary variable (low versus high) using any (more or less adequate) threshold. One may have to choose a technique to handle extreme values appearing as implausible, or a method that aims at estimating missing values so that the patients with incomplete data do not have to be excluded from the analysis. One may also have to choose whether to exclude part of the data for any reason or to consider relevant subgroups to be analyzed separately. These are only a few examples out of a long list. Data preparation steps performed in initial data analyses are particularly important and complex in the context of big molecular data. They also include, for example, data normalization, elimination of irrelevant variables, variable selection or correction for technical differences between parts of the data generated by different platforms or labs.

6.2   Researcher's degree of freedom

Choices regarding data preparation/initial data analyses may be driven by statistical considerations, by arguments from the consulting client or a mixture of both. For example, dichotomizing a measurement into two levels (low versus high) may make sense from a medical point of view. The choice of an imputation technique for missing values may be guided by statistical considerations regarding the number and type of missing values, as well as by the reason – described by the consulting client – why the values are missing. The choice to log-transform a measurement may have a statistical justification (typically, to make the measurement more normally distributed) and make sense from the point of view of the consulting client, for example for better comparability with previously published results – if this measurement is commonly log-transformed in the literature.

Beyond data preparation steps, the statistician has to choose the statistical method he/she will use to analyze the data. Note that this choice may interfere with the previous choices. For example, it makes sense to transform a measurement such that it is approximately normally distributed if one wants to apply a method assuming normality. Similarly, the choice of the statistical method may depend on whether the variables of interest have been dichotomized. In a sense, all preparation steps mentioned in the previous paragraph are part of the general statistical analysis approach and the separation between "initial data analysis" (data preparation) and "main analysis" is, in practice, blurred. In the remainder of this essay we will term the whole process as "statistical analysis approach".

From a naïve perspective one may feel that there is a unique "best approach" to analyzing the data at hand with respect to the question of interest. Firstly, scientific articles describing results obtained with specific statistical methods often carefully justify the chosen statistical analysis approach and may thus give the feeling that the chosen approach is clearly superior to other approaches. Secondly, statistics textbooks enumerating the assumptions underlying a statistical method implicitly suggest that it is straightforward to decide between the available methods just by theoretical arguments. In practice, however, things are much more complicated.

There are usually plenty of conceivable approaches to statistically analyze data that both make sense from a substantive point of view and are defensible from a theoretical point of view. To give a very simple example to readers familiar with statistics, it is often not so easy to decide whether to use the parametric t-test or the non-parametric Wilcoxon test. The consulting client often does not mind whether the null hypothesis concerns the median or the mean, and the sample size is often too small for tests of normality to reach an acceptable power, so the Wilcoxon test may be safer even if such a normality test supports the normal distribution. But the t-test may be appropriate as well. This question is by far not as trivial as it may look at first glance and testing for normality may yield substantial problems related to type-1 error (Rochon et al. 2012). This is a very simple example. The more complex the problem, the more difficult it becomes to decide between statistical methods. Often, practical situations require building complex models; and well-established guidelines to help decisions are not available. There are so many factors involved in the decision that it will never be possible to define guidelines covering all cases.

The researcher's degree of freedom may be directly observed through the application of a "split-team approach". This approach, which is common in some industrial contexts, consists of having the data analyses performed by two independent (groups of) experienced data analysts who are not allowed to communicate with each other before the end of the analysis. It is likely that the analysis strategies adopted by the two teams will be both suitable, but different from each other, thus reflecting the researcher's degree of freedom. Results may also differ.

6.3   Fishing for significance

On the whole, we can say that the results obtained with the chosen statistical analysis approach are but a point in the space of all possible analyses. This is not bad *per se*. However, this may lead to substantial problems if the statistical analysis approach is selected *a posteriori*, i.e. after answers to the questions of interest are obtained with several approaches. Indeed, researchers most often have expectations/hopes regarding the results of statistical analyses. For example, they sometimes hope that their results will yield completely new insights into a problem. They sometimes hope that the results will confirm previously obtained results. They sometimes hope that the results will fit nicely to substantive considerations. Now if several statistical analysis approaches are conceivable, both from a theoretical and practical point of view, researchers will naturally tend to try them successively and select those results that best fit their hopes. This mechanism, commonly termed "fishing for significance", is known to lead to substantial optimistic biases (Ioannidis 2005a, 2014). For example, consider question 2, i.e. whether exposure to a certain risk factor increases the risk of developing a disease or having a bad outcome. All other factors being fixed, imagine that in reality this exposure does *not* increase the risk. If one analyzes the data using a single statistical analysis approach, it may result in a false positive outcome, in the sense that the results suggest a risk increase although there is none. In statistical terms this would be a type-1 error. Type-1 errors happen, but it is possible to control their probability using a correct statistical test. Now if one analyzes the data in ten different ways, it is relatively likely that *at least one of the ten* analyses will yield such a type-1 error. The same problem occurs if one tests the difference between two groups using both a t-test and a Wilcoxon-test, although the number of analysis strategies is here two instead of ten. Similarly, the same mechanism is also at work in more complex situations beyond statistical testing, perhaps in a more subtle but not less dangerous way.

In provocative words, Ioannidis (Ioannidis 2005b, p454) illustrates this problem by claiming: "Give me information on a single gene and 200 patients, half of them dead, please. I bet that I can show that this gene affects survival ($p < 0.05$) even if it does not. One can do analyses: counting or ignoring exact follow-up, censoring at different timepoints, excluding specific causes of death, exploiting subgroup analyses, using dozens of different cut-off values to decide what constitutes inappropriate gene expression, and so forth. Without highly specified a priori hypotheses, there are hundreds of ways to analyze the dullest dataset."

6.4   Peculiarities in big data settings

The problem of fishing for significance and false research findings, outlined in the case of "a single gene" by Ioannidis (Ioannidis, 2005b), is enhanced in big data settings for various reasons. Firstly, the high dimensionality and the complexity of big molecular data essentially increase the researcher's degree of freedom, as already mentioned above for initial data analyses/data preparation. In particular, the methodological challenges arising from these complex data has attracted a lot of attention in the last years; as a consequence many statistical analysis approaches are available. In his essay on false research findings, Ioannidis (2005a) proposes a model for the probability of a research finding being true and derives several corollaries from this model. In particular, corollary 4 applies to big data settings which are characterized by a multiplicity of possible analysis strategies: "The greater the flexibility in designs, definitions, outcomes, and analytical modes in a scientific field, the less likely the research findings are to be true" (p698).

Secondly, in big data settings, guidelines supporting the choice of statistical approaches (to be paralleled to, e.g., normality tests telling us if the t-test can be applied) are in their infancy. Diagnostic tools are not well developed. The more complex the data are and the more methods are available, the less likely it is that useful practical guidelines or guidance documents will ever be available.

Thirdly and not least, the high dimensionality leads to a high variance of the results, in the sense that, for example, removing a few patients from the dataset may – much more dramatically than in low dimensional settings – lead to big changes in the results. Let us consider the example of the estimation of the error of prediction models. While the so-called "resubstitution error" obtained by applying the prediction model on the data it was trained on is a bit less critical in low dimensional settings (Dougherty and Bitter 2011), in big data settings one typically has to use cross-validation or a related method, which have a very high variance. Due to the resulting random fluctuations, the choices made in the statistical analysis approach can influence the results much more in comparison to classical data settings with few variables and a larger sample size. As a consequence, the "best-looking" result selected out of the results of several approaches in big data settings often tends to look better than its counterpart in low dimensional settings as demonstrated through an example below.

6.5   Example

To quantitatively illustrate the first and the third points concretely, we perform a small simulation study concerned with the construction and evaluation of classification models. The goal of the simulation is to demonstrate that the results (in our case, the estimated classification accuracies of the models) may look better than they actually are – just by fishing for significance, and that this over-optimism resulting from fishing for significance increases with the number of tried strategies (in our case, the number of classification algorithms) and the dimensions of the data.

We consider a real microarray gene expression dataset giving the expression levels of p=54,675 genes for n=200 patients, where 100 of these suffer from colorectal cancer and 100 are healthy controls. The disease status "cancerous" vs. "healthy" is considered as a class label. We are interested in classification models that predict the class label ("cancerous" or "healthy") based on the expression levels of the p=54,675 genes, and in their classification accuracies.

In our experiment, we artificially destroy the association between gene expression levels and class label by randomly permuting the class labels; the gene expression levels do not have any association with the disease status. Any classification algorithm is thus expected to perform no better than random guessing, i.e. to predict the correct class label in half of the cases only. Correspondingly, the classification error is expected to be approximately 0.5. We will see that, through fishing for significance, one can obtain an *apparently* smaller classification error estimate.

To see this, we apply A=10 different classification algorithms to this dataset and estimate their classification errors through a procedure called 10-fold cross-validation using the R package 'CMA' (Slawski et al. 2008). After having obtained error estimates for the A=10 classification algorithms, we "fish for significance" by selecting the smallest estimated error out of the A=10 estimated errors.

To illustrate that fishing for significance of this type has a less severe impact for a smaller number A of tried algorithms (see point 1 in Section 4), we additionally perform the same fishing for significance procedure with A=2 candidate algorithms instead of A=10. To demonstrate the impact of data dimensions (see point 3 in Section 4), we also apply the considered classification algorithms to 2, 10, 200 and 20000 randomly selected genes, successively, i.e. we artificially reduce the dimension of the dataset to different extents, to allow comparison between different dimensions in terms of fishing for significance. The numbers of 2 or 10 are typical for classical settings, in contrast to 200 or 20000 which represent big data settings. Moreover, to study the impact of the dataset's size, we repeat the whole analysis for a random selection of 50 patients out of 200 (25 "cancerous", 25 "healthy"). The whole process is repeated for 1000 random permutations of the class labels.

The results of this study are displayed in Fig. 6.1. This plot clearly shows that the smallest error is noticeably smaller than the error of random classification materialized by the green horizontal line: through fishing for significance, an over-optimistic result is produced. While over-optimism is weak or even negligible in all scenarios for A=2 (which is unsurprising, since a correct cross-validation procedure was used), it is substantial for A=10. Over-optimism is slightly enhanced by the high dimensionality of the data, i.e. increases with the number of variables nvar (it is more pronounced in the right panel than in the left panel). Lastly, the issue is seen to be more relevant for smaller datasets: the over-optimism is considerably stronger for the analysis involving only 50 patients than for the analysis with 200 patients. This is because the variance of the error estimates increases with decreasing sample size.

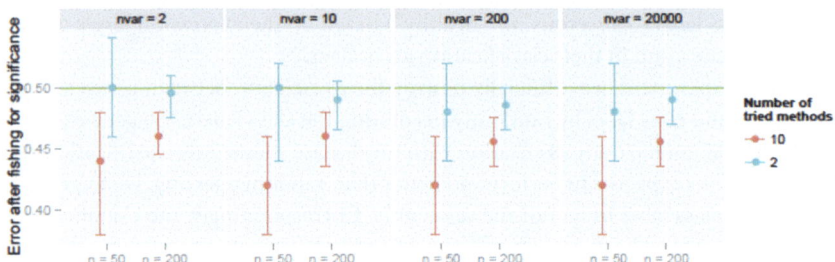

Fig. 6.1:   Results from the example study. Each bar shows the result of one scenario. The filled dots correspond to the medians over the 1000 values. The lower and upper ends depict the 25%- and 75%-percent quantiles of the 1000 values, respectively, i.e. the spans of the bars cover 50% percent of the 1000 values. nvar denotes the number of variables randomly selected in each repetition. The red bars correspond to the minimal error estimates out of all A=10 classification methods and the blue bars to the smaller errors out of pairs of classification methods (A=2). The green horizontal lines show the true classification error of the methods.

6.6   Potential approaches for improvements

6.6.1   Improvements through adjusting for multiple testing

Any statistician will probably think of adjustment for multiple testing when he/she hears of over-optimism resulting from the application of multiple tests. However, such approaches give a strong preference to the control of the type 1 error (i.e. probability of false positives) at the expense of a substantially increased type 2 error (i.e. probability of false negatives). In an example from low dimensional data, we will briefly outline the problem of fishing for significance and show that methods to adjust for multiple testing are available but that they do not solve the real practical problem.

To investigate the prognostic effect of a continuous marker a cut-off value is of-ten used to convert it into two groups with 'low' and 'high' values. For such a binary variable it is straightforward to test whether it has an effect on an outcome of inter-est. In the context of survival data and using the logrank test to compare survival times of two groups, Altman et al. (1994) illustrate that different cut-off values may be used to dichotomize the continuous marker and to conduct significance tests. Varying the cut-off value, the groups with 'low' and 'high' values overlap and tests are not independent. Searching for the cut-off value with the minimum p-value results in a severely increased type 1 error caused by multiple testing. For example, the type 1 error is about 40% instead of 5% in one of the scenarios considered by Altman et al. (1994). The authors provide a formula to correct for multiple testing and a suitable corrected p-value can then be calculated, but that does not answer the main question of assessing whether a continuous marker has a prognostic ef-fect. Dichotomization implies a substantial loss of information and therefore a loss of power if the marker has an effect; effect estimates of dichotomized variables are substantially biased and the 'optimal cut-off approach' usually selects different cut-off values in different datasets. For their specific example Altman et al. (1994) list 19 cut-offs used in the literature to determine whether a specific value of the continu-ous variable of interest is 'low' or 'high'.

6.6.2   Improvements through validation

After his statement on the possibility of finding that a "gene affects survival (p<0.05) even if it does not", Ioannidis (2005b) continues with: "Thus, no matter what my discovery eventually is, it should not be taken seriously, unless it can be shown that the same exact mode of analysis gets similar results in a different dataset. Valida-tion becomes even more important when datasets become complex and analytical options increase exponentially."

By "validation", he means here a validation step whose aim is to implicitly cor-rect for the multiplicity of the tried statistical analysis approaches. This is in con-trast with external or temporal validation of results (McShane et al. 2013), which generally aims at showing whether the results also hold in a different place (differ-ent hospital, different country, etc.) or at the same place but at a different time. See also König et al. (2007) for a discussion of these concepts in the context of predic-tion modeling. The motivation of the validation mentioned in the above citation by Ioannidis (2005b) is of statistical nature; it does not aim at generalizing the results to other settings.

Such a systematic validation approach for statistical purposes is described in Daumer et al. (2008) and has been adopted by this team in several research projects. The training data are used to perform all sorts of data analyses. At the end of the training phase the statistical analysis approach is conclusively chosen. It is implemented in such a way that it can be directly applied on the validation by so-called "data trusties" who are allowed to open the validation dataset. If the results on the validation dataset are sufficiently similar to those obtained from the training dataset (according to pre-defined criteria), they are considered as validated – no matter how much fishing for significance was done in the training phase.

Two important remarks should be made about this two-step procedure. Firstly, fishing for significance in the training phase does not call into question the validity of significant results in the validation phase, but the more fishing for significance that was done in the training phase, the less likely it is that the results will be as positive in the validation phase as in the training phase. Secondly, this procedure has a price in terms of power: the training analyses – whose results are ultimately reported as the final results – are performed on a dataset smaller than the total dataset consisting of training and validation dataset. This implies a decrease of power, which may be inacceptable in some cases (e.g., where the dataset is very small). In other words, the validation procedure reduces the risk of false positive results, but also increases the risk of false negative results.

6.6.3   Improvements through guidance

Another approach to reduce the occurrence of false positive research findings is to choose only one statistical analysis approach based on theoretical or practical considerations *before* seeing the results on the considered research questions. This is possible in some cases. For example, for the choice between the t-test and the Mann-Whitney test, guidelines are available; but even there things are not as easy as might be thought at first glance (Rochon et al., 2012). However, in most practical situations, especially in the context of big molecular data, there are no well-established procedures to guide these choices. We conjecture that it will remain so, considering the delay between the development of methods and the development of guidelines on these methods and, most importantly, the growing complexity of data and statistical analyses. Moreover, it may be possible to develop guidelines to steer the choice between two statistical tests, but decisions related to data preparation are harder to formalize since they are often highly dependent on the substantive context. Thus, recommending that researchers always choose the statistical analysis approach before seeing any results seems unrealistic.

However, we stress the importance of research that aims at developing guidance for the choice of methods. Even if such guidance documents cannot cover all issues, are subject to controversy and do not reflect the complexity of many analyses, they may contribute to a mitigation of the multiplicity problem discussed above and its consequences, such as the increase of false positives (in absence of validation) or the increase of false negatives (that result from the loss of power due to data splitting into training and validation data, if validation is performed).

Guidance documents may be developed in different ways; for example, by empirically investigating the behavior of the methods in real data sets or in simulations, by systematically reviewing the literature, by collecting experts' opinions and usually by a mix of these three approaches. An initiative called STRATOS (STRengthening Analytical Thinking for Observational Studies) was recently started in the context of classical medical statistics, i.e. without considering big data (Sauerbrei et al. 2014). Meanwhile, it has prompted a new topic group on high dimensional data. It aims at providing recommendations for the choice of analysis methods in practice. The concept of "evidence-based statistics", inspired from the well-known concept of "evidence-based medicine", could be used to describe the ultimate goal of such analyses. They are important to reduce the multiplicity problems outlined above by appropriate guidelines. The latter can namely not be formulated as part of an article describing a new method because such an article is most often biased in favor of the new method (Jelizarow et al. 2010; Boulesteix 2013; Boulesteix et al. 2013) and thus not neutral enough to provide reliable guidance.

Obviously, transparent and complete reporting is a pre-requisite to judge the usefulness of data and to interpret study results in the appropriate context. To improve on this issue, guidelines for the reporting of many types of studies have been developed during the last two decades (Begg et al. 1996; McShane et al. 2005) and the EQUATOR network (http://www.equator-network.org/) acts as an "umbrella" for developers of such guidelines (Simera et al. 2010). The necessity to follow reporting guidelines will also help to battle the problem of fishing for significance. For some years, some journals require researchers to follow the relevant reporting guideline when submitting a manuscript but the relevance of this topic is still severely underrated.

Note that the mitigation of the multiplicity problem (and thereby of fishing for significance opportunities) is merely a by-product of guidelines. The primary aim of guidelines is to ensure complete reporting and to reduce the use of inappropriate statistical methods by supporting researchers in their choices. But we claim this by-product is extremely important in practice, because false research findings due to fishing for significance are a major challenge in many scientific areas, not only in the context of big molecular data.

6.6.4   Improvements through study registration and publication of analysis plans

Several reasons were mentioned before to explain that results may be selectively reported by researchers. In the context of clinical trials the International Committee of Medical Journal Editors has emphasized that "... the research enterprise has an obligation to conduct research ethically and to report it honestly. Honest reporting begins with revealing the existence of all clinical studies, even those that reflect unfavorably on a research sponsor's product." They have requested registration in a public trials registry as condition of considering a trial for publication (De Angelis et al. 2004). As clinical trials have a protocol and a statistical analysis plan, fishing for significance is a less critical issue. However, it is well known that not all trials are reported as registered (Emdin et al. 2015; Dwan et al. 2014).

The principle from clinical trials to register a study and having a pre-specified analysis plan may also be transferred to observational studies but the realization of this is much more difficult. To improve on the transparency of reporting of observational studies, registration has been recently proposed (Riley et al. 2009; Andre et al. 2011, Peat et al. 2014; Altman 2014) but so far it hardly plays a role. Kasenda et al. (2014) is one of the rare exceptions.

Many researchers doubt that it is possible to pre-specify the analysis of an observational study and prefer having all degree of freedom to derive a suitable model data-dependently and by using the most suitable statistical approach. However, pre-specification of an analysis plan would not imply that the analysis has to be strictly followed (for example if an important assumption is seriously violated) but that changes and amendments have to be carefully justified (Dwan et al. 2011). Therefore, that may be an incentive not to fish for significance. The publication of analysis plans is subject to controversial discussions, can certainly not be recommended in all situations, and is not yet a mature concept in the context of explorative research, but it is certainly worth more attention than currently given.

6.6.5   Further incentives not to fish for significance

The main approaches discussed in Section 6 so far address the problems of fishing for significance and researchers' degree of freedom by correcting for the multiplicity of analyses after fishing for significance has been done and by reducing the multiplicity of choices (through guidance and through well-thought-out analysis plans), respectively. In this section we discuss further approaches that may be useful, although more indirectly, to reduce the publication of false research findings through fishing for significance.

Firstly, the increased publication of negative research findings in high impact journals may certainly reduce the incentive to fish for significance. Fishing for significance is indeed often performed in order to make a study more publishable, even if other aspects such as the hope for groundbreaking findings and fame may also play a role. An extreme strategy towards the publication of negative research findings would be to conduct review processes based solely on the description of the design and methods and *not* on the results. Such strategies are, however, subject to controversial discussions in the scientific community. While their beneficial effect with respect to fishing for significance seems unquestionable (among other advantages, e.g., regarding the publication bias), we do not claim that they should be applied everywhere without caution.

Secondly, the publication of analysis codes and data together with research articles promoted, for example, in a special section of the magazine Science (Peng 2011), may also reduce the temptation to fish for significance. The publication of analysis codes and data is known under the term "reproducible research". Among other advantages, it allows other researchers to check the results of a study, to better understand and interpret what has been done, to use the same analysis strategy on their own data, and to detect errors. As a side-effect, it may also discourage researchers from fishing for significance, since the availability of codes and data would make it relatively easy for other researchers to detect/suspect their fishing strategy.

6.7   Conclusion

The researcher's degree of freedom is high in practical data analysis, especially in big data settings. Often, multiple statistical analysis approaches are conceivable. This leaves room for fishing for significance, often leading to apparently interesting but false research findings. In this essay we discussed different strategies to address this major problem. A possible strategy, validation, is applied at the researcher level. The researcher may essentially apply as many approaches as he/she wants and possibly select one of them based on the results, as long as he/she validates these results using independent validation data. The price to pay is a loss of power. Another strategy, increased development of guidance documents, may be applied at the level of the scientific community. It calls for more studies whose aim is to help researchers choose their statistical analysis approach. Such studies may reduce the multiplicity problem beforehand, instead of correcting for it afterwards as in the validation strategy. In reality, guidance documents for analysis will probably never eliminate the researcher's degree of freedom, and validation will probably never be applicable or sensible everywhere. A mixture of both strategies may be a good compromise while reporting following guidelines, increased publication of negative research findings, pre-specified analysis protocols, and data and code for the purpose of reproducibility may also contribute to combating fishing for significance.

References

Altman, D. (2014). The Time Has Come to Register Diagnostic and Prognostic Research. *Clinical Chemistry, 60*(4), 580-582. doi:10.1373/clinchem.2013.220335.

Andre, F., McShane, L.M., Michiels, S., Ransohoff, D.F., Altman, D.G., Reis-Filho, J.S., Hayes, D.F., Pusztai, L. (2011). Biomarker studies: a call for a comprehensive biomarker study registry. *Nat Rev Clin Oncol, 8*(3), 171-176. doi: 10.1038/nrclinonc.2011.4.

Altman, D., Lausen, B., Sauerbrei, W., & Schumacher, M. (1994). Dangers of Using "Optimal" Cutpoints in the Evaluation of Prognostic Factors. *JNCI Journal Of The National Cancer Institute, 86*(11), 829-835. doi:10.1093/jnci/86.11.829.

Begg, C., Cho, M., Eastwood, S., Horton, R., Moher, D., Olkin, I., Pitkin, R., Rennie, D., Schulz, K.F., Simel, D., & Stroup, D.F. (1996). Improving the quality of reporting of randomized controlled trials. The CONSORT statement. *JAMA: The Journal Of The American Medical Association, 276*(8), 637-639. doi:10.1001/jama.276.8.637.

Boulesteix, A. (2013). On representative and illustrative comparisons with real data in bioinformatics: response to the letter to the editor by Smith et al. *Bioinformatics, 29*(20), 2664-2666. doi:10.1093/bioinformatics/btt458.

Boulesteix, A., Lauer, S., & Eugster, M. (2013). A Plea for Neutral Comparison Studies in Computational Sciences. *PLOS ONE*, *8*(4), e61562. doi:10.1371/journal.pone.0061562.

Daumer, M., Held, U., Ickstadt, K., Heinz, M., Schach, S., & Ebers, G. (2008). Reducing the probability of false positive research findings by pre-publication validation – Experience with a large multiple sclerosis database. *BMC Med Res Methodol*, *8*(1), 18. doi:10.1186/1471-2288-8-18.

De Angelis, C., Drazen, F.A., Haug, C., Hoey, J., Horton, R., Kotzin, S., Laine, C., Marusic, A., Overbeke, A.J., Schroeder, T.V., Sox, H.C., & Van Der Weyden, M.B. (2004). Clinical Trial Registration: A Statement from the International Committee of Medical Journal Editors. *Annals Of Internal Medicine*, *141*(6), 477. doi:10.7326/0003-4819-141-6-200409210-00109.

Dougherty, E., & Bittner, M. (2011). *Epistemology of the cell*. Piscataway, NJ: IEEE Press.

Dwan, K., Altman, D., Clarke, M., Gamble, C., Higgins, J., & Sterne, J. et al. (2014). Evidence for the Selective Reporting of Analyses and Discrepancies in Clinical Trials: A Systematic Review of Cohort Studies of Clinical Trials. *Plos Med*, *11*(6), e1001666. doi:10.1371/journal.pmed.1001666.

Dwan, K., Altman, D., Cresswell, L., Blundell, M., Gamble, C., & Williamson, P. (2011). Comparison of protocols and registry entries to published reports for randomised controlled trials. *Cochrane Database Of Systematic Reviews*, *19(1),MR000031*. doi:10.1002/14651858. mr000031.pub2.

Emdin, C., Odutayo, A., Hsiao, A., Shakir, M., Hopewell, S., Rahimi, K., & Altman, D. (2015). Association of Cardiovascular Trial Registration With Positive Study Findings. *JAMA Internal Medicine*, *175*(2), 304. doi:10.1001/jamainternmed.2014.6924.

Ioannidis, J. (2005a). Why Most Published Research Findings Are False. *Plos Med*, *2*(8), e124. doi:10.1371/journal.pmed.0020124.

Ioannidis, J. (2005b). Microarrays and molecular research: noise discovery?. *The Lancet*, *365*(9458), 454-455. doi:10.1016/s0140-6736(05)17878-7.

Ioannidis, J., Greenland, S., Hlatky, M., Khoury, M., Macleod, M., & Moher, D. et al. (2014). Increasing value and reducing waste in research design, conduct, and analysis. *The Lancet*,*383*(9912), 166-175. doi:10.1016/s0140-6736(13)62227-8.

Jelizarow, M., Guillemot, V., Tenenhaus, A., Strimmer, K., & Boulesteix, A. (2010). Over-optimism in bioinformatics: an illustration. *Bioinformatics*, *26*(16), 1990-1998. doi:10.1093/bioinformatics/btq323.

Kasenda, B., Sauerbrei, W., Royston, P., & Briel, M. (2014). Investigation of continuous effect modifiers in a meta-analysis on higher versus lower PEEP in patients requiring mechanical ventilation - protocol of the ICEM study. *Systematic Reviews*, *3*(1), 46. doi:10.1186/2046-4053-3-46.

König, I., Malley, J., Weimar, C., Diener, H., & Ziegler, A. (2007). Practical experiences on the necessity of external validation. *Statist. Med.*, *26*(30), 5499-5511. doi:10.1002/sim.3069.

McShane, L., Altman, D., Sauerbrei, W., Taube, S., Gion, M., & Clark, G. (2005). REporting recommendations for tumour MARKer prognostic studies (REMARK). *Br J Cancer*, *93*(4), 387-391. doi:10.1038/sj.bjc.6602678.

McShane, L., Cavenagh, M., Lively, T., Eberhard, D., Bigbee, W., & Williams, P. et al. (2013). Criteria for the use of omics-based predictors in clinical trials. *Nature*, *502*(7471), 317-320. doi:10.1038/nature12564.

Peat, G., Riley, R., Croft, P., Morley, K., Kyzas, P., & Moons, K. et al. (2014). Improving the Transparency of Prognosis Research: The Role of Reporting, Data Sharing, Registration, and Protocols. *Plos Med, 11*(7), e1001671. doi:10.1371/journal.pmed.1001671.

Peng, R. (2011). Reproducible Research in Computational Science. *Science, 334*(6060), 1226-1227. doi:10.1126/science.1213847.

Riley, R., Sauerbrei, W., & Altman, D. (2009). Prognostic markers in cancer: the evolution of evidence from single studies to meta-analysis, and beyond. *Br J Cancer, 100*(8), 1219-1229. doi:10.1038/sj.bjc.6604999.

Rochon, J., Gondan, M., & Kieser, M. (2012). To test or not to test: Preliminary assessment of normality when comparing two independent samples. *BMC Med Res Methodol, 12*(1), 81. doi:10.1186/1471-2288-12-81.

Sauerbrei, W., Abrahamowicz, M., Altman, D., le Cessie, S., & Carpenter, J. on behalf of the STRATOS initiative. (2014). STRengthening Analytical Thinking for Observational Studies: the STRATOS initiative. *Statist. Med., 33*(30), 5413-5432. doi:10.1002/sim.6265.

Simera, I., Moher, D., Hirst, A., Hoey, J., Schulz, K., & Altman, D. (2010). Transparent and accurate reporting increases reliability, utility, and impact of your research: reporting guidelines and the EQUATOR Network. *BMC Medicine, 8*(1), 24. doi:10.1186/1741-7015-8-24.

Simmons, J., Nelson, L., & Simonsohn, U. (2011). False-Positive Psychology: Undisclosed Flexibility in Data Collection and Analysis Allows Presenting Anything as Significant. *Psychological Science,22*(11), 1359-1366. doi:10.1177/0956797611417632.

Slawski, M., Daumer, M., & Boulesteix, A. (2008). CMA – a comprehensive Bioconductor package for supervised classification with high dimensional data. *BMC Bioinformatics, 9*(1), 439. doi:10.1186/1471-2105-9-439.

Xu, Y., Xu, Q., Yang, L., Ye, X., Liu, F., & Wu, F. et al. (2013). Identification and Validation of a Blood-Based 18-Gene Expression Signature in Colorectal Cancer. *Clin Cancer Res, 19*, 3039–3049. doi:10.1158/1078-0432.C

7. Big Data: Progress in Automating Extreme Risk Analysis

Nadine Gissibl[1], Claudia Klüppelberg[2], Johanna Mager[3]

Abstract

Big data can be a curse and a blessing for the statistician. We report in this paper about some positive effect of big data: big data may open the way for more reliable risk analysis, simply because more extreme data are available. However, big data also require a fully automated analysis. We present here a method, which can easily be implemented and used for large numbers of statistical attributes. We apply this method to safety issues at airplane landings.

1   Nadine Gissibl is a doctoral candidate at the Chair of Mathematical Statistics at the Center for Mathematical Sciences of the Technical University of Munich. Her research interests are extreme value theory and graphical models.

2   Claudia Klüppelberg holds the Chair of Mathematical Statistics at the Technical University of Munich. Her research interests combine various areas of applied probability and statistics with applications to technical and financial risk analysis. She has written numerous publications and written and edited books for extreme risk analysis. In 2008 Claudia Klüppelberg has been appointed Carl von Linde Senior Fellow at the Institute for Advanced Study of TUM. During her three years' period she has led a Focus Group on "Risk Analysis and Stochastic Modelling".

3   Johanna Mager was a master student at the Chair of Mathematical Statistics at the Center for Mathematical Sciences of the Technical University of Munich. At present she is a mathematical consultant in the department of airline insurance at Munich Re.

7.1 Introduction

There has been much debate about use and abuse of big data in recent years, Klaus Mainzer's excellent book [9] bears witness to this. Big data are often defined via the three Vs: *Volume* (data come in terabytes, are automatically recorded in files and tables, are often transactions), *Velocity* (they come in real time, are near time recorded or streamed) and *Variety* (they come structured or unstructured, or both).

On the negative side all of us are exposed to a supervision machinery unprecedented in history. Details of our lives are propagated to and within social networks without any respect or ethical conduct. Business concepts are based on masses of data which are either freely available or can be bought from illegitimate sources with the expressed goal of product placement.

On the positive side, however, scientists have never had so much information to study rare events like natural catastrophes (earthquakes, hurricanes, floods), technical risks like flight operation problems, nuclear power plant safety, or financial risks like the subprime crisis of 2007 or the ongoing European Sovereign-debt crisis.

Extreme value statistics has in the past always suffered from lack of data resulting in non-robust and inefficient estimators and prediction of rare events with often huge errors. As a consequence, extreme value statisticians performed each data analysis like a piece of art, supervising every step of the analysis with a catalog of provisions to avoid severely wrong and often disastrous conclusions. Other scientists took resort to standard statistical tools based on Gaussian models, which were often used for the whole data set, and extrapolated to extreme events, although extreme events can only be assessed statistically properly, when previous extreme events are analysed. Moreover, it has long been known that Gaussian models in almost all cases underestimate extreme events severly.

Now with huge data sets available and computers able to simulate and analyse masses of non-standard data originating from large networks or space-time measurements, risk assessment by extreme value statistics is again on the rise and in greater demand than ever before. New interesting problems arise, since environmentalists, engineers and economists ask for user friendly extreme value methods based on a high degree of automation.

One of the attractions of modern extreme value methods is the embedded dimension reduction of the data. Since risk events can only be found in a small amount of data, even in a huge data set, extreme value statistics automatically leads to a reduction of the dimensionality and complexity of the data.

Whereas there exist computer packages for extreme value statistics implemented in MATLAB and R, which allow for graphical assessment of the extreme value analysis, it still requires the eye of the analyst to take the decision, which events should be included in the analysis. This task is usually named "threshold selection" and is one of the most critical points in extreme value statistics. With hundreds or even thousands of different variables measured over grids, networks or simply as a high-dimensional vector, such a method is no longer feasible.

This has been recognized by a number of extreme value statisticians, who have suggested methods for automatic threshold selection. We shall show one of these methods at work for assessing the runway overrun risk at airplane landings based on operational flight data.

Our paper is organised as follows. In Section 2 we introduce extreme risk models, where in Subsection 2.1 we present some dimension reduction methods based on extreme value concepts. Subsection 2.2 is then devoted to the Peaks-over-Threshold method, giving a statistical model for all observations exceeding a high threshold. This sets the stage for Section 3, where we present the automated threshold selection method, which opens up the way to risk assessment based on big data. In Section 4 we show our method at work in a technical risk analysis exemplified for the safety of airplane landings.

7.2   Extreme risk models

7.2.1   Dimension reduction

Assume that we have a large number of observations for a high-dimensional vector $X = (X_1,..., X_d)$, which contributes to some risk in a system. In the simple case that all components measure the same kind of risk, for instance negative relative price changes of some financial assets, then, when we plot all such data together, only the most risky assets determine the high risk in the system. Note that throughout we think of risk as a positive quantity, and high risk corresponds to large values.

This knowledge allows us to perform an often substantial dimension reduction: We only need to consider the most risky assets, all the others contribute to the risk in the financial system only marginally. We can formulate this in mathematical terms. If extreme events correspond to large data values, like large financial losses for component i, then the behaviour of the distribution tail $\mathbb{P}(X_i > x)$ for large x matters only. For a common risk model this distribution tail is algebraically decreasing; i.e., $\mathbb{P}(X_i > x) = K_i x^{-\alpha}$ for large values of x. Now assume that among the components of X we have several with such algebraic tails, perhaps with different α's, and some may have even faster (e.g. exponentially) decreasing tails. When we now assess the risk of the vector X by the sum of its components, and $\mathbb{P}(X_j > x)/\mathbb{P}(X_i > x) \to 0$ as $x \to \infty$, for all $j \neq i$, then (regardless of the dependence structure between the components of X), $\mathbb{P}(\Sigma_{j=1}^{d} X_j > x)/\mathbb{P}(X_i > x) \to 1$ as $x \to \infty$. In other words, the risk with the slowest decreasing tail (which indeed generates the highest risk values) completely determines the distribution tail of the sum of extreme risks.

If there are several components (let's say components, or call them risk factors, with indices $i \in A \subset \{1,...,d\}$) with the same behaviour in the distribution tail (a statistical test could check this), then for every $i^* \in A$ we obtain $\mathbb{P}(X_i > x)/\mathbb{P}(X_{i^*} > x) \to C_i > 0$ as $x \to \infty$ for $i \in A$ and by Lemma A3.28 of [6] (for independent $X_1,..., X_d$),

$$\frac{\mathbb{P}\left(\sum_{j=1}^{d} X_j > x\right)}{\mathbb{P}(X_{i^*} > x)} \to \sum_{i \in A} C_i, \quad x \to \infty. \tag{2.1}$$

We observe a similar effect, when we have data classified with respect to some covariable Y, and want to calculate, for disjoint $B_1,..., B_n$ whose union covers all possible values for Y,

$$\mathbb{P}(X > x) = \sum_{i=1}^{n} \mathbb{P}(Y \in B_i)\mathbb{P}(X > x \mid Y \in B_i) =: \sum_{i=1}^{n} p_i \overline{F}_i(x), \tag{2.2}$$

where we have set $p_i := \mathbb{P}(Y \in B_i)$ and $\overline{F}_i(x) := \mathbb{P}(X > x \mid Y \in B_i)$. Then under the same condition as above, denoting $\overline{F}(x) := \mathbb{P}(X > x)$ for $x > 0$,

$$\frac{\overline{F}(x)}{\overline{F}_{i^*}(x)} \to \sum_{i \in A} p_i C_i, \quad x \to \infty. \tag{2.3}$$

Hence, for large values of x we can focus on all components with index in A and no others. This dimension reduction will also play a role in the aviation application of Section 4.

Example 2.1 (Sum of exponential and Pareto risks). Assume that we have two different risks, one being exponentially distributed with distribution tail $\mathbb{P}(X_1 > x) = e^{-x}$ and the other Pareto with distribution tail $\mathbb{P}(X_2 > x) = (1+x)^{-2}$ for $x > 0$. Figure 1 shows a simulation of a random sample of X_1 (left) and of X_2 (middle). Comparing the high peaks from the middle and right hand plot we see immediately that *high* risks originate mainly from the Pareto risk X_2.

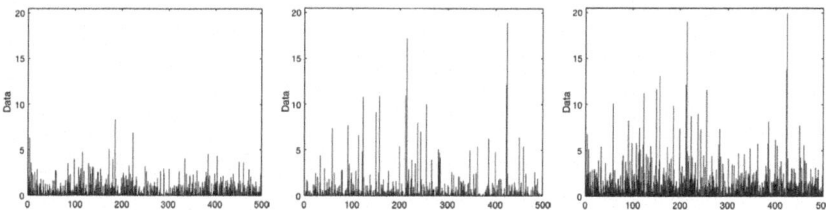

Fig. 7.1: 500 data points simulated from the distributions given in Example 2.1: exponential distribution (left) and Pareto distribution (middle). The right hand plot shows the respective sums of both data.

7.2.2   Let the tails speak for themselves: the POT method

This recipe goes back to William H. DuMouchel in the 1970ies, propagated by Richard Smith in the 1980ies, emphasising that a wrong model for the extreme data gives wrong answers for the far out tail estimation, leading to wrong risk estimation. Consider for instance, Figure 1: if a Pareto distributed risk (as in the middle plot) is wrongly modelled by an exponential distribution (left plot), then the high risk is grossly underestimated.

As we have indicated above, and know from Figure 1, risk based on large values can be assessed by distribution tails $\mathbb{P}(X_i > x)$ for large x. Consequently, we have to estimate this for large values of x regardless of the behaviour of the data around their mean.

Since extreme risk is present only in large observations, we use exactly those, and the method is called *Peaks-over-Threshold* or *POT method*. We start with a high threshold u and note that an observation larger than $u + y$ for some $y > 0$ is only possible, if the observation is in the first place larger than u; this means one needs an *exceedance* of u. The observation itself has then necessarily an *excess over the threshold u;* cf. Figure 2.

It is now important for the POT method that for a random variable X with distribution function F and a large threshold u the following approximation holds under weak regularity conditions by the Pickands-Balkema-de Haan Theorem (cf. [6], Theorem 3.4.13 and the succeeding Remark 6),

$$\mathbb{P}(X > u + y \mid X > u) = \frac{\overline{F}(u+y)}{\overline{F}(u)} \approx \overline{H}_{\xi,\beta}(y) = \mathbb{P}(Y > y) \quad \text{for } y > 0, \qquad (2.4)$$

where

$$H_{\xi,\beta}(x) = \begin{cases} 1-\left(1+\xi\frac{x}{\beta}\right)^{-\frac{1}{\xi}}, & \text{if } \xi \in \mathbb{R} \setminus \{0\} \\ 1-e^{-\frac{x}{\beta}}, & \text{if } \xi = 0 \end{cases} \text{ for } \begin{cases} x > 0, & \text{if } \xi \geq 0, \\ 0 < x < -\beta/\xi, & \text{if } \xi < 0. \end{cases} \quad (2.5)$$

Moreover, the shape parameter ξ is independent of the threshold u, whereas the scale parameter $\beta = \beta(u)$ may change with u. The class $H_{\xi,\beta}$ for $\xi \in \mathbb{R}$ and $\beta > 0$ is called *generalized Pareto distribution (GPD)*. In a general context, (2.4) says that excesses over a high threshold, which we denote by Y, are (for large enough u) approximately generalized Pareto distributed. (This may seem at first sight suprising. However, such results are prominent in probability theory. The central limit theorem is the most known example: if independent random variables have the same distribution with finite variance, then the partial sums properly centered and scaled converge to a normal distribution. A similar result holds for partial maxima of random variables (cf. [6], Chapters 2 and 3).)

For sufficiently large u we approximate the excess distribution by a GPD $H_{\xi,\beta}$ as in (2.4) and obtain the tail approximation

$$\mathbb{P}(X > x) = \mathbb{P}(X > u)\mathbb{P}(X > x \mid X > u) \approx \bar{F}(u)\bar{H}_{\xi,\beta}(x-u), \quad x > u. \quad (2.6)$$

The choice of the threshold is delicate and creates a so-called bias-variance problem in the following sense: from a statistical point of view more data give better parameter estimates (with smaller variance) suggesting to take a low threshold; however, a too low threshold may approximate observations by a GPD, which cannot yet be approximated by a GPD (giving high bias).

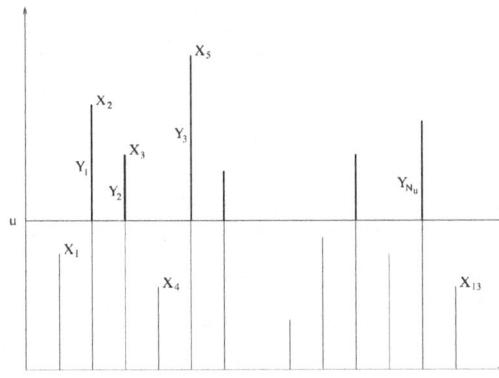

Fig. 7.2:   Data $X_1, \ldots, X_{13}$ with corresponding excesses $Y_1, \ldots, Y_{Nu}$.

For what follows we assume that we have independent and identically distributed (iid) observations $X_1, \ldots, X_n$ all with distribution function F. Let us also assume for the moment that we know the threshold u such that observations above it can be well approximated by a GPD. In order to identify the observations exceeding u we order the observations $X_1, \ldots, X_n$ as

$$\min\{X_1, \ldots, X_n\} = X_{n:n} < \cdots < X_{1:n} = \max\{X_1, \ldots, X_n\}. \tag{2.7}$$

Then observations larger than u correspond to the upper values $X_{k:n} < \cdots < X_{1:n}$ for some $k \in \{1, \ldots, n\}$, in particular, $X_{k+1:n}$ for $k < n$ can serve as a threshold u. Moreover, the conditonal distribution of the exceedances $Y_1, \ldots, Y_k$ given $X_{k+1:n} = u$ is approximated by the distribution function $H_{\xi, \beta}$ as in (2.5).

Since for all parameters ξ, β the GPDs are continuous and increasing functions, the integral transform yields for the ordered exceedances

$$H_{\xi, \beta}(0) = H_{\xi, \beta}(X_{k+1:n} - u) < H_{\xi, \beta}(Y_{k:n}) < H_{\xi, \beta}(Y_{k-1:n}) < \cdots < H_{\xi, \beta}(Y_{1:n}),$$

which are in distribution equal to ordered uniform random variables: $0 < U_{k:k} < U_{k-1:k} < \cdots < U_{1:k} < 1$.

This relation allows us to assess the goodness of fit of the upper order statistics by a GPD. To this end, note that, if the model is approximately correct, then the two random variables $H_{\xi, \beta}(Y_{i:n})$ and $U_{i:k}$ have the same distribution (denoted by $\overset{d}{=}$) resulting in

$$H_{\xi, \beta}(Y_{i:n}) = 1 - \left(1 + \xi \frac{X_{i:n} - u}{\beta}\right)^{-\frac{1}{\xi}} \overset{d}{=} U_{i:k} \quad \text{for } i = 1, \ldots \tag{2.8}$$

where we interpret the case of $\xi = 0$ as the limit for $\xi \to 0$ giving the exponential distribution function with mean β. If we replace the uniform order statistics by their expectations, i.e. $U_{i:k}$ by $\mathbb{E}[U_{i:k}] = 1 - \frac{i}{k+1}$ for $i = 1, \ldots, k$, and the parameters ξ and β by their maximum likelihood estimates (MLEs) $\hat{\xi}$ and $\hat{\beta}$ based on the upper k order statistics of the observations, we find from (2.8),

$$Y_{i:n} := X_{i:n} - u \approx \frac{\hat{\beta}}{\hat{\xi}}\left(\left(\frac{i}{k+1}\right)^{-\hat{\xi}} - 1\right) =: \hat{Y}_{i:n} \quad \text{for } i = 1, \ldots, k. \tag{2.9}$$

It is well-known that the asymptotic normality of the MLEs requires $\xi > -0.5$ (cf. [11]). So whenever we use the asymptotic normality of the MLEs we have to assume this.

7.3   Threshold selection

It is intrinsic to the problem that statistical estimation methods of tails and quantiles are as a rule very sensitive to substantial changes in the extreme data, since every data point is relevant among the not too many data points.

When only one or very few risk types have to be considered, then a useful graphical tool for threshold selection is based on the *mean excess function,* given for a positive random variable X by

$$e(u) = \mathbb{E}[X - u \mid X > u], \quad u > 0.$$

For X with distribution function $H_{\xi,\beta}$ such that $\xi < 1$ (only then $\mathbb{E}[X]$ exists) and $\beta > 0$, the mean excess function is given by

$$e(u) = \frac{\beta + \xi u}{1 - \xi}, \quad \beta + \xi u > 0.$$

The function is linear in u with slope $\frac{\xi}{1-\xi}$ and intercept $\frac{\beta}{1-\xi}$.

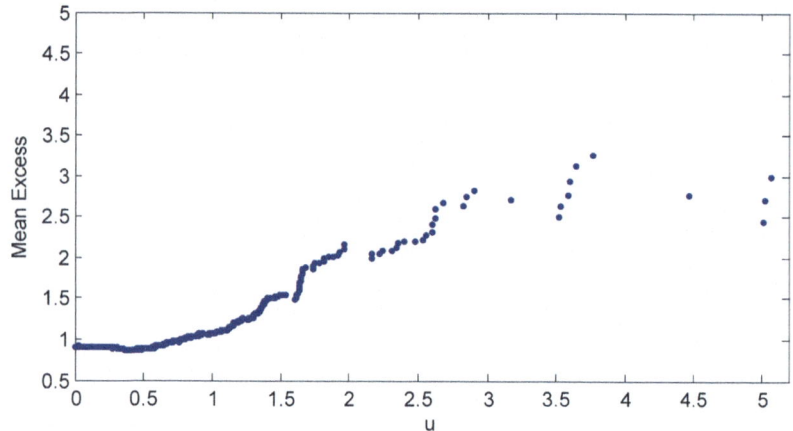

Fig. 7.3 Simulated independent random variables from the distribution function $\mathbb{P}(X \leq x) = \mathbf{1}(x \leq \Phi^{-1}(1 - p))\Phi(x) + \mathbf{1}(x > \Phi^{-1}(1 - p))(1 - p + pH_{\xi,\beta}(x - \Phi^{-1}(1 - p)))$ for $p = 0.1$, $\xi = 0.5$, $\beta = u\xi = 0.64$, where Φ denotes the standard normal distribution. Depicted is the mean excess plot. The true threshold (known by choice of the simulation method) is $u = z_{0.9} = 1.2816$ with $z_{0.9}$ being the 90% standard normal quantile.

When we define an empirical version $e_n(u)$ of the mean excess function and search for linearity beyond a high threshold, then from this threshold on a GPD makes a good model. Again choosing $u = X_{k+1:n}$, the empirical mean excess function is given by

$$e_n(u) = \frac{1}{k}\sum_{i=1}^{k}(X_{i:n} - u).$$

A mean excess plot consists of the points

$$\{(X_{i:n}, e_n(X_{i:n})) : i = 1,\ldots,n\} \in \mathbb{R}^2.$$

The mean excess plot should show an approximately linear behavior, when the underlying data follow a GPD. To choose an appropriate threshold for the POT method, we search for the smallest u where the plot is approximately linear. This guarantees that we use as many data points as possible to obtain GPD parameter estimates with minimal variance. This plot gives a first impression, whether a GPD approximation for high threshold excesses of the data is reasonable. For a simulation example see Figure 3.

As a second step the parameters ξ and β are estimated, where we focus on MLE (different estimation methods can be found in [6], Chapter 6). A meanwhile classical approach is to plot the estimates of the shape parameter $\xi = \xi(u)$ for different thresholds $u = X_{k+1:n}$ (equivalently, different k) and find the best threshold by eye inspection. Since ξ is independent of the threshold, we choose a threshold value (not too high to avoid high variance, and not too low to avoid a bias), where the estimates $\xi(u)$ are stable.

When we have, however, a large number of risk types, then an automatic reliable procedure for estimating an appropriate threshold is called for. Various methods have been suggested in the literature; cf. Beirlant et al. [2], Section 4.7. We extend a method proposed in [5] (cf. also [10]) for a Pareto distribution to the GPD.

The basic idea is to minimize the mean squared prediction error with respect to the threshold u; i.e. to minimize $\Gamma(k) := \Gamma(X_{k+1})$ given by

$$\Gamma(k) = \frac{1}{k}\sum_{i=1}^{k}\mathbb{E}\left[\left(\frac{\hat{Y}_{i:n} - \mathbb{E}[Y_{i:n}]}{\sigma_i}\right)^2\right] \tag{3.1}$$

with respect to k, where $\sigma_i^2 = \mathrm{Var}(Y_{i:n})$, and $Y_{i:n}$ and $\hat{Y}_{i:n}$ are as in (2.9). Since we cannot compute the moments $\mathbb{E}[Y_{i:n}]$ and $\mathrm{Var}(Y_{i:n})$ explicitly, we have to reformulate the problem. We reformulate $\Gamma(k)$ as follows:

$$\Gamma(k) = \frac{1}{k}\sum_{i=1}^{k}\mathbb{E}\left[\left(\frac{Y_{i:n}-\hat{Y}_{i:n}}{\sigma_i}\right)^2\right] + \frac{2}{k}\sum_{i=1}^{k}\frac{\mathrm{Cov}(Y_{i:n},\hat{Y}_{i:n})}{\sigma_i^2} - \frac{1}{k}\sum_{i=1}^{k}\frac{\mathrm{Var}(Y_{i:n})}{\sigma_i^2}$$

$$= \frac{1}{k}\sum_{i=1}^{k}\mathbb{E}\left[\left(\frac{Y_{i:n}-\hat{Y}_{i:n}}{\sigma_i}\right)^2\right] + \frac{2}{k}\sum_{i=1}^{k}\frac{\mathrm{Cov}(Y_{i:n},\hat{Y}_{i:n})}{\sigma_i^2} - 1.$$

(3.2)

The idea now is to replace $\Gamma(k)$ using some empirical versions of the moments.

Since tail estimation can be very sensitive to slight changes in the observations, we use classical methods of Robust Statistics to find an expression for the prediction error; cf. [7] and references therein. For simplification we set $\theta = (\beta,\xi)$ and, instead of working with the ideal GPD H_θ, in Robust Statistics, for some given value $\varepsilon > 0$, we define suitable distributional neighbourhoods about this ideal model. Here we restrict the method to neighbourhoods consisting of all distributions $H_\varepsilon = (1 - \varepsilon) H_\theta + \varepsilon\delta_x$, where δ_x is the Dirac measure in x; i.e., it puts mass 1 on the point x. The Influence Function (IF) of an estimator T at x specifies the infinitesimal influence of the individual observation on the estimator by

$$IF(x;T,H_\theta) = \lim_{\varepsilon\to 0}\frac{T(H_\varepsilon)-T(H_\theta)}{\varepsilon}.$$

When we estimate now the parameter $\theta = (\beta,\xi)$ via a functional T evaluated at the empirical distribution, this robust approach allows us to specify the infinitesimal influence of the individual observations on the estimator (cf. (2.8)). This idea is now extended to the minimization of $\Gamma(\cdot)$ in (3.2) by estimating σ_i^2 and $\mathrm{Cov}(X_{i:n},\hat{X}_{i:n})$ by robust methods based on different threshold values u corresponding to k. This means that for large k and $u = X_{k+1}$,

$$\sigma_i^2 = \mathrm{Var}(Y_{i:n}) \approx \frac{1}{k}\int_0^\infty IF(x;Y_{i:n},H_\theta)^2 H_\theta(dx),$$

(3.3)

$$\mathrm{Cov}(Y_{i:n},\hat{Y}_{i:n}) \approx \frac{1}{k}\int_0^\infty IF(x;Y_{i:n},H_\theta)IF(x;\hat{Y}_{i:n},H_\theta)H_\theta(dx).$$

(3.4)

In our model, for $\xi \neq 0$ it is possible to compute the influence functions $IF(x;Y_{i:n}, H_\theta)$ and $IF(x;\hat{Y}_{i:n},H_\theta)$, where $Y_{i:n}$ and $\hat{Y}_{i:n}$ are defined in (2.9). Influence functions corresponding to $\xi = 0$ can be obtained as limits for $\xi \to 0$.

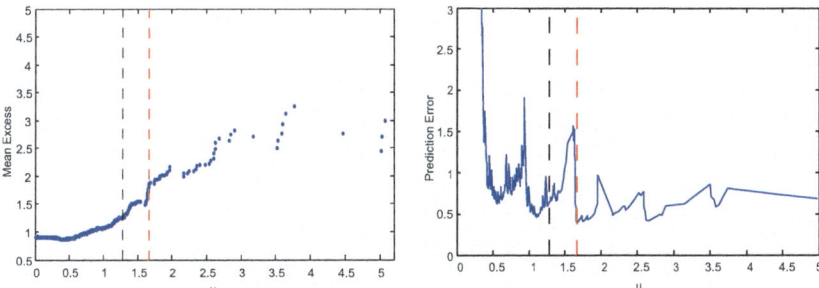

Fig. 7.4: Simulated independent random variables from the distribution function
$\mathbb{P}(X \leq x) = \mathbf{1}(x \leq \Phi^{-1}(1-p))\Phi(x) + \mathbf{1}(x > \Phi^{-1}(1-p))(1 - p + pH_{\xi,\beta}(x - \Phi^{-1}(1-p)))$
for $p = 0.1$, $\xi = 0.5$, $\beta = u\xi = 0.64$, where Φ denotes the standard normal distribution. The left plot shows the mean excess plot as in Figure 3 and the right plot depicts $\hat{\Gamma}$. The estimated threshold (red dashed line) is close to the true one (black dashed line, known by choice of the simulation method), which is $u = z_{0.9} = 1.2816$ with $z_{0.9}$ being the 90% standard normal quantile.

Empirical versions of the expressions of (3.3) and (3.4) can be obtained by introducing the MLE $\hat{\theta} = (\hat{\xi},\hat{\beta})$ for $\theta = (\xi,\beta)$ and the estimated order statistics using $H_{\hat{\theta}}(Y_{i:n}) = U_{i:k}$ (cf. (2.8)). Then we minimize the empirical prediction error $\Gamma(\cdot)$, which takes the form

$$\hat{\Gamma}(k) = \frac{10}{k}\sum_{i=1}^{k}\left(\frac{Y_{i:n} - \hat{Y}_{i:n}}{\hat{\sigma}_i}\right)^2 + \frac{2}{k}\sum_{i=1}^{k}\frac{\widehat{\mathrm{Cov}}(Y_{i:n},\hat{Y}_{i:n})}{\hat{\sigma}_i^2} - 1. \qquad (3.5)$$

This means we determine

$$k_{\mathrm{opt}} = \mathrm{argmin}_{k\in\{1,\ldots,n\}}\hat{\Gamma}(k). \qquad (3.6)$$

This procedure uses the MLEs $\hat{\xi}$, $\hat{\beta}$ as input parameters and their asymptotic normality; hence, we require that $\xi > -0.5$ as indicated in Section 2.2.

So far we have introduced statistical risk models, which are appropriate to model risk variables by distribution tails, giving the probability that some specific risk exceeds a certain threshold. For the estimation of this probability we start with a semiparametric tail approximation, which holds for most probabilistic risk models above a high threshold and yields to reliable approximations of high risks. We also have presented an automatic threshold selection method, which estimates the high risk probabilities for large numbers of different risk quantities in an automated way. In the next section we will show this method at work in a technical risk analysis.

7.4   Technical risk analysis

We apply extreme value statistics to the safety of airplane landings. One specifically risky event is the so-called *runway overrun* (RO), which describes the fact that an airplane is unable to stop before the end of the runway. Such an event happened for instance on December 29, 2012 in Moscow; cf. http://avherald.com/h?article=45b4b3cb for details. A case study can be found in [1].

We set extreme value statistics to work in order to estimate the risk of such serious incidents. From conversation with experts we know that not every variable defining a RO is appropriate for estimating the occurrence probability of a RO. As an appropriate risk variable Max Butter [4] proposed the *maximum deceleration needed to stop* (DNS) within the runway. At each time during the landing, the hypothetical deceleration needed to stop before the end of the runway is computed, and then the maximum over the whole landing process is taken. A RO occurs when the DNS is higher than the *maximal deceleration physically possible* (DPP), i.e. in this case the DNS value is physically not possible.

Obviously, the landing process depends on various variables, and one of the most relevant risk factors is the *runway condition* (RWY Cond), where we distinguish between a "dry" and a "wet" runway. Engineers calculate the DPP for a dry runway as lying between 0.54g and 0.60g, depending on the specific runway, and for a wet runway between 0.33g and 0.35g, again the precise number depends on the specific runway.

Our goal is to estimate the occurrence probability of a RO for dry and wet runways

$$\mathbb{P}(\text{RO} \mid \text{RWY Cond} = \text{dry}) \quad \text{and} \quad \mathbb{P}(\text{RO} \mid \text{RWY Cond} = \text{wet}). \quad (4.1)$$

7.4.1   The data and risk factors

Our data set contains around 500.000 operational flight data, recorded mostly from February 2013 to February 2015. Table 4.1 gives a short overview of the available parameters of each flight. One basic parameter is the *aircraft type*. Examples of aircraft types are the A320, one of the famous narrow-body jets from the Airbus family, and the A388, the world's largest commercial passenger aircraft.

The *runway ID* gives the airport, the degrees the runway is pointing to the north pole, and a letter defining the position of the runway (R=right, L=left, C=center). For example, an A320 landing at MUC 08L heads 80 degrees northeast on the left runway of the Munich Airport. MUC 26R is the same physical runway, but approached from the other side; i.e., 260 degrees southwest, the right runway.

The *landing distance available* (LDA) is the usable length of the runway for landing. Sometimes we observe different LDAs for the same physical runway, caused for instance by an obstacle on the runway.

The *inflight landing distance* (ILD) is the calculated landing distance which can be accomplished by an average line pilot adhering to standard technique. The ILD is being published by the manufacturer of the aircraft and takes into account the aircraft weight, the runway geometry, weather and runway condition and other environmental aspects.

First we classify our data to make sure that we work with identically distributed samples, when estimating the distribution tail. We do this step by step based on the different attributes from Table 4.1.

We start with the attribute aircraft type. Figure 5 shows the boxplots of the DNS for landings on a dry runway grouped by different aircraft types. The medians (red lines) as well as the first and the third quartiles (the bottom and top of the blue boxes) are clearly distinct for each aircraft type, indicating that the distributions differ for different aircraft types.

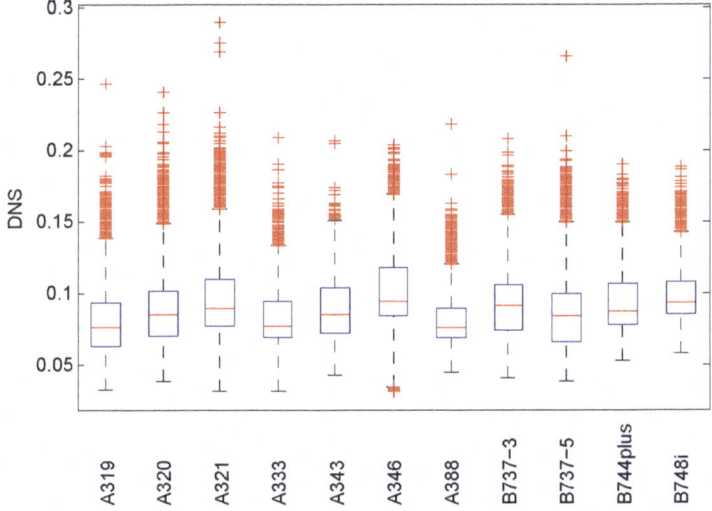

Fig. 7.5: RWY Cond = dry: boxplots of DNS grouped by aircraft type.

Pairwise Kolmogorov-Smirnov tests (e.g. [3]) confirm that we have to differentiate between all 11 aircraft types. Similar classifications apply to all given attributes. However, we have to be aware that too detailed classification may lead to so few data within classes that we cannot estimate the GPD parameters in a statistically reliable way.

Attribute	Description
Aircraft type	Aircraft type
Runway ID	Airport and runway identifier
DNS	Maximum deceleration needed to stop during roll-out measured in units of gravity ($1g = 9.81 \text{m/s}^2$)
LDA	Landing distance available in meter
ILD	Inflight landing distance in meter: an estimation of the required landing distance
RWY Cond	Runway condition (dry, wet)

Table 7.4.1:   Overview of the different parameters given for each flight.

Our data have to be classified by at least two of the most relevant attributes: The aircraft type as well as the runway ID have an influence on the distribution of the DNS. In this classification, however, the problem described above occurs and some classes would contain too few observations, in particular in case of wet runways. So finally we classify 11 aircraft types and 5 LDA categories (which summarize landings with similar LDAs).

7.4.2   Estimation

By definition, a runway overrun (RO) is an extreme event, and none was observed during the data records. So there is no way to estimate the occurrence probability of a RO with classical methods of statistics. It is, however, a typical problem to solve with extreme value statistics, which allows us to extrapolate risk assessment beyond the data range. We shall use the POT method as presented in Section 2 with the optimal threshold $u = X_{k+1:n}$ and k selected automatically within each data class by the optimization procedure from Section 3.

In the following we summarize the estimation algorithm and recall that the prediction error criterion leading to (3.6) is different for $\xi \neq 0$ and $\xi = 0$. This causes certain difficulties, which our algorithm has to deal with.

Algorithm

For a sample $X_1, \ldots, X_n$ of iid random variables and its corresponding order statistics $X_{n:n} < \cdots < X_{k+1:n} < X_{k:n} < \cdots < X_{1:n}$ perform the following steps.

1. **Find the MLEs**
   Find the MLEs $\hat{\xi} = \hat{\xi}(k)$ and $\hat{\beta} = \hat{\beta}(k)$ for all $k = k(n) \in [\min(40, \lfloor 0.02n \rfloor), \lfloor 0.2n \rfloor]$. This means, the percentage of the number of large sample values k taken into account ranges from 2% to 20%. If the upper 2% of the sample contains less than 40 values, for stability reasons, we take 40 extreme data points as the smallest k. On the other hand, we do not want to go too close to the center of the data. Hence, we do not consider more than 20% of the largest sample values. This is also motivated by the fact that for nearly all our data sets the optimal value for k was less than 20% of the upper data values. The estimator $\hat{\xi}$ will always give values different from 0, although close to 0 will be possible, indicating that the true ξ may be zero. We come back to this point below. We have also to ensure that the MLEs are asymptotically normal, requiring that $\xi > -0.5$. This is, however, not a problem for our data, since many estimates result in small positive or very small negative estimates.

2. **Estimate $\Gamma(k)$**
   Estimate the prediction error (3.1) under the assumption that $\xi \neq 0$: compute $\hat{\Gamma}(k)$ by (3.5) (details are given in [8], Theorem 4.9):

$$\hat{\Gamma}(k) = \sum_{i=1}^{k} \frac{1}{\hat{\beta}^2} \left(\frac{i}{k+1} \right)^{2\hat{\xi}} \left(\frac{k+1}{i} - 1 \right)^{-1} \left(X_{i:n} - X_{k+1:n} - \frac{\hat{\beta}}{\hat{\xi}} \left(\left(\frac{i}{k+1} \right)^{-\hat{\xi}} - 1 \right) \right)^2$$

$$+ \frac{2}{k} \sum_{i=1}^{k} \left(\frac{i}{k+1} \right)^{2\hat{\xi}} \left(\frac{k+1}{i} - 1 \right)^{-1} \frac{(1+\hat{\xi})^2(1+2\hat{\xi})}{\hat{\xi}^2} \left(\frac{\left(\frac{i}{k+1} \right)^{-\hat{\xi}} - 1}{\hat{\xi}} \right)^2$$

$$+ \frac{4}{k} \sum_{i=1}^{k} \left(\frac{i}{k+1} \right)^{\hat{\xi}} \left(\frac{k+1}{i} - 1 \right)^{-1} \frac{(1+\hat{\xi})\,(1+2\hat{\xi})}{\hat{\xi}^2} \frac{\left(\frac{i}{k+1} \right)^{-\hat{\xi}} - 1}{\hat{\xi}} \log\left(\frac{i}{k+1} \right)$$

$$+ \frac{2}{k} \sum_{i=1}^{k} \left(\frac{k+1}{i} - 1 \right)^{-1} \frac{(1+\hat{\xi})^2}{\hat{\xi}^2} \log^2\left(\frac{i}{k+1} \right) - 1.$$

3. **Find the optimal threshold**

   Find the number of excesses k such that the prediction error criterion (3.5) is minimal:

   $$k_{opt} = \underset{k}{\text{argmin}} \ \hat{\Gamma}(k) \quad \text{and set} \quad u_{opt} = X_{k_{opt}+1:n}.$$

   Denote by $(\hat{\xi}, \hat{\beta})$ the MLE of the GPD $H_{\xi,\beta}$ based on $X_{k:n}, \ldots, X_{1:n}$ for $\xi \neq 0$ and $k = k_{opt}$.

4. **Perform a likelihood ratio test for $\xi = 0$**

   Choose the $k = k_{opt}$ largest values of the sample and estimate the parameter β of the GPD $H_{0,\beta}$ based on $X_{k:n}, \ldots, X_{1:n}$. Denote the MLE by $\tilde{\beta}$.

   Perform a likelihood ratio test for $H_0 : \xi = 0$ with level $\alpha \in (0,1)$ on $X_{k:n}, \ldots, X_{1:n}$: reject H_0 if the likelihood ratio statistic

   $$D := 2 \left\{ -\left(\frac{1}{\hat{\xi}}+1\right) \sum_{i=1}^{k} \log\left(1 + \hat{\xi}\frac{X_{i:n}-u}{\hat{\beta}}\right) + \sum_{i=1}^{k} \frac{X_{i:n}-u}{\tilde{\beta}} + k\log\frac{\tilde{\beta}}{\hat{\beta}} \right\} > \chi^2_{1-\alpha},$$

   where $\chi^2_{1-\alpha}$ is the $(1 - \alpha)$-quantile of the χ^2 distribution with one degree of freedom. We choose the level $\alpha = 0.05$, which gives $\chi^2_{0.95} = 3.84$.

   We consider the two possible outcomes of the test:

   - If H_0 is rejected: Then $\hat{\xi}, \hat{\beta}$ are the MLEs obtained in step (1), based on the fitted GPD of the k_{opt} largest values and $u_{opt} = X_{k_{opt}+1:n}$ as in steps (2) and (3) above.

   - If H_0 is not rejected: Then we assume that $\xi = 0$ (likelihood ratio tests for k close to k_{opt} also do not reject H_0 for all our data sets). Finally, now for $\xi = 0$, we estimate the prediction error (3.1) $\Gamma^0(k)$ by (3.5) (details are given in [8], Theorem 4.14):

   $$\hat{\Gamma}^0(k) = \sum_{i=1}^{k} \frac{1}{\tilde{\beta}^2}\left(\frac{k+1}{i}-1\right)^{-1}\left(X_{i:n} - X_{k+1:n} + \tilde{\beta}\log\left(\frac{i}{k+1}\right)\right)^2$$
   $$+ \frac{2}{k}\sum_{i=1}^{k}\left(\frac{k+1}{i}-1\right)^{-1}\log^2\left(\frac{i}{k+1}\right) - 1.$$

   We find

   $$k^0_{opt} = \underset{k}{\text{argmin}} \ \hat{\Gamma}^0(k) \quad \text{and set} \quad u_{opt} = X_{k^0_{opt}+1:n}.$$

   Denote now by $\tilde{\beta}$ the MLE of the GPD $H_{0,\beta}$ based on $X_{k:n}, \ldots, X_{1:n}$ and $k = k^0_{opt}$.

5. Estimate the GPD for the excesses

Recall (2.4) and set

$$\hat{\xi} < 0 \qquad \overline{H}_{\hat{\xi},\hat{\beta}}(x) = \left(1 + \hat{\xi}\frac{x - u_{opt}}{\hat{\beta}}\right)^{-\frac{1}{\hat{\xi}}}, \quad x \in \left(u_{opt}, u_{opt} - \frac{\hat{\beta}}{\hat{\xi}}\right),$$

$$\hat{\xi} = 0 \qquad \overline{H}_{0,\hat{\beta}}(x) = \exp\left\{-\frac{x - u_{opt}}{\tilde{\beta}}\right\}, \quad x > u_{opt},$$

$$\hat{\xi} > 0 \qquad \overline{H}_{\hat{\xi},\hat{\beta}}(x) = \left(1 + \hat{\xi}\frac{x - u_{opt}}{\hat{\beta}}\right)^{-\frac{1}{\hat{\xi}}}, \quad x > u_{opt}.$$

6. Estimate the excess probability for the risk variable X

Recall (2.6), estimate $\mathbb{P}(X > u)$ by its empirical version and obtain

$$\widehat{\mathbb{P}(X > x)} = \hat{\overline{F}}(x) = \frac{k_{opt}}{n}\overline{H}_{\hat{\xi},\hat{\beta}}(x - u_{opt}) \quad \text{for} \quad x > u_{opt}.$$

This algorithm is applied to estimate (DNS is measured in units of gravity g):

$$\mathbb{P}(RO \mid RWY\ Cond = dry) = \mathbb{P}(DNS > 0.54) \text{ and}$$
$$\mathbb{P}(RO \mid RWY\ Cond = wet) = \mathbb{P}(DNS > 0.33),$$

as in (4.1) for all different classes determined in Section 4.1. The precise probabilities can be found in [8]. As to be expected, some classes lead to higher probabilities than others, which allows for an assessment of the riskiness of the specific values of the corresponding attributes.

Moreover, there are classes with less than 300 observations, which leads to unreliable parameter estimates. There are two possibilities to deal with this problem: consider only such classes with at least 300 observations, or merge such classes, whose distributions are not too different. Since we want to use most of the data, we use the classification by aircraft type and landing distance available (LDA) as already mentioned in Section 4.1. It is possible to summarize the results of the different classes by the theorem of total probability (cf. (2.2)) resulting in

$$\mathbb{P}(RO \mid RWY\ Cond) =$$

$$\sum_{i=1}^{n_T}\sum_{j=1}^{n_{LDA}} \mathbb{P}(RO \mid T_i \cap LDA_j \cap RWY\ Cond)\mathbb{P}(T_i \cap LDA_j \mid RWY\ Cond),$$

where T_i stands for aircraft type i with $i = 1,\dots, n_T = 11$ and LDA_j for LDA category j with $j = 1,\dots, n_{LDA} = 5$. RWY Cond is again either dry or wet. This way we use most of the data and obtain the estimates

$$\mathbb{P}(RO \mid RWY\ Cond = dry) = 3.72 \times 10^{-7}\ and$$
$$\mathbb{P}(RO \mid RWY\ Cond = wet) = 4.98 \times 10^{-6}.$$

When analysing these results more closely, we find that—for landings on a dry runway—the total probability is mainly driven by two classes for which relatively high probabilities have been estimated, such that (2.3) implicitly applies. These two classes involve the B744i-plus landings on runways with LDA between 3000m and 3500m and the B737-5 landings on runways with LDA between 3500m and 4000m.

7.4.3 Conclusion

We have applied extreme value statistics to assess the safety of airplane landings, focussing on the event of a runway overrun. Given the large number of data available we have developed an automated threshold selection to identify the relevant events, which we used for statistical estimation of the occurrence probability of a runway overrun. Although we had a large number of flight data available, we still had to merge certain data, which may not have been identically distributed. We have tried to keep track of the consequences of this simplification, but there is still room for improvement. Furthermore, we have used the maximum deceleration needed to stop during roll-out (DNS) as risk variable, which seems to be the relevant quantity to study in this context. This novel risk quantity is currently discussed among flight engineers. Further studies based on more detailed data will certainly help to clarify this notion. We also hope that this study helps to propagate the powerful statistical extreme value methods further in the engineering sciences.

Acknowledgements

We are grateful to Johan Segers for sharing his insight into different threshold selection methods with us. The first author acknowledges support from ISAM of TUM Graduate School at Technische Universität München. This research was funded by the German LUFO IV/4 project SaMSys—Safety Management System in Order to Improve Flight Safety.

References

[1] E.S. Ayra. Risk analysis of runway overrun excursions at landing: a case study. http://www.agifors.org/award/submissions2013/EduardoAyra.pdf, 2013.

[2] J. Beirlant, Y. Goegebeur, J. Segers, and J. Teugels. *Statistics of Extremes: Theory and Applications.* Wiley, New York, 2004.

[3] P.J. Bickel and K.A. Doksum. *Mathematical Statistics: Basic Ideas and Selected Topics, Volume I.* Chapman & Hall/CRC, New York, 2 edition, 2001.

[4] M. Butter. DNS as risk variable for a runway overrun. Personal Communication, 2015.

[5] D.J. Dupuis and M-P. Victoria-Feser. A robust prediction error criterion for Pareto modelling of upper tails. *Canadian Journal of Statistics,* 34(4):639–658, 2006.

[6] P. Embrechts, C. Klüppelberg, and T. Mikosch. *Modelling Extremal Events for Insurance and Finance.* Springer, Berlin, 1997.

[7] F.R. Hampel, E.M. Ronchetti, P.J. Rousseeuw, and W.A. Stahel. *Robust Statistics - The Approach Based on Influence Functions.* Wiley, New York, 1986.

[8] J. Mager. *Automatic Threshold Selection of the Peaks-Over-Threshold Method.* Master's thesis, Technische Universität München, 2015.

[9] K. Mainzer. *Die Berechnung der Welt.* C.H. Beck, München, 2014.

[10] P. Ruckdeschel and N. Horbenko. Robustness properties of estimators in generalized Pareto models. Bericht 182, Fraunhofer Institut für Techno- und Wirtschaftsmathematik, 2010.

[11] R.L. Smith. Estimating tails of probability distributions. *Annals of Statistics,* 15:1174–1207, 1987.

II
Berechenbarkeit

8. Computing in Space and Time

Andrei Rodin[1] [2]

Abstract

The Turing machine adequately accounts for the temporal aspect of real computing but fails to do so for certain spatial aspects of computing, in particular, in the case of distributed computing systems. This motivates a search for alternative models of computing, which could take such spatial aspects into account. I argue that a revision of the received views on the relationships between logic, computation and geometry may be helpful for coping with spatial issues arising in modern computing.

8.1 Introduction

Computing takes time. For practical reasons it is crucial how much time it takes. Processing speed (usually measured in FLOPS or more generically in cycles per second) is a basic measure of computer hardware performance. During the past several decades of continuing computer revolution the processing speed of hard-

---

1   Andrei Rodin is a senior researcher in the Institute of Philosophy of Russian Academy of Sciences in Moscow and an associated professor of philosophy and mathematics at Saint-Petersburg State University. His research interests include history and philosophy of mathematics, philosophy of science and philosophical logic. He is the author of „Axiomatic Method and Category Theory" appeared in 2014 in Springer (Synthese Library vol. 364).

2   The work is supported by Russian Foundation for Basic Research, research grant number 13-06-00515.
    I thank Sergei Kovalev for reading an earlier draft of this paper and very valuable advices.

ware increased dramatically. Computation time (aka running time) is equally crucial for evaluating the software performance: invention of faster algorithms is going along with building faster processors.

Computing also takes space. It equally matters how much space it takes. Today's digital processors are smaller in size than their early prototypes by several orders of magnitude; the minimization of physical sizes of computing devices is a continuing technological trend. In order to see the role of space in computing more clearly it will be helpful to consider first a more traditional computing device like a set of pebbles (Latin *calculi*). Manipulating with pebbles requires a space. A basic trick of traditional computing is that this required computation space is more "handy" and usually significantly smaller than the space where live the counted objects. Suppose for example that one wants to count sheep in a herd. A reason for replacing sheep by pebbles is that one can manipulate with pebbles in space by far more easily and more effectively than one can do this with the real animals. Using pebbles one can even count stars and other objects on which one has no real control at all. Thus if we consider computing as a process applied to certain external objects rather than an abstract procedure on its own rights, then we can observe that scaling the system of objects under consideration down (or up in the case of a microscopic system) to the spatial and temporal scale where humans may possibly provide for an effective control, is an essential part of this process. A spatio-temporal scaling is equally at work when calculations are done symbolically.

If we now turn back to the modern electronic computing then we remark that the spatial issues enter into the picture in a number of new ways. Spatial issues are dealt with by hardware engineers who seek to make handy human-scale computing devices as powerful as possible. The distributed computing made possible by the web technology involves hardware scattered over the globe, so spatial issues (along with temporal ones) become in this case more pressing. The case of remote control of spacecrafts where the time of signals' traveling between different parts of the same information system becomes an essential factor, makes a link (practical rather than only theoretical!) between computing and relativistic space-time. These are but the most apparent ways (randomly chosen) in which spatial and spatio-temporal issues can be relevant to modern computing; developers of information systems most certainly can specify more.

Theoretically spatial and spatiotemporal aspects of computing have been studied so far at least in the following two directions. In 1969 Konrad Zusse proposed a powerful metaphor of "computing space" [25] which gave rise to an area of research known as *digital physics*, which explores the heuristic idea according to which the physical universe can be described as a computing device. Reciprocally, the fact that computing proceeds in the real physical space and time led a number of authors to reconsidering standard theoretical models of computation [12] and more recently gave rise to a new field of research named *spatial computing* [9].

8.2   Time and space estimations

A commonly used theoretical model of computation is the *Turing machine*, which can be described informally as follows. It consists of an infinite strip of tape divided into equal cells, which moves forward and backward with respect to a writing device. The device writes and rewrites symbols into the cells according to a table of rules. Thus the Turing machine presents computation as a discrete process divided into discrete atomic steps. According to this model every accomplished calculation is characterized by a certain finite number n of such steps. This apparently naive model (which can be given a more precise formal presentation) turns out to be surprisingly effective for theorizing about computing in the context of modern information technology. The Turing machine model allows one to estimate the computing time straightforwardly. Given that an algorithm A requires n Turing moves for accomplishing a given task T, and given that the CPU of one's computer makes m operations per second, one can estimate the required running time t as $\frac{n}{m}$. Since the Turing machine is an ideal theoretical model but not a real computing hardware, the above calculation is by far *too* straightforward. In the real life the exact number n as above is undetermined, so one can only estimate how n varies with the variation of parameters T such as the number of elements in sorting. But notwithstanding these details the very fact that the Turing machine, theoretically described algorithms, algorithms realized in a software, and finally real CPUs all work step-by-step provides a firm ground for time estimations. The "internal time" of Turing machine measured in elementary moves of its tape turns out to be a good theoretical model for the running time of real computers.

The Turing machine also helps to estimate the space required for computing. This is done by the estimation of the number m of the required cells. If one knows how m depends on parameters of the given task, one can estimate the volume of required memory, which in its turn provides a reasonable estimation of size of the real computing device. In that respect the time estimation and the space estimation are similar. However this very structural similarity between the temporal and the spatial sides of the Turing machine makes a big difference in how the Turing model of computation relates to computations made in the real world. Let me for the sake of the following argument assume that the physical time and space are classical (Newtonian). The elementary moves of the Turing machine can be identified with ticks of physical clocks and its tape divided into equal cells can be used as a ruler. The number of ticks is all one needs for measuring a time span between two events and the number of cells is all one needs for measuring the distance between two points in space. What makes the two cases very different is this: while the arithmetic of natural numbers in a sense comprises the formal structure of classical time (as already Kant rightly acknowledged) this is not the case regarding the formal structure of classical Euclidean 3D space. A fundamental property of this space which remains unaccounted in this way is the number of its dimensions. The one-dimensional Turing tape may serve as a good instrument for *testing* various spatial structures—Euclidean and beyond—but it cannot, generally, *represent* such structures (including their global topological properties) in the same direct way in which it represents the running time in real computers. Thus one can remark a sharp difference between the temporal and the spatial relevance of the Turing model of computation: While this model adequately accounts for the temporal aspect of real computing (modulo usual reservations explaining the difference between a physical process and its theoretical model), it fails to do so with respect to certain spatial characteristics of modern computing devices.

Why this dissymmetry? Or perhaps it is more appropriate to ask why not? The success of the Turing machine and other related models of computing (such as the lambda calculus) suggests seeing the running time as an essential feature of computing and seeing all spatial aspects of computing as non-essential. Even if the Turing machine says nothing about spatial issues related to computing it is not obvious whether these issues should be taken into account by a theory of computing at the fundamental level. Perhaps these spatial issues can be better accounted for separately after the basic model of computation is already fixed. The idea to unify spatial and temporal aspects of computing within the same fundamental theory may appear tempting (and natural from the point of view of today's physics) but it certainly needs further arguments in its favor.

In what follows I shall try to provide such arguments. I shall start with some historical observations concerning the relationships between space, geometry and computing. Then I consider a recent theory, which reveals a deep link between computing and geometry in a modern mathematical context. Finally I discuss some related philosophical issues concerning the relationships between pure and applied mathematics.

8.3   Historical Forms of Computing

Examining the history of a subject unavoidably involves projecting of the present state of this subject onto its past. These days by a computer one understands a digital electronic device, which inputs and outputs sequences of 0s and 1s; a peripheral hardware translates between the 0-1 sequences and data of different types including the data that can be received and/or outputted by human users immediately (such as strings of symbols and imagery). Looking for a close historical analogue of modern computing one naturally points to arithmetical calculations in its various historical forms, some of which involve devices of abacus type [16]. However a closer examination shows that the historical forms of computing are more diverse. Suppose one needs to compute the height of an equilateral triangle knowing its side. Such geometrical problems are common in building construction and many other practical affairs. If one has an electronic calculator at hand then to compute a decimal fraction approximating $\frac{\sqrt{3}}{2}$ is a reasonable solution. Otherwise one may use a more traditional tool such as the ruler and the compass for solving the problem geometrically. If the size of the figure in question does not allow one to apply these instruments directly one first solves the problem on a sheet of paper or another appropriate support, and then uses a scaling technique (which typically but not necessarily involves arithmetical calculations) for applying this geometrical result in the given practical context.[3] There is a tendency dating back to Plato to overlook or underestimate the computational aspect of the traditional elementary geometry as presented in Euclid's *Elements*. Whatever may be the philosophical reasons behind it such an attitude is hardly appropriate when studying the history of computing.

The combination of ruler and compass works as a simple *analogue* computer. While modern *digital* computers use the idea of symbolic *encoding*, the analogue computers exploit the idea of *analogy* between different physical processes. This latter idea can be made more precise through the concept of *mathematical form* (which, of course, in its turn needs further specifications which I omit here). Different physical processes, including those having very different physical nature, happen to share the same mathematical form; in many cases, they may be adequately described by the same mathematical tools such as differential equations.

---

3   The Euclidean space is the only one among Riemannian spaces of constant curvature, which allows for a simple linear scaling. The importance of scaling in practical matters provides, in my view, a plausible explanation why the Euclidean geometry for many centuries was considered as the only "true" theory of space. The fact that the linear scaling property implies Euclid's Fifth Postulate (aka the Parallel Postulate) was first realized by Wallis in 1693 [2].

Let P be a class of processes sharing the same mathematical form F. Now the idea of analogue computing can be formulated as follows: choose in P an appropriate process C (for "computation"), which is artificially reproducible, well-controllable and conceptually transparent; then use C as a standard representation for F. What one learns about F through C applies to all other processes in P disregarding their specific physical nature. As an example, F could be the geometrical form of equilateral triangle and C the standard construction of such a triangle by the ruler and the compass.

Analogue computers have been largely superseded by their digital rivals at some time in the early 1960-ies (or earlier on some accounts [16]). A thorough discussion on digital and analogue computing is out of place here but I shall point to one advantage of digital computing which obviously contributed to its success. It consists in its *universality*. What we want to call a computer is not just a device that allows one to simulate physical processes and technological procedures of some particular type P as described above but rather a universal toolkit, which allows for simulating processes and procedures of many different sorts. The ruler and the compass meet this requirement only to a certain degree. These instruments can be used for solving a large class of geometrical problems but this class turns out to be limited in a way, which from the practical viewpoint may appear very strange and even arbitrary. Why the trisection of a given segment is doable by ruler and compass but the trisection of a given angle is not? Why a regular hexagon can be so easily computed but a regular heptagon cannot be computed by these means at all? Today we know good theoretical answers to these questions but they don't make the ruler and the compass more useful than they are.

Now consider the claim according to which all relevant mathematical procedures and mathematical structures serving as mathematical expressions of various physical "analogies" in the analogue computing as explained above, can *in principle* be encoded into (i.e., represented with) 0-1 sequences and operations with these things. This claim is problematic both from a theoretical and a practical point of view. Not all mathematical theories currently used in physics are constructive and moreover computable. A mathematical theory or structure, which is theoretically computable, may require unfeasible computations and thus be not computable in practice. The more computing power we get the more such limitations become visible. Nevertheless the idea of a single universal model of calculation appears so attractive and so promising that our technological development largely follows it anyway.

8.4   Geometrical Characteristic

Let me now turn from the history of computing to the history of ideas about computing. Leibniz is commonly and rightly seen as a forerunner of modern computing; his ideas about this subject he put under the title of *Universal Characteristic*, which he described as a hypothetical symbolic calculus for solving problems in all areas of human knowledge.

Although this idea sounds appealing in the modern context to reconstruct it precisely is a laborious historical task; moreover so since this idea never achieved in Leibniz' work a stable and accomplished form. I shall discuss here only one specific aspect of this general idea, which is relevant to my argument, namely the notion of *Geometrical* Characteristic [10], for partial English translation see [11].

Leibniz builds his idea of Geometrical Characteristic upon Descartes' *Analytic Geometry*. In its original form (unlike its usual modern presentations) this latter concept has little to do with the arithmetization of geometry through a coordinate system. It has been rather conceived by Descartes as a geometrical application of a general algebraic theory of magnitude. This general algebra of magnitudes was supposed to cover both arithmetic (the case of discrete magnitude) and geometry (the case of continuous magnitude). As Leibniz stresses in his Geometrical Characteristic paper the general algebra of magnitudes cannot be a sufficient foundation of geometry because this general algebra treats only metrical properties of geometrical objects while these objects also have relational *positional* properties (which we call today topological). Leibniz tries to push Descartes' project further forward by mastering a more advanced algebraic theory capable to account for positional properties of geometrical configurations along with their metrical properties. He conceives here of a possibility of replacing traditional geometrical diagrams with appropriate symbolic expressions and appropriate syntactic procedures on such expressions, which would express the positional properties directly, without using the Cartesian algebra of magnitudes. For this end Leibniz observes that the traditional geometrical letter notation (as in Euclid) is not wholly arbitrary but has a certain syntactic structure, which reflects certain positional properties. For example when one denotes a given triangle ABC the syntactic rules require A, B, C to be the names of this triangle's vertices, and AB, BC, AC be the names of its three sides. Leibniz' idea is to develop this sort of syntax into a full-fledged symbolic calculus similar (on its syntactic side) to Descartes' algebraic calculus.[4]

---

4   An attempt to develop geometry systematically on an algebraic basis (in the form of a general theory of magnitude) has been made by Descartes' follower Antoine Arnauld [1]. This work of Arnauld was carefully studied by Leibniz and contributed to his thinking

Leibniz's idea of Geometrical Characteristic is interesting because it directly links symbolic computation to geometrical reasoning on a fundamental theoretical level—while in the mainstream 20th-century theoretical works on computation by Church and others such a link appears to be wholly absent. However in the 19th century the idea of Geometrical Characteristic has a rich history, which involves works of Grassmann [5], Peano [17][18] and other important contributors. Even if in the 20th century this circle of ideas did not form the mainstream research in the theory of symbolic computing (which in this century was largely monopolized by logicians) it continued to develop during this century within other mathematical disciplines including algebraic geometry. Tracing this history of ideas continuously up to the present is a challenging historical task, which I leave for another occasion. In this paper I shall only briefly describe what I see as the latest episode of this history, which establishes a new surprising theoretical link between geometry and computing in today's mathematical and logical setting.

8.5   Univalent Foundations

The *Univalent Foundations* of mathematics (UF) is an ongoing research project headed by Vladimir Voevodsky and his collaborators at the Princeton Institute for Advanced Study; this project is closely related to the recently emerged mathematical discipline of *Homotopy Type theory* (HoTT). The backbone of UF/HoTT is a correspondence between a type calculus due to Martin-Löf (MLTT) [15] and a geometrical theory (in a broad sense of "geometrical") known as *Homotopy theory* (HT); see [19] for a systematic introduction and further references. A role in the discovery of this correspondence was played by the concept of infinite-dimensional *groupoid* first introduced by Grothendieck in 1983 [6]; more historical details are found in [21], Ch. 7.

For my present argument it is essential to take into account the specific character of correspondence between the type calculus MLTT and the geometrical theory HT, which gives rise to UF/HoTT. I leave now aside subtle model-theoretic issues of HoTT and disccuss only the semi-formal homotopical interpretation of MLTT as described in [19]. When one compares this interpretation of MLTT with standard

---

about geometrical matters. Another name for the same circle of Leibniz' ideas, which connects them to Descartes' work more directly, is Analysis Situs (Situational Analysis); under this latter name this circle of Leibniz' ideas plays a prominent role in the early history of modern topology [3]

examples of models such as Beltrami-Klein or Poincaré models of Hyperbolic geometry (HG), one immediately notices a striking difference. A standard axiomatic presentation of HG contains *non-logical* terms like "point", "lies between", etc, and certain *logical* terms like "and", "if then", "therefore" etc. As far as one thinks about formal axiomatic theories and their models along the pattern provided by Hilbert in his classical [7], one assumes that the meaning of logical terms is fixed (and commonly understood), while the non-logical terms are place-holders, which get definite semantic values only under this or that possible *interpretation*; when such an interpretation turns the axioms of the given theory into true statements this interpretation qualifies as a model of that theory. The distinction between logical and non-logical terms is of a major epistemic significance here because it usually (and certainly in Hilbert's works) goes along with the assumption according to which logical concepts are more epistemically reliable than mathematical (and in particular geometrical) ones. The assumption about the epistemic primacy of logic provides a ground for the claim that a Hilbert-style formal axiomatic presentation of a given mathematical theory is a genuine epistemic gain rather than just one's favorite style of writing mathematical textbooks.

In UF/HoTT the above familiar pattern of axiomatic thinking does not apply. For HoTT provides geometrical interpretations for those terms of MLTT, which by all usual accounts qualify as *logical*. The most interesting (both mathematically and philosophically) case in point is the concept of *identity* (as in MLTT without the additional axioms of extensionality which makes the identity concept in this theory trivial) and its homotopical interpretation. By all usual accounts (including Frege's classic [4]) the concept of identity is logical. In HoTT it receives a highly non-trivial geometrical (homotopical) interpretation in the form of *fundamental groupoid* of a topological space, which is the groupoid of *paths* between points of this space. This construction of "flat" fundamental groupoid of paths is further extended onto that of infinite-dimensional *higher homotopy groupoid*, which accounts for *higher identity types* appearing in MLTT. This geometrical interpretation makes intelligible the complexity of the identity concept as in MLTT, which otherwise may appear as unnecessarily technically complicated and conceptually opaque. So in this case a logical concept is analyzed and clarified by geometrical means rather than the other way round. The reciprocal epistemic impact of logic onto geometry in UF/HoTT is also significant (see below) but it would be clearly wrong to see the impact as one-sided. What brings an epistemic gain in this case is a cross-fertilization of logic and geometry rather than a one-sided influence.

UF/HoTT has a special relevance to computing, which I am now going to describe. MLTT has been designed from the outset as a formal calculus apt for computer implementations. It is a constructive theory in the strong sense of being

Turing computable. The homotopical interpretation of MLTT makes possible to see MLTT (possibly with some additional axioms such as the Univalence Axiom) as a computable version of HT and use program languages based on MLTT for computing in HT. The fact that MLTT has been designed as a general formal constructive framework rather than a formal version of any particular mathematical theory suggests that UF/HoTT may serve as a foundation of all mathematics and that its computational capacities can be used also outside HT, ideally everywhere in mathematics and mathematically-laden sciences. The realization of this project remains a work in progress.

Like Leibniz's Geometrical Characteristics UF/HoTT can be seen as a theoretical means for reducing geometrical constructions to symbolic expressions, which can be managed by the Turing Machine. However the link between geometry and computing established in UF/HoTT can also be explored in the opposite direction and provide a theoretical ground for attributing to computations a geometrical (topological) structure. I submit that such a notion of internal geometrical structure of computing may be used for designing distributed computing systems and coping with the other spatial aspects of modern computing mentioned above. This is, of course, nothing but a bold speculation *à la* Leibniz, which I cannot support by any specific technical argument. Instead I shall discuss certain related philosophical issues. One's stance towards these issues can make the above guess appear more reasonable or, on the contrary, less reasonable and direct one's technical efforts accordingly. Leibniz's example demonstrates that in the past philosophical speculations played a role in later technological developments. I cannot see a reason why this should not work today and in the future.

8.6   Geometrical thinking

Is it reasonable to expect that geometrical methods may help one to cope with spatial issues arising in engineering (including IT engineering)? Two centuries ago the answer in positive would have been a matter of course. However today we live with a very different received view on the nature and the subject-matter of mathematics. This modern vision has been strongly influenced by Hilbert's notion of axiomatic theory and stabilized at some point in the mid-20th century. A concise presentation of this received view is found in Professor Mainzer's recent monograph [13], where he describes a "mathematical universe" of "proper worlds and structures the existence of which is thought of solely in terms of accepted axioms and logical proofs", (*op. cit.*, p. 280, my translation from German). I shall call this view the *standard pic-*

ture (SP) for further references. It should be understood that SP is not a description of what mathematicians are doing in their everyday work but rather a judgement on what pure mathematics really *is* in the proper philosophical analysis. Elementary arithmetical calculations like $7 + 5 = 12$ at the first glance do not look like logical inferences. In order to fit $7 + 5 = 12$ into SP one needs to make a judgement like the following: this calculation is ultimately justified by a logical inference, which is made explicit by a logical reconstruction of arithmetic, i.e., by presenting this traditional mathematical discipline in the modern axiomatic form of Peano Arithmetic or similar. Such a gap between the current mathematical practice and SP exists in all areas of today's mathematics including mathematical logic itself. It is a controversial matter among philosophers whether or not such a gap is tolerable.

SP implies that there is no direct connection between the "proper worlds" of mathematical structures and the material world in which we live, act and develop our technologies. How it happens that some of these structures play a significant role in natural sciences and technologies constitutes a philosophical puzzle famously called by Wigner [24] the "unreasonable effectiveness" of mathematics. This puzzle has a number of plausible solutions compatible with SP (including one explained in Professor Mainzer's book, ch. 14), which I shall not discuss here. Instead I shall try to revise SP and briefly present a different understanding of modern mathematics, which establishes (or rather re-establishes) a stronger conceptual connection between mathematics, natural sciences and technology. Such a link was taken for granted by many philosophers, mathematicians and scientists in the past but was later lost of sight in popular 20th century accounts of the so-called "non-Euclidean revolution" [23] of the mid 19th century. Without going into a thorough historical discussion of this matter I shall try to show here that the results of this alleged revolution have been largely misconceived and somewhat exaggerated.

SP comes with the following assumption, which at the first glance may look merely technical but in fact is epistemically important: an axiomatic presentation of mathematical (and in fact also all other) theories involves a definite *symbolic syntax*. So in addition to the ideal existence "in terms of accepted axioms and logical proofs" all mathematical objects and structures enjoy within SP a more palpable form of existence, namely, the existence in the form of symbolic representations. Hilbert, who was a pioneer of formal axiomatic method, described this double form of mathematical existence explicitly. He qualified mathematical symbols as the only "real" mathematical objects, while the rest of mathematical objects on his account were merely "ideal" [8]. Accordingly, he exempted a part of mathematics from SP and called this special part *metamathematics*. Hilbert conceived of metamathematics as a foundational discipline, which allows one to develop the rest of mathematics safely using symbolic logical methods. Hilbert hoped that metamathematics would reduce to a theoretically transparent and wholly unproblematic fragment of *finitary* mathematics.

Thanks to Gödel and others we know today that Hilbert was seriously mistaken here; for this reason mathematicians and logicians today usually feel free to apply in mathematical logic and in metamathematics any sort of mathematics that may prove useful, i.e., that may prove some non-trivial results. Yuri Manin expresses this changed attitude by saying that "good metamathematics is a good mathematics rather than shackles on good mathematics" ([14], p. 2). As we have seen HoTT applies the Homotopy theory (HT) for a similar purpose: it provides a new geometrical semantics for a symbolic calculus (MLTT) the intended semantics of which is logical (in a broad sense of the word). A logical inference in HoTT is a different name for a geometrical construction. The "existence of mathematical structures" in HoTT is as much logical as it is geometrical. It is clear that this feature of HoTT does not square with SP.

As we have seen, Hilbert in 1927 believed that finite strings of symbols are privileged mathematical objects, which serve as a unique join between abstract mathematics and the concrete material world. Even if modern presentations of SP don't make the same point explicitly, they need to use this assumption tacitly because it is enforced by the current standard of formal logical rigor, which requires using symbols. But since Hilbert's project of building mathematical foundations on the basis of finitary mathematics is given up, I can see no further reason to justify the aforementioned assumption either. Mathematically speaking, the combinatorics of symbols is important but it does not play a distinguished role in mathematical matters—whether one provides it with one's favorite logical semantics or not. Epistemologically speaking, there is no reason to consider symbols as the sole tool, which connects human cognition with the outer world. Geometrical intuition is another obvious candidate.

One should keep in mind that the implementation of mathematical ideas in physics and technology is never a straightforward matter. It is not straightforward even in the Euclidean case, and it is by far less straightforward in the case of modern geometry. Nevertheless I cannot see that modern geometry differ drastically in this respect from the traditional Euclidean geometry, as proponents of the non-Euclidean Revolution often tend to say. Mathematics in general and geometry in particular is a cognitive activity rooted in human material practices and experiences, which on this basis explores further theoretic possibilities by modeling them conceptually. Even if the testing of such newly discovered theoretic possibilities against new experiences and new practices belongs not to pure mathematics but rather to science and technology, there is no reason, in my view, to think of mathematics as a genuinely independent discipline exploring its own "proper world". Human experiences and practices do not, generally, simply guide one's "choice of axioms" for developing on this basis some useful mathematical theories, as SP suggests, but rather help one to build conceptual frameworks, in which certain axioms and certain inferences from these axioms can be later established.[5]

On this—admittedly merely speculative—ground I suggest that HoTT indeed qualifies as a reasonable candidate for a theory of spatial computing or at least for a fragment of such a theory. In fact, it appears as the only such candidate since no other mathematical theory treating the concept of computing *geometrically*, to the best of my knowledge, is presently known.

8.7   Conclusion

Computing is an old and very important channel, which connects the research in pure mathematics (when such an activity is practiced in a society) with the society's economy, administration, political institutions, technology, and natural science. Professor Mainzer [13] provides a detailed account of how this channel functions in today's information societies. In particular, he shows how today's standard picture of mathematics fits contemporary ideas about computing and its implementation in the existing computing technology.

On my part, I tried to suggest a revision of this standard picture and offer a different view on mathematics and computing, which, as I believe, may help one to cope with some technological challenges related to the spatial aspect of computing technologies.

In this context I argued that Hilbert's view on what is real and what is ideal in mathematics is biased. However important is the historical impact of symbolic writing techniques on mathematics, it is certainly not the only thing, which connects mathematics to the material world and to human material practices. However impressive is the implementation of these techniques in modern digital computing it would be wrong to isolate these specific techniques from other mathematically-laden material practices and technologies and think of symbolic techniques (possibly providing it with one's favorite logical semantics) as a unique and exceptional channel that links mathematics to the material world. Among other things such an ideological focus on symbolic processing and on the Turing model of computation artificially isolates the temporal aspect of computing from the spatial one

---

5   For the notion of geometrical intuition in modern mathematical contexts see [20]. For more details concerning the role of geometrical modeling in axiomatic theories see [22].

and thus makes it more difficult to theorize mathematically about spatial aspects of computing. As a possible remedy I pointed to the ongoing research in Univalent Foundations and Homotopy Type theory, which provide a surprising conceptual link between geometry and computing. Whether or not this theory may indeed help one to cope with distributed information systems and long-distance control at the present stage of research is wholly unclear, and in any event there would be a very long to go to it. However I tried to demonstrate using this example that contemporary mathematics—by which I here mean the very edge of the ongoing mathematical research—can be more friendly to technological implementations in general and to computer implementations in particular than suggests the popular picturing of this mathematics as exceedingly abstract and wholly detached from all other human affairs.

References

[1]   A. Arnauld. *Nouveaux Elements de Geometrie*. Guillaume Desprez, Paris, 1683.

[2]   R. Bonola. *Non-Euclidean Geometry*. Dover Publications, 1955.

[3]   V. de Risi. Geometry and Monadology: *Leibniz's Analysis Situs and Philosophy of Space*. Birkhauser, 2007.

[4]   G. Frege. Über Sinn und Bedeutung. *Zeitschrift für Philosophie und philosophische Kritik*, 100:25-50, 1892.

[5]   H. Grassmann. *Geometrische Analyse, geknüpft an die von Leibniz erfundene Geometrische Charakteristik*. Leipzig: Weidmann, 1847.

[6]   A. Grothendieck. *Pursuing Stacks (Letter to Quillen)*. Unpublished, 1983.

[7]   D. Hilbert. *Grundlagen der Geometrie*. Leipzig, 1899.

[8]   D. Hilbert. Grundlagen der Mathematik. *Abhandlungen aus dem mathematische Seminar der Hamburgischen Universität*, 6(1/2):65-85, 1927.

[9]   O. Michel J.-L. Giavitto and A. Spicher. Unconventional and nested computations in spatial computing. *International Journal of Unconventional Computing*, 9(1-2):71-95, 2013.

[10] G. Leibniz. Characteristica geometrica. in G. I. Gerhardt (ed.). *Leibnizens Mathematische Schriften*. Halle. 1849-1863, 5:141-168, 1679.

[11] G. Leibniz. Characteristica geometrica. *in M. Dascal, Leibniz: Language, Signs and Thought, John Benjamins B.V.* 1987, p. 167-174, 5:141-168, 1987.

[12] G. Longo and Th. Paul. The mathematics of computing between logic and physics. *in B. Cooper and A. Sorbi (eds.) Computability in Context, Imperial College Press*, pages 243–274, 2009.

[13] K. Mainzer. *Die Berechnung der Welt: Von der Weltformel zu Big Data*. C.H. Beck, 1899.

[14] Yu. Manin. *Foundations as Superstructure*. arXiv:1205.6044, 2012.

[15] P. Martin-Löf. *Intuitionistic Type Theory (Notes by Giovanni Sambin of a series of lectures given in Padua, June 1980)*. Napoli: BIBLIOPOLIS, 1984.

[16] G. O'Regan. *A Brief History of Computing*. Springer, 2012.

[17] G. Peano. *Calcolo Geometrico secondo l'Ausdehnungslehre di H. Grassmann, preceduto dalle operazioni della logica deduttiva*. Torino: Fratelli Bocca Editori, 1888.

[18] G. Peano. *Geometric Calculus according to the Ausdehnungslehre of H. Grassmann*, translated by Lloyd C. Kannenberg. Birkhäuser, 2000.

[19] Univalent Foundations Program. *Homotopy Type Theory: Univalent Foundations of Mathematics*. Institute for Advanced Study (Princeton); available at http://homotopytypetheory.org/book/, 2013.

[20] A. Rodin. How mathematical concepts get their bodies. *Topoi*, 29(1):53–60, 2010.

[21] A. Rodin. *Axiomatic Method and Category Theory (Synthese Library vol. 364)*. Springer, 2014.

[22] A. Rodin. On Constructive Axiomatic Method. Forthcoming in Logique et Analyse.

[23] R.J. Trudeau. *The non-Euclidean Revolution*. Birkhäuser, 1986.

[24] E. Wigner. The unreasonable effectiveness of mathematics in the natural sciences. *Communications on Pure and Applied Mathematics*, 13:1–14, 1960.

[25] K. Zuse. *Rechnender Raum*. Friedrich Vieweg & Sohn, Braunschweig, 1969.

9. Physik, Biologie und Mathematik: Grundbegriffe, Skalen und Allgemeingültigkeit

J. Leo van Hemmen[1]

Abstract

Die Mathematisierung biologischer und insbesondere neurobiologischer Wirklichkeit erzeugt a priori, aber wie hier gezeigt wird, unberechtigt, einen starken Widerstand. Auf Basis dreier Arbeitshypothesen bzw. Thesen wird nun das Tor zur Mathematisierung geöffnet. Erstens, eine mathematische Beschreibung physikalischer oder biologischer Realität braucht geeignete Grundbegriffe, ohne die sie nicht funktionieren kann. Zweitens, jede mathematische Formulierung experimentell vorgegebener Fakten gilt auf einer begrenzten Skala in Raum und Zeit. Drittens, universelle Gültigkeit mathematischer Beschreibung ist in der Physik zwar allgegenwärtig, ist aber auch z.b. in der Neurobiologie möglich und gibt es bereits.

A mathematization of natural phenomena never happens by itself but needs the fulfillment of two preconditions, which are specified here. First, appropriate key concepts must be found that are intimately connected with the phenomena one wishes to describe and explain mathematically. Second, the scale on, and not beyond, which a specific description can hold must be specified. Different scales in

---

1 J. Leo van Hemmen ist Emeritus des Lehrstuhls für Theoretische Biophysik neuronaler Informationsverarbeitung an der Technischen Universität München. Er ist Zweitmitglied der mathematischen Fakultät der TUM und befasst sich eingehend mit der Frage, wie Perzeption verschiedener Sinnesorgane wie Sehen (normal wie Infrarot bei Schlangen), Hören (Schallortung bei Vögeln und Reptilien) und, allgemeiner, Mechanosensorik (z.B. Beuteortung des Wüstenskorpions, Seitenlinien-System des Krallenfrosches *Xenopus* und der Fische) sowie multimodale Integration dieser Sinne auf Basis der vorgegebenen, sensorischen Biophysik begrifflich verstanden und mathematisch beschrieben werden kann.

space and time allow for different conceptual and mathematical descriptions. This is the scaling hypothesis. Furthermore, the question is analyzed as to whether a mathematical description can be universally valid and, if so, how? As an illustration we put forth the argument that universals exist not only in physics, a generic example, but also in theoretical neuroscience, that evolution proves the rule there, and that theoretical neuroscience is a domain with still lots of space for new developments initiated by an intensive interaction with experiment.

9.1   Vorrede

Gibt es ein Geist-Gehirn-Problem? Dies ist eine Frage, die Philosophen seit Jahrhunderten fasziniert. Falls es ein Geist-Gehirn-Problem gibt, was ist dann eigentlich das Problem? Das Gehirn. Ein wichtiger Aspekt der Gehirn-Geist Problematik und des Anspruchs der Neurowissenschaften, den Menschen in seinem Erleben und Verhalten zumindest im Ansatz erklären zu können, hängt mit der Frage zusammen, ob die Neurowissenschaft mathematische Beschreibungen erlaubt, und, ob erwartet werden kann, dass eine Wechselwirkung zwischen experimenteller und theoretischer Neurowissenschaft für beide vorteilhaft ist. In der Tat existiert eine unübersehbare Vielzahl an Daten, deren Struktur nur durch mathematische Verfahren aufgedeckt werden kann.

Es wird hier argumentiert, dass eine Mathematisierung natürlicher Phänomene niemals von alleine kommt. Zunächst muss man nämlich geeignete Grundbegriffe finden, die mit dem Phänomen, das man mathematisch beschreiben und erklären möchte, eng verbunden sind. Zweitens muss man die geeignete Skala festlegen, auf der eine bestimmte Beschreibung gelten kann und jenseits derer sie nicht gilt. Unterschiedliche Skalen lassen unterschiedliche begriffliche und mathematische Beschreibungen zu. Dies ist die Skalenhypothese. Drittens, kann eine mathematische Beschreibung allgemeingültig sein, und, wenn ja, wie? Hier bringen wir das Argument vor, dass Universalien auch in der theoretischen Neurowissenschaft existieren, dass Evolution die Regel bestätigt, und, dass es sich um ein Gebiet handelt, in dem noch viel Platz ist für neue, mathematisch initiierte Begriffsbildung, die durch eine intensive Wechselwirkung mit dem Experiment eingeleitet wird. Schließlich erhält man einen tiefen Einblick durch eine sorgfältige Analyse der Weise, in der bestimmte Gehirnstrukturen auf Wahrnehmungs-Input antworten und damit eine Aktion in der Umgebung eines Tieres veranlassen.

9.2 Einführung: Wie lauten die Fragen?

Die Biologie, insbesondere die Neurobiologie, auf die wir hier unsere Aufmerksamkeit richten werden, ist eine ziemlich facettenreiche Wissenschaft mit einer überwältigenden Menge an Fakten und Begriffen, aber nur wenigen allgemeingültigen Leitprinzipien. Eine noch *geringe* Rolle spielt die Mathematik. Die grundlegende Frage, die wir im vorliegenden Aufsatz betrachten wollen, ist, ob allgemeingültige Prinzipien existieren und, falls ja, ob sie durch mathematische Ausdrücke formuliert werden können. Des Weiteren lohnt es sich, darüber nachzudenken, ob die vorherige Frage in einem derart allgemeinen Kontext gestellt werden kann. Wir werden sehen, dass Konzeptualisierung, Skalierung und Allgemeingültigkeit die drei zur Orientierung nötigen Eckpfeiler sind. Wie sich herausstellen wird, beschränken diese auch den Gültigkeitsbereich unserer Argumente, und zwar wesentlich.

Jedes Gebiet der Wissenschaft besitzt seine eigenen Grundbegriffe basierend auf einer gewaltigen Fülle an Tatsachen. Die Wissenschaftsgeschichte kann uns darüber aufklären, warum und wie diese Grundbegriffe zustande kamen und worauf sie hinaus laufen. Wir alle wissen, dass Mathematik existiert, und viele von uns wissen sogar um ihre Stärken. Aber können wir und, wenn ja, *wie* können wir beim Aufzeigen ihrer Bedeutung die Stärke ersichtlich werden lassen? Falls die Natur für quantitative Analysen zugänglich ist, so ist Mathematik der *einzige* Weg, um die Natur zu quantifizieren. Das heißt, sie ist der einzige Weg, um quantitative Theorien zu formulieren, die beschreiben oder gar vorhersagen, was bei geeignet gewählten Anfangs- und Randbedingungen geschehen wird. Im Grunde bedeutet „quantitativ" den Gebrauch von Zahlen, um den Wert der gemessenen oder zu messenden Größen festzulegen, und Zahlen sind naturgemäß bereits wesentlicher Bestandteil der Mathematik. Man kann 1/7 als natürlich betrachten, da 1 durch eine positive ganze Zahl dividiert wird, nicht aber die Quadratwurzel von 2.

Das Ziel des vorliegenden Aufsatzes ist, zu zeigen, dass die Quantifizierung der Natur nicht von alleine kommt. Wir werden die Physik, insbesondere die Mechanik als konkretes Beispiel nehmen, um zu verdeutlichen, dass man zuerst passende *Grundbegriffe* finden muss, ehe man Naturphänomene in Form einer konkreten mathematischen Beschreibung quantifizieren kann. Dabei werden wir auf verschiedene Größenordnungen der Phänomene in Raum und Zeit, auch *Skalen* genannt, stoßen. Es macht einen großen Unterschied, ob wir einen Fußball als Ansammlung von Atomen und Molekülen beschreiben wollen oder als (normalerweise) runde elastische Hülle. Ist des Weiteren die Mechanik eine universelle Theorie, um sowohl Fußbälle als auch Kanonenkugeln auf allgemeingültige Weise zu behandeln? In anderen Worten, was bedeutet Allgemeingültigkeit und gilt sie immer und überall?

Das heißt, können wir uns *Allgemeingültigkeit* als eine der mathematischen Beschreibung innewohnende Eigenschaft vorstellen oder hängt sie von der Natur und Ausdehnung des Untersuchungsgegenstandes ab?

Noch bevor wir wirklich loslegen, haben wir bereits drei Begriffe kennengelernt, die unserer sorgsamen Aufmerksamkeit bedürfen. Zuerst den des *Grundbegriffs* und wie er geprägt wird, dann die *Skalen*, auf denen wir bestimmte Phänomene analysieren, und schließlich müssen wir die Frage beantworten, ob *Allgemeingültigkeit* existiert und, wenn ja, was sie bedeutet. Nachdem wir durch Klärung unserer Ideen bezüglich der Rolle von Grundbegriffen, Skalen und Allgemeingültigkeit den Boden bereitet haben, werden wir uns der Analyse von Verträglichkeit zwischen Mathematik und Neurowissenschaft zuwenden. Am Ende dieser Abhandlung sind Schlussfolgerung und Ausblick zu finden.

9.3 Mathematisierung der physikalischen Wirklichkeit: Prägen von Grundbegriffen und Unterscheidung von Skalen

Wissenschaft ist ein Streben, eine aufklärende Expedition, um „logische" Erklärungen für Phänomene zu finden, die in der uns umgebenden Welt auftreten. Solch ein Streben ist wie die Suche nach Orientierungspunkten und sodann nach Aussichten auf eine noch unbekannte Landschaft. Man kann falsche Richtungen einschlagen, die zwar Erkenntnisse versprechen, aber ins Nichts führen. Dennoch weiß man erst im Nachhinein, dass sie „falsch" waren. Was wir hier nicht analysieren werden, aber was man immerwährend im Kopf behalten sollte, ist, dass viele gelehrsame Streitpunkte, die im Laufe der Wissenschaftsgeschichte auftauchten, wie etwa der *horror vacui*, Epizyklen, *minima naturalia*, Phlogiston etc., hitzig debattiert wurden und sich dann früher oder später als irrelevant auflösten. Was die Physik betrifft, so sollte der Leser die Literatur (Dijksterhuis 1956, Simonyi 2012, Smolin 2006) konsultieren, um sich zu informieren, was *nicht* funktionierte. Im vorliegenden Kontext können wir aber nicht anders, als uns auf das zu konzentrieren, was *schon* funktionierte.

Wie erhält man die mathematische Beschreibung eines natürlichen Phänomens? Diese Frage faszinierte die Griechen schon 500 v. Chr. mit Pythagoras und fand ihre Krönung in den tiefgründigen Ergebnissen Archimedes' (Syrakus, 287-212 v. Chr.). Die Mathematisierung der Natur ist seither eine faszinierende Frage und daher scheint es angemessen, die Geschichte der Wissenschaft und insbesondere die der Physik sorgfältig zu analysieren. Denn die Physik weist klar einen wesentlichen Aspekt auf, der für sich genommen eine sorgfältige Analyse verdient: Die Prägung geeigneter Begriffe in Verbindung mit der zugehörigen Mathematik. Bei unserer Analyse werden wir von der fundamentalen Studie über die Entwicklung der Mechanik ausgeführt von E.J. Dijksterhuis (1956) Gebrauch machen. Die klassische Mechanik – im Gegensatz zur Quantenmechanik (1924-8); siehe dazu z.b. Simonyi (2012) – war das erste und lange Zeit führende Gebiet der Physik und zeigt in beispielhaftem Ausmaß, wie Mathematik in einem physikalischen Bereich zur Anwendung kommt, um Geschehnisse in der Natur, d.h. in „natürlichen" Phänomenen zu quantifizieren. Aus diesem Grund war Dijksterhuis' Wahl ein ausgezeichneter Schritt und machte seine wichtigste Arbeit (Dijksterhuis 1956) zu einem Klassiker. Anstelle einer Definition des Begriffs „natürlich", über die man gut einen gesonderten Aufsatz schreiben könnte, die aber weitgehend vom Geschmack des Autors abhängt und damit praktisch keine Bedeutung hat, konzentrieren wir uns darauf, was „klassisch" im Sinne der Newtonschen Mechanik bedeutet. Für alle Details zu den untenstehenden Argumenten, einschließlich Stevin, wird der Leser auf Dijksterhuis (1956) verwiesen.

Um das sogenannte zweite newtonsche Gesetz zu verstehen, müssen wir auf Stevin zurückgehen, der im 16. Jahrhundert als erster den Vektorcharakter von Kräften wie der Schwerkraft klar erkannt hat. Grob gesprochen bedeutet das, dass in dem dreidimensionalen Raum, in dem wir leben, jeder Vektor drei Komponenten besitzt, so dass, wenn $\mathbf{v} = (v_1, v_2, v_3)$ und $\mathbf{w} = (w_1, w_2, w_3)$ zwei Vektoren sind, die Summe der Vektoren $\mathbf{v} + \mathbf{w} = (v_1 + w_1, v_2 + w_2, v_3 + w_3)$ ist. Das heißt, wir addieren Vektoren komponentenweise. Was jetzt vielen offensichtlich erscheint, war alles andere als das, als Stevin 1586 seine *Grundlagen der Kunst des Wiegens* veröffentlichte und den Vektorcharakter von Kräften klar herausstellte, der für ihn eine Hypothese war, mit der er die physikalische Welt um ihn herum erklären konnte; siehe Abb. 9.1.

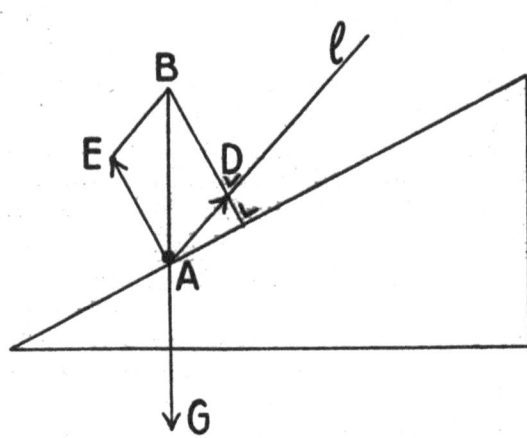

Abb. 9.1: Simon Stevins (1586) Illustration des Vektorcharakters einer Kraft, die aufgrund
des Gewichts **G** einer Masse auf einer schiefen Ebene auftritt. Wie angezeigt kann
eine Kraft in zwei Komponenten zerlegt werden. Vektoren wie **D** und **E** sowie **B** als
deren Zusammensetzung können entweder komponentenweise entlang der karte-
sischen horizontalen und vertikalen Achse addiert werden oder man benutzt al-
ternativ die Parallelogrammregel für die Pfeile **D** und **E**, um **B** zu erhalten. Somit
liegt Stevins Zeichnung in einer zweidimensionalen Ebene, wobei **B** in die zwei
Vektoren **D** und **E** zerlegt ist.

Nun kommt der Grundbegriff, auf den die Physik zwei Jahrtausende (Dijksterhuis 1956) warten musste, ehe er es Newton erlaubte, sein zweites Gesetz zu formulieren. Wir beginnen mit dem Geschwindigkeitsvektor **v** = d**r**/dt, wobei **r** der
Ortsvektor ist, der im Allgemeinen von der Zeit t abhängt, also **r** = **r**(t). Um die
Geschwindigkeit zu erhalten, benötigte Newton einen neuen, diesmal mathematischen Begriff, nämlich den der Differentiation d**r**/dt nach der Zeit, welche er und
Leibniz unabhängig voneinander erfanden. Die Einheit der Geschwindigkeit ist,
sagen wir, Meter/Sekunde (m/s). Der Begriff des Vektors und der Vektoraddition,
d.h. komponentenweiser Addition, existierte bereits für Kräfte und war von Stevin
klar vermerkt und veröffentlicht worden. Wir multiplizieren dann die Geschwindigkeit eines Teilchens mit der Teilchenmasse m, um so den Impuls **p** = m**v** zu erhalten. Der Begriff des Impulses war für die Wissenschaft zu Newtons Zeiten völlig
verblüffend. Und ebenso im Rückblick: Warum soll man eine Größe der Einheit
m/s mit einer anderen der Einheit Kilogramm (kg) multiplizieren, um so **p** = m**v** zu
erhalten? Das hat kaum einen Sinn, bevor wir uns Newtons zweitem Gesetz zuwen-

den: $F = dp/dt$ mit F als die Kraft. So einfach das aussieht, warum sollte das so sein? Bis heute weiß das niemand, aber, und das ist die einzige Erklärung, es funktioniert und zwar seit Jahrhunderten. Sowohl Impulse als auch Kräfte sind Vektoren. Warum? Sie sind es einfach. Simon Stevin schloss schon 1586, dass Kräfte so geartet sind; siehe Abb. 9.1.

Es gibt natürlich auch eine andere Antwort auf die Warum-Frage. Der Mythos von Newtons Entdeckung, dass der Impuls der Grundbegriff ist, um mathematisch ein Gesetz zur Beschreibung der Kraft zu formulieren, ist nicht nur anmutig, sondern enthält auch die „richtige" Idee: Als Newton unter einem Apfelbaum sitzt, wird er von einem herab fallenden Apfel getroffen. Da dessen Impuls innerhalb kurzer Zeit vernichtet wird, muss der Apfel eine Kraft auf Newtons (oder unseren eigenen) Kopf ausüben. Bei $F = dp/dt$ anzukommen, ist ein gewaltiger Schritt, aber beim gründlichen Studium zum Beispiel Dijksterhuis' sorgfältiger geschichtlicher Analyse der Entwicklung der Mechanik in den vorangegangenen zwei Jahrtausenden, insbesondere des Jahrhunderts vor Newton, erkennt man die zugrunde liegende Logik. Auf gut Deutsch, Newtons Entdeckung war keineswegs eine „creatio ex nihilo". Dank ihrer mathematischen Natur öffnete sie jedoch auch die Tür zu einer *mechanistischen* Analyse der Phänomene, bei denen Kräfte eine wichtige Rolle spielen, und dies sogar schon zu Newtons Zeiten.

Was lernen wir von $F = dp/dt$ aus Newtons Sicht? Zunächst stellte Newton als Hypothese auch sein drittes Gesetz auf: *actio = -reactio*. Wenn sodann zwei Körper 1 & 2 auf einer flachen horizontalen Ebene ohne Reibung kollidieren, so gibt es keine Nettokraft, da die Gravitationskraft in vertikaler Richtung wirkt und die Summe aller Kräfte in der horizontalen Ebene senkrecht dazu zu allen Zeiten verschwindet: $F_1 + F_2 = 0$. Die Impulse pi befinden sich ebenso in genau derselben Ebene. Wir verwenden nun $F_i = dp_i/dt$ für jede der Massen, $i = 1, 2$, und, indem wir das dritte newtonsche Gesetz anwenden, können wir gar nicht anders, als zu schließen, dass $d(p_1 + p_2)/dt = 0$ und somit $p_1 + p_2$ erhalten bleibt, was Experimente in der Tat schon vor Newton gezeigt hatten. Wenn man über dieses Ergebnis eine Minute nachdenkt, dann kann man erkennen, dass es für Beobachter im 17. Jahrhundert verblüffend war, und eigentlich immer noch ist, dass der *Gesamtimpuls* erhalten bleibt. Aber immerhin können wir jetzt den zugrunde liegenden „Mechanismus" sehen, der zur Impulserhaltung führt, falls es keine äußeren Kräfte gibt.

Mehrere Aspekte des vorangegangenen klassischen Arguments sind bemerkenswert. Zunächst sollen die Begriffe Impuls und Kraft auf mechanistische Art interpretiert werden. Nur, wenn eine Kraft wirkt, ändert sich der Impuls und zwar gemäß $F = dp/dt$. Zugegeben, Impuls und Kraft spielen in der Mechanik eine wesentliche Rolle, aber das ist hier nicht gemeint. Wir wagen nicht, eine Kraft zu definieren, sondern verweisen einfach auf Abb. 9.1, um zu zeigen, dass sie real ist, weil

sie verwendet, zerlegt und mit unseren Sinnen erfasst werden kann. Die Gewichts-
kraft aufgrund einer Masse kann gewogen werden. Wenn wir sie fallen lassen, so
wie Galileo es angeblich getan hat, können wir ihre Beschleunigung messen, ihre
Geschwindigkeit und damit ihren Impuls bestimmen, der vernichtet wird, wenn sie
am Boden aufschlägt, so dass sie eine Kraft ausübt. Auf diese Weise ist Gravitation
der Mechanismus, um Beschleunigung zu erzeugen, so wie der Impuls der relevante Begriff ist, der die Wirkung einer Kraft bei der Erzeugung von Beschleunigung
zu quantifizieren hat.

Um unsere Argumente auf das Wesentliche zu fokussieren, ohne den Grundbegriff des *Impulses*, einem Vektor, hätte Newton sein allgemeingültiges zweites Gesetz nicht formulieren können. Wir werden bald auf seine „Allgemeingültigkeit"
zurückkommen, akzeptieren sie im Moment aber und erinnern uns einfach zum
Beispiel an Architektur, die uns tagtäglich die Gültigkeit des zweiten newtonschen
Gesetzes vor Augen führt, vorausgesetzt, dass die Praktiker ihre Hausaufgaben
richtig gemacht haben. Wie Dijksterhuis (1956) überzeugend gezeigt hat, benötigte
die Physik in der Tat zwei Jahrtausende, ehe sie zu der Einsicht kam, dass Impuls
der „richtige" Grundbegriff ist: $F = dp/dt$. Nebenbei bemerken wir, dass die meisten
Leute das newtonsche Gesetz in der Form $F = ma$ kennen, wobei F die Kraft auf ein
Teilchen der Masse m ist und $a = dv/dt$ die Beschleunigung mit v als Geschwindigkeit. Da die Masse m in der klassischen Mechanik eine Konstante ist und $p = mv =$
$m dr/dt$, bleibt uns noch $F = ma$.

Konzeptualisierung in Form prägender Grundbegriffe. Die Geschichte des Kon-
struierens von Grundbegriffen lässt sich faktisch unbegrenzt fortführen. Sie fand
einen zwischenzeitlichen Höhepunkt bei dem Entwurf der Quantenmechanik in
den zwanziger Jahren des letzten Jahrhunderts. Es war Dirac (1930, 1958), der
ihr eine ausgeprägte Formulierungen gab, in der Observablen wie etwa der Impuls $p = (p_x, p_y, p_z)$ und der Ort r, üblicherweise geschrieben als $r = (q_x, q_y, q_z)$,
wobei p_x, q_x, usw. Zahlen sind, jetzt zu Operatoren werden, die eine nichttriviale
Kommutator-Relation $[q_j, p_j] = q_j p_j - p_j q_j = ih/2\pi$ erfüllen, wobei $j = x, y, z$ und h
das Plancksche Wirkungsquantum ist, während alle anderen Kommutatoren verschwinden; siehe auch Messiah (1961, Ch. 4). Aufbauend auf einer Darstellung
der Kommutator-Relationen hat man die Wellenfunktion à la Schrödinger, die der
Schrödinger-Gleichung gehorcht, so dass man eine dynamische Entwicklung in der
Zeit erhält und ihre äußerst erfolgreiche Wahrscheinlichkeitsinterpretation insbesondere des Messprozesses, die jetzt „Kopenhagener Deutung" genannt wird (Born
in Göttingen als Vorgänger und insbesondere Bohr und seine Kollegen an der Universität von Kopenhagen). Grundbegriffe und ihre innewohnende mathematische
Formulierung durchziehen somit die gesamte Physik.

Bevor wir weitergehen, ist es vielleicht ganz gut, die obenstehende Idee des Prägens von Grundbegriffen, die eine neue theoretische, d.h. mathematische Beschreibung anstoßen, Heisenbergs (1931) „Folge abgeschlossener Theorien" gegenüberzustellen. Wissenschaftsgeschichte zeigt, dass Theorien, um erfolgreich zu sein, zwar nicht abgeschlossen sein müssen, aber durch Verwendung ihrer Grundbegriffe eine vollständige mathematische Beschreibung der physikalischen und im vorliegenden Fall biologischen Realität an die Hand geben und dabei keine Widersprüche enthalten sollten. Durch ein stetiges Zusammenspiel von mathematischer oder, in anderen Worten, theoretischer Beschreibung mit experimenteller Verifikation reifen physikalische Theorien, bis sie an die Grenzen ihrer Gültigkeit stoßen, wie etwa Skalen in Raum und Zeit, außerhalb derer sie ihre Bedeutung verlieren. So wie in Diracs und Heisenbergs Fall, in dem sich klassische Mechanik auf atomarer Ebene als ungeeignet herausstellte.

Allgemeingültigkeit. Wie allgemeingültig ist, sagen wir, das (zweite) newtonsche Gesetz? In der klassischen Mechanik der Architektur, bei Kanonenkugeln à la Stevin und in der makroskopischen Physik im Allgemeinen hat sich das Gesetz immer als gültig erwiesen. Auf atomarer Skala, die acht Größenordnungen (10^8 mal) kleiner ist, gilt es allerdings nicht. Stattdessen müssen wir mit Quantenmechanik arbeiten, welche über einen völlig anderen Formalismus läuft, der in den Goldenen Zwanzigern des letzten Jahrhunderts aufgedeckt wurde, aber nach wie vor extrem nützlich ist. Das heißt, die Wissenschaftsgeschichte lehrt uns, dass zumindest in der Physik mathematische Formulierungen nur auf einer bestimmten Skala in Raum und Zeit gelten. Quantenmechanik kann *nicht* von der klassischen Mechanik abgeleitet werden. Ihr mathematischer Formalismus einschließlich der Feinheiten ihrer experimentellen Interpretation, ohne welche sie nicht bestehen kann, existiert mit eigener Berechtigung. Lediglich umgekehrt lässt sich die klassische Mechanik eines makroskopischen Körpers in gewissem Umfang aus der Quantenmechanik herleiten, aber die zugrunde liegende Mathematik ist in hohem Maße nichttrivial, um es sachte auszudrücken.

Das bedeutet, wir müssen verschiedene *Skalen* unterscheiden: Für makroskopische Körper die der klassischen Mechanik und für Atome die der Quantenmechanik. Für Elementarteilchen müssen wir noch mal acht Größenordnungen heruntergehen, so dass wir bei Quanten*feld*theorie (QFT) enden. Und wieder betreten wir ein anderes Regime mit verschiedenen Regeln, die nicht aus der Quantenmechanik hergeleitet werden können. Was den Ursprung des Universums betrifft, müssen wir andere Zeitskalen unterscheiden als die, mit denen wir im täglichen Leben vertraut sind. Auf jeder Skala begegnet man neuen Regeln, die von neuen Grundbegriffen, welche sich nicht von den „gröberen" ableiten lassen, herrühren und dennoch eng mit diesen verbunden sind. In der Regel kann man, wenn man von feineren zu gröberen Skalen in der Physik übergeht, einige der mathematischen Gesetze, die auf der „gröberen" Skala gelten, herleiten, aber nicht mehr als das und zwar trotz der gewaltigen Literatur, die verschiedene Aspekte des Übergangs von Quantenfeldtheorie zu Quantenmechanik und von Quantenmechanik zur klassischen Mechanik behandelt. Nur am Rande sei bemerkt, dass der Begriff des „Funktional"-Integrals (Dirac 1958), der damit zusammenhängt, dass es über einem Funktionenraum anstatt des üblichen dreidimensionalen Raums definiert ist, physikalisch extrem nützlich ist, aber mathematisch noch viele lose Enden hat. Kurzum, es existieren Verbindungen, aber es gibt sozusagen keine breite Brücke zurück in Richtung gröberer Skalen.

9.4   Kann man Neurobiologie mathematisieren und, wenn ja, wie?

Wenn wir bedenken, warum die Physik so erfolgreich war, dann können wir von ihrer reichen Erfahrung über die Jahrhunderte lernen (Dijksterhuis 1956): Eine Theorie muss nicht ausschließlich auf experimentell verifizierten Tatsachen bauen, sondern kann auch mathematische Prinzipien aufdecken, was für den Moment eine Hypothese aufzustellen bedeutet, welche zu einer konsistenten Beschreibung von Experimenten führen. Das heißt, vom hier vertretenen Standpunkt aus sollte sie einen Vorhersagewert haben, so dass ein Teil einer Theorie durchaus *prae facto* anstatt *post factum* sein darf und somit zu experimenteller Verifikation auffordert; vgl. Dijksterhuis (1956), Mach (1904) und insbesondere Smolin (2006). Es war genau dieser konstruktive Austausch zwischen Theorie und Experiment, der (wohl) die Physik zur Vorzeigeunternehmung des zwanzigsten Jahrhunderts gemacht hat. Wer könnte bestreiten, dass es für praktisch die gesamte Biologie, die auf quantitative Beschreibung der natürlichen Welt abzielt oder davon Gebrauch macht, keine ähnliche Geschichte geben wird?

Gegeben eine bestimmte Skala, sagen wir, die der klassischen Mechanik oder Quantenmechanik, gelten die physikalischen Gesetze ohne Ausnahme. In der Biologie gibt es allgemeingültige Regeln und Mechanismen, aber man muss mit Ausnahmen leben, welche die Regel „bestätigen" (engl. exceptions proving the rule[2]), denn die Evolution mag Lösungen finden, die in einer bestimmten Situation „bessere" Arbeit leisten als die „allgemeine" Lösung. Um die Aussage zu erläutern, dass

---

2   Dieses "proving" stammt von lat. "probare", was prüfen bedeutet; siehe Rall, J.E. 1994. Proof positive. *Nature* 370: 322. Seine Schlussfolgerung ist jedoch falsch.

in der Neurobiologie auf einer geeigneten Skala in Raum und Zeit allgemeingültige Gesetze existieren, wenden wir uns drei anschaulichen Beispielen zu, die wir der Literatur (van Hemmen 2007) entnehmen.

Erstens, Aktionspotentiale oder, kurz, Spikes werden durch koordinierte Aktivität vieler Ionenkanäle erzeugt. Das Resultat ist ein Spannungsimpuls mit einer Amplitude von 1/10 Volt (V) und einer Dauer von ungefähr einer Millisekunde (ms). Es gibt kaum Zweifel daran, dass einzelne Ionenkanäle erstaunlich detailliert im Kontext biologischer Physik beschrieben werden können. Wie man dann mathematisch präzise das Spike-erzeugende Verhalten einer Ansammlung von hunderten Ionenkanälen erfasst, ist nach wie vor außer Reichweite der theoretischen Neurobiologie und biologischen Physik. Dementsprechend ist die relevante Skala die neuronale und nicht die der Ionenkanäle und wir richten unser Augenmerk auf ein Neuron als *Schwellenelement*, was bedeutet, dass es nur dann ein Aktionspotential generieren kann, wenn sein Membranpotential einen Schwellenwert überschreitet. Dieser Begriff erwies sich als extrem fruchtbar. Er führte nicht nur zu formalen oder McCulloch-Pitts (1943) Neuronen, welche in diskreten Zeitschritten von 1 ms arbeiten und entweder 1 für aktiv, d.h. Spike-Erzeugung, oder 0 für den inaktiven Zustand ausgeben, sondern auch zu Hodgkin and Huxley (1952), deren Werk ihnen den Nobelpreis einbrachte und eine überwältigende Fülle an hoch-detaillierten Neuronenmodellen anstieß. Diese Modelle beschreiben viele unterschiedliche Situationen, aber alle weisen effektiv einen Schwellenwert auf und die meisten von ihnen spiegeln auf die eine oder andere Weise die mathematische Struktur wider, welche von Hodgkin und Huxley entwickelt wurde. Die beiden hatten ihr Gleichungssystem zur Beschreibung von Aktionspotentialen im Riesenaxon des Tintenfisches nicht aus Grundprinzipien hergeleitet, sondern es sich schlichtweg anhand einer komplizierten numerischen Passung (fit) ausgedacht.

Zweitens, Lernen geschieht im Allgemeinen an den Synapsen im Kontext neuronaler Dynamik. Die sogenannte *Spike-timing-dependent plasticity* (STDP) erwies sich als allgemeingültiger Mechanismus, um synaptisches Lernen zu erklären. Seine Schlüsselidee (Gerstner et al. 1996, Markram et al. 1997) ist das *Lernfenster*. Für eine erregende Synapse bedeutet das, dass, wenn das postsynaptische Neuron feuert und der präsynaptische Spike geringfügig früher ankommt, die Synapse ihre Arbeit richtig macht und abhängig von der Zeitdifferenz zwischen dem Auftreten der zwei Spikes wird sie mehr oder weniger verstärkt. Wenn andererseits der präsynaptische Spike „zu spät" kommt, d.h. nachdem das postsynaptische Neuron feuerte, dann ist die Synapse zu schwächen; „wer zu spät kommt, den bestraft das Leben". Der wesentliche Bestandteil ist das Lernfenster als eine *Funktion*, welche die Zu- oder Abnahme der synaptischen Übertragungsstärke in Abhängigkeit der Ankunftszeiten von prä- und postsynaptischem Spike beschreibt. Die einzige Sache, die sich von einem Fall, zum Beispiel Typ der Synapse, Gehirnbereich oder Spezies, zum nächsten ändert, ist das Lernfenster. Eine gewaltige Menge an experimentellen Nachweisen hat inzwischen die große Fruchtbarkeit der Idee mit dem Lernfenster gezeigt.

Schließlich wenden wir uns dem dritten Begriff zu, der sowohl die Existenz von Allgemeingültigkeit in der Neurobiologie als auch die Relevanz von Skalen unterstreicht. Es ist die *Populationsvektorkodierung* (Georgopoulos et al. 1986; Velliste et al. 2008) als Mechanismus zur Erklärung, wie Populationen von Neuronen im motorischen Kortex Bewegungsrichtungen der Muskeln und somit der Gliedmaßen kodieren. Man kann dies wohl das „zweite newtonsche Gesetz für kortikale Motoneuronen" nennen. Wie Newtons Gesetz handelt es sich um eine experimentelle Erkenntnis und basiert auf dem mathematischen Begriff des Vektors. Man ordnet jedem Motoneuron i eine Vorzugsrichtung, den Einheitsvektor $\mathbf{e}_i$ zu. Die resultierende Bewegung, welche durch die neuronale Population kodiert wird, ist dann die Vektorsumme (wie in Abb. 9.1) der Vorzugsrichtungen der einzelnen Neuronen multipliziert mit ihrer (momentanen) Feuerrate f_i. Die Richtung ist also gegeben durch die Summe $\Sigma_i f_i \, \mathbf{e}_i$. So einfach das aussieht, die Vorhersagekraft dieser Regel ist beeindruckend und ebenso ihr Nutzen für mathematische Modellierung, das heißt theoretische Neurowissenschaft und computergestützte Anwendungen. Abbildung 9.2 zeigt eine simple, wenn auch hochkomplexe Demonstration seiner Stärke.

Im vorliegenden Kontext werden die beiden Grundbegriffe der Vorzugsrichtung und der momentanen Feuerrate gepaart und die resultierenden Vektoren in einer Vektorsumme kombiniert. Wir dürfen uns dies als ein mathematisches Bindeglied vorstellen, welches die Auswirkung einer Population von kortikalen Motoneuronen beschreibt. Bei einem multiplikativen Faktor von 1000 sind wir mindestens drei Größenordnungen höher und treffen auf ein Gesetz, das nicht aus Grundprinzipien hergeleitet werden kann. Deshalb hat das niemand je hergeleitet, aber tatsächlich gilt es auf der Skala motorischer Aktion, welche die neuronale Skala um mehrere Größenordnungen übertrifft.

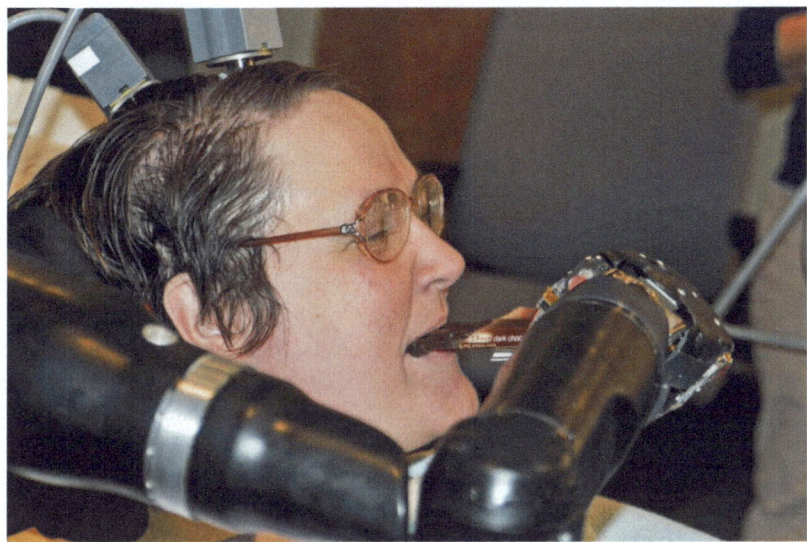

Abb. 9.2:   Eine querschnittgelähmte Frau nutzt Populationsvektoren, um sich mit Schokolade zu füttern. Dieser Erfolg (Collinger et al. 2013) moderner Neurowissenschaft durch den Populationsvektor-Algorithmus macht deutlich, dass ein Gehirn nicht für sich alleine existiert, sondern sich während der Evolution in enger Wechselwirkung mit seiner Umgebung entwickelte. Foto mit freundlicher Genehmigung von Prof. Andrew B. Schwartz (Motor Lab, University of Pittsburgh, PA, USA).

Daher ist es angebracht, die obige Kodierung als zweites newtonsches Gesetz für kortikale Motoneuronen zu bezeichnen. Wir können dies mit dem Verhältnis zwischen klassischer und Quantenmechanik vergleichen, da beide eng miteinander zusammenhängen, wir aber die Wirkung nicht aus Grundprinzipien des jeweils gröber- oder feiner-skaligen Gegenstücks aus herleiten können. Während sich aber in der Quantenmechanik die Skalengröße im Vergleich zur gewöhnlichen, sogenannten klassischen Mechanik verringert, vergrößert sie sich in der Neurowissenschaft, wenn wir von den Ionenkanälen über ihre Gesamtwirkung der Spike-Erzeugung (Feuern) weitergehen zur motorischen Aktion kodiert durch Populationsvektoren, deren Skala die einzelner Neuronen um mehrere Größenordnungen übertrifft. Als ein Algorithmus und wie in Abbildung 9.2 veranschaulicht, kann die Populationsvektorkodierung nicht losgelöst werden vom dem Kontext, für den sie geschaffen wurde: Antriebssteuerung in normalerweise feindlicher Umgebung.

9.5   Was bedeutet Allgemeingültigkeit?

Das zweite newtonsche Gesetz beschreibt die Wirkung irgendeiner Kraft **F** auf irgendein Teilchen mit Masse m und Impuls $\mathbf{p} = m\mathbf{v}$ durch $\mathbf{F} = d\mathbf{p}/dt$. In der Mechanik, Quantenmechanik, Optik, Elektromagnetismus, kurz, in der gesamten Physik sind Naturgesetze allgemeingültig. In der Biologie funktioniert Quantifizierung der Natur geringfügig anders. Obwohl ein Neuron als ein (näherungsweises) Schwellenelement ein allgemeingültiger Begriff ist, gibt es einen Zoo mathematischer Neuronenmodelle (Koch 1999; Ermentrout und Terman 2010) beginnend mit Hodgkin und Huxleys bahnbrechendem Werk 1952, dem nur die verblüffende Arbeit K.F. Bonhoeffers 1948 vorausging, der die meisten seiner Analysen einschließlich einer im zweidimensionalen Phasenraum in Leipzig während der frühen 1940er Jahre durchführte. In einem nächsten Schritt gelangen wir zu einem Aktuator-Algorithmus (Handlungsalgorithmus), der über die Populationskodierung für kortikale Motoneuronen zur Verfügung steht. Was steuert dann diese Motoneuronen? Die aktivierende Aktuator-Geometrie der kortikalen Motoneuronen weist auf eine Hierarchie hin. Wenn man den Hirnstamm hinuntergeht, findet man niedere Motoneuronen (engl. lower motoneurons, LMNs), höhere Motoneuronen (engl. upper motoneurons, UMNs), … Was kommt als nächstes?

Auch in der Biologie ist die Gültigkeit jeder mathematischen Beschreibung auf eine bestimmte Skala in Raum und Zeit begrenzt. Was Neuronen betrifft, so handelt es sich dabei um Schwellenelemente, um sie aber mathematisch zu beschreiben, gibt es sozusagen einen Zoo von mathematischen Modellen für einen Zoo von Ionenkanälen in einem Zoo von Tieren. Die Wirklichkeit ist facettenreich und, um es mit einem Ausdruck der Maßtheorie zu umschreiben, ein mathematisches Gesetz gilt nun „fast immer" anstatt „immer", wobei – für die Experten – das Maß der Ereignisse von der Evolution geeicht wird.

Angesichts all der oben aufgeführten Tatsachen erscheint es sinnvoller, bescheiden zu sein und an der Hypothese festzuhalten, dass zwischen Populationsvektorkodierung und Psychologie mehrere Beschreibungsebenen zu unterschiedlichen Skalen liegen. Kurzum, das ist die *Skalenhypothese* und im Moment wissen wir noch nicht, worum es sich bei diesen Schichten handelt und welche die relevanten Begriffe zur Beschreibung ihres Verhaltens sind. Geschweige denn, welche Mathematik, wenn überhaupt, diese Beschreibungsebenen bestimmt. Nichtsdestotrotz scheint es eine relativ sichere Sache, dass sie existieren[3]. Für ein einzelnes Neuron ist seit Hodgkin

---

3   Hier sollte vielleicht noch Bezug auf Sandra Mitchell's Arbeit *Komplexitäten. Warum wir erst anfangen, die Welt zu verstehen* (2008, Frankfurt am Main: Suhrkamp) genommen werden. Zentral bei dieser Autorin steht der nicht-definierte Begriff „Komplexität"

und Huxley (1952) und nach dem frühen Werk von Bernstein (1908) und Bonhoeffer (1948) bekannt, dass Aktionspotentiale auf die *kollektive* Wirkung von Ionenkanälen zurückgehen, die auf die Spannung, die sie erfahren, reagieren, und, dass es einen reichhaltigen Zoo von mathematischen Modellen gibt, der dem Zoo von Ionenkanälen (Koch 1999; Ermentrout und Terman 2010) entspricht, die in einem bestimmten Neuron existieren. In anderen Worten, ein Aktionspotential wird auf einer Skala erzeugt, die mindestens zwei Größenordnungen über der von Ionenkanälen liegt.

Für die kollektive Wirkung vieler kortikaler Motoneuronen bei der Erzeugung einer motorischen Handlung in einem Muskel wissen wir auch, wie wir ihre Aktion mathematisch beschreiben können, so überrascht wir auch sein mögen, jedem kortikalen Motoneuron eine Vorzugsrichtung, einen Einheitsvektor $\mathbf{e}_i$ zuzuweisen, dies mit der momentanen Feuerrate f_i zu multiplizieren, um $f_i\,\mathbf{e}_i$ zu erhalten, und die Richtung, die ein Muskel veranlasst, durch Summation aller Vektoren $f_i\,\mathbf{e}_i$ zu $\Sigma_i\,f_i\,\mathbf{e}_i$ vorherzusagen. Im Rückblick sieht das alles vernünftig aus, aber warum sollte es so sein? Tatsächlich kann die Populationsvektorkodierung bereits auf der Wahrnehmungsebene (van Hemmen und Schwartz 2008) gefunden werden, was die Konsistenz von Grundbegriffen auf Wahrnehmungs- und Handlungsebene gewährleisten würde.

9.6 Schlussfolgerung: Warum Analogie sehr fruchtbar sein kann

In philosophischen Diskussionen der Neurowissenschaft spielt der Begriff der Skalen noch keine Rolle, obwohl ich behaupte, dass er wichtig, ja sogar wesentlich ist. Wir werden uns nun mit der Frage befassen, inwieweit eine Analogie mit den Skalen der Physik uns beim Verständnis der „inneren" Struktur der Neurobiologie und

---

zusammen mit den beiden Wörtern „integrativer Pluralismus." Da die Welt komplex sei, gäbe es eine Vielzahl von begrifflichen Ebenen, die kontinuierlich verschachtelt die Wirklichkeit erfassen. Als Wissenschaftler solle man versuchen, diese vielschichtige Beschreibung zu integrieren. Die Beziehung zur im vorliegenden Aufsatz ponierten Skalenhypothese ist gar keine. Nur macht die Skalenhypothese mit ihren verschiedenen Skalen und zugehörigen Beschreibungen, die sich nicht eindeutig aus einander herleiten lassen, sofort verständlich, weshalb Mitchell Komplexität eher als verworrenes Netzwerk erfährt und darstellt. Der Begriff Komplexität ist auf diese Art wenig hilfreich. Hier sei eher auf Mainzers (2005) Buch zur Komplexität verwiesen.

ihrer mathematischen Beschreibung weiterhilft. Das vorherige Jahrhundert hat gezeigt, wie man in der Physik immer kleinere Skalen und Theorien entdeckte, die nur funktionieren konnten, weil ihre Mathematik in enger Verbindung mit zugehörigen physikalischen Grundbegriffen ersonnen wurde. Diese Theorien existieren eigenständig. Für die Neurowissenschaft behaupte ich das Gegenteil, dass man sich immer größere anstatt immer kleinere Skalen vornimmt und sowohl feststellt, dass es für die neuronale Arbeitsweise auf kleineren Skalen ein subtil funktionierendes chemisches Substrat gibt, als auch, dass gleichzeitig ein Evolutionsdruck am Werk ist, um Ausnahmen zur Optimierung bestimmter Randbedingungen zu finden.

Im vorliegenden Aufsatz hätten wir auf molekularer Ebene beginnen können, aber stattdessen nahm unsere Analyse die Ionenkanäle als ihren Ausgangspunkt. Als nächstes kommt die synaptische und neuronale Ebene. Ein Lernfenster beschreibt die dynamische Entwicklung synaptischer Übertragungsstärken auf Basis der Ankunftszeiten eines präsynaptischen Spikes und der Feuerzeiten des postsynaptischen Neurons und wir bleiben hier beim einfachst möglichen Kontext. Wir erhalten dann eine allgemeingültige mathematische Beschreibung des Lernens, die auch ein detailliertes Verständnis vieler daraus folgender Lernprozesse erlaubt, wie etwa die Kartenbildung. Eine *Karte* ist eine neuronale Darstellung der sensorischen Außenwelt und ist normalerweise in einem einzelnen anatomischen Kern (Nukleus) geortet. Sie besteht aus vielen Neuronen, wobei „viele" acht im Fall des Wüstenskorpions bedeuten kann und, sagen wir, 10.000 für eine azimutale Schallortungskarte im laminaren Nukleus der Schleiereule, eines der berühmtesten Beispiele (Konishi 1993). Karten unterschiedlicher Modalität werden (im Tectum opticum der Wirbeltiere oder dem Colliculus superior der Säugetiere) integriert und veranlassen Bewegung. Wie wir sehen, vergrößert sich die experimentelle Skala stetig. Es gibt keinen Zweifel daran, dass die neuronale Skala von zentraler Wichtigkeit ist, und, dass sie die Grundlage praktisch aller Überlegungen in der Neurowissenschaft darstellt.

Ebenso gibt es keinen Zweifel daran, dass sowohl qualitatives als auch quantitatives Verständnis der Ionenkanäle auf physikalischen Gesetzen basiert. Indem wir in der Skala aufsteigen, verlieren wir die physikalische Einsicht und gewinnen neue neurowissenschaftliche Begriffe wie etwa Populationsvektorkodierung. Durch stetig größer werdende Skalen können wir nicht anders, als letztlich die Ebene zu erreichen, auf der wir denken und argumentieren. Aber können wir letzteres vom Standpunkt der heutigen Neurowissenschaft aus verstehen? Nein. Bis jetzt sind weder die geeigneten Grundbegriffe noch die entsprechenden mathematischen Beschreibungen verfügbar. Man könnte auf dem Zusatz „bis jetzt" herumreiten, aber der Gegenstand des vorliegenden Aufsatzes ist, dass dieses Herumreiten bedeutungslos ist, während das Finden der „richtigen" Grundbegriffe eine wahre Herausforderung darstellt. Durch die Skalenhypothese gewinnen wir auch einen Einblick in die Natur dessen, was noch fehlt, und, wie wir uns die fehlenden Verknüpfungen vorstellen dürfen. Vom jetzigen Standpunkt aus ist das Geist-Gehirn-Problem irrelevant. Ein Gehirn liefert sozusagen die Hardware für die Gedanken, die zu dem gehören, was wir Geist nennen, aber die Neurowissenschaft bietet noch kein fundamentales oder mechanistisches Verständnis dafür, was Gedanken sind und wie sie entstehen. Das heißt, unser neurowissenschaftliches Verständnis ist davon noch einige Ebenen entfernt und, über den „Geist" zu spekulieren, ist ebenfalls fraglich.

9.6.1   Wahrscheinlichkeitstheoretisch basierte Beschreibungen

Im vorliegenden Kontext stoßen wir auch auf eine andere, probabilistische Beschreibung, die häufig unter dem Namen *Bayes'sche Wahrscheinlichkeit* läuft, benannt nach dem Pastor Thomas Bayes (1702-1761), der als einer der ersten mit bedingten Wahrscheinlichkeiten arbeitete. Nicht mehr und nicht weniger. Es ist aber wichtig, zu erkennen, dass man durch das Einbeziehen von Wahrscheinlichkeiten ausdrücklich *mangelndes Wissen* bezüglich des betrachteten Systems zulässt. Dieses Wissen kann auch nicht erworben werden, da man sonst genau das machen und die Wahrscheinlichkeiten glücklich weglassen würde. Anders ausgedrückt bieten Wahrscheinlichkeiten einen häufig verwendeten Weg, unser Wissen oder vielmehr unseren Mangel an Wissen bezüglich des betrachteten Systems quantitativ darzustellen (de Finetti 1974). Indem man so verfährt, lässt man auch die mechanistische Herangehensweise fallen und ersetzt sie durch eine quantitative Beschreibung, was das Gleiche ist wie beim Würfeln oder Münzenwerfen, und für Bayes'sche Probleme um eine Bedingung erweitert ist, wie für den Münzwurf mit einem österreichischen Euro, bei dem Kopf und Zahl offenbar nicht gleich wahrscheinlich auftreten, die Münze also aufgrund ungleicher Gewichtsverteilung "biased" ist, und man somit Vorwissen bräuchte, um auf Dauer nicht zu verlieren.

Kurz, auf der Grundlage der Geschichte der Physik und einer angemessenen Interpretation[1] der Art und Weise, auf die Mathematik zur Quantifizierung natürlicher Phänomene benutzt wird, kann man durchaus eine oft detaillierte und quantitative Erklärung der biologischen Wirklichkeit erwarten. Das heißt, eine Erklärung der Teile der Biologie, die einer quantitativen Beschreibung zugänglich sind. Die große Verheißung der Zukunft ist nicht die „Mathematisierung" der Biologie als solche, sondern die schöpferische *Wechselwirkung* zwischen experimenteller Biologie und das, was man in Analogie zur Physik einfach theoretische Biologie oder theoretische Neurowissenschaft bezeichnen mag, so dass aus dieser schöpferischen Wechselwirkung neue Grundbegriffe hervorgehen. Die Wissenschaftsgeschichte sagt uns, dass genau das der Schlüssel zum Erfolg ist, nämlich das Finden der richtigen Grundbegriffe, die mathematische Formulierung ihrer „allgemeingültigen" Gesetze und die Bestimmung des Gültigkeitsbereichs in Raum und Zeit. Durch ihren Vorhersagewert laden sie zu neuen Experimenten und experimentellen Paradigmen ein, um ihre Gültigkeit herauszufordern – ein Ursprung wissenschaftlichen Fortschritts so alt wie der in der Mechanik; vgl. z.B. Dijksterhuis (1956).

Bevor wir zum Ausblick kommen, wäre eine Bemerkung dazu angebracht, was die jetzigen Argumente *nicht* bezwecken möchten. Wir argumentieren nicht im Sinne des Paradigmenwechsels von Thomas Kuhn (1962). Ein schönes Beispiel für Letzteres und von Kuhn extensiv besprochen ist die Weise, wie Kopernikus die Sonne anstelle der Erde als Mittelpunkt des Universums behandelte. Es gibt keinen Zweifel, dass dies ein Paradigmenwechsel ist, aber das hat nichts zu tun mit dem Prägen neuer Grundbegriffe wie im Fall des zweiten newtonschen Gesetzes. Es war Newton, der – mechanistisch gedacht – mit seinem zweiten Gesetz die Keplerschen Gesetze herleiten konnte, die dank Kopernikus entstanden waren.

9.6.2   Ausblick

Wie ich an anderer Stelle (van Hemmen 2007) im Detail diskutiert habe, ist der Ausblick in der theoretischen Neurowissenschaft mindestens so gut wie der in der theoretischen Physik. Die Reichweite der Neurowissenschaft in Richtung eines Verständnisses logischer Prozesse ist nach wie vor ziemlich beschränkt, so dass Bescheidenheit hierbei mehr als angebracht ist. Es ist, als ob wir zu Newton zurückkehrten, während er unter dem Apfelbaum sitzt und nachdenkt. Der Apfel fällt und Newton bemerkt ihn. Die Neurowissenschaft kann uns heutzutage viel über Sehen und Greifen gesteuert vom Populationsvektor-Algorithmus (Georgopoulos et al. 1986, Gerstner et al. 1996, van Hemmen und Schwartz 2008) erklären; vgl. Abb. 9.2. Mit anderen Worten, wie Newton den Apfel wahrnimmt und seine Greifbewegung steuert, ist mittlerweile ziemlich gut verstanden. Wie die theoretische Neurowissenschaft die Fülle der Phänomene in der experimentellen Neurobiologie jenseits, sagen wir, der hier behandelten Beispiele mathematisch erfassen kann, ist ihre entscheidende Herausforderung. Inspiriert durch die Geschichte der Physik haben wir nun zumindest eine Vorstellung, worauf wir Wert legen sollten, auch und gerade wenn nicht alles funktioniert, wie es sollte, sowohl in der Biologie als auch in der Physik.

9.6.3   Perspektive

Was können wir dann zum Beispiel über *Bewusstsein* sagen? Ist das ein Problem? Nein, es ist überhaupt keines, lediglich eine Frage der Definition im Auge des Betrachters. Im Kontext der Phänomenologie könnte man den Begriff der Definition durch Beschreibung ersetzen. Man kann einen Satz verwenden (siehe unten), eine Seite, einen Aufsatz (Chalmers 2000) oder ein Buch (Koch 2004; Damasio 2010). Hier ist eine Ein-Satz-Definition: Bewusstsein ist die Fähigkeit sich in einer (üblicherweise) feindlichen Umgebung als autonome Einheit zu handhaben. Natürlich könnte man sich beschweren, dass wir jetzt „sich handhaben" definieren müssen. Autonom agierende Staubsauger zum Beispiel können sich mit Sicherheit nicht handhaben, denn der Besitzer zieht den Stecker aus der Steckdose – und das war's mit der Autonomie. Hier also ein zweiter Satz, falls man ihn wirklich benötigt (der derzeitige Autor hält ihn für überflüssig): „Sich handhaben" bedeutet, auf jede Aktion von Außen angemessen zu reagieren, so dass die Unabhängigkeit gewährt bleibt. Eine unmittelbare Konsequenz ist die Definition von *Kognition*: Kognition ist die Fähigkeit sich in einer (üblicherweise) feindlichen Umgebung als autonome Einheit zu handhaben, indem man sich Erfahrungen aus der Vergangenheit zu Nutze macht. Das lateinische „cognoscere" bedeutet gerade „sich Erfahrung aus der Vergangenheit zu Nutze machen". Man muss sozusagen aus Erfahrung klug werden. Indem wir diesen Weg gehen, haben wir zumindest zwei Probleme weniger.

Danksagung

Der Autor bedankt sich herzlich bei drei Personen: Bei seinem Freund und Kollegen Andy Schwartz, seinem Doktoranden Matthias Krippner und bei seinem Kollegen Felix Tretter, der mit seiner großartigen Tagung zum Thema *Homo neurobiologicus* (München, 21.-22. Juni 2012) und seinem ständigen Engagement und Interesse den Anstoß zu diesem Aufsatz gegeben hat. Außerdem zeigt er sich der Hanns Seidel Stiftung für Ihre großzügige Unterstützung dieses wissenschaftlich herausfordernden und spannenden Unterfangens sehr erkenntlich. Schließlich bedankt sich der Autor bei der Hanns Seidel Stiftung für die Genehmigung diese Überarbeitung seines Essays „Neurowissenschaft und Mathematik: Grundbegriffe, Skalen und Allgemeingültigkeit", der in *Homo neurobiologicus* (Höfling und Tretter 2013) erschien, hier zu veröffentlichen.

Literatur

Chalmers, D.J. 2000. What is a neural correlate of consciousness? In *Neural Correlates of Consciousness: Empirical and Conceptual Questions*, Hrsg. T. Metzinger, 17-40. Cambridge, MA: MIT Press.

Collinger, J.L., B. Wodlinger, J.E. Downey, W. Wang, E.C. Tyler-Kabara, D.J. Weber, A.J.C. McMorland, M. Velliste, M.L. Boninger, A.B. Schwartz. 2013. High-performance neuroprosthetic control by an individual with tetraplegia. *Lancet* 381: 557-564

Damasio, A. 2010. *Self comes to mind: Constructing the Conscious Brain*. London: Heinemann.

Dijksterhuis, E.J. 1956. *Die Mechanisierung des Weltbildes*. Berlin: Springer. Die englische Übersetzung *The Mechanization of The World Picture* erschien 1961 in Oxford bei Oxford University Press. Für das niederländische Original *De mechanisering van het wereldbeeld* (Meulenhoff, Amsterdam, 1950) erhielt der Autor 1952 den niederländischen Staatspreis für Literatur (P.C. Hooft Preis). Die Sicht von Dijksterhuis unterscheidet sich prägnant von klassischen Analysen wie der von Mach (1904). Auf diese Unterschiede brauchen wir hier nicht näher einzugehen, da sie am Ziel vorbeiführen.

Dijksterhuis, E.J. 1970. *Simon Stevin: Science in the Netherlands around 1600*. Den Haag: Martinus Nijhoff.

Dirac, P.A.M. 1st Ed., 1930; 4th Ed. 1958. *Quantum Mechanics*. Oxford: Oxford University Press.

Ermentrout, G.B., D.H. Terman. 2010. *Mathematical Foundations of Neuroscience*. New York: Springer.

de Finetti, B. 1974. *Theory of Probability*, vol. I. London: Wiley.

Gerstner, W., R. Kempter, J.L. van Hemmen, H. Wagner. 1996. A neuronal learning rule for sub-millisecond temporal coding. *Nature* 383: 76-78.

Georgopoulos, A., A.B. Schwartz, R.E. Kettner. 1986. Neuronal population coding of movement direction. *Science* 233: 1416-1419.

Heisenberg, W. 1931. Kausalgesetz und Quantenmechanik. *Erkenntnis* 2: 172-182.

van Hemmen, J.L. 2007. Biology and mathematics: A fruitful merger of two cultures. *Biological Cybernetics* 97: 1-3.

van Hemmen, J.L., A.B. Schwartz. 2008. Population vector code: a geometric universal as actuator. *Biological Cybernetics* 98: 509-518.

Hodgkin, A.L., A.F. Huxley. 1952. A quantitative description of membrane current and its application to conduction and excitation in nerve. *Journal of Physiology* 117: 500-544.

Höfling, S., F. Tretter (Hrsg.) 2013. *Homo neurobiologicus – Ist der Mensch nur sein Gehirn?* München: Hanns Seidel Stiftung, Argumente und Materialien zum Zeitgeschehen 87.

Koch, C. 1999. *Biophysics of Computation*. New York: Oxford University Press.

Koch, C. 2004. *The Quest for Consciousness: A Neurobiological Approach*. Colorado: Roberts.

Konishi, M. 1993. Listening with two ears. *Scientific American* 268(4): 34-41.

Kuhn, T.S. 1962, 1979, 1996, 2012. *The Structure of Scientific Revolutions*. Chicago: University of Chicago Press.

Mach, E. 1904. *Mechanik in ihrer Entwicklung. Historisch-kritisch dargestellt*, 5. Aufl. Leipzig: F.A. Brockhaus.

Mainzer, K. 2005. *Symmetry and Complexity: The Spirit and Beauty of Nonlinear Science*. Singapore: World Scientific.

Markram, H., J. Lübke, M. Frotscher, B. Sakmann. 1997. Regulation of synaptic efficacy by coincidence of postsynaptic APs and EPSPs. *Science* 275: 213-215.

McCulloch, W.S., W.H. Pitts. 1943. A logical calculus of ideas immanent in nervous activity, *Bulletin of Mathematical Biophysics* 5: 115-133.

Messiah, A. 1961. *Quantum Mechanics*, Vol. 1. Amsterdam: North-Holland; Dover, Mineola, NY, 1999.

Simonyi, K. 2012. *A Cultural History of Physics*. Boca Raton, FL: CRC Press.

Smolin, L. 2006. *The Trouble with Physics*. Houghton Mifflin Harcourt. Der Autor ist ein ausgesprochener Kritiker der modernen Stringtheorie. Als solcher ist sein Buch ziemlich kontrovers, unterstreicht aber die entscheidende Bedeutung einer offenen wissenschaftlichen Diskussion.

Stevin, Simon. 1586. *De beghinselen der weeghconst*, Leyden: François van Raphelinghen. Siehe auch Dijksterhuis (1950, 1957, 1961) Abb. 27, und sein Buch zu Stevin (1970).

Velliste, M., S. Perel, M.C. Spalding, A.S. Whitford, A.B. Schwartz. 2008. Cortical control of a prosthetic arm for self-feeding. *Nature* 453: 1098-1101.

10. Warum ist überhaupt etwas und nicht vielmehr nichts? – Ansätze und Perspektiven der Physik und Kosmologie

Harald Lesch[1]

Abstract

Die moderne Naturwissenschaft geht von einer großen Hypothese aus: Die Natur ist ein Ganzes, es gibt keinen besonders ausgezeichneten Ort im Kosmos. Mit anderen Worten, die Gesetzlichkeiten der Natur, die wir auf der Erde, durch den Wettbewerb von Theorie und Experiment, als nicht falsch identifiziert haben, gelten immer und überall im ganzen Universum. Die entsprechenden Schlussfolgerungen für den Anfang von Allem lautet: Er muss sich mit der Theorie beschreiben lassen, die die elementaren Bausteine und Prozesse der Materie erklärt. Es muss möglich sein, das Allergrößte, das Universum in seinem Beginn mit dem Allerkleinsten zusammenfallen zu lassen. Das Universum hat eben klein angefangen und wir können das verstehen.

Modern science is based on a grand hypothesis: Nature is a whole, there is no distinguished place in the cosmos. In other words, the laws of nature, which we on earth have identified as not wrong through an interplay of theory and experiment, hold always and everywhere in the entire universe. Correspondingly, we can state for the beginning of everything: It must be describable by a theory that explains the fundamental building blocks and processes of matter. It must be possible to let

1 ZDF-Wissenschaftsmoderator; Physiker, Astronom, Naturphilosoph, Autor; seit 1995 Professor für theoretische Astrophysik an der Ludwig-Maximilians-Universität in München; seit 2002 Lehrbeauftragter Professor für Naturphilosophie an der Hochschule für Philosophie (SJ) in München; seit 2008 Moderator der Sendungen „Abenteuer Forschung", „Leschs Kosmos" und „Terra X-Faszination Universum" beim ZDF.

the greatest, the universe, at its beginning coincide with the smallest. The universe started small and we can understand that.

Die Frage: „Warum ist etwas und nicht vielmehr nichts?" gehört zu den zentralen Problemen der Philosophie und kann als metaphysische Grundfrage bezeichnet werden. Ein metaphysischer Systementwurf untersucht in seiner klassischen Form die zentralen Probleme der theoretischen Philosophie: die Beschreibung der Fundamente, Voraussetzungen, Ursachen oder „ersten Gründe", die Gesetzlichkeiten und Prinzipien sowie den Sinn und Zweck der gesamten Wirklichkeit bzw. allen Seins.

Kurt Gödel, der sich mit den Grenzen mächtiger, formaler Systeme beschäftigte, konnte 1931 zeigen, dass es unmöglich ist, ein System von Aussagen aus sich selbst heraus zu begründen. Entsprechend diesem wichtigen nach ihm benannten Unvollständigkeitssatz gibt es in hinreichend mächtigen Systemen, wie sie beispielsweise die Metaphysik darstellt, Aussagen, die weder formal zu beweisen noch zu widerlegen sind. Das heißt: Jedes hinreichend mächtige formale System ist entweder widersprüchlich oder unvollständig beziehungsweise kann nicht zum Beweis seiner eigenen Widerspruchsfreiheit verwendet werden. Damit ist klar, dass die Frage nach den Gründen des Seins nicht zu lösen ist. Aber sie hat ihre Faszination nicht verloren und kann gerade und vor allem in der Physik als Leitfaden dienen und das aus mehreren Gründen. Pointiert formuliert: Es gibt keine Physik ohne Metaphysik.

Das Scheitern dieser ontologischen Kernfrage macht klar, dass auch die Physik als Wissenschaft von der Natur nur unter Voraussetzungen durchgeführt werden kann, die sie selbst nicht begründen kann. Das soll an drei Stichpunkten erklärt werden: Der vermuteten Ordnung in der Natur, den Naturgesetzen und dem ontologischen Realismus.

Die ersten beiden Themen behandeln implizit die Berechenbarkeit der Natur. Die vermutete Ordnung mündet ja geradezu zwingend in ein Netz von Gesetzlichkeiten, deren mathematische Formulierung unmittelbar die quantitative Berechenbarkeit von Naturprozessen nach sich zieht. Allerdings wird sich zeigen, dass fundamentale Grenzen dieser Berechenbarkeit vorliegen, die vor allem in den quantenmechanischen Unbestimmtheitsrelationen formuliert werden. Es liegt von Anfang an, also direkt im Anfang von Allem, die Unbestimmtheit aller physikalischen Grundgrößen vor. Sie ist geradezu die conditio sine qua non, denn ohne die entsprechenden kontingenten Fluktuationen wäre eine Bildung von gravitativ instabilen materiellen Strukturen überhaupt nicht möglich gewesen. Mit anderen Worten, die Existenz aller Planeten, Sterne und Galaxien hing empfindlich von den Anfangs- und Randbedingungen des Anfangs ab und die eben nicht berechenbar waren, sondern mussten fluktuieren. Gleiches gilt für die Entstehung der chemischen Elemente in Sternen durch die Verschmelzung von leichteren Kernen zu immer schwereren Kernen. Nur der kontingente Charakter des sogenannten Tunneleffekts garantiert die verschwindende Möglichkeit einer Annäherung zweier gleichnamigen Ladungen, die sich daraufhin zu einem neuen, schwereren Kern formieren. Angesichts der Bedeutung der Existenz schwerer Kerne wie Sauerstoff, Kohlenstoff usw. für die Möglichkeit, dass sich Leben auf einem Planeten entwickelt, wird auch hier die Stochastizität der Welt zu einer unbedingten Bedingung. Diese und viele anderen kontingenten Beziehungen zwischen Naturkonstanten in der Physik der Elementarteilchen und möglichen Wechselwirkungen zwischen bottom-up Kausalitätsketten (vom Elementaren zum Komplexen) und top-down Vernetzungen (globale Eigenschaften definieren Möglichkeitsräume des lokal Elementaren) machen die grundsätzliche Unberechenbarkeit der Natur zu einer ihrer hervorstechendsten Eigenschaften. Im Übrigen, zeigt sich bereits im einfachsten mechanischen Umfeld die prinzipielle Unkenntnis über zukünftige Abläufe und Wirkungszusammenhänge, wie sie kennzeichnend für Mehrkörper-Probleme sind. Abschließend sei noch daraufhin hingewiesen, dass sich aus der Nichtgleichgewichtsthermodynamik, die sich mit der grundsätzlichen Irreversibilität aller realen physikalischen Prozessketten auseinandersetzt, die prinzipielle Unberechenbarkeit als grundlegend für die komplexe Wirklichkeit ergibt und zwar zwingend. Man kann sozusagen die Unberechenbarkeit berechnen.

10.1 Die drei metaphysischen Säulen der Physik

10.1.1 Die Ordnung in der Natur

Die physikalische Erforschung der Vorgänge in der Natur ist eine quantitative Tätigkeit, d. h. es wird gemessen und gerechnet. Es wird vorausgesetzt, dass sich Mess- und Rechengrößen aus der unendlichen Vielfalt der natürlichen Möglichkeiten so definieren und isolieren lassen, dass man sie einzeln behandeln und darstellen kann. Das setzt voraus, dass es in der Natur, zumindest in Teilen, eine Ordnung gibt, die sich in quantitativer Form wiedergeben lässt. Physik setzt Ordnung voraus, kann sie aber nicht begründen.

Das Modell einer geordneten Natur, in der sich Elemente ineinander verwandeln, in der Kräfte am Werk sind, die unserer Erkenntnisfähigkeit grundsätzlich zugänglich sind, ist der Beginn der griechischen Philosophie, also der vorsokratischen Naturphilosophen. Dieses Modell steht am Anfang der Philosophie und ist bis heute der harte Kern der Physik. Dies gilt insbesondere auch für die Astrophysik, die physikalische Untersuchungsmethode der Strukturelemente des Universums und für die Kosmologie, die Lehre vom ganzen Universum als Untersuchungsobjekt.

10.1.2 Die Naturgesetze

Die Physik geht nicht von einer beliebigen Ordnung in der Natur aus, sondern von einer durch mathematische Gesetzmäßigkeiten strukturierten Ordnung. Die Physik sucht explizit nach diesen Naturgesetzen. Nach Gerhard Vollmers Definition (2003, S. 144ff.) sind Naturgesetze zunächst einmal Beschreibungen von Regelmäßigkeiten im Verhalten realer Systeme. In einem ersten Schritt wird niemand ein Naturgesetz formulieren. Es wird schrittweise, mit immer höherer Präzision auf ein Phänomen angewandt und dabei immer genauer bestätigt. Zumeist zeichnen sich die Naturgesetze dadurch aus, dass sich ihre Anwendung auf Bereiche der Wirklichkeit ausdehnen lässt, an die man anfänglich nie gedacht hätte. Aber auch hier gilt, die Forderung nach solchen universellen, bedingten, synthetischen, relationalen Aussagen, die als wahr akzeptierbar sind, irreale Konditionalsätze stützen können und Notwendigkeitscharakter tragen, lässt sich aus der Physik selbst nicht begründen (Vollmer 2003, S. 164ff.). Die Physik setzt diese Gesetze voraus und hofft, dass sie sich bestätigen lassen.

Die Naturgesetze, die wir auf der Erde entdecken, sollen überall im Universum gültig und auf die Entwicklung des gesamten Kosmos anwendbar sein. Wobei als Einschränkung zu beachten ist, dass es noch weitere Gesetzmäßigkeiten geben kann, die wir noch nicht kennen, die wir aber durch die Erforschung des Universums, mit seinen teilweise extremen Materie- und Energiestrukturen, vielleicht noch entdecken werden. Diese neuen Naturgesetze dürfen denen, die wir bereits kennen und mit Hilfe hoch präziser Experimente bestätigen konnten, nicht widersprechen. Unser Bestand an Naturgesetzen ist die „Minimalausrüstung"; es mag noch andere geben, aber prinzipielle Grenzen wie die Lichtgeschwindigkeit oder das Plancksche Wirkungsquantum müssen auch von neuen Naturgesetzen als Grenzen wiedergegeben werden. Damit ist auch die Anwendung und Erweiterung der Naturgesetze auf den gesamten Kosmos gemeint.

10.1.3　Der ontologische Realismus

Der ontologische Realismus geht von der Grundannahme aus, dass es eine Welt „da draußen" gibt, die in ihrer Existenz und in ihren Eigenschaften unabhängig von unserem Bewusstsein ist. Das Mobiliar dieser Welt ist äußerst sparsam: Raum-Zeit, Materie (Teilchen und Felder) und Energie. Die reale Welt ist evolutionär, vergänglich, zusammenhängend, separabel und quasi-kontinuierlich, d. h., absoluter Zufall ist möglich. Einige Wechselwirkungen sind schwächer als andere, sodass man Dinge zerlegen und voneinander trennen kann oder wenigstens der Fehler nicht gar zu groß ist, wenn man sie als getrennt betrachtet und behandelt, obwohl sie – wie die Quarks – „eigentlich" gar nicht trennbar sind.

Für den ontologischen Realismus spricht vor allem das Scheitern von Theorien. Denn woran sollte eine Theorie scheitern, wenn nicht daran, dass es tatsächlich eine Welt gibt, die anders ist als die Theorie behauptet? Zudem gibt es eine bemerkenswerte Konvergenz physikalischer Forschung, dass nämlich Personen aus völlig unterschiedlichen Kulturen bei ihren Experimenten zu den gleichen Ergebnissen kommen. Die elektrische Elementarladung eines Elektrons hat in indischen Laboratorien exakt den gleichen Wert wie in brasilianischen Forschungseinrichtungen oder im Large Hadron Collider am europäischen Forschungszentrum CERN in der Schweiz. Dies gilt selbstverständlich auch für andere Planeten und andere Sterne in anderen Galaxien. Es gibt keine „Harry-Potter-Inseln" im Universum, in der andere Realitäten existieren, mit anderen Ordnungsparametern und Naturgesetzen als bei uns.

Damit haben wir die wesentlichen metaphysischen Grundlegungen für die Untersuchung der Natur mit physikalischen Mitteln zusammengefasst: Natur – Ordnung – Gesetz.

In einer ersten kurzen Zwischenbilanz kann man also bereits den Wert metaphysischer Spekulationen, obschon grundsätzlich unentscheidbarer Natur, erkennen. Sie liefern nämlich das Fundament, auf dem sich das Unternehmen „Physik als Grundlagenwissenschaft" abspielt. Für die Physik als angewandte Forschung, die zur Entwicklung von Technologien führt, sind metaphysische Aussagensysteme zunächst weniger relevant, weil es sich bei der Technik um ein vom Menschen gemachtes Design handelt. Die Frage nach dem Urgrund technischen Seins lässt sich deshalb auf triviale Weise beantworten.

10.2   Die Bedingung der Möglichkeit, Fragen zu stellen

Die Frage nach dem Urgrund des Seins selbst liefert aber noch weitere Perspektiven, die zu einer ausführlichen Diskussion einladen. Sie führt zu einer interessanten Variante der Kantschen Fragetechnik, nämlich nach den Bedingungen dafür, dass überhaupt etwas möglich ist. Hier ist es die unabdingbare Voraussetzung, dass jemand die Frage nach dem Urgrund des Seins stellen kann. Wer fragt und wonach? Die Frage setzt den Fragesteller voraus und dass diese Frage sinnvoll nach etwas fragt, was vor ihm schon da war.

Wir fragen also nach den Voraussetzungen, die erfüllt sein müssen, damit die Person existieren kann, damit er oder sie die metaphysische Grundfrage stellen kann. Die Antwort lautet: Die Materie, aus der die Person aufgebaut ist, muss stabil sein, d. h., sie darf nicht zerfallen. Außerdem muss die Person mit sich selbst und der Materie um sie herum in dauerhafte und für sie selbst nachvollziehbare Wechselwirkungen treten können. Dies bedeutet, die Person muss sich selbst als Individuum wahrnehmen und ihre Erfahrungen mit sich und der Welt in einem logischen Rahmen einordnen können. Hierzu sind Sinneswahrnehmungen und deren mentale Verarbeitung unerlässlich. Den entsprechenden Verarbeitungsapparat stellt das menschliche Gehirn dar. Es sorgt für die Korrespondenz des Äußeren mit den individuellen Konstruktionen, indem es das Äußere rekonstruiert. Wir wollen hier keine evolutionäre Erkenntnistheorie betreiben, sondern den Bedingungen der Möglichkeit der Existenz einer wahrnehmenden Person nachgehen. Stabilität und Wechselwirkungen der Materie stehen dann auf dem Programm.

Eine der wichtigsten Erkenntnisse in der gesamten Menschheitsgeschichte ist die Entdeckung, dass unsere Welt aus Atomen besteht. Auf dem Planeten Erde hat sich im Rahmen der planetaren Evolution tote Materie in lebendige Materie verwandelt. Die genauen Umstände sind bis heute immer noch Gegenstand der Forschung, es gibt eine ganze Reihe verschiedener Theorien dazu. Unbestritten aber ist: Lebewesen bestehen aus Molekülen, sehr großen Molekülen, deren Grundbestandteile Atome sind. Die Frage nach den materiellen Grundlagen der Person, die die metaphysische Grundfrage nach dem Grund des Seins stellt, lässt sich also übersetzen in das Problem der Stabilität der atomaren Bausteine, die die Person aufbauen. Und mit dieser veränderten Perspektive landen wir bei der wichtigsten Theorie der Physik, der Quantenmechanik. Sie klärt eines der mysteriösesten Geheimnisse der Natur: Warum sind Atome stabil? Dieser Tatsache verdanken wir unser aller Existenz. Sterne, Planeten und Lebewesen bestehen ausschließlich aus den 92 stabilen chemischen Elementen.

Zu Beginn des zwanzigsten Jahrhunderts entdeckte man den inneren Aufbau der Atome: Sie bestehen aus negativ geladenen, sehr leichten Elektronen und positiv geladenen, schweren Protonen, die zusammen mit elektrisch neutralen Neutronen den vergleichsweise winzigen Atomkern aufbauen. Niemand konnte erklären, warum die Atome stabil bleiben und nicht sofort zusammenbrechen, denn schließlich werden die negativ geladenen Elektronen ja von den positiv geladenen Protonen elektrisch angezogen. Eigentlich müssten die Elektronen in den Atomkern stürzen, denn diese elektromagnetische Kraft, die sie in Richtung des positiv geladenen Kerns zieht, würde sie zwangsläufig beschleunigen. Beschleunigte Ladungen aber geben elektromagnetische Strahlung ab. Die Elektronen müssten demnach fortwährend Energie in Form von Strahlung verlieren und langsam aber sicher auf immer tiefere Bahnen um den Kern fallen. Der Sturz in den Kern wäre unvermeidbar. Die klassische Physik wiederspricht damit völlig unserer Erfahrung, dass sich ein Großteil unserer makroskopischen Welt aus stabilen Atomen zusammensetzt.

Die Lösung bietet die Quantenmechanik. Sie stellt eine mathematische Struktur dar, die die Wechselwirkung von Materie und Licht im Rahmen grundlegender Bedingungen beschreibt. Zu diesen Bedingungen gehört die Forderung, dass Teilchen Eigenschaften besitzen, die man sonst nur Wellen zuordnet. Mit anderen Worten, Elektronen verhalten sich unter gewissen Umständen so, als entspräche ihnen eine Welle mit einer bestimmten Wellenlänge. Sie weisen also keine definitive Lokalisierbarkeit auf. Die Quantenmechanik beschreibt die Dynamik der Wellenfunktion von Elektronen, vergleichbar mit der Ausbreitung und Entwicklung von Wellen, die sich auf einer schwingenden Membran ausbilden können. Die Form und Größe dieser schwingenden Fläche grenzt die möglichen Wellenlängen ein. Die Wellen müssen auf der Membran Platz finden. Analog lässt sich das Verhalten der Elektronen im elektrischen Feld eines Atoms verstehen. Dieses Feld hat eine bestimmte Form, die die Dynamik der Elektronen im Atom sehr stark einschränkt, weil sie eben wellenartige Eigenschaften besitzen. Eine Welle ist charakterisiert durch die Frequenz ihrer Schwingung bzw. durch ihre Wellenlänge und ihre Amplitude, d. h. die Auslenkung der Welle aus der Nulllage. Die Aufenthaltswahrscheinlichkeit für das Elektron ist dort am größten, wo die Amplitude der Welle am größten ist. Die Stellen, an denen die Auslenkung beziehungsweise die Wellenamplitude gleich Null ist, bezeichnet man als Knoten. Dort ist die Aufenthaltswahrscheinlichkeit null. Für das Elektron als Teilchen bedeutet das, wir werden es mal hier und mal dort messen, mit entsprechenden Wahrscheinlichkeiten. Nur an den Knoten kann es sich nicht befinden. Der positiv geladene Kern des Atoms stellt eine solche Verbotszone dar. Elektronen *können* sich gar nicht im Bereich des Atomkerns aufhalten! Deshalb sind Atome stabil!

Zusätzlich ergeben sich noch weitere Verbotszonen um den Kern herum, auch dort darf das Elektron nicht sein. Sofort erkennt man das Modell der Energieniveaus der Elektronenhülle. Wird einem Elektron Energie durch elektromagnetische

Strahlung zugeführt, dann wird das Atom angeregt, das Elektron kann dann eine höhere Energiestufe annehmen und die Energie später wieder in genau definierten Portionen abgeben.

Der quantenmechanische Charakter der Materie erklärt also nicht nur die Stabilität, sondern auch die Emission und Absorption von Strahlung durch die Atome. Alle Vorhersagen der Quantenmechanik haben sich mit einem Höchstmaß an Präzision bestätigt. Materie, wie wir sie um uns und in uns haben, verhält sich grundsätzlich quantenmechanisch, d. h., die Welt ist in ihrem physikalischen Grund geprägt von nicht genau zu lokalisierenden, energetischen Strukturen unterschiedlicher elektrischer Ladung und Masse. Ort und Impuls der materiellen Konstituenten sind ebenso grundsätzlich unbestimmt wie Energie und Drehimpuls. Alles schwankt mit einem winzigen, aber endlichen Betrag, geprägt durch das Plancksche Wirkungsquantum. Dieser fluktuierende Charakter der physikalischen Welt ist die fundamentale Bedingung für die Stabilität und Dynamik der Materie und damit auch der Person, die die Grundfrage der Metaphysik stellt.

Bereits die Existenz eines Fragestellers, der aus stabilen Atomen besteht, setzt die Regeln und Prinzipien der Quantenmechanik voraus. Damit sind die notwendigen Vorbedingungen für die Existenz von ‚Etwas‘, als materiell-energetischem ‚Etwas‘, definiert. Denn nach ‚Etwas‘ kann nur gefragt werden, wenn etwas anderes ist, und sein kann es nur, wenn seine Wechselwirkung mit sich und der es umgebenden Welt beschrieben und erklärt werden kann. Die Quantenmechanik bietet uns hierfür ein Instrumentarium.

Es gibt also zumindest eine empirische Teilantwort auf die metaphysische Frage nach dem Urgrund des Seins: Das physikalische Sein ist fundamental quantenmechanisch.

10.3   Von der besten aller möglichen Welten zur Quantenmechanik

Kehren wir nochmals zurück zur Ausgangsfrage: „Warum ist etwas und nicht vielmehr nichts?" Gottfried Wilhelm Leibniz leitete daraus ab, dass wir in der besten aller möglichen Welten leben. Das Beste ist ein extremer Begriff, er bedeutet ‚äußerst‘ oder ‚unübertrefflich‘. Für die Entwicklung der Mechanik hatte die Äußerung Leibniz' die Konsequenz, dass man eine höchst ergiebige Forschungsrichtung in der theoretischen Untersuchung physikalischer Phänomene und Experimente einschlug. Man stellte in verschiedenen Varianten Extremalprinzipien auf, d. h. Prinzipien, die einen Extremwert zum Erwartungswert eines physikalischen Ablaufs machen. Beispielsweise das Prinzip, dass Licht in einem Medium zwischen zwei Punkten Wege nimmt, auf denen seine Laufzeit bei kleinen Veränderungen, mathematisch korrekt: ‚infinitesimalen Variationen‘, des Weges stationär ist. Mit anderen Worten, Licht nimmt zwischen zwei Punkten immer den Weg mit der kürzesten Laufzeit. Die Lichtgeschwindigkeit ist dabei abhängig vom Medium, das durchflogen wird. Stellt man den Photonen unterschiedliche Materialien in den Weg, so wählen sie durchaus einen Umweg, falls damit eine höhere ‚Reisegeschwindigkeit‘ und somit eine kürzere Laufzeit ermöglicht wird. Dieses Prinzip wurde experimentell hervorragend durch das Brechungs- und Reflexionsgesetz bestätigt.

Dieses grundlegende Verständnis zur Ausbreitung des Lichts war sehr hilfreich bei der Überprüfung der Allgemeinen Relativitätstheorie (ART). Diese postuliert, dass Lichtwege durch die Anwesenheit von Massen verändert, soll heißen, verbogen werden. Mit anderen Worten, eine der wichtigsten Vorhersagen der ART, dass die Lichtwege um große Massen verbogen werden, fußt auf einem Extremalprinzip der Physik, das sich mit Hilfe der metaphysischen Frage von Leibniz ableiten ließ. Wieder einmal wurde die Metaphysik zum Motor für die theoretische Physik.

Noch wirkungsvoller wurde dieses Verhältnis von Metaphysik und Physik bei dem berühmten Prinzip von der kleinsten Wirkung. Pierre Maupertuis formulierte dieses in einer ersten Form 1744 (also 30 Jahre nach Leibniz’ Tod).[2] Physikalische Felder und Teilchen nehmen danach für eine bestimmte Größe den kleinsten der möglichen Werte an. Knapp 90 Jahre später brachte William Hamilton das Prinzip der kleinsten Wirkung auf seine moderne Form (vgl. Hamilton 1835). Aus dem Hamiltonschen Prinzip folgen bei geeignet gewählter Wirkung die Newtonschen Bewegungsgleichungen, aber auch die Gleichungen der relativistischen Mechanik, die Maxwellgleichungen der Elektrodynamik, die Einsteingleichungen der Allgemeinen Relativitätstheorie und die Gleichungen, mit denen man die anderen elementaren Wechselwirkungen beschreibt. Vor allem ergibt sich mit Hilfe der auf dem Prinzip der kleinsten Wirkung fußenden Hamiltonschen Dynamik die Grundgleichung der Quantenmechanik, die sogenannte „Schrödinger-Gleichung“.

Die Frage nach dem Urgrund des Seins lässt sich bis zur Entwicklung der Theorie über die Struktur der Materie und ihrer Wechselwirkung mit Licht zurückverfolgen. Beide haben wir bereits als Grundvoraussetzungen dafür identifiziert, dass diese Frage überhaupt gestellt werden kann. Ein naturphilosophischer Hochseilakt!

Nachdem wir damit also den Rahmen der physikalischen Seinsformen (entweder durch Felder und Teilchen oder in Form von Energie, Materie und Strahlung) vor allem mit Hilfe der Quantenmechanik erklärt haben, kommen wir zur größ-

---

2 Vgl. Maupertuis (1744, S. 250).

ten Seinsstruktur, die noch von der Physik untersucht werden kann, dem Kosmos. Auch hier wird die Quantenmechanik eine herausragende Rolle spielen. Doch bevor wir beginnen, noch eine Abschlussbemerkung.

10.4   Abschlussbemerkung vor dem Anfang

Man könnte uns vorwerfen, wir hätten uns mit einem philosophischen Kunstgriff aus der Affäre gezogen. Denn schließlich haben wir die Ausgangsfrage nach dem Urgrund allen Seins für prinzipiell unbeantwortbar erklärt und zugleich seine Bedeutung für die Erlangung empirischen Wissens mittels der Physik betont. Wir haben damit die Metaphysik im Wesentlichen als Erkenntnistheorie identifiziert und uns der Interpretation Immanuel Kants angeschlossen, der in der *Kritik der reinen Vernunft* mit der transzendentalen Methodenlehre die didaktischen und argumentativen Verfahren, die an die Stelle der älteren und dogmatischen Metaphysik treten, ausführlich untersucht hat.

Kant macht Raum und Zeit zu Formen des sinnlichen Anteils der Erkenntnis und damit zu Grundlagen der Mathematik als apriorischer Wissenschaft, aber auch der Naturwissenschaft und der Alltagserkenntnis. Er unterscheidet nicht zwischen einem idealen Raum der Mathematik und einem realen Raum der physischen Wechselwirkung oder zwischen einem realen Raum der Physik und einem phänomenalen Raum des Erlebens. Alle Anschauungen sind nach Kant Empfindungen in einer räumlichen und zeitlichen Ordnung, die den objektiven Beziehungen zwischen den Gegenständen, so wie wir sie erfahren, zu Grunde liegt.

Kant führt aus, dass zur Erkenntnis aber auch bestimmte reine Begriffe, die Kategorien, notwendig sind. Nur durch sie können aus dem sinnlich Gegebenen Gegenstände der Erfahrung werden. Diese Begriffe findet Kant am Leitfaden der möglichen logischen Verknüpfung von Vorstellungen. Durch Anwendung der Kategorien auf Raum und Zeit ergibt sich ein System von Grundsätzen, die a priori gewiss sind und allgemeine Bedingungen für erfahrbare Objekte darstellen, wie z. B. die kausale Verknüpfung aller Erscheinungen. Damit ist die Möglichkeit von Mathematik und Naturwissenschaften gegeben. Der Preis, den wir nach Kant hierfür bezahlen müssen, liegt darin, dass unsere Begriffe nicht auf die Dinge, wie sie „an sich" sind, anwendbar sind, sondern nur, sofern sie die Sinnlichkeit betreffen und somit Vorstellungen in der Ordnung von Raum und Zeit im individuellen Bewusstsein erzeugen.

Zugleich haben Kants Erkenntnisse aber auch Folgen für die Möglichkeit metaphysischer Spekulationen, wie sie die Frage nach dem Grund für das Etwas und für das Nichts darstellt. Denn in dem Versuch der menschlichen Vernunft, dieses Unbedingte des Seins zu erkennen und damit die sinnliche Erkenntnis zu übersteigen, verwickelt sie sich in Widersprüche, da jenseits der Erfahrung keine Kriterien für Wahrheit mehr zugänglich sind. Und dennoch hat die Vernunft ganz offenbar ein notwendiges Bedürfnis, diese Versuche immer wieder aufs Neue zu unternehmen. Kant spricht *expressis verbis* von der „Naturanlage zur Metaphysik", da nur ein solcher Versuch zwischen Erfahrungswelt und dem Subjekt eine sinnvolle Verbindung stiftet. Wir kommen also nicht umhin, immer wieder metaphysische Spekulationen anzustellen, wohlwissend, dass wir keine klaren Antworten auf unsere Fragen an die Welt erhalten können. Warum? Weil wir das Wesen sind, das nach dem Sinn sucht. Und jetzt zum Anfang des Universums.

10.5   Das Rätsel des Anfangs

10.5.1 Kosmologische Fakten

Das Universum hatte einen Anfang. Drei Beobachtungen begründen diesen Standpunkt: 1. Das Universum expandiert. 2. Das Universum ist gleichmäßig in allen Richtungen von Strahlung erfüllt. 3. Die leichtesten chemischen Elemente, Wasserstoff und Helium, finden sich im intergalaktischen Raum, weit entfernt von allen Galaxien, im Verhältnis 3 zu 1.

Alle drei Beobachtungen lassen sich widerspruchsfrei nur in einem Modell zusammenfassen, das einen heißen Anfang des Universums voraussetzt – dem sogenannten Urknall. Die Expansion des Universums ergibt sich nun ebenso zwanglos wie die Strahlung als Überrest der hohen Anfangstemperaturen, abgekühlt auf den heute beobachteten Wert von 2,71 K. Die Existenz und die Zahlenverhältnisse von Wasserstoff- und Heliumkernen ergeben sich im Urknallmodell auf natürliche Weise, sie sind das Ergebnis der Kernfusionsprozesse in einem sich durch Expansion abkühlenden Medium. Rund drei Minuten nach dem Beginn waren die Elemente Wasserstoff und Helium entstanden. Jede weitere Synthese zu schwereren Elementen war unmöglich, weil einerseits die Neutronen fehlten bzw. zerfallen waren und andererseits die Temperatur zu niedrig waren, um noch Elemente mit größeren Kernen wie z.B. Kohlenstoff oder Sauerstoff zu bilden. Kurzum, es gibt genügend empirische Befunde, die sich kaum anders als mit einem sehr heißen Anfang des Universums erklären lassen. Das Universum hatte also einen Beginn.

Was wissen wir über die ersten Entwicklungsstufen des Universums? Wie nahe kommt man dem Anfang mit den Methoden der Physik?

10.5.2 Grenzen physikalischer Erkenntnis

Anfänge definieren immer Grenzlinien. Sie begrenzen und unterscheiden eindeutig das Danach und Davor. Aber was war vor dem Urknall? Was war seine Ursache? Diese Fragen quälen uns schon lange und schon Aristoteles hat folgendes logisches Problem angesprochen: Kann man sich eine Ursache vorstellen, die selbst keine Ursache hatte? Gibt es einen unbewegten Beweger? Muss für die Ursache des Urknalls ein ‚höheres Wesen', als Schöpfer tätig gewesen sein oder müssen wir uns den Kosmos als eine rein zufällige Schwankung vorstellen? Aber was genau soll da geschwankt haben?

Die Physik bietet hier eine ganz einfache Antwort: Wir können niemals wissen, was sich am Anfang des Universums abgespielt hat. Unserer Erkenntnismöglichkeit sind Grenzen gesetzt, die sich aus zwei inzwischen durch kaum noch zählbare Experimente in jeder bis heute denkbaren Hinsicht überprüften Theorien ergeben: der Relativitätstheorie und der Quantentheorie. Mit anderen Worten: Die grundsätzlichsten physikalischen Modelle definieren ebenso grundsätzliche, weil nicht überschreitbare Schranken des möglichen empirischen Erfahrungshorizonts.

Die Anfangsbedingungen des Universums sind mit nichts vergleichbar, was sich auch nur im Entferntesten mit der menschlichen Erfahrungswelt in Verbindung bringen ließe. Das Urknallmodell jedoch stellt zumindest zwei unverzichtbare Bedingungen: Das Universum muss zu Beginn sehr klein und sehr heiß gewesen sein.

Räumliche Kleinheit aber ist eine wichtige Eigenschaft quantenmechanischer Systeme. Hohe Temperaturen entsprechen hohen Geschwindigkeiten, das Maximum ist die Lichtgeschwindigkeit. Ergo beschreiben die Quantenmechanik und die Relativitätstheorie den Anfang des Universums.

Letztlich kann nur eine Vereinigung von Quantenmechanik und Relativitätstheorie die Physik des Urknalls theoretisch behandeln. Leider sind wir noch sehr weit von einer solchen Theorie entfernt.

Aber wir können die Eigenschaften der kleinsten physikalisch sinnvollen kausalen Struktur angeben – die Eigenschaften der Planck-Welt. Interessanterweise definierte Max Planck bereits 1899 ein universell gültiges System von Einheiten, das nur aus verschiedenen Kombinationen der Gravitationskonstanten G, der Lichtgeschwindigkeit c und dem von ihm eingeführten Wirkungsquantum h besteht. Das folgende Zitat aus seiner Publikation „Über irreversible Strahlungsprozesse"

vermittelt einen Eindruck von dem Stellenwert, den Planck diesen Einheiten ein-
räumte: „...*diese Einheiten werden ihre Bedeutung für alle Zeiten und für alle, auch
außerirdische und außermenschliche, Kulturen nothwendig behalten und können daher als ‚natürliche Maßeinheiten‘ bezeichnet werden...*" (1899, S. 479).

Erst später erkannte man die tiefere physikalische Bedeutung dieser ‚Spielerei
mit Naturkonstanten‘, als nämlich Relativitätstheorie und Quantenmechanik als die
wichtigsten Theorien der modernen Naturwissenschaften entwickelt waren. Eine
Zusammenschau der grundsätzlichen Begriffe beider Theorien liefert die gleichen
Ergebnisse wie Plancks Dimensionsanalyse, wie im Folgenden erklärt wird.

Die Relativitätstheorie kennt den Begriff des Ereignishorizontes. Alles, was
hinter ihm verborgen ist, hat keinerlei kausale Verbindung mit dem Geschehen
diesseits des Horizontes. Für einen Körper der Masse M ist der Ereignishorizont
eine wohldefinierte Größe, der so genannte Schwarzschildradius: $R_{SW}=2GM/c^2$. c
stellt die Lichtgeschwindigkeit dar (c=300.000 km/sec) und G ist die Gravitationskonstante (G=6.67·10^{-11} m^3/(sec kg)). Der Schwarzschildradius der Sonne beträgt
drei Kilometer. Angesichts ihres heutigen Radius von 700.000 km wird klar, wie
dramatisch Prozesse sein müssen, die Materie so stark verdichten, dass ein Stern bis
zu seinem Ereignishorizont schrumpft. Verantwortlich für einen solchen Kollaps
ist die Gravitation als einzige physikalische Wechselwirkung, die nur anziehend
wirkt. Wenn ihr keine Druckkräfte mehr entgegenwirken können, kollabiert ein
Stern unter seinem eigenen Gewicht. Dieser Prozess ist beobachtbar, bis der Stern
so klein geworden ist, dass er den Ereignishorizont unterschritten hat. Die meisten
Sterne erreichen dieses Stadium nicht, sondern werden durch gewisse Druckkräfte
letztendlich stabilisiert.

Nur sehr schwere Sterne von einigen zehn Sonnenmassen beenden ihr Da-
sein in einem finalen gravitativen Kollaps. Übrig bleibt ein Schwarzes Loch, von
dem keinerlei Information mehr in den umgebenden Raum gelangen kann. Alles,
was sich innerhalb eines Schwarzen Loches abspielen mag, ist für den Beobachter
grundsätzlich nicht beobachtbar. Wir können also nicht wissen, was sich hinter
dem Ereignishorizont ereignet.

Während sich die eine Grenze empirischer Forschung aus der klassischen Relativitätstheorie ergibt, wird die andere Grenze durch eine nichtklassische Theorie, die Quantenmechanik, definiert. Die Quantenmechanik ist die physikalische
Beschreibung des Verhaltens von Licht und Materie im atomaren Bereich. Hier
verhalten sich die physikalischen Prozesse auf eine Art, wie man es aus der All-
tagserfahrung niemals erwarten würde. Der Ort eines Teilchens mit definiertem
Impuls wird in der Quantenmechanik durch eine ebene Welle seiner Aufenthaltswahrscheinlichkeit beschrieben. Dies bedeutet aber, dass der Ort eines solchen
Teilchens grundsätzlich unbestimmt ist, denn Wellen breiten sich aus und können

sich sogar überlagern. Aufgrund des Wellencharakters der Aufenthaltswahrscheinlichkeit von Teilchen gelten in der Quantenmechanik Impuls und Ort eines Teilchens grundsätzlich als komplementäre Eigenschaften, die niemals gleich genau
bestimmt werden können. Man kann nicht gleichzeitig beide Seiten einer Medaille
betrachten.

Formal drückt sich die Komplementarität in der Unbestimmtheitsrelation aus,
die 1927 von Werner Heisenberg aufgestellt wurde. Sie besagt, dass die Bestimmt-
heit zweier komplementärer Eigenschaften folgender Ungleichung genügen muss:

$$\Delta x \cdot \Delta p \geq h/2\pi$$

wobei die beiden Ausdrücke auf der linken Seite die Unbestimmtheit von Ort und
Impuls bezeichnen. Der Ort von irgendetwas kann also nur bis auf

$$\Delta x \geq h/(2p\pi)$$

bekannt sein. Der maximale Impuls $p=mc$ entspricht der kleinsten Länge $\Delta x=h/(2\pi mc)$. Dies ist keine Grenze, die etwa durch Mängel an experimenteller Messge-
nauigkeit gegeben ist, sie ist vielmehr von prinzipieller Natur. Kommt ein Messge-
rät während eines Experimentes dem materiellen Teilchen zu nahe ($\Delta x \cong h/(2p\pi)$),
wird sein Impuls unbestimmt, kann der Impuls sehr genau bestimmt werden, ist
der Ort völlig unbestimmt.

Wir haben jetzt die Möglichkeit, die zwei Grenzen empirischer Erkenntnis
– Schwarzschildradius $R_{SW}=2GM/c^2$ und Unbestimmtheitslänge $\Delta x \geq h/(2p\pi)$ –
gleichzusetzen und erhalten als eine Masse:

$$m_{Planck} = (hc/4\pi G)^{0.5} = 1.5 \cdot 10^{-5} \text{ Gramm.}$$

Dies entspricht der Masse eines Staubkorns. Wir setzen diese Masse wieder in die
Unschärfelänge oder den Schwarzschildradius ein und erhalten die kleinste physikalisch sinnvolle Länge, die so genannte Plancklänge:

$$l_{Planck} = (Gh/\pi c^3)^{0.5} = 10^{-33} \text{ cm.}$$

Sie gibt die kleinste Ausdehnung eines physikalischen Systems an, von dem man
überhaupt noch irgendeine Information im Sinne einer Ursache-Wirkungsbezie-
hung erhalten kann. Entsprechend lässt sich eine kleinste physikalisch gerade noch
sinnvolle Zeiteinheit durch

$$t_{Planck} = l_{Planck}/c = 5 \cdot 10^{-44} \text{ Sekunden}$$

definieren.

Neben den drei Grundgrößen Masse, Länge und Zeit werden auch folgende abgeleitete Größen verwendet:

$$\text{Planck-Fläche: } A_{Planck} = l^2_{Planck} = 10^{-66} \text{ cm}^2$$

$$\text{Planck-Energie: } E_{Planck} = m_{Planck}c^2 = 1.9 \; 10^9 \text{ J} = 10^{19} \text{ GeV}$$

$$\text{Planck-Temperatur: } T_{Planck} = 10^{32} \text{ K}$$

$$\text{Planck-Dichte: } \rho_{Planck} = m_{Planck}/l^3_{Planck} = 10^{93} \text{ Gramm/cm}^3$$

Diese Werte charakterisieren die kleinste Raumeinheit, die mit Relativitätstheorie und Quantenmechanik gerade noch vereinbar ist. Die Planck-Welt ist die kleinste kausale Struktur in unserem Universum, in dem Lichtgeschwindigkeit, Gravitationskonstante und Plancksches Wirkungsquantum wohl definierte Naturkonstanten darstellen. Wären ihre Zahlenwerte anders, würde sich auch die kausale Grundstruktur ändern.

Wie weit weg die Planck-Welt selbst von den exotischsten Materieformen entfernt ist, veranschaulicht schon die Tatsache, dass die Planck-Länge ca. 10^{20}mal kleiner ist als der Durchmesser des Protons und damit weit jenseits einer direkten experimentellen Zugänglichkeit liegt. Wollte man die Planck-Welt mit einem Teilchenbeschleuniger untersuchen, so müsste die Wellenlänge der Strahlung oder der Teilchen (die sog. de Broglie Wellenlänge) vergleichbar mit der Planck-Länge sein, bzw. ihre Energie vergleichbar mit der Planck-Energie von 10^{19} GeV. Die über $E=mc^2$ zugeordnete Masse ist über 10^{16} mal größer als die Masse des schwersten bekannten Elementarteilchens, des Top-Quarks. Ein entsprechender Teilchenbeschleuniger müsste astronomische Ausmaße besitzen.

10.5.3 Symmetrisch war der Anfang, sehr heiß und sehr ordentlich

Was bedeuten die Eigenschaften der physikalisch kleinsten gerade noch eine Kausalstruktur enthaltenden Planck-Welt? Zuerst einmal ist die Planck-Zeit als Elementareinheit der Dimension Zeit eine wichtige Grenze für die empirischen Wissenschaften, die Zeit t=0 gibt es in den Naturwissenschaften nicht. Gleiches gilt auch für die räumliche Ausdehnung des Universums. Die Planck-Länge, als kleinstmögliche empirische Struktur, verbietet Längen, die gleich Null sind. Mit anderen Worten: Der gedankliche Beginn des Kosmos mit t=0 und l=0 kann nicht Gegenstand der Naturwissenschaften sein. Deshalb ist Kosmologie immer Innenarchitektur des Kosmos. Nur innere Eigenschaften wie Strahlung und Materie des Universums können Thema der Astrophysik sein. Der eigentliche Anfang wird uns immer ein Rätsel bleiben; Fragen nach dem Davor und dem Draußen sind naturwissenschaftlich sinnlos.

Nehmen wir diese unabänderlichen Einschränkungen hin, so haben wir nur noch die Möglichkeit, den Anfangszustand des Universums in Begriffen der Planck-Welt zu erklären.

Zur adäquaten Beschreibung physikalischer Vorgänge muss jedoch noch eine weitere physikalische Theorie hinzugenommen werden – die Thermodynamik. Sie ist eine sehr allgemeine Beschreibung von Systemeigenschaften. Ein besonders wichtiger thermodynamischer Begriff ist ‚Entropie'. Die Entropie gibt an, welche Realisierungsmöglichkeiten physikalische Prozesse im Universum haben. Im Allgemeinen scheint ein Zustand niedriger Entropie einer sehr hohen Ordnung zu entsprechen, während hochentropische Zustände sehr ungeordnet sein sollten. Hohe Temperaturen würden dann hoher Entropie und niedrige Temperaturen einer niedrigen Entropie entsprechen. Dem ist aber nicht immer so. Reversible oder auch ideale Prozessketten können ihren thermodynamischen Anfangszustand wieder vollständig erreichen, d.h. Energie wird nicht dissipiert – die Entropie ändert sich nicht.

Alle realen Prozesse hingegen sind irreversibel, denn immer wird Energie teilweise in Wärme umgewandelt und immer nimmt die Entropie zu.

Die Entropie hängt von der räumlichen Größe des Systems ab. Damit ist sofort zweierlei klar:

1. Die Entropie des Universums im kleinstmöglichen räumlichen Zustand war sehr klein, obwohl eine sehr hohe Temperatur herrschte. Durch die Expansion hat sich die Entropie des Universums ständig erhöht, obwohl es sich währenddessen abkühlte.

2. Die Zunahme der Entropie im Universum definiert eine Zeitrichtung: vom Zustand niedriger Entropie hin zum Zustand höherer Entropie. Insbesondere dieser Zusammenhang erklärt die Entstehung von kleinen räumlichen Strukturen wie Galaxien in einem homogenen, isotropen und expandierenden Kosmos. Denn während der Raum expandierte, haben sich aufgrund von Instabilitäten kleine Raumbereiche von der allgemeinen Expansion entkoppelt und sind unter ihrem eigenen Gewicht zusammengestürzt. Dort hat sich die Entropie erniedrigt, die dafür nötige Energie kam aus dem Gravitationsfeld der in sich zusammenstürzenden Gaswolken. Gleiches gilt auch für Sterne. Auch sie sind im Vergleich zur Ausdehnung einer Galaxie winzige Inseln der Ordnung in einem Meer zunehmender Unordnung und damit zunehmender Entropie. Alle Strukturen im Kosmos, vor allem Lebewesen, sind auf externe Energiequellen ange-

wiesen, um dem allgemeinen Trend zum Zerfall bis hin zur völligen Unordnung und damit hohen Entropie wenigstens für eine gewisse Zeit zu entgehen.

Doch zurück zum Anfang: Obwohl also das Universum in seinem Anfangszustand eine sehr hohe Temperatur besaß und damit eigentlich eine hohe Entropie geherrscht haben sollte, ist die fast verschwindende räumliche Größe der dominierende Faktor, der die Entropie des Universums am Anfang minimiert hat.

Diese hohe Temperatur des Anfangs, immerhin 10^{32} K, hat eine ganz wichtige Bedeutung für die weitere Entwicklung des Kosmos gehabt. Ein Beispiel: Nehmen wir an, wir seien Lebewesen in einem ferromagnetischen Material. Solange unsere ‚Welt' kühl genug ist, ist sie magnetisiert. Bewegungen entlang der magnetischen Feldlinien fallen uns leicht, während Bewegungen senkrecht zu den Feldlinien durch die magnetischen Kräfte sehr erschwert werden. Unsere Welt ist also nicht symmetrisch, sie enthält eine Kraft, verursacht durch das Magnetfeld. Eine Raumrichtung, nämlich die parallel zu den Feldlinien, ist ausgezeichnet. Eine klare Orientierung liegt vor, deshalb spricht man in der Physik davon, dass die Symmetrie ‚gebrochen' ist. Einer gebrochenen Symmetrie entspricht eine Kraft. Wird nun das ferromagnetische Material über eine bestimmte Temperatur hinaus erhitzt, verschwindet plötzlich das Magnetfeld. Die physikalischen Ursachen dafür sind in diesem Zusammenhang nicht wichtig. Ohne Magnetfeld jedoch gibt es keine ausgezeichnete Richtung mehr, unsere Welt ist symmetrisch geworden. Wir haben es in unserem Beispiel mit einer ‚versteckten' Symmetrie zu tun, die sich erst oberhalb einer gewissen Temperatur zeigt. Bei Ferromagneten nennt man diese Temperaturschwelle ‚Curie-Temperatur'.

Für unsere Fragestellung nach dem Anfang des Universums lässt sich dieses Beispiel einfach übertragen. Die Planck-Temperatur entspricht der Curie-Temperatur des Universums. Wird sie erreicht, dann ist das Universum völlig symmetrisch. Kühlt sich der Kosmos aufgrund seiner Expansion unter die Planck-Temperatur ab, tauchen Kräfte auf, die gewisse Eigenschaften des Universums auszeichnen. Jede Kraft ist mit einem Symmetriebruch verbunden. In unserem Universum gibt es vier fundamentale Kräfte: die Gravitation (Schwerkraft), die elektromagnetische Wechselwirkung (Ströme, Magnetfelder, elektromagnetische Wellen), die starke Kernkraft (Zusammenhalt der Atomkerne) und die schwache Wechselwirkung (radioaktiver Zerfall). Jede dieser Wechselwirkungen entspricht also einem Symmetriebruch während der frühen Phasen des Kosmos. Wann genau die jeweiligen Wechselwirkungen ‚ausfrieren', hängt von der Energie bzw. Masse der Teilchen ab, die die jeweilige Wechselwirkung vermitteln. Auch hier sollen uns die Details nicht interessieren. Die Experimente, die in den großen Teilchenbeschleunigern durchgeführt wurden, haben in den letzten 20 Jahren zu einem recht klaren Bild der ganz frühen Entwicklung des Kosmos geführt. Sie konnten zeigen, dass oberhalb einer Energie von ca. 100 GeV, was einer Temperatur von 10^{15} K entspricht, die elektromagnetische und die schwache Wechselwirkung zu einer elektroschwachen Wechselwirkung verschmelzen. Dies muss überraschen, denn die schwache Wechselwirkung hat eine sehr kleine Reichweite (kleiner als der Radius des Atomkerns – 10^{-14} m) und wird von Teilchen mit einer sehr großen Ruhemasse von ca. 80-90 GeV/c^2 vermittelt. Demgegenüber hat die elektromagnetische Wechselwirkung eine prinzipiell unendlich große Reichweite und wird von Teilchen vermittelt, die keine Ruhemasse besitzen, den Photonen. Also sind beide Kräfte sehr verschieden und auch die entsprechenden Symmetriebrüche sind sehr unterschiedlich. Oberhalb der Temperatur von 10^{15} K allerdings verschwinden diese Unterschiede, hier haben wir es wieder mit einer verborgenen Symmetrie zu tun.

Der erreichte Energiebereich von 100 GeV stellt die Bedingungen des Universums dar, als es ca. 10^{-14} Meter groß war und nur wenige Billionstel Sekunden alt.

Die experimentelle Elementarteilchenphysik hat sich zum Ziel gesetzt, den Zustand des Universums zu rekonstruieren, in dem die elektroschwache Wechselwirkung mit der starken Wechselwirkung verschmolz. Würden diese Versuche gelingen, wäre eine neue Symmetriestufe erreicht. Die Krönung jedoch wären Experimente, die auch noch die Gravitation mit den bereits verschmolzenen Wechselwirkungen vereinigen. Dafür wären allerdings Energien von der Größenordnung der Planck-Energie (10^{19} GeV) notwendig. Dass das im Labor jemals gelingen wird, ist unwahrscheinlich.

10.6 Von Strings und Membranen – die Theorien für alles und gar nichts

Theoretische Modelle vereinigter Wechselwirkungen gibt es schon, man spricht von 'Grand Unified Theories' (GUT) für den Fall der drei Wechselwirkungen Elektromagnetismus, starke und schwache Wechselwirkung und von der 'Theory of Everything' (TOE) für die Fusion aller vier Wechselwirkungen.

Bei all diesen Theorien handelt es sich um mathematisch ausgesprochen komplexe Modelle, deren physikalische Interpretation sehr schwierig und deren experimentelle Überprüfung bis jetzt unmöglich ist. Ein Beispiel für die TOE ist die Superstring-Theorie, die die Vereinigung aller Kräfte in einer Supersymmetrie beschreibt. Sie will die beiden Hauptpfeiler der heutigen Physik vereinigen: die allgemeine Relativitätstheorie, welche bei Strukturen im Großen gültig ist, und die Quantenfeldtheorie, die im Mikrokosmos angewendet wird. Dabei erscheinen sozusagen als Nebenprodukt alle Elementarteilchen und ihre Wechselwirkungen. Die primäre Aussage der Stringtheorie ist: Die Elementarteilchen manifestieren sich als unterschiedliche Anregungszustände der so genannten Strings. Das sind eindimensionale Fäden, die wie Saiten (englisch: *string*) in einem vieldimensionalen Raum schwingen. Je nach ‚Frequenz' (Energie) und Raumdimension stellen sie unterschiedliche Varianten von Elementarteilchen dar. Elektronen oder Quarks entsprechen nahezu masselosen Anregungszuständen (‚Nullmodi') der Strings. Besonders Erfolg versprechend ist, dass einer dieser masselosen Zustände genau die Eigenschaften des hypothetischen Gravitons hat, das als Wechselwirkungsteilchen der Gravitation gilt. Die Superstringtheorie beschreibt die Schwerkraft adäquat, ebenso wie alle anderen Teilchen, die Wechselwirkungen vermitteln.

Daneben gibt es ein Vibrationsspektrum von unendlich vielen Schwingungsmodi, welche aber zu hohe Massen (Energien) haben, um direkt beobachtet werden zu können. Denn aus theoretischen Überlegungen sollten Strings eine Ausdehnung in der Größenordnung der Planck-Länge besitzen, somit müssten die Vibrationsmodi Massen besitzen, die um ein Vielfaches größer wären als 10^{19} GeV größer wären. Das liegt um viele Größenordnungen über dem, was man experimentell beobachten kann. Daher wird man auf einen direkten Nachweis dieser Vibrationsmodi verzichten müssen und stattdessen versuchen, im Sektor der fast masselosen Teilchenanregungen Eigenschaften zu finden, die spezifisch für die Stringtheorie und gleichzeitig experimentell beobachtbar sind. Dies stößt aber auf die Schwierigkeit, dass gerade der zugängliche masselose Sektor in nur geringem Maß von der zugrunde liegenden Stringtheorie bestimmt wird, zumindest nach heutigen Erkenntnissen. Superstringtheorien werden nur in 10 oder 11 Dimensionen formuliert und können nur in diesen Dimensionen ein mehr oder weniger eindeutiges Spektrum haben. Um auf unsere vierdimensionale Raum-Zeit zu kommen, muss man eine sogenannte. Kompaktifizierung (in etwa: Aufwicklung) der 6 bzw. 7 ‚überschüssigen' Dimensionen postulieren, die der direkten Beobachtung nicht zugänglich sind. Der Prozess der Kompaktifizierung ist bei weitem nicht eindeutiger, er führt zu einer Überfülle von möglichen vierdimensionalen Theorien.

Bislang hat man keine Eigenschaften des masselosen Sektors finden können, welche spezifisch für die Stringtheorie und in naher Zukunft experimentell überprüfbar wären. Deshalb ist ein großer Teil der Forschung mehr mit theoretischen und konzeptionellen Fragen beschäftigt, z. B. mit Problemen, die im Zusammenhang mit der Anwendung der Quantenfeldtheorien in der Nähe von Schwarzen Löchern stehen.

Die Stringtheorie wurde ursprünglich rein mathematisch aus Symmetrieprinzipien abgeleitet. Anfangs wurden fünf Stringtheorien entwickelt, die sich später

als unterschiedliche Näherungen einer umfassenden Theorie (M-Theorie) heraus-
stellten. Der Nachweis wurde durch Aufzeigen von Dualitäten zwischen den einzelnen Stringtheorien erbracht. Ein interessantes Ergebnis dieser Vereinigung der
Teiltheorien war, dass die elfdimensionale Supergravitation als weiterer Grenzfall
der M-Theorie erkannt wurde. Diese enthält aber keine Strings, sondern ist eine
Teilchen-Approximation von zwei- und fünfdimensionalen Membranen. Tatsächlich hat sich in den letzten Jahren gezeigt, dass höherdimensionale Membranen
(D-branes) eine sehr wichtige Rolle in der Stringtheorie spielen. Ein neues kosmo-
logisches Modell nutzt diese Membranen, um die theoretischen Unzulänglichkei-
ten des Urknallmodells zu umgehen.

10.7   Von dunklen Energien

Irdische Experimente können die physikalischen Bedingungen des ganz jungen
Kosmos nicht klären, die theoretischen Modelle bieten keine realisierbaren experimentellen Tests an. Vielleicht können kosmologische Beobachtungen hier Abhilfe
schaffen. Leider hat sich auch in der beobachtenden Kosmologie die Erkenntnisla-
ge in den letzten Jahren ziemlich verschlechtert.

Früher war man der Meinung, das Universum bestehe aus Strahlung und Materie, die elektromagnetische Strahlung abgibt. Heute wissen wir, dass der größte
Teil des Kosmos mit Dunkler Materie angefüllt ist, die keinerlei Wechselwirkung
mit Strahlung besitzt und nicht aus den uns bis heute bekannten Teilchensorten besteht. Es gibt rund fünfmal mehr dunkle als leuchtende Materie. Die zahllosen Galaxien, die wir mit unseren Teleskopen beobachten, sind nur ein winziger Bruchteil
der gesamten Materie des Kosmos. Darüber hinaus hat sich in den letzten Jahren
herausgestellt, dass das Universum sich nicht gleichmäßig ausbreitet, sondern seit
rund 8 Milliarden Jahren beschleunigt expandiert. Offenbar gibt es eine Energieform, die für diese beschleunigte Expansion verantwortlich ist, man spricht von
sog. Dunkler Energie. Diese Energie ist nicht mit einer Masse gleichzusetzen, wie
sie mit der berühmten Formel $E=mc^2$ beschrieben wird. Vielmehr muss es sich um
eine von Masse gänzlich unabhängige Energieform handeln, die mit der Energie
des Vakuums in Zusammenhang gebracht wird. Nach den allerneuesten Messungen der Planck-Mission, macht sie 68,3% der Energiedichte des Universums aus.
Die Dunkle Materie trägt 26,8% bei und die gesamte leuchtende Materie ist nur mit
4,9% beteiligt.

Weder die Natur der Dunklen Materie noch der Dunklen Energie ist uns bekannt. Sicher ist nur, dass ohne diese beiden Bestandteile weder die Entstehung

von Milchstraßen noch die Expansion des Universums möglich gewesen wäre. Nachgerade als katastrophal erweist sich, dassdie gemessene Energiedichte der Dunklen Energie um 120 Größenordnungen kleiner ist, als die Vakuumenergiedichte, die sich aus den heute gängigen Quantenfeldtheorien ergibt, die exakt der Planck-Energie hoch vier entspricht. Eine größere Diskrepanz ist in der Physik zwischen Messwert und Theorie nicht bekannt. Man erkennt die Unzulänglichkeit moderner kosmologischer Theorien.

10.8   Paralleluniversen oder Uhrmacher?

Natürlich werfen diese Inkonsistenzen zwischen Theorie und Beobachtung grundlegende Fragen zum Ursprung des Universums auf. Hat ein Designer die Anfangsbedingungen so exakt eingestellt, dass sich die Entwicklung des Universums mit Galaxien, Sternen und Planeten genau so hat abspielen können? Musste die kosmische Entwicklung intelligente Lebensformen hervorbringen, die auf einem bewohnbaren Planeten existieren, der sich um einen Stern dreht, der gerade die richtige elektromagnetische Strahlung produziert, die als Grundlage aller Lebensvorgänge dient? War das zwangsläufig? Oder sind das alles, der Kosmos und seine bemerkenswert fein aufeinander abgestimmte Innenarchitektur, nur ein Zufall? Die feine Abstimmung zeigt sich nicht nur in den kosmischen Dimensionen mit ihren merkwürdigen dunklen Energie- und Materieformen. Vielmehr sind auch die bereits erwähnten fundamentalen Wechselwirkungen, die für die Stabilität der Materie (Elektromagnetismus und starke Wechselwirkung), aber auch für die Kernprozesse in den Sternen (schwache und starke Wechselwirkung, Gravitation und Elektromagnetismus) verantwortlich sind, extrem genau aufeinander abgestimmt. Bedenkt man zudem, dass wir als Lebewesen aus Elementen wie Kohlenstoff, Stickstoff, Phosphor, Calcium, Eisen etc. bestehen, die aus der Verschmelzung ganz kleiner Atomkerne wie Wasserstoff und Helium in den Sternen entstanden sind, wird die enge Verflechtung des Menschen mit den Grundgesetzen des Kosmos deutlich. Wir sind Sternenstaub! Jede noch so kleine Veränderung der Naturkonstanten hätte verhindert, dass Sterne entstehen und damit auch die Synthese schwerer chemischer Elemente unmöglich gemacht. Unsere Existenz hängt von der Genauigkeit der Naturgesetze und der Naturkonstanten ab – wir konnten uns nur in diesem Universum entwickeln. Wir sind Kinder dieses Universums und deshalb wollen wir wissen, wie alles angefangen hat. War es Zufall oder Notwendigkeit?

Als Alternative zum Designer-Modell diskutieren Naturwissenschaftler die Möglichkeit von Paralleluniversen. Das Problem der unwahrscheinlichen Feinabstimmung wird umgangen, indem für den Anfang sehr viele 'Versuche' angenommen werden, die jeder zu einem Universum führten mit je unterschiedlichen physikalischen Naturgesetzen und Konstanten. Aber nur einer davon war erfolgreich in der Hervorbringung von Galaxien, Sternen, Planeten und Lebewesen. Nach diesem Modell müssten wir uns überhaupt nicht wundern, dass die Zusammenhänge so sind wie sie sind, denn wenn es anders wäre, gäbe es uns ja nicht. Allerdings müssten ca. 10^{57} Paralleluniversen entstanden sein, damit eines mit den richtigen Parametern darunter sein konnte.

Dieser Ansatz ist ziemlich fragwürdig, weil es keinerlei Möglichkeit gibt, die Existenz der Paralleluniversen zu überprüfen. Es gibt keine kausale Verbindung zwischen den angenommenen Universen. Damit wird aber gegen eine der Grundregeln naturwissenschaftlichen Tuns verstoßen: Theorien müssen grundsätzlich falsifizierbar sein.

10.9   Summa summarum

Von der tatsächlichen Geburt des Universums kann die Physik nichts berichten. Diese Aussage kommt aus der Physik selbst. Ihre erfolgreichsten Theorien liefern grundsätzliche Grenzen empirischer Forschung. Wir wissen weder, warum noch wie das Universum sich in das Abenteuer seiner Existenz gestürzt hat. Es könnte auch Nichts sein.

10.10   Eine Nachbetrachtung

Eines der letzten Worte des vorherigen Abschnitts lautet „könnte" – verweist auf das auch Mögliche. Möglichkeiten sind die zur Wahl stehenden Varianten, sie bezeichnen die Potenziale dessen, was eintreten könnte. In der Philosophie wird der Ausdruck Kontingenz, mit dem Adjektiv ‚kontingent', verwendet, und bezeichnet den Status von Tatsachen, deren Bestehen gegeben und weder notwendig noch unmöglich ist. Insofern steht unsere Ausgangsfrage „Warum ist überhaupt Etwas und nicht vielmehr Nichts?" in engstem Zusammenhang mit der Kontingenz des Seins. Ist das Sein notwendig oder ist es kontingent?

Wenn wir hier nicht in der reinen Konjunktivfassung bleiben wollen, dann müssen wir uns auch in einer Nachbetrachtung positionieren und Annahmen machen, deren Tragfähigkeit sich erweisen muss. Dabei müssen wir auch weiterhin den schon anfangs zitierten Gödelschen Beweis der Unvollständigkeit von Aussagensystemen als Zensor akzeptieren.

Es ging uns um eine saubere Darstellung des Urknalls als Transformation des physikalischen Seins von der Möglichkeit in die Wirklichkeit. Das Sein der Wirklichkeit – so zeigt sich mittels eines ausgeklügelten Erkenntnisprozesses, der durch den Wettbewerb von empirischen Hypothesen charakterisiert ist, dessen Ausgang durch die schärfste Waffe der Kritik, das Experiment bzw. die Beobachtung, entschieden wird – stellt sich in seiner Urform als reine Energie dar, deren Schwankungen sich in Feldern und Teilchen manifestieren. Bedingung der Möglichkeit dieser Manifestation ist die Stabilität dessen, was entstand. Unser Universum ist das Ergebnis einer Abfolge ineinandergreifender Prozesse, die so abliefen, weil die Teilchen und Kräfte exakt die Eigenschaften haben, die sie haben. Gedankenexperimente mit den fundamentalen Wechselwirkungen und ihren Parametern zeigen, dass das Universum genau so sein muss, wie es ist, ansonsten wären die Seinsstrukturen so sicher nicht entstanden. Aus diesen physikalischen Spekulationen lassen sich bemerkenswerte metaphysische Spekulationen entwickeln.

10.11 Die Solidarität des Universums – der Kosmos als Prozess

Kommen wir zurück an den Anfang. Wir fragten nach den Voraussetzungen, die erfüllt sein müssen, damit ein Wesen die metaphysische Grundfrage stellen kann. Wenn es nur eine Wirklichkeit gibt, an der Wahrnehmender und Wahrgenommenes teilhaben, dann müssen die unterschiedlichen Sichtweisen in irgendeiner Form aufeinander bezogen sein.

Der Mensch ist samt seinem Erkenntnisvermögen das Resultat der Evolution. Er partizipiert an der Natur und beeinflusst sie durch seine Handlungen, Ängste, Visionen und Hoffnungen. Die Position eines außerweltlichen Beobachters ist ihm deshalb verwehrt. Aber die Naturwissenschaften beschreiben nur die kausalen Zusammenhänge empirischer Daten, sie sind nach Thomas Nagel „ein Blick von nirgendwo" (1986). Als vollständiger Teil dieses Universums machen Menschen ihre Erfahrungen zum Fundament ihrer Modelle über das Universum. Dieses Weltganze, das Universum, zeigt sich aber nur unter der durch die Sinneswahrnehmungen dargestellten leiblichen Konstitution und unter der durch die besondere Form des erkennenden Bewusstseins vermittelten Perspektive.

Und was erkennt der Mensch, der das Universum erforscht, es nach Alternativen durchsucht? Ganz einfach: Das Universum hat, zumindest auf unserem Planeten, eine materielle Transformation durchgemacht. Aus unbelebter Materie ist Leben entstanden. Leben ist eine Form des Seins, die sich vor allem durch das Bestreben auszeichnet, sich vor den natürlichen Gegebenheiten und Widrigkeiten zu schützen und wenn möglich zu befreien. Eingebettet in einen kosmischen Energiestrom, der, ausgehend von der Sonne, die Erde erwärmt und erhellt, entwickelte sich in sehr langen Zeiträumen eine immer komplexere Flora und Fauna. Physikalisch definiert ist ein Lebewesen ein sich selbst organisierendes, dissipatives Nichtgleichgewichtssystem. Es verteilt und verarbeitet die bereits aufgenommene und in molekular Form verwandelte Sonnenenergie immer wieder aufs Neue. In vernetzten Kreisläufen und in engster Wechselwirkung mit den äußeren Umweltbedingungen hat das Phänomen Leben Bedingungen geschaffen, die seine Weiterentwicklung in immer komplexere Individuen überhaupt erst ermöglichten.

Nach der über fast neun Milliarden Jahre langen, alleine durch die kosmische Entwicklung bedingten Transformation der Materie in schwere Elemente, bildeten sich das Sonnensystem und die Planeten. In dieser Kulisse vollzieht sich auf der Erde, vermutlich in den Urmeeren, die allmähliche Verwandlung von einfachsten Molekülen zu immer komplexeren molekulare Strukturen, die sich als Selbstorganisationsphänomen in zahllosen „Testreihen" zu allerersten Vorformen einfachster Zellen entwickeln. Die Zelle ist das Atom des Lebens. Die biologische Evolution geht ans Werk und erschafft immer neue Möglichkeiten der Befreiung von der Lebensumwelt. In immer verfeinerten Verfahren schaffen es die Lebewesen, sich in kooperativen Strukturen gegenseitig zu erhalten und zu unterstützen. Alfred North Whitehead schreibt hierzu: „Nature is a theatre for interrelations of activities. All things change, the activities and their interrelations" (Whitehead, S. 35f.). Das Zauberwort der Evolution ist „Koevolution", sie verbindet die unterschiedlichen Organismen und ihre Umwelt. Umgebung und Spezies sind nicht voneinander zu trennen, beide entwickeln sich. Die erfolgreichsten Organismen sind die, die ihre Umwelt so verändern, dass sie einander gegenseitig unterstützen. Man könnte von einer Gemeinschaft aller Prozesse und Teile der Welt sprechen, weit ausgeholt, von der „Solidarität des Universums".

Wäre es so, dann wäre der Kosmos ein einziger, im Grunde unberechenbarer Selbstorganisationsprozess, dessen Inhalt und Zweck die ständige und kontinuierliche Schaffung neuer Möglichkeiten darstellt: Zunächst die Materie als Substrat, aus dem sich überhaupt etwas „machen lässt", dann durch zunehmende Strukturierung in Galaxien, Sterne und Planeten. Je kleiner die Strukturen, umso höher wird ihr Organisationsgrad. Das Organisationsphänomen Leben ist dann nur ein notwendiger Schritt unter den besonders günstigen Bedingungen eines Planeten im richtigen Abstand um einen nicht zu heißen und nicht zu kalten Stern. Nach viereinhalb Milliarden Jahren Werden und Vergehen auf der Erde taucht dann eine ganz neue Organisationsform auf: das menschliche Gehirn mit der besonderen Eigenschaft des Selbstbewusstseins und der Selbstreflexion. Diese neue Lebensform erhöht das Tempo der Selbstorganisation durch die Entwicklung von Technik und Wissenschaft im Rahmen sich ständig verändernder kultureller Organisationsstrukturen. Immer geht es um mehr Freiheit von der Umwelt durch immer neue Systemvernetzungen und Kooperativen. Das lebendige Sein kann sich wehren, kann sich neue Chancen und neue Möglichkeiten für die Zukunft verschaffen. Das Sein, mit seiner Ausformung ‚Universum‘, hat sich für die Chance der vielen Möglichkeiten entschieden und nicht für die einzige Möglichkeit des Nichtseins.

Kein Wesen kann zu Nichts zerfallen!
Das Ew'ge regt sich fort in allen,
Am Sein erhalte dich beglückt!
Das Sein ist ewig: denn Gesetze
Bewahren die lebend'gen Schätze,
Aus welchen sich das All geschmückt.
(J. W. von Goethe: Das Vermächtnis)

Literatur

Ellwanger, U. 2015. *Vom Universum zum Elementarteilchen*. Berlin: Springer Verlag.

Gil, F. J. S. 2014. *Philosophie der Kosmologie*. Frankfurt: Verlag Peter Lang.

Hamilton, W. 1835. Second Essay on a General Method in Dynamics. *Philosophical Transactions of the Royal Society of London* 125, 95-144.

Kanitscheider, B. 1984. *Kosmologie*. Ditzingen: Reclam.

Lesch, H. 2015. In *Quanten 3 (Schriften der Heisenberg-Gesellschaft)*, hrsg. K. Kleinknecht. Hirzel Verlag.

Maupertuis, P.-L. 1744. Der Einklang verschiedener Naturgesetze die bislang als unvereinbar angesehen wurden. Lu à L'Academie des Sciences le 14 avril 1744.

Nagel, Thomas. 1986. *The View From Nowhere*. Oxford: OUP.

Planck, M. 1899. Über irreversible Strahlungsprozesse. *Sitzungsberichte der Preußischen Akademie der Wissenschaften* 5.

Treichel, M. 2000. *Teilchenphysik und Kosmologie*. Göttingen: Springer Verlag.

Vollmer, G. 2003. Wieso können wir die Welt verstehen? Stuttgart: Hirzel.

Whitehead, A.N. 1934. *Nature and Life*. New York: Greenwood Press.

11. Ist zukünftiges Klima berechenbar?

Konrad Kleinknecht[1]

Abstract

Die Vorhersagen über die Entwicklung des globalen Klimas, die das International Panel for Climate Change (IPCC) veröffentlicht, beruhen auf komplexen Differenzialgleichungsystemen mit vielen unbekannten Parametern. Durch die Korrelationen zwischen verschiedenen physikalischen Größen sind diese Systeme nichtlinear, sie werden in der Mathematik „chaotisch" genannt. Die Ergebnisse solcher Klimamodellrechnungen sind nicht zuverlässig. So widerspricht die beobachtete Entwicklung der mittleren Oberflächentemperatur der Erde für die letzten 17 Jahre den Vorhersagen des IPCC. Klima-Modellrechnungen über Zeiträume von hundert Jahren sind mit so großen Unsicherheiten behaftet, dass sie nicht zur Grundlage politischer Entscheidungen gemacht werden können.

Model calculations about climate change rely on complex coupled differential equations. From such models, the International Panel for Climate Change (IPCC) tries to predict an increase in the global surface temperature. However, these predictions fail to agree with the observed temperatures during the last 17 years which show no increase. The complexity of the problem is due to the correlations between the physical properties and a great number of unknown parameters. Predictions for periods as long as hundred years are unreliable and cannot serve as basis for political decisions.

---

1   Konrad Kleinknecht ist em. Professor für Experimentalphysik, er arbeitete am CERN in Genf, an der Universität Dortmund, am California Institute of Technology, an der Harvard University, an den Universitäten in Mainz und München (LMU). Seine Forschungsgebiete sind die Physik der elementaren Bausteine der Materie und die Energieversorgung. Für seine Arbeiten erhielt er viele Preise, u. a. den Leibniz-Preis der DFG und die Stern-Gerlach-Medaille der DPG. Neben Forschungsarbeiten veröffentlichte er Bücher über Teilchendetektoren und die Materie-Antimaterie-Asymmetrie und war Klimabeauftragter der DPG. Er ist Präsident der Heisenberg-Gesellschaft.

„Die langfristige Voraussage des zukünftigen Klimas ist nicht möglich." (IPCC-Bericht 2001)

Als Klima bezeichnet man die Beschreibung der über längere Zeiträume gemittelten physikalischen Eigenschaften der Erdatmosphäre, also der statistischen Mittelwerte der Temperatur an der Erdoberfläche, des Luftdrucks oder der Zusammensetzung der Atmosphäre über etwa dreißig Jahre. In der Geschichte der Erde hat sich das Klima auf natürliche Weise häufig stark geändert.

Seit 200.000 Jahren gab es mehrere Eiszeiten und Warmzeiten. Sie hatten verschiedene Ursachen. Ein Teil der Variationen war verursacht durch periodische Änderungen der Erdumlaufbahn um die Sonne und der Neigung der Erdachse zur Umlaufebene, wie der ungarische Bauingenieur Milutin Milankovic Anfang des 20. Jahrhunderts herausfand. Die zyklischen Schwankungen haben verschiedene Perioden. Die Erdbahn ist derzeit kein Kreis, sondern eine Ellipse. Dies wissen wir, seitdem Johannes Kepler aus den Messungen von Tycho Brahe in Prag die genauen Bahndaten berechnete. Durch den Einfluss der anderen Planeten verändert sich die Bahn mit einem Zyklus von 100.000 Jahren von einer Ellipse zum Kreis und wieder zurück. Das hat durch die Änderung des Abstandes zwischen Erde und Sonne eine geringe Variation der Sonneneinstrahlung zur Folge.

Einen größeren Einfluss hat die Neigung der Erdachse zur Umlaufbahn. Sie verursacht den Wechsel von Sommer und Winter. Da die Erde keine perfekte Kugelform hat, sondern an den Polen abgeplattet und am Äquator ausgebeult ist, ähnelt sie einem Kreisel. Wie wir von der Beobachtung des Kinderkreisels wissen, taumelt er unter dem Einfluss der Schwerkraft, d.h. seine Achse führt eine kreisförmige Bewegung um ihre zentrale Lage aus. Es gibt für einen Kreisel zwei solcher Taumelbewegungen, die in der klassischen Mechanik Präzession und Nutation genannt werden. Beim Kreisel Erde taumelt die Drehachse relativ zur Ebene der Erdbewegung um die Sonne mit den zwei Perioden von 41.000 und 22.000 Jahren. Der Winkel zwischen der Äquatorebene und der Erdbahnebene schwankt so zwischen 21,5 Grad und 24,5 Grad. Dadurch werden verschiedene Teile der Erde unterschiedlich mit Sonnenenergie bestrahlt. Die Unterschiede zwischen Winter und Sommer verändern sich.

Das Wasser der Ozeane absorbiert die Wärme stärker als die Kontinente oder schneebedeckte Eiskappen. Die unterschiedliche Reflexionsfähigkeit der Erdoberfläche wird Albedo genannt, nach dem lateinischen Wort „albus" für weiß. Die Albedo für schneebedeckte Gletscher beträgt 95 Prozent, die für offenes Meer unter wolkenfreiem Himmel nur weniger als 30 Prozent. So beeinflusst die Stellung der Erdachse und die Ausdehnung der Eiskappen die Temperatur der Atmosphäre.

Die Wärmeaufnahme der Erde ändert sich auch dann, wenn sich die tektonischen Platten, auf denen die Kontinente liegen, verschieben. Die Sonneneinstrahlung am Äquator ist viel intensiver als an den Polen. Ein Kontinent am Äquator absorbiert mehr Wärme als an den Polkappen. Während der beiden letzten Eiszeiten war es um etwa fünf Grad kälter als heute. Das kann man aus Bohrkernen im Festlandeis von Grönland und der Antarktis erschließen. Zwischen der damals herrschenden Temperatur und der Kohlendioxid-Konzentration ergibt sich eine Parallelität: wenn die Temperatur durch die veränderten Bahnparameter der Erde anstieg, folgte mit einer Verzögerung von 50 bis 100 Jahren eine Erhöhung der CO_2-Konzentration. Die CO_2-Konzentration war also damals eine Folge der höheren Temperatur, keineswegs die Ursache.

Seit dem Ende der letzten Eiszeit und dem Beginn unserer stabilen Warmzeit vor 11500 Jahren betrug die mittlere Oberflächentemperatur der Erde 15 Grad Celsius, und damit wurde die Entwicklung der menschlichen Zivilisation möglich. Diese Temperatur verdanken wir dem natürlichen Treibhauseffekt der Atmosphäre.

Wenn die Erde keine Atmosphäre hätte, würde sich die Temperatur der Oberfläche auf -18 Grad einstellen, was für menschliches Leben nicht ausreichend warm wäre. Es ist die Atmosphäre, die dafür sorgt, dass auf der Erdoberfläche die angenehme mittlere Temperatur von plus 15 Grad herrscht, die unser Leben ermöglicht. Worauf beruht nun dieser Effekt? Die Ursache liegt in den sogenannten Treibhausgasen, an erster Stelle Wasserdampf, daneben auch Kohlendioxid und andere nur in Spuren vorhandenen Gase wie Stickstoffoxid, Methan, Ozon u.a. Diese Gase lassen die einfallende sichtbare und kurzwellige Strahlung ungehindert durchtreten, absorbieren aber die langwellige vom Boden kommende Wärmestrahlung. Diese hat Wellenlängen zwischen 3 und 100 Mikrometer, und in diesem Bereich können die Treibhausgase die Strahlung absorbieren. Dabei erwärmen sie sich und tragen zum Treibhauseffekt bei.

Der natürliche Treibhauseffekt der Erdatmosphäre beträgt etwa 33 Grad Celsius. Hinzu kommt seit dem Beginn der Industrialisierung ein kleiner Anteil durch das chemisch inerte Spurengas Kohlendioxid aus der Verbrennung von Kohle, Erdöl und Erdgas. Der daraus folgende Anstieg der Kohlendioxid-Konzentration in der Atmosphäre verstärkt den natürlichen Treibhauseffekt. Wie groß wird dieser Effekt sein? Wird sich die Erde wirklich um mehr als zwei Grad erwärmen, wie manche Rechnungen andeuten? Eine große Zahl von Atmosphärenphysikern versucht durch Modellrechnungen die Wirkung der anthropogenen Treibhausgase zu analysieren. Sie versuchen auch, aus den Modellen die zukünftige Entwicklung des Klimas und insbesondere der mittleren Oberflächentemperatur der Erde zu berechnen. Das International Panel for Climate Change (IPCC) macht keine eigenen Forschungen, sondern wählt veröffentlichte Ergebnisse aus, fasst sie zusammen und bildet daraus eine Mehrheitsmeinung. Wie zuverlässig sind diese Vorhersagen? Die Berichte des IPCC basieren auf sehr komplexen Modellrechnungen der Klimaentwicklung. Sie beschreiben die Entwicklung durch partielle Differentialgleichungssysteme, die durch gegenseitige Rückkopplungen nichtlinear sind.[2] Solche Systeme nennt man in der Mathematik chaotisch. Zusätzlich tritt ein weiteres Problem auf: die Stärke der Rückkopplungseffekte ist zum großen Teil unbekannt, bei manchen Effekten ist sogar umstritten, ob sie verstärkend oder abschwächend wirken. Alle im Bericht des IPCC im Jahre 2007 zitierten Modellrechnungen sagten für den Zeitraum von 1997 bis 2014 einen merklichen Anstieg der mittleren Oberflächentemperatur der Erde von 0,2 bis 0,6 Grad voraus.[3] Im Bericht des Jahres 1995 (SAR) gingen die für diesen Zeitraum vorhergesagten Temperaturanstiege sogar bis zu 0,8 Grad. Die Messungen widersprechen aber diesen Vorhersagen, die Temperatur hat sich in 17 Jahren um weniger als 0,05 Grad erhöht, obwohl die Konzentration des Spurengases Kohlendioxid sich stark erhöht hat. Am Nordpol wird es zwar wärmer – das sehen wir am langsamen Abschmelzen des Grönlandeises -, jedoch wird es an verschiedenen Stellen der Südhalbkugel kälter, sodass im Mittel keine Erwärmung festzustellen ist. Dabei ist zwischen den Temperaturmessungen von Satelliten aus und den Messungen von Bodenwetterstationen ein kleiner Unterschied festzustellen. Die Bodenmessstationen haben Standorte auf den Kontinenten, nicht auf dem Meer. Außerdem stehen viele Wetterstationen in Städten oder sind durch das Wachstum der Siedlungen in deren Nähe geraten. In Städten ist es wärmer als auf dem Lande. Das ergibt eine Tendenz zur Temperaturerhöhung durch Urbanisierung. Wahrscheinlich sind die Satellitendaten zuverlässiger.

Eines der weltweit führenden Zentren für solche Modellrechnungen ist das Max-Planck-Institut für Meteorologie in Hamburg. In einer neueren Publikation erläutert ein Team von Autoren um Thorsten Mauritsen aus diesem Institut, wie sie die ungewissen oder sogar unbekannten Parameter in diesem komplexen Gleichungssystem „tunen", d.h. solange verändern, bis die sich gewünschten Resultate einstellen (Mauritsen et al. 2012). Sie betonen, dass eine beträchtliche Unsicherheit in der Wahl dieser Parameter besteht, und beschreiben das Problem sehr offen: „Die Wahl, die wir machen, hängt ab von unseren vorgefassten Meinungen, Präferenzen und Zielen." Umso erstaunlicher ist es, dass die Resultate dann doch als gültig betrachtet werden.

---

2   Grundlegendes über den Treibhauseffekt und die Klimamodellrechnungen findet sich
    z. B. in dem Standardwerk „Klimatologie" von Christian-Dietrich Schönwiese (2008).
3   Der Bericht IPCC AR4 (Assessment Report4) stammt aus dem Jahr 2007.

Der Widerspruch zwischen den vorausgesagten und den beobachteten mittleren Temperaturen in den letzten 17 Jahren führt nun aber nicht dazu, dass nach Fehlern oder Fehlannahmen in den Modellen gesucht wird, sondern die Diskrepanz wird auf „Fluktuationen" zurückgeführt oder in der Sprache des Potsdam Instituts für Klimafolgenforschung als unerklärter „Hiatus" bezeichnet. Wenn die Klimaforscher die Ursache der seit 17 Jahren andauernden Erwärmungspause kennen würden, könnten sie die fehlerhaften Annahmen korrigieren und neue Berechnungen anstellen. Dann würden sich natürlich manche dramatischen Vorhersagen des Temperaturanstieges über 100 Jahre ändern. Auf die Ergebnisse kann man gespannt sein.

Das grundlegende Problem der Klimamodellrechnungen besteht in ihrer Komplexität. Wenn man isoliert nur die Auswirkung einer Verdopplung des Kohlendioxid-Gehaltes auf die Temperatur berechnet, ergeben die meisten Gleichgewichtssimulationen einen Anstieg bis zum Jahr 2100 um 1,2 Grad mit einer kleinen Unsicherheit (Schönwiese 2008). Die Prognose wird erheblich unsicherer, wenn man die Wirkung dieses geringen Anstiegs auf andere Messgrößen, d.h. die Rückkopplung innerhalb der Gleichungssysteme, berücksichtigt. Insbesondere ist hier zu berücksichtigen, dass wärmeres Wasser an der Oberfläche der Meere mehr Wasserdampf in die Atmosphäre entlässt. Die IPCC-Modelle nehmen an, dass durch den zusätzlichen Wasserdampf der Treibhauseffekt erhöht wird, und kommen so zu einem Temperaturanstieg um mehr als 2 Grad, allerdings mit einer großen Unsicherheit.

Es gibt aber Wissenschaftler, die den Fehler der Klimamodelle in der unzureichenden Berücksichtigung der abkühlenden Wirkung von Wolken sehen. Bei höherer Konzentration von Wasserdampf in der Atmosphäre bilden sich vermehrt Wolken, anderseits wirkt Wasserdampf auch als Treibhausgas. Die Frage ist, welcher der beiden Effekte stärker wirkt. Der Atmosphärenphysiker Richard S. Lindzen vom Massachusetts Institute of Technology (MIT) in Cambridge (USA) hat dazu Messungen mit Satelliten und Rechnungen durchgeführt. Lindzen und sein Ko-autor Yong-Sang Choi benutzten Messdaten des Satelliten ERBE (Earth Radiation Budget Experiment). Sie kommen zu dem Ergebnis, dass der kühlende Effekt dominiert, mathematisch ausgedrückt, dass der Rückkopplungsterm in den nichtlinearen Differentialgleichungen negativ sein muss (Lindzen und Choi 2009, 2011). In allen Klimamodellen des IPCC ist diese Rückkopplung als positiv angenommen.

Die Ergebnisse von Lindzen und Choi führen dazu, dass die Klimasensitivität auf CO_2 nur halb so groß ist wie in den anderen Klimamodellen. Bei einer Verdopplung der CO_2-Konzentration in der Atmosphäre ergäbe sich dann nur ein Temperaturanstieg um weniger als ein Grad Celsius. Wenn das richtig ist, gibt es keinen Grund, den Klimawandel zu dramatisieren. Lindzens Ergebnisse werden aber unter Klimatologen nicht offen diskutiert. Stattdessen wird er ausgegrenzt und persönlich angegriffen. Dabei gilt doch, was der Klimaforscher Mike Hulme vom Kings College in London sagte: „Eigentlich gibt es in der Wissenschaft nur Fortschritt, wenn sich Wissenschaftler nicht einig sind."[4]

Die Modellrechnungen über Zeiträume von hundert Jahren sind mit so großen Unsicherheiten behaftet, dass die vorhergesagten Entwicklungen der Temperatur mit Vorsicht zu betrachten sind. Die Modelle müssen wesentlich genauer werden, wenn man sie ernst nehmen will. Es ist voreilig, aus den Ergebnissen politische Konsequenzen zu ziehen.

Literatur

Lindzen, Richard S. und Yong-Sang Choi. 2009. On the determination of climate feedbacks from ERBE data. *Geophysical Research Letters* 36, *L16705; doi:10.1029/2009GL039628;* http://onlinelibrary.wiley.com/doi/10.1029/2009GL039628/abstract.

Lindzen, Richard S. und Yong-Sang Choi. 2011. On the observational Determination of Climate Sensitivity and Its Implications. *Asia-Pacific J. Atmos. Sci.* 47(4), 377-390. http://link.springer.com/article/10.1007%2Fs13143-011-0023-x/lookinside/000.png.

Mauritsen, Thorsten et al. 2012. Tuning the climate of a global model. *Journal of Advances in Modeling Earth Systems* 4 M00 A01.

Schönwiese, Christian-Dietrich. 2008. *Klimatologie.* 3. Aufl. Stuttgart: Verlag Eugen Ulmer.

---

4   Das Interview mit Mike Hulme findet man in der Süddeutschen Zeitung vom 1.9.2014.

12. Are We Living in a Computable World? Answers and Controversies from Chinese Scholars

Chadwick Wang[1]

Abstract

The ongoing controversies in China whether "society can or cannot be computed" are both in and about science and they are far from closure. It is found that this problem originates in China's historical division between the natural and the social sciences, as well as in its long established culture regarding a "differential mode of association", which is present even in academia. Scientific proposals in this context, such as that "the natural sciences and the social sciences must closely work together", are politically improper and somehow incompatible with the overall ideology. A consensus on the computability of society problem is not to be expected under this fragmentation.

1   Dr. Chadwick (Chengwei) WANG is an assistant professor at Tsinghua University, Institute of Science, Technology and Society and a visiting scholar at Harvard University, Department of the History of Science. His principal research interests include discursive history of science, technology and innovation, as well as non-human actors in the institutional entrepreneurship. This research is funded by the Chinese Scholarship Council (CSC).

12.1   Prologue

Can society be computed, or not? That is the question.

On June 13-14, 2008, a special academic salon was held by the China Association for Science and Technology (CAST) in Beijing. A number of scientists and social scientists met together (although only one from management science, one professor of obstetrics and gynecology and one government officer could be counted as representatives of the soft sciences[2]) to discuss one of the most popular terms back then and still today, "social computing", with the aim of providing a proper answer to the question being asked at the very beginning of this paper: Can the society we live in and inhabit be or not be computed?

Surprisingly, their viewpoints were as diverse as they were fascinating[3]: "Society is definitely computable, with no doubt. And social computing should serve the development of our harmonious society with regard to its complex nature", one contributor said. "My team has already been practicing this for years....", another professor echoed. Others were not so optimistic, however. "Our primary job here is to clarify the definition of 'social computing', as dozens of versions may exist," said an authoritative expert on social computing, for example, politely pushing aside the "definitely can" idea. "A new scientific system must be established however, to compute the incomputable problems in the complex social systems", he added, "and for this greater good, the natural sciences and social sciences must closely work together". Unfortunately, the ambition to bridge the natural and social sciences did not attract much attention. On the contrary, the issue of "how to compute the incomputable" turned out to be the focus of discussion. One professor, for example, proposed that society is always computable if "computation is further broadly defined as information processing," while a senior academician from the Chinese Academy of Sciences (CAS) argued from a more operable perspective: "Whether or not society can be computed indeed heavily depends on how we form hypotheses for the models ... and probably, we may always assume that the 'society' in models is computable".

---

2   "Hard science" and "soft science" are terms frequently used in China to compare and also label scientific fields on the basis of perceived methodological rigor, exactitude, and objectivity. Roughly speaking, the natural sciences are considered "hard", whereas the social sciences are usually described as "soft" (Price 1970).

3   The live discussion in this salon was recorded and published by: Academic Department of CAST. *Is Society Computable?* Beijing: China Science and Technology Press, 1999. (in Chinese)

Then another question emerged, "Which 'society' are we computing?" Some of the discussants believed that the models in "social computing" should be examined by the empirical data, which is the only way to "eliminate the false and retain the true". Others refuted this idea by referring to Plato's Utopia and Popper's three worlds[4], which implies that it is only the parallel "artificial society" they can really compute and that is "inspiring enough" at this stage, and so on. Beyond the definition as a result, there was actually no consensus on proper solutions for the paradox of computing the incomputable or for filling the gaps between computing models and social reality. The "can or cannot be" question that was asked is indeed still unanswered.

These are neither the first nor last controversies on this topic, and it is not even helpful to expose them all. Accordingly, this paper will modestly take a step back to focus on how and why the question is framed and responses in certain social contexts.

12.2 What's past

12.2.1 The Helplessness of STS Controversies Studies

Controversies characterize intellectual change and developments within and about science and technology (Engelhardt and Caplan 1987). As a distinguished feature and necessary condition for the so-called paradigm shift (Kuhn 1962), controversies have long been considered to be one of the essential topics in science and technology studies (STS). Especially in the 1970s and 1980s, there was a tendency to open up different scientific and technological controversies, and expose the "soft social inside filled with seeds of everyday thought" (Collins and Evans 2002, p. 248). Although scientific and technological controversies are usually seen to be the sorts of disputes that are to be resolved by appeal and to rigorous reasoning concerning facts and artifacts, many (if not all) of these controversies have heavy social and political overlays (Engelhardt and Caplan 1987). And moreover, these overlays are twofold.

---

4   Popper split the world into three categories: World 1: the world of physical objects and events, including biological entities; World 2: the world of mental objects and events; World 3: objective knowledge. See: Popper, Karl R. *Objective Knowledge: An Evolutionary Approach*. Oxford: Clarendon Press, 1972.

First of all, only a few controversies can truly lead to the paradigm shift, as long as they do, however, the fact and artifact along with the context that produced them will evolve through revolutions. Take the famous controversies on *plenists/vacuists* for instance: Not only was the knowledge of the "air's *elasticity*" deepened, but also, and more importantly, the tradition of experimental science was also established when deduction was previously assumed to be the only way of (re)producing scientific knowledge (Shapin and Schaffer 1985). In the debates over several candidate technological solutions for the modern bicycle as well, more than just a pure selection process was implemented. New social norms and habitus, such as the attitude toward women wearing trousers, were changed and the love of outdoor life was fostered in the pursuit the personal mobility of human beings (Pinch and Bijker 1987; Herlihy 2004, p. 264).

Nevertheless, compared with the social influence of the controversies, STS is more inclined to explore the mechanism of the "closure" of controversies, that is, the conclusion, ending, or resolution of a controversy. And more than fact or truth, it is believed that the representation of artifacts themselves and the interpretation by relevant social groups really matter with respect to the end of controversies. The acceptance of "the air's *elasticity*" for example, was positively correlated to the diffusion of scientific equipment, in particular the air pump, as "inscription devices" (Latour and Woolgar 1986; Latour 1986,1983). And the groups with higher intellectual faculties, such as "Oxford professors" or upper class fans of "smoke and mirrors", were considered to be more credible and reliable as witnesses than the "Oxfordshire peasants" (at least in the early stage) (Shapin and Schaffer 1985, p. 58). In the case of the bicycle, however, the dominant design was historically shaped by different social groups with conflicting technical requirements (generally speed for men and safety for women) and solutions (Pinch and Bijker 1987). And the only design[5] that succeeded in the end is precisely the one which can and does coordinate and mediate all requirements: "...*to youths it gave speed; to women, freedom; and to many ordinary citizens it was simply a source of great pleasure and utility. To all, it offered exercise and adventure...*" (Herlihy 2004, p. 264).

After all, STS has developed remarkably effective and insightful perspectives on investigating controversies. For example, the distinction between controversy *in*

---

5   Tricycles were once regarded as the machines permitted for women. But engineers and producers anticipated the importance of women as potential bicyclists (Pinch and Bijker 1987). As a result, "A sudden desire awoke in the feminine mind to ascertain for itself, by personal experience, those joys of the two-wheeler which they had so often heard vaunted as superior, a thousand times, to the more sober delights of the staid tricycle" (Herlihy 2004, p. 244). One possible reason for this is the desire for the "dignity and equality of women" (p. 266).

science, which refers to a disagreement (assumed to be) about "a matter of belief, of knowledge claim" by the predominant scientists (McMullin 1987), and controversies *about* science, where there may not be a common understanding about the problem at stake and a focus on diverse moral, political, and economic concerns (Nelkin 1995, 1979); the symmetrical approach to open up the "black box" of controversies, aiming to explore the reasons and forces for the debates to disperse the air of inevitability, even addressing arguments that failed in the end (Sismondo 2004); and the way it is terminated, i.e.: (1) *Resolution*: A controversy is resolved when an agreement is reached on the merits of the case in terms of what the participants take to be standard epistemic factors; (2) *Closure*: A controversy reaches closure when it is terminated on the basis of non-epistemic factors, such as the authority of the state, the pride, ambition, or laziness of a controversialist, or the withdrawal of publication facilities; (3) *Abandonment*: Controversies may terminate through participants losing interest (McMullin 1987; Engelhardt and Caplan 1987).

The problem in applying these perspectives, however, is that the "can or cannot be computed society" issue is both *in* and *about* science, or to use a more formal word, *technoscience* (Latour 1987; Haraway 1997), both epistemological and methodological. Even trickier, these are the ongoing controversies that have not reached a closure yet, thus it is even impossible to apply the symmetrical analysis where the "right" or "wrong" arguments must be established but re-unlabeled. What we are only sure about is that the controversies surrounding this question are continuous episodes in construction, and the only question that could be answered is how these disagreements are shaped by society (including cultures and events) and vice versa (Sismondo 2008).

12.2.2   The Monodrama of the "Theory of Everything"

Historically, the definition of social computing referred specifically to the nature of a computing application "in which software serves as an intermediary or a focus for a social relation". "If software were developed by individuals for their own use at their own behest, the social nature of computing might not exist", it was argued, while "the social nature of software is inescapable" (Schuler 1994). The primary argument, as a matter of fact, is made from the technological point of view, as computing and computers at that time were an ongoing source of change in the way we conduct our lives (Nissenbaum 1994). The only reasons "social" aspects were addressed was because the computing technologies along with the artifacts were built *for* "social" (inter)actions. Besides, there were already certain controversies on the economic and social role of computation, such as work life, class divisions in society, human safety and critical computer systems, democratization, employment, education, gender biases, military security, health, computer literacy, privacy and encryption, scholarship and so on (Kling 1996). No doubt these controversies were essentially *about* computing, with the main concern of getting out of the "social" control or even putting society at harm or risk (Nissenbaum 1996,1994).

Later, in the era of Web 2.0, a large number of new applications and services came to dominate the Internet, where collective action and social interaction online were facilitated with the rich exchange of multimedia information and evolution of aggregate knowledge. Also, in such a way, the meaning of social computing shifted from the professional computation job performed by experts to the everyday use by ordinary people with no expertise (as users), enabling them to "manifest their creativity, engage in social interaction, contribute their expertise, share content, collectively build new tools, disseminate information and propaganda, and assimilate collective bargaining power" (Parameswaran and Whinston 2007; Charron et al. 2006). Exactly as addressed in the *Science* article "The coming age of computational social science", the advance of computer and network technology as well as the capacity to collect and analyze massive amounts of data have been unambiguously enhanced, which dramatically transformed the society we inhabit (Lazer et al. 2009). The technology-dominant way of thinking, or we can say the technological determinism outlook, however, had not changed at all. The nature of social computing in the Web 2.0 is still expected and assumed to make computing or computer-related technologies more sociable *for* society, and even *for* social sciences research in the sense of enriching the approach and infrastructure for collecting data. The "can or cannot be computed society" question, as before, is put aside.

There are only few exceptions in the definition, actually, where balanced social-technical viewpoints are implemented. One of the conceptualizations is from the very Chinese scholar who ambitiously intends to bridge natural and social sciences as mentioned in the prologue. According to his interpretation, social computing should be better understood as a combination of "computational facilitation of social studies" and "human social dynamics" with "the design and use of ICT technologies that consider social context" (Wang et al. 2007, p. 79). That is to say (as shown in Figure 12.1), not only the technological infrastructure and applications of social computing are included. The "theoretical underpinnings", where natural and social sciences "join together"[6] to identify the cross-disciplinary nature of social

---

6   This argument makes Wang's theory essentially different from a one-way transplant of ideas from the Chinese version of "social physics". By emphasizing the Chinese version, it is implied that the non-Chinese actually define this term in another way. For

computing, are also listed as an important and fundamental part. Moreover, due to the "difficulties of testing real systems that are inherently open, dynamic, complex, and unpredictable", simulation methods of computing artificial society, such as the agent modeling techniques, as well as the computational experiments with artificial systems are employed to incorporate "social theories into technology development". At the same time, real and virtual data will be collected and co-analyzed in parallel using data mining and text mining tools in order to "to seek effective solutions" (Wang et al. 2007, p. 80).

Finally, in Wang's "theory of everything" model, the "can or cannot be computed society" question is answered: Society is too complex to be perfectly computed. What we can do accordingly is to simulate the artificial society based on the theoretical underpinnings, and test as well as modify the solutions with the real-world data. Thus he not only wants to cross the gap between natural and social sciences, but also between the method of simulation and hypothesis testing. This idea is not brand new of course. For example, agent-based modeling and virtual experimentation are also frequently used in "computational sociology"[7] (Hummon and Fararo 1995; Macy and Willer 2002; Squazzoni 2012). Wang himself also admitted in his later works that his contribution primarily lies in "the combination and intersections of the information technologies and social studies, especially their forms on the Web" (Wang 2008), for "dealing with issues of cyber/physical interactions" (Wang 2009). Though far less than a paradigm revolution, his distinguished viewpoint can definitely be considered a Schumpeterian sense of innovation (Schumpeter 1934).

---

example, MIT media lab conceptualized social physics as "a quantitative social science that describes reliable, mathematical connections between information and idea flow on the one hand and people's behavior on the other. Social physics helps us understand how ideas flow from person to person through the mechanism of social learning and how this flow of ideas ends up shaping the norms, productivity, and creative output of our companies, cities, and societies". See: Pentland, A. (2014). *Social Physics: How Good Ideas Spread—The Lessons From a New Science*. New York, The Penguin Press.

7   Computational sociology (sometimes called "social simulation") is an outstanding method for modeling and building explanations of social processes from a "bottom up" approach, based on ideas about the emergence (and "self-organization") of complex behavior from simple activities in the social sciences (Salgado and Gilbert 2013).

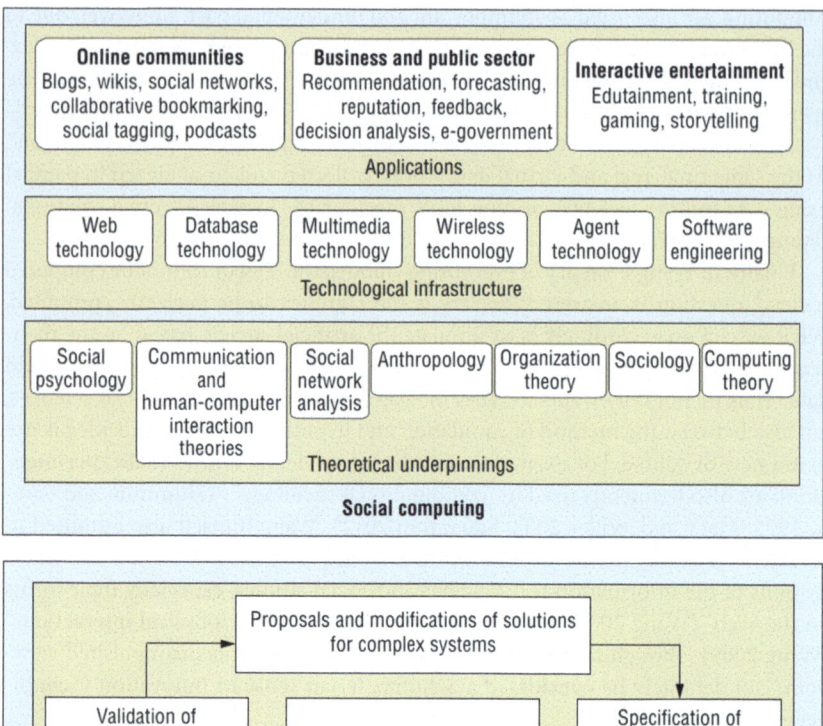

Fig. 12.1:   A New Definition of Social Computing from a Chinese Scholar (Source: Wang, F. and K. M. Carley, et al. 2007. Social computing: From social informatics to social intelligence. *Intelligent Systems, IEEE* 22 (2): 79-83.)

Wang was not the only one who had such an ambition. As explicitly mentioned in his framework, complex system theory had proposed a similar approach long before, although it may not be easily applicable. Complexity scholars believe that "complexity and nonlinearity are prominent features in the evolution of matter, life, and human society … and the complex system approach may be a method of bridging the gap between the natural sciences and the humanities" (Mainzer 1994, p. VII, 10). Especially for computing a society, where it is usually difficult to provide empirical tests and confirmation of the complex models, "the computer-assisted simulation of complex cultural systems has become a crucial instrument for providing new insights into their dynamics which may be helpful for our decisions and actions" (Mainzer 1994, pp. 267-8). Actually, in the 5th edition of Mainzer's (2007) book, the idea of "ubiquitous computing"[8] was introduced to encompass much broader computing by means of what we today call the "web of things". As a natural scientist, Wang was inspired and took a further step beyond philosophical thinking by providing a more technical specification for interoperability (TSI).

	SOCIAL COMPUTING	COMPUTATIONAL SOCIOLOGY	WANG'S FRAMEWORK	COMPLEXITY APPROACH
COMPUTE *FOR* THE SOCIAL	1	0	1	1
COMPUTE THE SOCIAL'	0	1	1	1
BRIDGE NATURAL/ SOCIAL SCIENCES	0	0	1	1
TSI	1	1	1	0

Note: Social' refers to the artificial society that was computed, which may or may not coincide with the "real" social.

Table 12.1: A Comparison of the Related Theories

---

8   Ubiquitous computing is a concept in software engineering and computer science where computing is made to appear everywhere. In contrast to desktop computing, ubiquitous computing can occur using any device, in any location, and in any format. It is said that Mark Weiser coined the phrase "ubiquitous computing" around 1988 (or 1991), during his tenure as Chief Technologist at the Xerox Palo Alto Research Center (PARC).

In the sense of a "new combination", Wang's social computing model excellently builds up the alliance with the traditional social computing and computational sociology theories, and also implements the complexity approach (as shown in Table 12.1). If these ideas were taken by the Chinese academia, the controversies at the very beginning would never have happened. Since the focal controversies occur in the Chinese scientific arena and not many influential works by Chinese scholars have been published in English journals, the status quo Chinese literature on related topics will be mapped out in the next section.

12.3   Show more than thou speakest

12.3.1   Research Method and Strategies for Data Collection

There is too much related literature to be reviewed, however, which makes it impossible to represent the controversies with words. Fortunately, controversies are mainly the "argumentation in print" (Keith and Rehg 2008), and figures can be used to visualize these disagreements.

The method for visualization that will be used in the following is called "co-word analysis", an established method[9] that has been utilized in scientometrics since the 1980s for exploring and elucidating the structure of scientific knowledge (as well as ideas, problems, etc.) represented in appropriate sets of documents (Whittaker 1989; Callon et al. 1983; Law et al. 1988). Co-word analysis draws on the assumption that the keywords for a paper (or other words, such as titles) constitute an adequate description of its content or of the links that the paper established between problems. Two or more keywords co-occurring within the same paper are an indication of a link/strategic alliance between the topics/research themes to which they refer (Cambrosio et al. 1993). With this approach, co-word analysis reduces and projects the data into a specific visual representation while maintaining the essential information contained in the data. It also reveals patterns and trends in a specific discipline by measuring the association strengths of terms representative of relevant publications produced in this area. It is based on the meaning of words, which are the important carriers of scientific concepts, ideas and knowledge (Van Raan and Tijssen 1993; Su and Lee 2010; Ding et al. 2001).

---

9   We should be aware, however, that a more traditional bibliometric technique is the author/journal co-citation analysis, although its main problem is its "inability" to picture the actual content of scientific topics.

In this section specifically, a social network analysis and bibliometric keyword analysis were integrated in order to draw a picture for the "controversies map" for studies related to "social computing" in China. The processes for this research method are: (1) literature retrieval and filtering, (2) keyword revision and basic statistical description, (3) visualization of a 2-dimensional keyword network, (4) network properties calculation and subgroup analysis, as well as (5) comparison of domestic and international network structure.

To retrieve sufficient "social computing" related papers, the CNKI (China National Knowledge Infrastructure) database, a recognized platform for domestic publications, is initially used for the data collection. Furthermore, to avoid the so-called "indexer effect"[10], the symmetry principles described in Section 2.1 are implemented. Specifically, all the terms in the official controversies[11] as well as all theories reviewed in Section 2.2 are queried: (1) social computing, (2) computational sociology, (3) social physics[12], (4) artificial society, (5) social complexity, and (6) social system(s), from 2001 to 2014.

---

10   "Indexer effect" refers to the phenomena of results that were, in any case, influenced by the way in which the indexers who chose the keywords conceptualized the scientific fields with which they were dealing, so that the pictures which emerged were more akin to their conceptualizations than to those of the scientists whose work was to be studied (Whittaker 1989).

11   The special salon held by CAST is one of the top-level academic events in China, with all of the speakers regarded as the academic authority selected by government officials in CAST.

12   "Social physics" was included in the search strategy because this idea was influential in Chinese academia for many reasons. One of the academicians is the advocator for this idea, who also participated the "can or cannot be computed society" controversies organized by the CAST, has a political identity as the consultant of the State Council, and is a member of The World Academy of Sciences (TWAS) for the advancement of science in developing countries. Interestingly, one of his former PhD students, who also works for CAST, was invited to this special academic salon and gave a presentation entitled "Getting to know social computing from the lens of social physics".

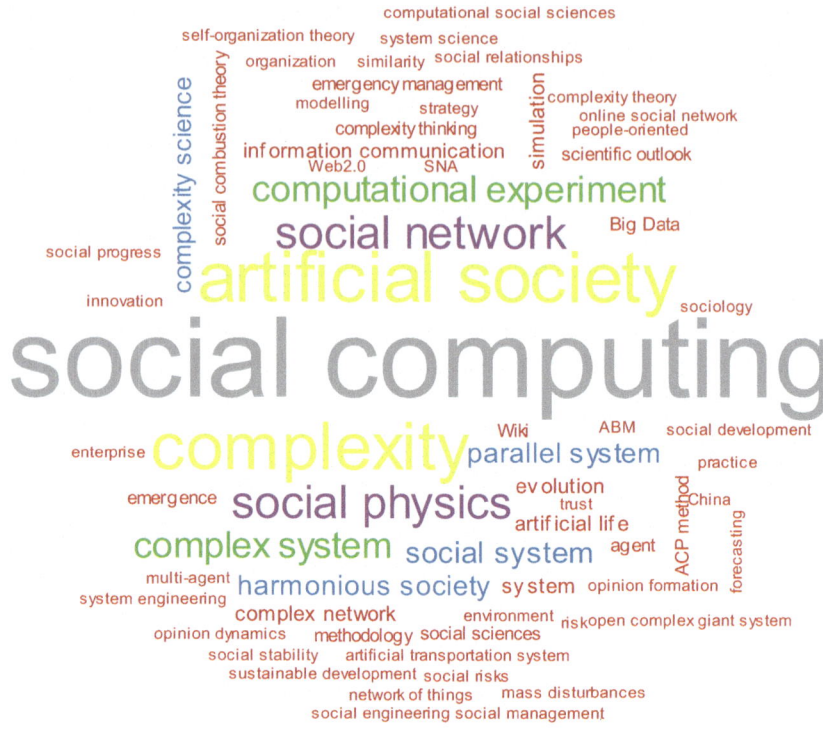

Fig. 12.2: Keyword cloud from the papers in the "social computing" controversies (appearance frequency3, generated with R "word cloud" package)

402 papers were finally selected as the bibliometric sample after examining the paper titles and abstracts and manually deleting all irrelevant and non-academic publications (identified by the lack of keywords provided by the authors). The keywords were extracted by the *Statistical Analysis Toolkit for Informetrics* (SATI)[13] software. 1,182 keywords were generated from all these papers, with "social computing" appearing 82 times as the keyword with the highest frequency (6.93%)[14], and "artificial society" in second place with 50 occurrences (4.23%), while the other three in the top five are "complexity" (46, 3.89%), "social network" (29, 2.45%), and "social physics" (28, 2.37%). It should also be noted that the attenuation of the keywords

---

13   See: http://liuqiyuan.com/#sati.
14   % refers to the ratio with which keywords appear, applies to the same below.

frequency spectrum is quite sharp (as shown in Figure 12.2, which is drawn in line with the positive correlation relationship between the words' size and their appearance frequency). The first keyword with an appearance frequency of two is ranked 68th, while the first of one-frequency keywords is ranked 165th, which means the majority of keywords (86.1%) in these papers show up only once. It is clear however, that this distribution should not be regarded as a sign of disagreement between the scholars. Some of the keywords, such as those with extraordinary high appearance frequency, might play the role of the "hard core" [15] of an established "social computing" paradigm, which further makes the other keywords merely do the "puzzle-solving" (Kuhn 1962, p. 35-42) jobs. The simple frequency statistics method thus could not help with the identification of controversies. More linkage data should be supplemented in order to map out the relationships among the keywords.

Again, to avoid the "indexer effect", the linkage we intend to explore is based on the authors' original keywords. For example, although "social complexity" is logically a kind of subset of "complexity", they are still regarded as different keywords, because they are treated differently in the original studies by the authors themselves.

12.3.2   Visualization of Controversies in Vivo

Social network analysis (SNA) is used to map and measure relationships among components in these controversies. A network in SNA consists of a set of nodes and links. The nodes represent the components, and the links stand for relationships between the nodes. In this section specifically, we structure the keyword network of research on "social computing", where the nodes are the keywords provided by the authors and the links represent the co-occurrence of those keywords. The co-occurrence matrix is generated by the SATI software, and the cutoff threshold is set to "co-occurrence frequency≥2" (13.7 as a share of the total number of keywords).

A primary objective of this co-word analysis is to explore the substructures of the controversies network, as stated before, a top-down approach will be used accordingly (Carolan 2014; Hanneman and Riddle 2011). In particular, the Girvan-Newman iterative algorithm[16], which is widely accepted in social computing

---

15   "Hard core" theoretical assumptions (auxiliary hypotheses) that cannot be abandoned or altered without abandoning the program altogether (Lakatos 1978).

16   The GN-algorithm seeks to create clusters of nodes that are closely connected within, and less connected between clusters via calculating the edge betweenness centrality of all the edges and then deletes the edge or edges with the highest value as a typical "block

and complex system studies, will be applied to the binary co-occurrence matrix, which solely counts whether or not associations between keywords exist, to detect the subgroups (communities, which are further represented by different shapes and colors of nodes in figures). Moreover, since the communication between two nodes in a network can be facilitated, blocked, distorted or falsified by a node falling between them, the *node betweenness centrality*[17] as the medial measure (Freeman 1977,1980; Borgatti and Everett 2006) is also demonstrated in the figure in terms of the nodes' sizes. That is to say, the bigger the size of the nodes in the figure, the more the other nodes tend to rely on them to communicate.

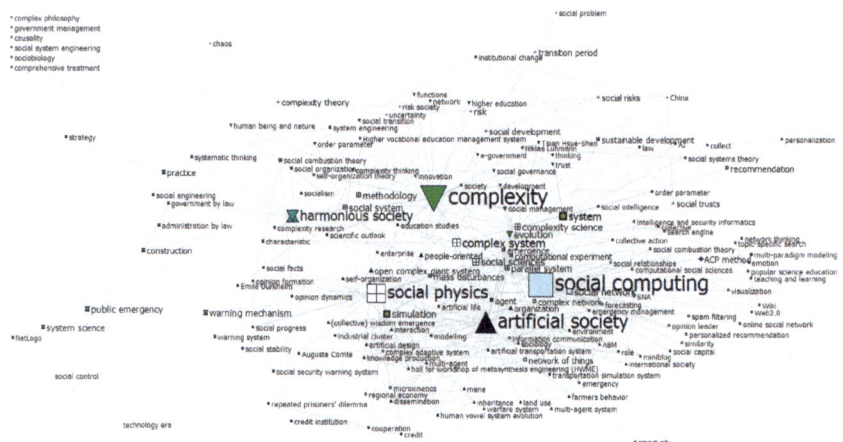

Fig. 12.3: Subgroups in the "social computing" controversies co-words network (Note: co-occurrence frequency≥2 as the cutoff condition)

modeling" method in SNA (Girvan and Newman 2002). The blocks are confirmed by the goodness-of-fit of the configuration with a maximum of Q-value; where Q-value or in other words Newman and Girvan's modularity Q as a measure of the quality of a particular division of a network, means the fraction of edges that fall within the partition minus the expected such fraction if the edges were distributed at random (Newman and Girvan 2004). All calculation is provided by the standard package in UCINET and NetDraw software (Borgatti et al. 2002).

17   Loosely described, the betweenness centrality of a node is the number of times that any actor needs a given actor to reach any other actor.

Generally speaking, the most prominent feature of this co-words network of "social computing" controversies (as shown in Figure 12.3) is fragmentation. The network is partitioned from 2 to 100 blocks via the GN algorithm. And the height of a peak with $Q=0.510$ is the best community division we can obtain with the blocks number of 10. More importantly, even beyond the isolated (degree=0) and pendant (degree=1) subgroups, which are already marginal in the controversies, there has not yet been any sign of the emergence of dominated subgroups (in number of nodes covered) among the remain 8 modularities. The main connections between the fragmented subgroups at the same time are via the general concept mediators with high node betweenness centrality, such as "complexity" (24.211), "social computing" (22.501), "artificial society" (19.613), "social physics" (16.680) and, very surprisingly, "harmonious society" (9.219), which is definitely a political formulation (Schoenhals 1991, *tifa*, 提法) but not an academic term, as the top 5. Although many papers in this sample use the same term as keywords as a result, they may discuss completely different things or even could not manage to get into the same page. Actually, all of these mediators in the controversies network are "big words" with highly interpretive flexibility (Pinch and Bijker 1987), akin to a huge umbrella, under which one can talk about almost everything according to their cognition and interests. Just as with the dispute in the CAST salon, there are a thousand "social computings" in a thousand scholars' eyes. Or in other words, to some extent, the controversies even do not exist, since everyone is so focused on telling their own stories.

The strength of the association is then added to the visualization as a resolution parameter of the GN modularity clustering (Waltman et al. 2010), a clustered density map is thus obtained as shown in Figure 12.4. Obviously, the fragmentation and "self-serving" still exists. And even worse, the "topic islands" begin to appear: The non-academic ones with a political complexion, such as "harmonious society" and "inherence" (*chuancheng*, 传承), are apparently partially isolated. Strangely, some of the key concepts in Wang's "theory of everything" model, like the "ACP method"[18], as well as "artificial design" and "multi-agent", partially fall into these "islands", too.

---

18 ACP stands for "artificial systems, computational experiments and parallel execution", according to Wang's model.

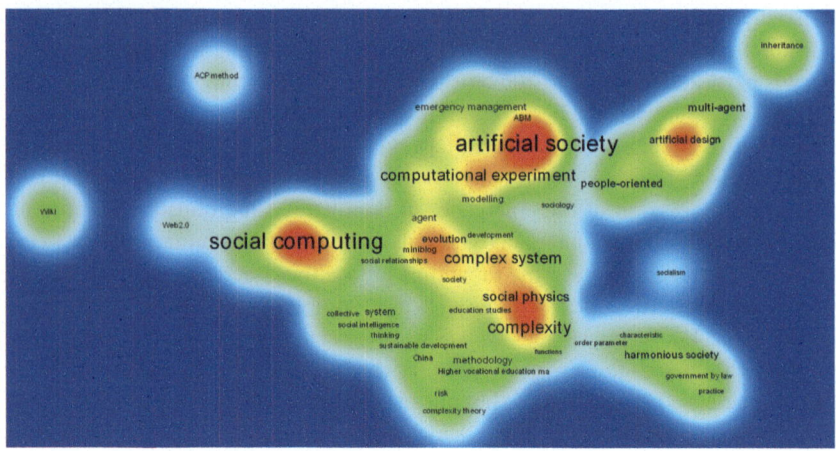

Fig. 12.4: Density map of the "social computing" controversies co-words network (Note: co-occurrence frequency2 as the cutoff condition, via the VOSviewer software)

12.4   The game is up

12.4.1   History of China's Social Commuting Literature Revisited

So what truth will come out from the fragmented "social computing" controversies in China? Let's go back to the historical context to find some answers, since "history matters" (North 1990, p. vii). The sampling in the last section starts from 2000, when the first Chinese papers with "social computing" and "social physics" as their essential ideas were published. If we further trace other related keywords back to the 1980s, a totally different picture emerges.

Comte's "social physics"[19] term, for example, was introduced by a Chinese academic journal for the first time in 1984[20]. Very interestingly, the article logically started by introducing Comte's idea of "social physics". And then rapidly switched to Marx's critique of Comte with regard to his position as a "spokesperson for capitalism". "There is a sociology for capitalism, and a sociology for socialism", the authors argued, "thus the sociology as a discipline in China should be rebuilt... to help solve the social problems in the 'Four Modernizations' (*Sihua*, 四化)[21]". Tough "modern science and technology were embedded in the legitimating ideologies and governing structures of the regime" (Greenhalgh 2008, p. 5), sociology in the early post-Mao era was nearly suspended for ideological reasons (Arkush 1981). And actually, all other social sciences were "either abolished or forcefully transformed into fields of Marxian inquiry and annexed by the land of ideology/politics" (Greenhalgh 2008, p. 28). There were only a few publications which introduced or implemented related ideas, but all intended to serve ideology with no exception, since it was well accepted at that time that "calculation and supervision is the essential condition for making communism work" (Малышев 1963 [1960], p. 1). In the very book cited, the term "social calculation" was used for the first time in Chinese literature, in which it was regarded as an "essential instrument to manage the planned social economy". Beyond translation, the calculation concept was eventually practiced, for example because there was an urgent need for China's GDP indicators to correspond with the western world (mainly the U.S.) after China got its "legitimate position back in the United Nations" (Qian 1982).

Under the shadow of ideology, none of the social scientists' work could reach the power and influence of the natural scientists. Scientists in the field of cybernetics (as part of complex system theory), for instance, managed to change one "basic state policy" (*jiben guoce*, 基本国策), the one-child policy, by computing the society using the then cutting-edge technologies, computer via mathematization and visualization (Greenhalgh 2008). Thus the advocators for "social physics" were indeed natural scientists who devoted themselves to applying the principles, methods

---

19 Many people attribute the origin of using the term "social physics" to Auguste Comte, when he tried to bring in the positivism idea to social sciences (*science sociale*) research, where the negative, metaphysical speculation was supposed to be avoided. According to Comte, the social sciences field should work similarly to the natural sciences, thus this field was also referred by him earlier as "social physics".

20 This article was published in "*Dongyue Tribune*" (东岳论丛) 1984, (01). To prevent any political or ethical issues, no author information was provided.

21 The Four Modernizations as a political *tifa* were goals first set forth by Zhou Enlai in 1963, and enacted by Deng Xiaoping, starting in 1978, to strengthen the fields of agriculture, industry, national defense, and science and technology in China.

and ideas of the natural sciences to social research, which could never escape from the trace of ideology. As stated in one article published in a reputable STS journal titled "Studies of Science of Science" in 1986[22], "the subject of social physics is the natural properties of society, it includes mechanic, thermodynamic, electromagnetic and systematic properties of society, etc. The social physics of today, which can serve as fodder for other social sciences as well as Marxism, ... is different from the social physics of Comte". The author, who was then an official in CAST, made the same arguments in many other subsequent papers, which represented the approval from the government as well as CCP. As a matter of fact, it was frequently proposed at that time that the surplus of talents in physics should turn and "down" to the sociology research to accommodate the global trend. The consultant of the State Council who participated in the CAST salon is just one of the early practitioners.

For ideological reasons as well, then, people totally believed that social physics or any other forms of social computing, like population prediction and control, could work perfectly, because society must have some (if not all) "natural attributes", which could be treated as "objects" and certainly be computed in the way that was done in natural sciences, no matter how complex it is. After all, the intellectual and positioning divisions of natural and social sciences had been established. And the path dependency determines that Wang's ambition to bridge the two would fail, or at least be extremely difficult.

12.4.2   Boundary Drawn Academically

This is not to intentionally propagate the historical determinism. As long as the communication and interaction between the two sides persisted, there could be some new mechanism other than isolation and fragmentation emerging. Unfortunately, the truth is just the opposite.

Let's continue the "social physics" story. The first and probably the last "social physics" seminar to which both natural and social scientists (actually even more humanists) were invited, was held in July 1986. During the meeting, complex system theories with similar ideas were brought forward to back up "social physics" as an alternative intellectual foundation[23]. But their gallantry was never appreciated by the natural scientists. One reason is because the humanists (mainly philosophers) over-emphasized the "dialectical unity" (of regularity and occasionality) although

---

22   See: "Studies of Science in Science" (科学学研究), 1986, (04).
23   See: "Journal of Dialectics of Nature" (自然辩证法通讯)1986, (05).

they also tended to admit "the complexity of society is determined by the 'Law of History'"[24], while in natural science only contingency was welcomed. And the second reason is their simple symmetry positioning between nature and society, although in some natural philosophy articles, it was proposed that scientific theories like the "second law of thermodynamics" be applied in the analysis of the complexity of society[25], which is believed far beyond the capability to do themselves.

For the sake of desperately seeking certainty as well as escaping the metaphysical guidance, natural scientists had a deep, heartfelt hope to truly practice the "social physics" idea like engineers. As a result, the Systems Engineering Society of China (*Zhongguo Xitong Gongcheng Xuehui*, 中国系统工程学会) was founded in 1980 with the support of Hsue-shen Tsien (钱学森) and many other famous natural scientists. Just 2 years after the society was founded, a professional board for social and economic systems engineering research was established, and numerous "social"-related papers were published in the proceedings of the Society, titled "Systems Engineering: Theory and Practice". From 1981 to 2000, 90 "social"-related[26] academic articles were published in that journal, with few of them contributed by humanists or social scientists, except for a few management scholars, who imitated natural scientists by applying natural sciences theories and techniques to social and, most importantly, economic research. Those who could or would just deal with philosophical issues were eventually isolated.

For most humanists and social scientists, however, the distinctions in terms of journals, methods, disciplines or even paradigms never bothered them at all. In one article published in the "Journal of Systemic Dialectics", it was delightedly, but rather naively argued that the concept of complexity, which was traditionally the domain of social sciences, was gaining its own popularity in natural science[27]. They seemed to never learn the lesson from the case of the population issues, when the natural scientists really came "down", there would be nothing left for them to study. Fortunately, there were some who did foresee the risk of coupling their tiny forts, a society called the "China Institute of the System Science" (*Zhongguo Xitong Kexue Yanjiuhui*, 中国系统科学研究会) as well as a journal titled "Chinese Journal of Systems Science" were thus established in 1993, to play their own metaphysical games on their academic "territories". They were in favor of and really proud of

---

24 See: "Studies on Marxism" (马克思主义研究) 1986, (03).

25 See: "Qinghai Social Sciences" (青海社会科学) 1988, (04).

26 Technically, the "social"-related articles were identified by either topics, abstracts or keywords mentioning "social" or "society".

27 See: "Journal of Systemic Dialectics" (系统辩证学报) 1994, (01).

the "science" in their society's title, where to them, science means something more "fundamental" as compared with the practical engineering approach.

Inside each camp, actually, the splitting up continued by establishing their own professional committees in the *Society* or *Institute*, founding new existence journals or occupying existing ones, forming a citation alliance with their colleagues, friends, students or others, and so on. Each work done was to ensure they "stand at the center of the circles produced by his or her own social influence" in the "differential mode of association" (*chaxu geju*, 差序格局) (Fei 1992 [1947], pp. 62-63). The Consultant of the State Council mentioned above published the "Bulletin of Chinese Academy of Sciences" in 2001[28]. A definition of "social physics" was provided along with his own proposed "social combustion theory" as a means to formulate an "early warning system" for maintaining the stabilization (*weiwen*, 维稳). His political identity and the ideology-favorable topic made sure he attracted lots of followers domestically, although the theory was developed to "simulate and forecast" the national stable situation, while the method for performing the simulation and forecasting was not provided for, except for using the metaphor of combustion. Moreover, 8 years later, the expanded theory[29] was finally published in English in the proceedings of an international conference, but has not been cited at all until now. Therefore, the way of Chinese academia working could never break up the historical path dependence, but rather empowers the distinctions by all means, which eventually leads to the fragmentation of controversies, or to put it more properly, more self-serving (as well as self-protection) in the small cliques.

With respect to the "can or cannot be computed society" question at the same time, very few care about it. The focus is on "have to", "how to", and "be able to". And the worldwide specialization of disciplines echoes with the local "differential mode of association" culture. No wonder Wang's ambition could not get its deserved attention and response in China. Nevertheless, on the international stage, Wang's (2007) paper attracted 120 citations in the Web of Science Citation database by 2014, which is a remarkable achievement for Chinese publications. 28 of them, however, are from the two journals Wang serves on as a chief editor, and the main part of the co-author citation network is Chinese or more precisely Wang's kinship network (as shown in Figure 12.5).

---

28   See: "Bulletin of Chinese Academy of Sciences" (中国科学院院刊) 2001, (01).

29   The main expansion extends into another concept called "Social Harmony Equation" (SHE), which is believed to lead the "social system to the evolution direction of social by accumulation of 'social combustion substances'".

Fig. 12.5: Wang's (2007) Co-author Citation Network in the Web of Science Citation Database (Note: co-occurrence frequency≥2 as the cutoff condition)

12.5   What's done is done

Whether or not society is computable is truly a tough question that cannot be quickly and easily answered. A number of studies in disciplines and specialized fields also illustrate that cooperation can proceed *without* consensus (Clarke and Star 2008). Unfortunately, no sign of cooperation has been found, instead, only the isolation, fragmentation and small cliques are left and sustained, which makes the controversies in both the CAST salon and academic journals act more like a salad bowl rather than a supposed-to-be melting pot. We should not expect more from the simple co-word analysis unless we intend to over-interpret it. But the inconvenient truth is that the problem of China's fragmented controversies on the "can or cannot be computed society" is deeply rooted in the historical division of natural and social sciences, as well as in the long established "differential mode of association" culture.

Is it going to be changed, or still be out of control? Let's jump out of the academic box to see what is happening in society.

As a matter of fact, the paradox of "compute the incomputable" is already there. The metaphor of "society as a complex system" was used as early as the beginning of the Reform and Opening-up to refer the complex situation China was facing, so that one should be aware of the difficulties and risks lying ahead[30]. Political formulations like this were actually taken for granted in many ways. One paper later published in a top academic journal in 1988 was even titled "On Reform as an Arduous Social System Project". But very few words were used to substantiate its argument, except the general declaration that "this is the order of history" and "similar developments are happening in other countries"[31]. Like many other articles, the metaphor of reform as a "complex project" was the starting point of almost any related academic discussion at that time, and frequently showed up in the "introduction" part to gain legitimacy by catering to the ideology / national interests. At any rate, the system perspective and approach were applied crazily and freely to "planned immunization", "athlete cultivation" and so on, especially after the "construction and perfection of a socialist market economy as an arduous and complex system project" was made a "fixed expression" by the 14th National Congress of the Communist Party of China held in 1992. The society under this framework is more likely to be incomputable, but that could never be the excuse for us to stop reforming society. Otherwise the practical spirit emphasized by Mao and the "color of the cat" theory proposed by Deng[32] (Pye 1986), and "experimentation under hierarchy" as an advantage for institutional adaptation (Heilmann 2008) could not be that important. The ideology of the whole society is thus actively formed by this pragmatic "complex engineering" discourse (Schoenhals 1992; Ko 2001) and many of its evolved versions, such as "deep-water zone" (*shenshuiqu*, 深水区).

Wang's assertion that "natural sciences and social sciences must work closely together" is scientifically correct, as the history of complexity studies has demonstrated. But this ambitious thought is also politically improper, since it is far from the historical trajectory and association culture for Chinese society, and even somehow incompatible with the overall ideology. It is a shame that the true mutual commutation and understanding was abandoned, when Chinese scholars tried to understand the computability of society. We should feel more ashamed that the academic discussion somehow inhabits politics, which leads to fragmentation and "abusive" usage and interpretation of the popular words at the same time.

What's done is done, and history is repeating itself over and over again.

---

30   See: "Study and Research" (学习与研究)1986, (03).
31   See: "Jilin University Journal Social Science Edition" (吉林大学社会科学学报) 1988, (05).
32   The whole theory is expressed as "a good cat should be good at catching rats". Similar *tifa* also includes "fumbling the way to cross the river".

References

Arkush, RD. 1981. Fei Xiaotong and Sociology in Revolutionary China. Cambridge, Mass.: Council on East Asian Studies, Harvard University: Distributed by Harvard University Press.

Borgatti, S, M Everett, L Freeman. 2002. UCINET for Windows: Software for Social Network Analysis. Boston: Harvard Analytic Technologies. Retried from

Borgatti, SP, MG Everett. 2006. A Graph-theoretic perspective on centrality. Social Networks, 28(4), 466-484. doi: 10.1016/j.socnet.2005.11.005

Callon, M, J Courtial, WA Turner, S Bauin. 1983. From translations to problematic networks: An introduction to co-word analysis. Social Science Information, 22(2), 191-235

Cambrosio, A, C Limoges, J Courtial, F Laville. 1993. Historical scientometrics? Mapping over 70 years of biological safety research with coword analysis. SCIENTOMETRICS, 27(2), 119-143. doi: 10.1007/BF02016546

Carolan, BV. 2014. Groups and Positions in Complete Networks. In BV Carolan (Ed.), Social Network Analysis and Education: Theory, Methods & Applications (111-137). Los Angeles: SAGE. (Reprinted.)

Charron, C, J Favier, C Li. 2006. Social Computing: How Networks Erode Institutional Power, And What to Do About It. Forrester Customer Report

Clarke, AE, SL Star. 2008. The Social Worlds Framework: A Theory/Methods Package. In EJ Hackett, O Amsterdamska, M Lynch, J Wajcman (Eds.), The Handbook of Science and Technology Studies (113-137). Cambridge, Mass.: MIT Press. (Reprinted.)

Collins, HM, R Evans. 2002. The Third Wave of Science Studies: Studies of Expertise and Experience. SOC STUD SCI, 32(2), 235-296

Ding, Y, GG Chowdhury, S Foo. 2001. Bibliometric cartography of information retrieval research by using co-word analysis. INFORM PROCESS MANAG, 37(6), 817-842

Engelhardt, HT, AL Caplan. 1987. Introduction: Patterns of controversy and closure: the interplay of knowledge, values, and political forces. In Jr. HT Engelhardt, AL Caplan (Eds.), Scientific controversies: Case studies in the resolution and closure of disputes in science and technology (1-24). Cambridge: Cambridge University Press. (Reprinted. doi: 10.1017/CBO9780511628719.002.)

Fei, X. 1992. From the Soil, The Foundations of Chinese Society: A Translation of Fei Xiaotong's Xiangtu Zhongguo, With an introduction and epilogue. Berkeley: University of California Press.

Freeman, LC. 1977. A Set of Measures of Centrality Based on Betweenness. Sociometry, 35-41

Freeman, LC. 1980. The gatekeeper, pair-dependency and structural centrality. Quality and Quantity, 14(4), 585-592

Girvan, M, ME Newman. 2002. Community structure in social and biological networks. Proceedings of the National Academy of Sciences, 99(12), 7821-7826

Greenhalgh, S. 2008. Just One Child: Science and Policy in Deng's China. Berkeley: University of California Press.

Hanneman, RA, M Riddle. 2011. Concepts and Measures for Basic Network Analysis. In J Scott, PJ Carrington (Eds.), The Sage Handbook of Social Network Analysis (340-369). London; Thousand Oaks, Calif: SAGE. (Reprinted.)

Haraway, DJ. 1997. Modest_Witness@Second_Millennium. FemaleMan_Meets_Onco-Mouse: Feminism and Technoscience. New York, NY [u.a.]: Routledge.

Heilmann, S. 2008. From Local Experiments to National Policy: The Origins of China's Distinctive Policy Process. The China Journal, 59 (January), 1-30

Herlihy, DV. 2004. Bicycle: The History. New Haven: Yale University Press.

Hummon, NP, TJ Fararo. 1995. The emergence of computational sociology. J MATH SOCI-OL, 20(2-3), 79-87

Keith, W, W Rehg. 2008. Argumentation in Science: The Cross-Fertilization of Argumentation Theory and Science Studies. In EJ Hackett, O Amsterdamska, M Lynch, J Wajcman (Eds.), The Handbook of Science and Technology Studies (211-239). Cambridge, Mass.: MIT Press. (Reprinted.)

Kling, R. 1996. Social Controversies about Computerization. In R Kling (Ed.), Computerization and Controversy: Value Conflicts and Social Choices (10-15). San Diego: Academic Press. (Reprinted.)

Ko, S. 2001. China's pragmatism as a grand national development strategy: Historical legacy and evolution. Issues & Studies, 37(6), 1-28.

Kuhn, TS. 1962. *The Structure of Scientific Revolutions*. Chicago: University of Chicago Press.

Lakatos, I. 1978. The methodology of scientific research programmes. Cambridge; New York: Cambridge University Press.

Latour, B. 1983. Give Me a Laboratory and I will Raise the World. In K Knorr-Cetina, MJ Mulkay (Eds.), Science Observed: Perspectives on the Social Study of Science (141-170). London; Beverly Hills: Sage Publications. (Reprinted.)

Latour, B. 1986. Visualization and Cognition. Knowledge and Society, 6, 1-40

Latour, B. 1987. Science in Action: How to Follow Scientists and Engineers Through Society. Cambridge, Mass.: Harvard University Press.

Latour, B, S Woolgar. 1986. Laboratory Life: The Construction of Scientific Facts. Princeton, N.J.: Princeton University Press.

Law, J, S Bauin, J Courtial, J Whittaker. 1988. Policy and the mapping of scientific change: A co-word analysis of research into environmental acidification. SCIENTOMETRICS, 14(3-4), 251-264. doi: 10.1007/BF02020078

Lazer, D, AS Pentland, L Adamic, S Aral, AL Barabasi, D Brewer, N Christakis, N Contractor, J Fowler, M Gutmann, T Jebara, G King, M Macy, D Roy, M Van Alstyne. 2009. Life in the network: the coming age of computational social science. SCIENCE, 323(5915), 721-723. doi: 10.1126/science.1167742

Macy, MW, R Willer. 2002. From Factors to Actors: Computational Sociology and Agent-based Modeling. Annual Review of Sociology, 143-166

Mainzer, K. 1994. Thinking in Complexity: The Computational Dynamics of Matter, Mind, and Mankind (1st Edition) (1st ed.). Berlin; Heidelberg; New York: Springer.

Mainzer, K. 2007. Thinking in Complexity: The Computational Dynamics of Matter, Mind, and Mankind (Fifth Revised and Enlarged Edition) (Fifth Revised and Enlarged Edition.). Berlin; New York: Springer.

Малышев, ИС. 1963. She hui zhu yi zhi du xia lao dong de she hui ji suan he jia ge (Общественный учет труда и цена при социализме, Social Computing and Price of Labor Under the Socialist System)(社会主义制度下劳动的社会计算和价格). Beijing: Sheng huo, du shu, xin zhi san lian shu dian (DX Joint Publishing Company).

McMullin, E. 1987. Scientific controversy and its termination. In HTHT Engelhardt, AL Caplan (Eds.), Scientific controversies: Case studies in the resolution and closure of disputes in science and technology (49-91). Cambridge; New York: Cambridge University Press. (Reprinted.)

Nelkin, D. 1979. Controversy, Politics of Technical Decisions. Beverly Hills, Calif.: Sage Publications.

Nelkin, D. 1995. Science Controversies: The Dynamics of Public Disputes in the United States. In S Jasanoff, GE Markle, JC Petersen, T Pinch (Eds.), Handbook of Science and Technology Studies (444-456). Thousand Oaks, Calif.: Sage Publications. (Reprinted.)

Newman, ME, M Girvan. 2004. Finding and evaluating community structure in networks. PHYS REV E, 69(2), 26113

North, DC. 1990. Institutions, Institutional Change, and Economic Performance. Cambridge; New York: Cambridge University Press.

Nissenbaum, H. 1994. Computing and Accountability. COMMUN ACM, 37(1), 72-80

Nissenbaum, H. 1996. Accountability in a Computerized Cociety. SCI ENG ETHICS, 2(1), 25-42. doi: 10.1007/BF02639315

Parameswaran, M, AB Whinston. 2007. Social Computing: An Overview. Communications of the Association for Information Systems, 19(1), 37

Pinch, TJ, WE Bijker. 1987. The Social Construction of Facts and Artifacts: Or How the Sociology of Science and the Sociology of Technology Might Benefit Each Other. In WE Bijker, TP Hughes, TJ Pinch (Eds.), The Social Construction of Technological Systems: New Directions in the Sociology and History of Technology (17-50). Cambridge, Mass.: MIT Press. (Reprinted.)

Price, DJ. 1970. Citation measures of hard science, soft science, technology, and nonscience. In CE Nelson, DK Pollock (Eds.), *Communication Among Scientists and Engineers* (3-22). Lexington, MA: Heath Lexington Books. (Reprinted.)

Pye, LW. 1986. On Chinese Pragmatism in the 1980s. The China Quarterly, 106, 207-234

Qian, B. 1982. Guo min jing ji zong he ping heng tong ji xue (Statistics for Understanding the Balance of National Economy)(国民经济综合平衡统计学). Beijing: Zhongguo cai zheng jing ji chu ban she (China Financial and Economic Publishing House).

Salgado, M, N Gilbert. 2013. Emergence and Communication in Computational Sociology. Journal for the Theory of Social Behaviour, 43(1), 87-110. doi: 10.1111/jtsb.12004

Schoenhals, M. 1991. The 1978 Truth Criterion Controversy. The China Quarterly, 126, 243-268

Schoenhals, M. 1992. Doing Things with Words in Chinese Politics: Five Studies. Berkeley: Center for Chinese Studies, Institute of East Asian Studies, University of California.

Schuler, D. 1994. Social Computing. COMMUN ACM, 37(1), 28-29

Schumpeter, JA. 1934. The Theory of Economic Development: An Inquiry into Profits, Capital, Credit, Interest, and the Business Cycle (R Opie, Trans.). Cambridge, Mass.: Harvard University Press.

Shapin, S, S Schaffer. 1985. Leviathan and the Air-Pump : Hobbes, Boyle, and the Experimental Life: including a translation of Thomas Hobbes, Dialogus physicus de natura aeris by Simon Schaffer. Princeton, N.J.: Princeton University Press.

Sismondo, S. 2004. An Introduction to Science and Technology Studies. Malden, MA: Blackwell Pub.

Sismondo, S. 2008. Science and Technology Studies and an Engaged Program. In EJ Hackett, O Amsterdamska, M Lynch, J Wajcman (Eds.), The Handbook of Science and Technology Studies (Third Edition) (13-30). Cambridge, MA; London: The MIT Press. (Reprinted.)

Squazzoni, F. 2012. Agent-based Computational Sociology. Chichester, West Sussex: Wiley & Sons. (Reprinted.)

Su, H, P Lee. 2010. Mapping knowledge structure by keyword co-occurrence: a first look at journal papers in Technology Foresight. SCIENTOMETRICS, 85(1), 65-79. doi: 10.1007/s11192-010-0259-8

Van Raan, A, R Tijssen. 1993. The neural net of neural network research. SCIENTOMETRICS, 26(1), 169-192. doi: 10.1007/BF02016799

Waltman, L, NJ van Eck, ECM Noyons. 2010. A unified approach to mapping and clustering of bibliometric networks. Journal of Informetrics, 4(4), 629-635. doi: 10.1016/j.joi.2010.07.002

Wang, F, KM Carley, D Zeng, W Mao. 2007. Social Computing: From Social Informatics to Social Intelligence. Intelligent Systems, IEEE, 22(2), 79-83

Wang, F. 2008. (Social Computing: Fundamentals and Applications. IEEE International Conference on Intelligence and Security Informatics, 2008. ISI 2008., 2008. IEEE, p xxxv-xxxviii

Wang, F. 2009. Beyond x 2.0: Where Should We Go? IEEE INTELL SYST, 24(3), 2-4

Whittaker, J. 1989. Creativity and Conformity in Science: Titles, Keywords and Co-word Analysis. SOC STUD SCI, 19(3), 473-496

13. Computational Social Science and Big Data: A Quick SWOT Analysis

Theodor Leiber[1]

Abstract

Do computational social science and Big Data constitute a methodological revolution of the complex, data-intensive sciences? This question is approached by means of a *quick* analysis of strengths, weaknesses, opportunities and threats (SWOT analysis) of the two approaches. It is concluded that computational social science and Big Data do mark an important methodological improvement but should probably not be qualified as "scientific revolution" or "paradigm change". From the SWOT analysis it also follows that further research is necessary for a coherent development of computational social science and Big Data, in particular with respect to the ethics of privacy; balancing the low explanatory power of computational models; developing an epistemological position between naïve realism and radical constructivism; integrating computer science and social science.

---

1   Theodor Leiber is Associate Professor of Philosophy at University of Augsburg (Germany). He received doctorates in theoretical physics and philosophy. His main areas of interest are philosophy of science and technology, epistemology and ethics of nature. Leiber is also a higher education researcher focusing on models of teaching and learning, governance and impact analysis of quality management in higher education.

13.1   Introduction

It is undeniable that our present time is confronted with a number of basic and high-risk problems which are certainly in need to be tackled (although they are very likely unsolvable for complexity reasons). And it is perhaps a truism that all these problems are correlated to the functioning (or malfunctioning) of human-kind's societies and, therefore, require social science understanding. In the words of a recent "Manifesto of Computational Social Science":

> "In a world of demographic explosion, global crisis, ethnic and religious distur-bances and increasing crime the understanding of the structure and function of society, as well as the nature of its changes, is crucial for governance and for the well-being of people" (Conte et al. 2012, p. 325).

There are a number of authors who seem to believe that contemporary computational social science and Big Data approaches could provide a main contribution, if not a complete solution, to these problems (cf. Anderson 2008; Cioffi-Revilla 2010; Conte et al. 2012; Epstein 1999; Lazer et al. 2009). Even the 19th century's dream of "so-cial physics" is still present when it is assumed that through the computational social science approach "sociology in particular and the social sciences in general would undergo a dramatic paradigm shift, arising from the incorporation of the scientific method of physical sciences" (Conte et al. 2012, p. 341; see also Pentland 2012; Chang et al. 2014). How so? The basic assumptions of these rather optimistic authors are that computational complexity models provide computer-based quantitative methods and measures—not available before the advent of digital computing machines of very large power—which allow for treating social science problems of considerable com-plexity by "objective" scientific methods which, as a matter of course, are equated with mathematised quantitative models such as in physics. Moreover, there are proponents of "Big Data science" who think—or just want to provoke others?—that "the end of theory" (Anderson 2008) has come and any why-questions should be avoided further on and theories abolished and causal inquiries omitted. An example is the following enthusiastic interpretation of Big Data:

> "This is a world where massive amounts of data and applied mathematics re-place every other tool that might be brought to bear. Out with every theory of human behaviour, from linguistics to sociology. Forget taxonomy, ontology, and psychology. Who knows why people do what they do? The point is they do it, and we can trace and measure it with unprecedented fidelity. With enough data, the numbers speak for themselves" (Anderson 2008).

It seems to be quite obvious that what Chris Anderson is saying here is overly reductionistic, extremely hypothetical and relying on an exaggerated reliability of the "novel" big data.

But there are also others who, while surely acknowledging its methodological potential, clearly refer to deficits of computational modelling in social science. Here is an example:

> "Powerful computational resources combined with the availability of massive social media datasets has given rise to a growing body of work that uses a combination of machine learning, natural language processing, network analysis, and statistics for the measurement of population structure and human behaviour at unprecedented scale. However, mounting evidence suggests that many of the forecasts and analyses being produced misrepresent the real world" (Ruths and Pfeffer 2014, p. 1063).

This suggests that, in contrast to the extremist position cited before, a serious social sciences approach relying on computational modelling should preferably respond to the challenge how social scientists can "best use computational tools to analyse such data, problematic as they may be, with the goal of understanding individuals and their interactions within social systems?" (Shah et al. 2015, p. 6).

Thus, for the purposes of this essay and neglecting the before mentioned exaggerating optimism and methodological reductionism, computational social science and Big Data can be characterised in the following way: computational social science uses computational computer-based methods to deal with complex problems of 21st century social science. Thus, computational social science is conceived as the "interdisciplinary investigation of the social universe on many scales, ranging from individual actors to the largest groupings, through the medium of computation" (Cioffi-Revilla 2013, p. 2). Big Data enters the scene when processed data become "big" in terms of volume (amount of data), velocity (of data production) or variety (of data sources and types) (cf. Russom 2011, p. 6) and datasets' "size is beyond the ability of typical database software tools to capture, store, manage, and analyse" (Manyika et al. 2011, p. 1). For example, this is typically the case when social media and the underlying networks are analysed in the social sciences (see, e.g., Ch'ng 2014; Twombly 2011). Furthermore, it is assumed that the general goal of computational social science and Big Data (like that of the social sciences in general) is to improve humankind's understanding and planning and implementation of quality enhancing interventions in social systems. For overviews of the state of the art and also some critical reflection of computational social science and Big Data the reader is referred to (Cioffi-Revilla 2013; Mainzer 2014; Marr 2015; Reichert 2014).

In this situation, it is quite plausible to undertake an interim assessment of computational social science and Big Data: Do they revolutionize epistemology and methodology of the complex sciences? Are they helpful to establish quantitative approaches of relevance in the social sciences and the humanities and thus make them (more of) an "objective" enterprise? Do they contribute to a better understanding of the social dynamics of individuals and organisations? Based on literature review, this essay attempts to approach these questions by means of a quick analysis of some strengths, weaknesses, opportunities and threats (SWOT analysis).[2]

13.2   What is a SWOT Analysis?

A SWOT analysis is a structured assessment method which evaluates the strengths (S), weaknesses (W), opportunities (O) and threats (T) involved in a process or a structure in the most general sense of these terms. A SWOT analysis involves specifying the objectives of the process or structure, identifying the internal and external influences with regard to the degree of achievement of these objectives and, finally the core element, characterising the strengths, weaknesses, opportunities and threats of the process or structure under scrutiny. In general, a SWOT analysis can help developing the assessed entities for further rounds of improved goal achievement, and it usually has an exploratory dimension bringing to the fore aspects which have not been noticed by other means of analysis. This exploratory force originates from the requirement to identify and distinguish explicitly the four different categorisation dimensions of processes or structures.

Typical areas of SWOT are strategy formation and organisational and human resources development. A SWOT analysis aims to identify the key internal and external factors seen as important to achieving an objective. For analytical purposes a SWOT analysis may group key pieces of information into two main categories: internal factors—the *strengths* and *weaknesses* internal to the organisation; external factors—the *opportunities* and *threats* presented by the environment external to the organisation. In general, a SWOT analysis can be used to identify barriers that will limit goals/objectives; decide on direction that will be most effective; reveal possibilities and limitations for change; revise strategies and plans.

---

2   After writing this essay the author came to a publication by Chirag Rabari (2013) which follows similar ideas and reasoning (though in somewhat more detail, with different emphasis and with a practical background in the social science area of urban planning).

13.3   A Quick SWOT Analysis of Computational Social Science and Big Data

13.3.1   Epistemological and Methodological Aspects

13.3.1.1 Strengths

Computational social science provides methodologies for the complex sciences and humanities where system law hypotheses (Leiber 2007, pp. 201ff.) with broad application ranges, few exceptions to the laws and relatively simple law functions (such as, e.g., in physics) are not available. In such situations explaining by subsuming as many phenomena as possible under as few concepts and regularities as possible (hierarchical modelling) does not work. Instead, a more phenomenological approach is necessary looking at structures and processes which are not reducible to mathematical models that are more condensed with regard to their information content (horizontal modelling). Here, methods of data-intensive simulations and data-induced algorithms apply (also cf. Pietsch 2015). For example, computational social science models can exploit the advantage computers have against humans since computers do not have to be very restrictive with respect to storing and handling large amounts of collected data.

Computational complexity models of the horizontal modelling type are methodologically very flexible as compared to axiomatically formulated theory: Since their algorithms are just based on huge sets of data and do not contain content-specific knowledge (such as, e.g. parametrized laws), they represent templates which may be applied in various contexts. Computational complexity models can also create macro-level patterns which can simulate observable real-world patterns.

Some computational social science approaches—such as agent-based modelling (Conte and Paolucci 2014; Epstein 1999)—allow for modelling agents characterized by bounded rationality, i.e., agents who "do not fully optimize their utility functions given their information"; who are "not forward-looking in being able to predict accurately the probability of certain outcomes in the future given their behaviour"; who are "myopic in that they do not have information outside of some defined local 'zone' of interaction" (Kollman 2012, p. 372). This is clearly an advantage of computational social science making some of its quantitative (and qualitative) models more realistic.

13.3.1.2 Weaknesses

Causal laws that are established by data-intensive ("Big Data") computational complexity models usually are system law hypotheses applicable only to very few systems and also small systems, and they are characterised by tentatively very complicated functional dependencies or no such explicit functions at all. Therefore,
general claims based on the corresponding computational models and Big Data can
only be justified on the basis of large numbers of computer program runs, because
there are only (numerous, locally valid) computer algorithms but no theory-based
generalisation (e.g. law-like hypotheses of a wider range of applicability) beyond
the large amounts of computer-generated data. One immediate consequence is that
data-intensive computational complexity models have less explanatory power as
compared to "traditional" hierarchically nested systems of information condensing
law hypotheses and conceptual unification. Thus, by definition computational complexity models provide less understanding but instead focus very much on prediction and manipulation (cf. Pietsch 2015, p. 10 and elsewhere).

A further weakness seems to be that many endeavours and research papers in
computational social science are very much focused on computer science, and,
therefore, are of limited relevance to social science (cf. Watts 2013, p. 8; Wallach
2015, p. 2). Here, more interdisciplinary, transdisciplinary, more integrative approaches and cooperation efforts are needed. In particular, there is further need for
clarification of the application ranges and limits of certain approaches and models and their potential epistemological and methodological complementarity with
each other. This should help to avoid giving, sometimes, the impression of qua-
si-omnipotent reductionist models which are in fact very abstract and sometimes
probably very close to toy problems which are not applicable empirically.

Another weak point concerns the following: As already mentioned in the Introduction it seems that major proponents of computational social science are motivated by the "big" questions of social science research, e.g. risk management in financial systems; dynamics of epidemics or social movements; decision making and
collective action; organisational problem solving; the relationship between deliberation, governance, and democracy (Watts 2013, pp. 5, 7). However, when we look
at some of the probably hundreds or thousands of published papers and books on
computational social science, it seems very plausible to assume that many of them
represent valuable (mainly methodological) developments but it also seems to be
the case that obvious progress on the "big" questions of social science has not been
achieved to a relevant extent. Duncan J. Watts makes three reasonable suggestions
why the bigger questions have not been answered by computational social science

so far (although they are very often mentioned as the most motivating research goals and authors often also insinuate their accessibility and solution):

"First, social science problems are almost always much more difficult than they seem. Second, the data required to address many problems of interest to social scientists remain difficult to assemble. And third, thorough exploration of complex social problems often requires the complementary application of multiple research traditions—statistical modelling and simulation, social and economic theory, lab experiments, surveys, ethnographic fieldwork, historical or archival research, and practical experience—many of which will be unfamiliar to any one researcher" (Watts 2013, pp. 5-6).

One reason for the difficulty to assemble relevant data is that the required data are generally recorded and stored separately, e.g., in the context of single online platforms such as Facebook, Twitter, Ebay, LinkedIn. A "social supercollider" would be helpful that is "a facility that combines multiple streams of data, creating richer and more realistic portraits of individual behaviour and identity, while retaining the benefits of massive scale" (Watts 2013, p. 8).

13.3.1.3 Opportunities

Insofar as data-intensive computational social science relies on horizontal modelling there is an opportunity (but also a need and, probably, even a threat) to develop "new data analysis tools, as well as specific data representation and visualisation tools" (Conte et al. 2012, p. 341) which could take the place of more hierarchical modelling and information reduction by theoretical and mathematical models which are characterised by only few model parameters. In other words, here is an opportunity to further develop and improve the phenomenological approaches of computational social science.

The big opportunity and, at the same time, the major threat of data-intensive computational social science, however, is to materialise computational models that help in answering "bigger" questions of social science (such as those mentioned above). In that sense, it is a hope for the future computational social science "that many more socially minded computer scientists will start turning their attention to bigger-picture social questions, most likely in collaboration with computationally minded social scientists" (Wallach 2015, p. 7).

In this context it is a big opportunity that computational social science with Big Data may help to overcome the Runkel and McGrath's (1972) so-called three-horned dilemma for research methods. This dilemma consists in the fact that large amounts of process data, high data precision and realistic (i.e. complex) modelling are competing dimensions which cannot usually be optimised altogether at the same time. Large computers and Big Data, however, allow to improve on all three dimensions at once and thus overcome the dilemma (cf. Chang et al. 2014, p. 70).

13.3.1.4 Threats

The mathematisation and algorithmisation inherent in computational modelling seems to offer new ways for the social sciences and the humanities to claim the status of quantitative science and "objective" methodology. This may be regarded as a strength and/or a weakness and/or an opportunity depending on one's episte-mological stance towards the various scientific disciplines: For example, typically physicists and engineering scientists of course will welcome the intrusion of quan-titative modelling into the social sciences and humanities, while hermeneutical interpretationists, for example from qualitative social science research, anthropo-geography or philosophy would usually complain about an indefensible reduction-ism. Be that as it may, the basic threat seems to be the adoption of an epistemolog-ically balanced concept of "objectivity" (in relation to "subjectivity") taking into account the unavoidable interweaving of quantitative and qualitative reasoning and research methods (triangulation; mixed-method). In the words of Boyd and Craw-ford (2012, p. 667): "All researchers [across all subject cultures, disciplines, methods and theories of knowledge] are interpreters of data [be they quantitative or quali-tative in nature]." Most probably, there is no such thing as pure data, which is just "given" ("datum" in Latin) together with their immediate and completely obvious understanding (cf. e.g. the long-standing epistemological debate about the "Myth of the Given" and the realism/antirealism debate in philosophical epistemology). In particular, this implies that there is no pure quantitative data (numbers) that, without any qualitative conceptual comparative semantics, could have any definite empirical meaning. Moreover, a glimpse into humankind's as well as the human individual's cognitive genesis and history reveals that numbers, abstract entities as they are, are "most probably" not among the very first things that come to humans' minds. For example, from a coherent empirical point of view it seems to be more plausible that the basic forms of geometry were somehow quasi-inductively or di-alectically invented in a step-by-step process by empirical man in his attempts to "measure the earth" (geo-metry), and not just discovered in a pure trans-empirical mind or in the heavens of ideas (with whatever the latter terms may be correlated). Furthermore, any theoretical modelling and experimental modelling involves design decisions, e.g., determined by definitions; recognized axioms; accepted measurements and their underlying theoretical models and (quasi-)experimental procedures; etc. These primary decisions co-determine what will be (or can be) measured—and they are types of interpretations, i.e., they are not just "given" by or analytically deducible from putatively "purely objective data". Quite to the contrary, the decision for a methodological design and measurement design implies, e.g. "decisions about what attributes and variables will be counted, and which will be ignored" (Boyd and Crawford 2012, p. 667; also cf. Bollier 2010, p. 13).

Moreover, in data-intensive computational approaches there is the general issue of systematic data errors stemming from (unrecognized) biases and peculiarities of the data samples. In particular, if large data sets are available the risk is high that the users of that big data assume a sort of statistical representativeness of the sample while erroneously neglecting relevant specificities of the sample and its typical members. Systematic errors may also occur because of the fact that the control over data acquisition and data presentation is often held by biased owners who are following their specific stakeholder agendas (such as, e.g. internet companies and providers), which are often not transparent to others. Again in the words of Boyd and Crawford:

> "Interpretation is at the centre of data analysis. Regardless of the size of a data, it is subject to limitation and bias. Without those biases and limitations being understood and outlined, misinterpretation is the result. Data analysis is most effective when researchers take account of the complex methodological processes that underlie the analysis of the data. [...] [Therefore,] understanding sample [...] is more important now than ever" (Boyd and Crawford 2012, p. 668).

Another threat is to counteract the low or even missing explanatory capacity of data-intensive computational social science. For example, according to Conte et al. "[t]he construction of plausible generative models is a challenge for t[h]e new computational social science" (Conte et al. 2012, p. 340), meaning that computational models should be supplemented, if possible, by causal mechanism hypotheses, i.e. by formation of causal theory models. Such mechanisms would, among other things, build a connecting bridge between extreme horizontal modelling (e.g. establishing customer profiles from data acquired in the world wide web) and extreme hierarchical modelling (e.g. in theoretical physics). Note that, in general, causal social mechanism hypotheses are a valuable ingredient of reconstructing and understanding the functioning of complex social systems, in particular their reaction to interventions (cf. e.g. Hedström and Ylikoski 2010; Leiber et al. 2015; Little 2011; Steel 2011).

Finally, it is an overarching threat to further develop computational social science as a truly inter- and transdisciplinary enterprise. In the words of Hannah Wallach: "first and most importantly, we need to understand each other's disciplinary norms, incentive structures, and research goals" (Wallach 2015, p. 8). Furthermore, for making computational social science a genuinely interdisciplinary progressive research field, it is necessary to implement successful interdisciplinary collaborations in the education system long before people graduate from higher education.

13.3.2 Ethical Aspects: Weaknesses/Threats

In the context of Big Data, in particular if collected *via* the social e-networks or by large internet companies or secret services, it is to a large extent unclear and unregulated ethically and by law which agent has which moral and legal rights to access, process and use the data (e.g. just think of the NSA spy scandal). In particular, it is often unclear how individual and organisational privacy can be guaranteed (also cf. Boyd and Crawford 2012, pp. 671-673; Martin 2015). At first glance it might seem that Big Data "offers easy access to massive amounts of data" (Boyd and Crawford 2012, p. 673). Closer inspection reveals, however, that Big Data quite easily contributes to creating new forms of digital divide, because it is not at all transparent and decided participatively, "who gets access, for what purposes, in what contexts, and with what constraints" (ibid.). Quite to the contrary, business, researchers and higher education institutions which are well-resourced will usually get access much easier than the under-resourced ones. For example, it is "possible that engaging in platform-driven research will become increasingly difficult [...] for researchers in academic positions", "who do not have close ties to industry" (Wallach 2014, p. 7). In other words: "The current ecosystem around Big Data creates a new kind of digital divide: the Big Data rich and the Big Data poor" (Boyd and Crawford 2012, p. 674). This undeniable weakness of data-intensive social sciences is at the same time a major threat and challenge: to clarify the ethics and juridical law of the socio-technological phenomenon of Big Data.

13.3.3   Political and Societal Aspects: Threats

Finally, in addition to epistemological, methodological and ethical threats there are also general political and societal threats and challenges: For example, in order to be successful as an interdisciplinary and transdisciplinary endeavour computational social science will require

> "also new institutions that are designed from the ground up to foster long-term, large-scale, multidisciplinary, multi-method, problem-oriented social science research. To succeed such an institution will require substantial investment, on a par with existing institutes for mind, brain, and behaviour, genomics, or cancer, as well as the active cooperation of industry and government partners" (Watts 2013, p. 9).

Thus, it is important as well that educational programs for future computational social scientists will be (further) developed and offered in higher education institutions.

13.4   Concluding Remarks

From the above presented *quick* SWOT analysis it follows that computational social science is an exciting field still under development. As often, in this case it is also not easy to decide whether computational social science and Big Data mark a "scientific revolution" or a "paradigm change". One source of this difficulty is that the two concepts are notoriously underdetermined and therefore controversial. At the same time, both concepts have their revolutionary aura and motivating spirit which often makes them attractive, especially when the old shall be ousted and a new era be ushered. In contradistinction to such approaches and to avoid the difficulties of a final concept definition, in this short essay a conceptually more modest (some may even call it conservative) attitude is predominant which does not see a "revolution" since there is no novel theory involved nor is it the case that theory as such can be abolished and replaced by computational social science and Big Data analysis. Furthermore, no methodology of data handling which is completely new from the ground up has appeared with computational social science and Big Data, but what we witness is a true and effective methodological development which offers new analytical options and allows and promises insights not available before. The core structure and elements of social science research, however, seem to remain untouched: qualitative and quantitative approaches and data acquisition from various sources and methodologies, big data as well as small data. Which of these we need in any particular case depends on the research problem and the resources available. What usage of these shall be allowed follows from human rights, in particular ethics of self-determination and ethics of data protection and privacy rights.

Referring to the often stated putative "revolutionary" character of computational social science and Big Data it should be noted that computational social science is a phenomenon similar to computational physics or computational biology. In all these cases the novelty arises from the availability of powerful computers (excessing human capabilities) for acquiring, storing and processing digitized data and their automated modelling. This is a severe methodological change but not a "revolution" which would, e.g., replace traditional theories and methodologies as dispensable. No traditional "theoretical core" (Imre Lakatos) is replaced by a more progressive and novel one. Rather, it is to be expected that in the long term the computational methods will become part of the established standard research method tool box of the social sciences, complementing the well-established methods. The latter will remain and will not be replaced, e.g. because social science problems will stay that can neither be treated as data-intensive problems nor be captured by more hierarchical law approaches (e.g. organisational development and learning in smallsize and midsize organisations). In other words: Bigger data are not in each case necessarily more useful data because

> "[r]esearch insights can be found at any level [of investigation and data acquisition], including at very modest scales. [...] The size of data should fit the research question being asked; in some cases, small is best" (Boyd and Crawford 2012, p. 670).

Furthermore, it seems at least not very helpful to claim that computational complexity models and Big Data—such as pursued in computational social science—establish, as some say, a "fourth paradigm" in addition to theory, experiment, and simulation because data-intensive computational complexity models are incapable of surviving epistemologically and methodologically without the explanatory capacities and empirical cross-check of the "three traditional paradigms".

Computational complexity models and Big Data science are, however, an approach very appropriate for the social sciences because these are usually not that accessible to unificationist and law-based explanation since system law hypotheses with considerably large application ranges, few exceptions and simple law functions cannot be justified (basically for complexity and dynamicity reasons). Nevertheless, the statement seems clearly exaggerated that

"Big Data [as well as computational social science in general] reframes key questions about the constitution of knowledge, the processes of research, how we should engage with information, and the nature and the categorisation of reality" (Boyd and Crawford 2012, p. 665).

This exaggeration has two basic dimension: first, it remains usually unclear or at least underspecified which key questions are to be reframed; second, the thematic fields of announced reframing seem to be very broad and basic so that it is difficult to see at which object the intended reframing is aimed. However, if contextualised in a relevant and balanced way the above citation can be interpreted as indirectly indicating areas where further reflection and investigation seems necessary for a coherent further development of the related fields of computational social science and Big Data. Among these areas are the ethics of privacy; balancing the low explanatory power of computational models; developing a plausible intermediate epistemological position between naïve realism and radical constructivism; and integrating computer science and social science.

Thus, the developing and exciting computational social science can be seen as a methodologically pluralistic (mixed-method) social and computational science enterprise which defines

"a larger umbrella under which different approaches might coexist and somehow feel legitimate. Hence, generative ABM [agent-based modelling] might be practiced by a subset of social scientists, while others might prefer a purely [?] quantitative approach, based on data-mining and numerical simulation, and still others might continue to formulate abstract theories of social action in elegant equations and deduce their macro-level consequences" (Conte and Paolucci 2014, p. 7).

One subarea then of computational social science is big data (i.e. data-intensive) computational models which are basically data-driven, where horizontal modelling dominates and where powerful computing machines are indispensable.

References

Anderson, C. 2008. The end of theory. Will the data deluge makes the scientific method obsolete? *Edge.* http://edge.org/3rd_culture/anderson08/anderson08_index.html. Access: June 28, 2015.

Bollier, D. 2010. *The promise and peril of Big Data.* Washington: The Aspen Institute. http://www.aspeninstitute.org/sites/default/files/content/docs/pubs/The_Promise_and_Peril_of_Big_Data.pdf. Access: June 28, 2015.

Boyd, D., and Crawford, K. 2012. Critical questions for Big Data. Provocations for a cultural, technological, and scholarly phenomenon. *Information, Communication & Society 15(5)*, pp. 662-679.

Chang, R. M., Kauffman, R. J., and Kwon, Y. O. 2014. Understanding the paradigm shift to computational social science in the presence of big data. *Decision Support Systems 63*, pp. 67-80.

Ch'ng, E. 2014. The value of using Big Data technologies in Computational Social Science. 3rd ASE Big Data Science Conference, Tsinghua University Beijing, 3-7 August 2014, 4 pages. http://arxiv.org/ftp/arxiv/papers/1408/1408.3170.pdf. Access: January 2, 2016.

Cioffi-Revilla, C. 2010. Computational social science. *Wiley Interdisciplinary Reviews: Computational Statistics 2(3)*, pp. 259–271.

Cioffi-Revilla, C. 2013. *Introduction to Computational Social Science.* New York: Springer.

Conte, R., Gilbert, N., Bonelli, G., Cioffi-Revilla, C., Deffuant, G., Kertesz, J., Loreto, V., Moat, S., Nadal, J.-P., Sanchez, A., Nowak, A., Flache, A., San Miguel, M., and Helbing, D. 2012. Manifesto of Computational Social Science. *The European Physical Journal Special Topics 214*, pp. 325-346.

Conte, R., and Paolucci, M. 2014. On agent-based modeling and computational social science. Frontiers in Psychology 5 (art. 668), 9 pages. http://www.ncbi.nlm.nih.gov/pmc/articles/PMC4094840/pdf/fpsyg-05-00668.pdf. Access: January 01, 2016.

Epstein, J. M. 1999. Agent-based computational models and generative social science. *Complexity 4(5)*, pp. 41-60.

Hedström, P., and Ylikoski, P. 2010. Causal mechanisms in the social sciences. *Annual Review of Sociology 36*, pp. 49-67.

Kollman, K. 2012. The potential value of computational models in social science research. In H. Kincaid (Ed.), *The Oxford Handbook of Philosophy of Social Science* (pp. 355-383). Oxford: Oxford University Press.

Lazer, D., Pentland, A., Adamic, L., Aral, S., Barbási, A.-L., Brewer, D., Christakis, N., Contractor, N., Fowler, J., Gutmann, M., Jebara, T., King, G., Macy, M., Roy, D., and van Alstyne, M. 2009. Computational social science. *Science 323*, pp. 721-723.

Leiber, T. 2007. Structuring nature's and science's complexity: System laws and explanations. In T. Leiber (Ed.), *Dynamisches Denken und Handeln. Philosophie und Wissenschaft in einer komplexen Welt* (pp. 193-212). Stuttgart: Hirzel.

Leiber, T., Stensaker, B., and Harvey, L. 2015. Impact evaluation of quality assurance in higher education: methodology and causal designs. *Quality in Higher Education 21(3)*, pp. 288-311.

Little, D. 2011. Causal mechanisms in the social realm. In P. McKay Illari, F. Russo, & J. Williamson (Eds.), *Causality in the Sciences* (pp. 273-295). Oxford: Oxford University Press.

Mainzer, K. 2014. *Die Berechnung der Welt. Von der Weltformel zu Big Data.* München: Beck.

Manyika, J., Chui, M., Brown, B., Bughin, J., Dobbs, R., Roxburgh, C., and Byers, A. H. 2011 *Big Data: the next frontier for innovation, competition, and productivity.* McKinsey Global Institute Report, New York (May 2011), 20 pages.

Marr, B. 2015. *Big Data: using SMART Big Data, analytics and metrics to make better decisions and improve performance.* New York: Wiley & Sons.

Martin, K. E. 2015. Ethical issues in the Big Data industry. *MIS Quarterly Executive 14(2),* pp. 67-85.

Pentland, A. 2012. Reinventing society in the wake of big data. *Edge.* http://edge.org/conversation/reinventing-society-in-the-wake-of-big-data. Access: June 28, 2015.

Pietsch, W. 2015. The causal nature of modelling with Big Data. *Philosophy and Technology.* http://link.springer.com/article/10.1007/s13347-015-0202-2. Access: June 28, 2015.

Rabari, C. 2013. Big Data: The promise of a "Computational" social science. *Critical Planning. A Journal of the UCLA Urban Planning Department 20,* pp. 27-44.

Reichert, R. (Ed.). 2014. *Big Data. Analysen zum digitalen Wandel von Wissen, Macht und Ökonomie.* Bielefeld: transcript.

Runkel, P. J., and McGrath, J. E. 1972. *Research on human behaviour: a systematic guide to method.* New York: Holt, Rinehart and Winston.

Russom, P. 2011. *Big Data analytics.* Best practices report, fourth quarter 2011. Renton (WA): The Data Warehouse Institute. http://www.sas.com/content/dam/SAS/en_us/doc/research2/big-data-analytics-105425.pdf. Access: January 30, 2016.

Ruths, D., and Pfeffer, J. 2014. Social media for large studies of behaviour. *Science 346,* pp. 1063-1064.

Shah, D. V., Cappella, J. N., and Neuman, W. R. 2015. Big Data, digital media, and computational social science: Possibilities and perils. *Annals, American Academy of Political and Social Science 659,* pp. 6-13.

Steel, D. 2011. Causality, causal models, and social mechanisms. In I. C. Jarvie & J. Zamora-Bonilla (Eds.), *The SAGE Handbook of the Philosophy of Social Sciences* (pp. 288-304). Thousand Oaks: Sage Publications.

Twombly, M. 2011. Introduction: challenges and opportunities. *Science 331(6018),* pp. 692-693.

Wallach, H. 2015. Computational social science: toward a collaborative future. http://dirichlet.net/pdf/wallach15computational.pdf. Access: June 28, 2015.

Watts, D. J. 2013. Computational social science. Exciting progress and future directions. *The Bridge 43(4),* pp. 5-10.

14. Computability and Instability in Sociodynamics

Wolfgang Weidlich[1]

Abstract

About Sociodynamics (abbreviated SD) there already exist several survey articles (Weidlich 2003, 2005, 2006b; Weidlich and Huebner 2008) and books (Weidlich and Haag 1983; Weidlich 2006a) explaining its principles and describing a few seminal models. Therefore we will here give a short summary of SD only and instead focus on its methodological problems and implications. These include a comparison of the structures of SD and physics in view of the scope and limits of calculability in both sciences. Of particular interest is the role of stability versus instability in the human society treated in terms of SD.

14.1   Intentions of SD

Sociodynamics (SD) aims at providing a theoretical concept for an *integrated strategy* of mathematical modelling in the social sciences. Its method should e.g. be applicable to problems of demography, sociology, political science, economics, and regional science. SD can be considered as an *open frame* for designing models in these sectors of the social sciences.

Since social systems include elements of chance as well as quasi-deterministic structures the preferred mathematical method of SD is the *theory of the dynamics of stochastic, i.e. probabilistic systems*. It consists of universally applicable mathematical theorems.

---

1   Wolfgang Weidlich (*14. April 1931 in Dresden; †21. September 2015 in Stuttgart) was a German theoretical physicist and a pioneer in the field of sociophysics.

14.2   The Model Design of SD

The procedure of designing models in SD consists of three steps:

- The first step consists in finding—in cooperation with social scientists—*appropriate macrovariables*, also called key variables or order parameters, for the social sector under consideration.
- The second step introduces the *elementary dynamic processes* taking into account their specifically *social nature* (differing from processes in physics)
- The third step consists in setting up equations for the evolution of the key variables by inserting their elementary dynamics, and by solving the equations either on the probabilistic or the quasi-deterministic level of description.

Let us explain the three steps in more detail.

14.2.1   The Macrovariables

Different kinds of collective key variables must be introduced:

Collective material variables, such as found in economics (commodities, prices, ...) or in regional science (buildings, quarters, cities, ..) set up the *material configuration* m.

Collective personal variables namely the numbers n(a,i) of members of subpopulations P(a) having a certain (social, or political, or economic) attitude "i" set up the *personal configuration* n.

Collective trendparameters set up the *trend configuration* k. The trends k(a) characterize the collective behavioral trends in a population P(a). They appear in the construction of the elementary dynamics.

The set of variables comprised by (m,n,k) constitutes the *socioconfiguration* C. The variables of C describe the condition of societies on a more or less coarse-grained level. There exists an analogy between the space Y of statistical physics and the space S expanded by the components of the socioconfigurations: Each point of Y represents a physical system, and each point of S represents a concrete society. The number of components of the multiple (m,n) span the dimension of S. Fine-grained socioconfigurations lead to a high dimension, and coarse-grained socioconfigurations to a low dimension of S.

14.2.2 The Elementary Dynamics

The elementary steps of change of a concrete society—which corresponds to a concrete socioconfiguration **S**—consist in changes of one component of (**m,n**), by one unit, e.g. m(k) → (m(k)+1) or m(k) → (m(k) - 1), and n(a,i) → (n(a,i) +1) or n(a,i) → (n(a,i) - 1).

The core of the modelling procedure now consists in the choice of *probabilistic transition rates per unit of time* between the components of a socioconfiguration. They are the "driving forces" behind the evolution of the society. Their properties are:

They are engendered by human individuals, i.e. on the microlevel of the society. Therefore they initiate the *bottom-up* interaction from its micro- to macro-level.

On the other hand they depend on the *macrovariables* before and after the transition and on certain *trendparameters*. This fact implies the *top-down* interaction from macro- to micro-level.

The origin of the transition rates are *decisions of humans*. These depend on considerations about the *utility of the transition*. Fortunately these transition rates can be cast into a standard form: They are the product of a *mobility term* and a *utility term*. The latter contains the difference between the utility of the state before and after the transition. The choice of *probabilistic* instead of *deterministic* transition rates expresses the fact, that there exists a *freedom of decision* for the individual in spite of certain preferential trends showing up in the transition rates.

14.2.3 Equations of Evolution for the Macrovariables

14.2.3.1 The Probabilistic Description

The probabilistic transition rates are now the decisive construction elements in setting up the central equation—called *masterequation*—for the evolution with time of the quantity **P(m,n,t)**. This quantity **P(m,n,t)** is by definition the probability to find the set of macrovariables (**m,n**) realized at time t. It gives the most detailed description of the probabilities of possible evolutions of a society in terms of its variables (**m,n**) in the course of time.

14.2.3.2 The Deterministic Description

The necessity of going over to an approximate deterministic description is obvious: Only the evolution of *one society* (or at best of a few comparable societies) instead of a statistical ensemble of samples is empirically available. Therefore one should compare the data of the only one *realized society* with the *most probable trajectory* within the theoretical ensemble of probabilistically evolving *virtual societies!*

The *most probable trajectory* is obtained as follows:

Starting from a given initial configuration $(\mathbf{m}(t(0)),\mathbf{n}(t(0)))$ this will in the small interval of time dt develop into the "preferred" configuration $(\mathbf{m}(t(0)+dt),\mathbf{n}(t(0)+dt))$, which is defined as the set of probabilistically weighted *quasi-meanvalues* of the components of $(\mathbf{m},\mathbf{n})$ reached at time $t(0)+dt$ via the transition rates. The sequence of preferred configurations is thereupon (in the limit dt $\to$ 0) composed to find the preferred configurations $(\mathbf{m}(t),\mathbf{n}(t))$ for all future times $t>t(0)$. This procedure leads to a set of coupled nonlinear deterministic differential equations. Their solutions constitute the *most probable trajectory* developing via temporal quasi-mean-values.

It is obvious, that this *smoothly evolving most probable trajectory* is only an approximation. Due to the stochastic nature of the elementary steps of change the real society will instead traverse a *stochastic trajectory*, at best in the vicinity of the most probable trajectory.

The *utility* of so constructed SD-models as compared to purely qualitative descriptions of social systems will consist in the following: Their equations contain trend parameters $\mathbf{k}$ describing the behavior of the individuals. Of course, only *one* set of $\mathbf{k}$ is realized in the *one* existing society. However, the solutions of SD-models can *simulate virtual societies* in the vicinity of the existing one by varying the numerical values of the trend parameters. The consequences of this variation for the dynamics of the social system can thereupon be studied *analytically*.

14.3   Analogies and Differences between Physics and SD

The idea of SD consists in a *transfer of mathematical methods* (transition rates, masterequation, systems of nonlinear differential equations) which were successfully applied so far to problems of physics, to the treatment of a different layer of reality, namely the human society. This leads to analogies and differences between both fields of science, but not at all to the shortcut but misleading argument, SD be "nothing but physicalism". ("Physicalism" is by definition the *direct application* of results in physics to phenomena in the society.)

In order to disprove this argument, let us go into some detail:

Indeed there appear *on the macrolevel* of the physical and social reality also *structural analogies*. They consist in certain *macro-events*, such as e.g. *phase transitions*, i.e. relatively sudden transitions between different global states of the system. Such events are describable by similar mathematical equations in physics and in SD. However this does not mean that the systems of physics and the society are identifiable on the *microlevel*. Instead there takes place on the way from microlevel to macrolevel an *information compression* wiping out many detailed differences between both systems of reality. These differences do appear if the "blind unconscious" action of laws of nature in physics are compared with the "conscious decision making" of humans in the society.

Let us now discuss the effect of the differences to physics in the structures of SD-models. We begin with the smoothly evolving "most probable trajectories". They are solutions of nonlinear differential equations for the key-variables depending on trendparameters k. According to the choice of the latter the solutions may exhibit several kinds of attractors, namely stationary points, limit cycles or even "strange" chaotic attractors. But here begins the difference: Whereas the fundamental equations of motion in physics are dominated by *constants of nature*, such as G = constant of gravitation, h = Planck's constant, c = velocity of light, the equations of SD are dominated by *trendparameters* only. These k are much slower varying than the key-variables, but they are *no constants!* Instead they are rooted in the historical and cultural traditions of a society, or, even more slowly changing, subject to the evolution of the genetic outfit of a society. The effect of this slow but not negligible change of the trends k is the higher frequency of dramatic dynamic events in the society as compared to the ultrahigh stability of the "laws of nature" detected in physics.

14.4 Analysis of Instabilities in SD

Since there is "no end of history" of such dramatic global evolutions and revolutions in societies, the question arises, whether there can be a special contribution of SD to their analysis. This seems indeed to be the case, because the core of SD is a probabilistic, not deterministic philosophy of model construction. Its consequences must now be applied to the analysis of originally stable stationary situations threatening to become unstable.

Whenever a stochastic trajectory, accompanied by a smooth approximate most probable trajectory, traverses through the space **S** of the societies, there will exist a normal "forward bias" towards some direction among the forth and back pointing transition rates between adjacent states of **S**. Correspondingly, the accompanying most probable trajectory moves forward in this direction. But it can happen that the forward and backward probabilistic fluctuations of all key-variables simultaneously come to an equilibrium around some point of **S**. Thereby they establish this state of **S** as a state of *stationarity*. This state remains *stable* if the amplitude of the fluctuations around it is small and symmetrical without bias.

However, due to a slow change of the trends **k** the amplitude of fluctuations may increase and evolve towards *critical fluctuations*. Then they become *indicators* of a *threatening instability* of the stationary state and a possibly arising preferential direction may eventually lead to the breakdown of this state!

Of course, the *meaning of fluctuations in a society* is very different from that in physics. In the latter fluctuations are accidental deviations from a mean value by pure chance only. But in a society fluctuations are expressions of consciously motivated individual intentions and activities disagreeing with the established situation. They may actively initiate the instability. Correspondingly, their sociological analysis must include their ideological and socio-political background.

But in all cases the inclusion of fluctuations via the probabilistic model design of SD can be used for forecasting stable as well as unstable situations.

References

Weidlich, Wolfgang. 2003. Sociodynamics – A Systematic Approach to Mathematical Modelling in the Social Sciences. *Fluctuation and Noise Letters.* 3: L223.

Weidlich, Wolfgang. 2005. Thirty Years of Sociodynamics. An Integrated Strategy of Modelling in the Social Sciences. *Chaos, Solitons and Fractals* 24(1): 45-56.

Weidlich, Wolfgang. 2006a. *Sociodynamics – A Systematic Approach to Mathematical Modelling in the Social Sciences.* Dover.

Weidlich, Wolfgang. 2006b. Intentions and Principles of Sociodynamics. *Evol. Inst. Econ. Rev.* 2(2): 161-165.

Weidlich, Wolfgang and Günter Haag. 1983. *Concepts and Models of a Quantitative Sociology.* Springer.

Weidlich, Wolfgang and Heide Huebner. 2008. Dynamics of Political Opinion Formation Including Catastrophe Theory. *Journal of Economic Behavior & Organisation* 67: 1-26.

15. Komplexität, Berechenbarkeit und Big Data in der Psychologie

Günter Schiepek[1]

Abstract

Komplexität und nichtlineare Dynamik sind für das Verständnis psychischer und sozialer Phänomene essentiell. Dabei geht es nicht nur um Komplexitätsreduktion, sondern um die adäquate Erfassung und Analyse komplexer Systeme in Forschung und Praxis. Mit der Einführung von innovativen Technologien des internet-basierten Prozessmonitorings hat sich gerade die Praxis der Psychotherapie grundlegend geändert. Mit einem kontinuierlichen Prozessfeedback lassen sich Precursors von Ordnungsübergängen und andere nichtlineare Phänomene auf der Höhe des Geschehens („real-time") erfassen und für die Therapiesteuerung nutzbar machen. Die Einführung dieser Technologie in die Routinepraxis generiert umfassende Datensätze (big data), deren theoriegeleitete Analyse nun auch unser Verständnis von therapeutischen Prozessen und Effekten verändert. Diese Datensätze können auch zur Validierung mathematischer Modelle der nichtlinearen Interaktion und Dynamik therapeutischer Wirkfaktoren herangezogen werden. Die Mathematisierung der Psychotherapie eröffnet aktuell neue Wege der systemischen Theoriebildung und der Theorie-Praxis-Verschränkung.

---

1   Prof. Dr. phil. Günter Schiepek. Leiter des Instituts für Synergetik und Psychotherapieforschung und Professor an der Paracelsus Medizinischen Privatuniversität Salzburg. Professor an der Ludwig-Maximilians-Universität München. Mitglied und Senatsmitglied der Europäischen Akademie der Wissenschaften und Künste. Ehrenmitglied der Systemischen Gesellschaft (Deutscher Dachverband für systemische Forschung, Therapie, Supervision und Beratung). Mitglied des wissenschaftlichen Direktoriums der Deutsch-Japanischen Gesellschaft für integrative Wissenschaft.

Complexity and nonlinear dynamics are crucial for the understanding of mental, behavioral, and social systems. The aim in reserach and practice is an adequate assessment and analysis of the network dynamics of these stystems, not only complexity reduction for pragmatic reasons. Innovative technologies of internet-based process monitoring have changed the routine practice of psychotherapy. Continuous feedack with integrated methods of nonlinear time-series analysis allow for the identification of precursors of pattern transitions and other nonlinear phenomena in the functioning of mental and behavioral systems. Together with the client this real-time monitoring opens new ways for a continuous cooperative process control and creates big data sets in the routine practice of clinical psychology. Theory and the analysis of big data sets are converging to create a new understanding on how psychotherapy works and how treatment effects can be assessed. Actual developments in the mathematical modelling of psychotherapy processes are featuring nonlinear and self-organizing change processes and can be validated by those process-outcome data sets. The mathematization of psychotherapy is a promising way in systemic theory development and the integration of practice and research.

15.1   Komplexität und Komplexitätsreduktion

Es sollte nicht überraschen, dass der Begriff „Komplexität" selbst komplex ist. Es gibt eine Vielzahl von Bedeutungen dieses Begriffs mit dazu korrespondierenden Operationalisierungen und Messansätzen (Mainzer, 2007a, 2008), was zur Folge hat, dass die Vorschläge zur Komplexitätsreduktion ebenso vielfältig sind. Im sozialen Zusammenleben beispielsweise wurde auf Vertrauen als eine Möglichkeit der Komplexitätsreduktion hingewiesen (Luhmann, 1984). Wenn ich darauf vertrauen kann, dass das, was ein anderer sagt (z.B. Informationen, die er gibt, Zusagen, die er macht, Beschlüsse, die er fasst) auch stimmt, eingehalten oder umgesetzt wird, erspare ich mir jede Menge Recherche, Kontroll- und Steuerungsaufwand. Komplexität wird damit reduziert und die Dinge werden vorhersehbar(er). Ähnliches gilt für rollenkonformes Verhalten, Gehorsam und andere Formen der (mehr oder weniger freiwilligen) Selbst-Trivialisierung. Komplexitätsreduktion bedeutet in diesem Zusammenhang so etwas wie Aufwandsreduktion und Zugewinn an Verlässlichkeit und Vorhersehbarkeit.

Ein Spezialfall der Selbst- und Fremdtrivialisierung ist die Anwendung einfacher Entscheidungsregeln: Wenn die Bedingung oder Bedingungskombination a vorliegt, dann tue x, wenn b vorliegt, dann tue y. Trivialisierung bedeutet hier im

Sinne Heinz von Foersters (1985), dass Systeme nicht lernen, also ihr Verhalten nicht von ihrer eigenen Funktionsgeschichte oder ihren bisherigen System-Umwelt-Interaktionen abhängt. Ein Computer, eine Kaffeemaschine oder ein Auto sollen unter identischen Bedingungen das tun, was sie immer tun, nämlich anspringen, Kaffee produzieren oder eine Datei aufrufen. Wer das Ding bedient oder wer ins Auto steigt, sollte (bei fachgerechter Bedienung) keine Rolle spielen. Anders z.b. bei einem Pferd, da mag es durchaus eine Rolle spielen, welche Erfahrungen es mit dem Reiter gemacht hat.

Eine fundamentale Möglichkeit der Komplexitätsreduktion hat die Natur selbst eingeführt, nämlich durch die Erzeugung von selbstorganisierten Strukturen und Mustern in Systemen, also überall dort, wo Teile (Atome, Moleküle, Neuronen, Zellen jeder Art, Menschen, usw.) miteinander interagieren. Sobald in Systemen aus diesem Zusammenwirken von Teilen Strukturen und Muster („Ordner" im Sinne der Synergetik) emergieren, „versklaven" sie die Teile, d.h. die Freiheitsgrade ihres Verhaltens reduzieren sich drastisch. Die Teile werden in kollektive Moden (Ordner) eingebunden und bringen diese Ordner zugleich auch hervor (Kreiskausalität) – ein Vorgang, der von der Synergetik in vielen Anwendungsfeldern erklärt und beschrieben wurde (Haken, 2004). Mit dem „Kollaps" der Freiheitsgrade ist eine enorme Informationskompression verbunden, da es nun ausreicht, das raum-zeitliche Verhalten der Ordner zu beschreiben anstatt des Verhaltens von möglicherweise extrem vielen Teilen. Damit einher geht meist auch eine deutliche Zunahme des Wirkungsgrades der jeweiligen Systeme. In pathologischen Fällen (d.h. bei neurologischen oder psychischen Erkrankungen) lassen sich allerdings auch Übersynchronisationen in neuronalen oder psychischen Systemen finden, die mit massiven Funktionsbeeinträchtigungen einhergehen (z.B. Tass & Hauptmann, 2007; Schiepek et al., 2016d).

Die Entstehung kollektiver Ordnung ist eine universelle Form der Komplexitätsreduktion in Systemen jedweder Art, aber insofern janusköpfig, als damit neue Formen von Ordnung und Ordnern entstehen können, die über die Eigenschaften der Teile hinausgehende Qualitäten haben (emergente Eigenschaften, Mainzer, 2007b) und sich auch vielgestaltig und nicht vorhersehbar (chaotisch) verhalten können. Die Komplexität verschwindet auf einer Ebene, nämlich auf der Ebene der Teile, und tritt auf der Ebene der Ordnung und der Ordner wieder auf. Besonders interessant wird es, wenn wir von einer Ordnungsparameterdynamik ausgehen, die aus der Wechselwirkung der Ordner entsteht. Sind diese Wechselwirkungen durch Nichtlinearität und durch gemischtes Feedback (also ein Zusammenspiel von positiven und negativen Rückkopplungen) charakterisiert, so kann Chaos auftreten (Mainzer, 2007a,b; Strunk & Schiepek, 2006, 2014). Dies bedeutet, dass ein Systemverhalten irregulär und wie zufällig aussieht, aber eine innere Ordnung aufweist, die erkennbar wird, wenn man die Dynamik des Systemverhaltens in einem Phasenraum darstellt. Weiterhin bedeutet dies, dass das Verhalten eines Systems langfristig nicht vorhersehbar ist, auch wenn das System völlig deterministisch funktioniert, und dass kleinste Unterschiede in den Ausgangsbedingungen und minimale Einflüsse von außen oder Fluktuationen von innen zu deutlich geänderten Prozessen führen können („Schmetterlingseffekt"). Nichtlineare Systeme mit gemischtem Feedback scheinen in der Natur der Normalfall zu sein, weshalb in natürlichen Systemen ein breites Spektrum komplexer Dynamiken realisiert wird, mit anderen Worten, vielfältige Varianten des Chaos beobachtbar sind (Mainzer, 2007a).

Gemäß einer kürzlich vorgeschlagenen Einteilung in einfache, komplizierte, komplexe und zufällige Systeme (Strunk & Schiepek, 2014) wäre das Spektrum der Komplexität vor allem durch vielfältige Formen des Chaos abgedeckt. Diese Formen manifestieren sich in unterschiedlichen Attraktoren, in unterschiedlicher fraktaler Dimensionalität (Anzahl der das Systemverhalten hervorbringenden Variablen), in unterschiedlichen Vorhersagehorizonten des Systemverhaltens (Ausprägung der Lyapunov-Exponenten des Systems) und schließlich in den Übergangsszenarien zwischen Attraktoren. „Einfach" bedeutet dagegen, dass wenige Elemente eines Systems durch mehr oder weniger lineare Wirkungen aufeinander verbunden sind und vorhersehbares Verhalten erzeugen. Systeme können zunehmend komplizierter werden, wenn immer mehr Elemente hinzukommen sowie die Verschaltungen oder Verbindungen anspruchsvoller werden, wie die Mechanik einer Uhr oder die Schaltpläne von Elektrogeräten oder die Verdrahtungen auf der Festplatte eines Rechners. Solche Systeme sind dann kompliziert, aber insofern (noch) nicht komplex, als ihre Struktur verstehbar und ihr Verhalten vorhersehbar ist. Beim Zufall dagegen ist gar nichts verstehbar, keine Regel erkennbar, keine Ordnung rekonstruierbar und schon der nächste Schritt in der Abfolge von Systemzuständen ist weder berechenbar noch vorhersehbar. Das (deterministische) Chaos liegt dazwischen, wobei es in der realen Welt auch ein Zusammenspiel von Zufall (z.B. dynamisches Rauschen) und Chaos gibt (Mainzer, 2007b; Schiepek et al., 2017). Hinzu kommt die bereits angesprochene Möglichkeit, dass sich chaotische Dynamiken selbst über die Zeit verändern, also Übergänge zwischen Attraktoren stattfinden. Anpassungs-, Lern- und Entwicklungsprozesse biologischer, psychischer und sozialer Systeme sind eben dadurch charakterisiert (Abb. 15.1).

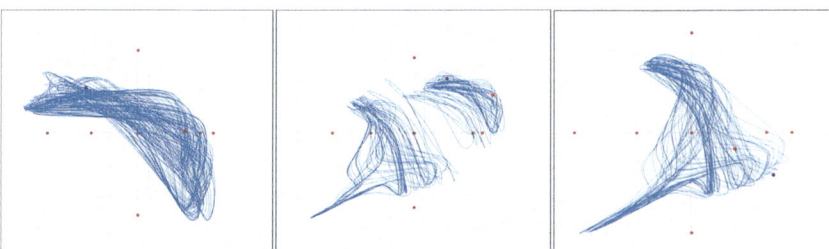

Abb. 15.1: Übergang zwischen zwei verschiedenen dynamischen Mustern (Attraktoren, links und rechts) in einen computersimulierten System (s. unten, Abb. 15.6). Die Transiente (Mitte) enthält das Muster beider Attraktoren.

In unserer Welt ist Komplexität dieser Art allgegenwärtig. Physiologische (z.b. neurobiologische), psychologische und soziale Systeme (vom Kleinen wie in dyadischen Interaktionen bis zu gesamtgesellschaftlichen Entwicklungen) erzeugen und verändern spontan dynamische Muster und Strukturen – ein Prozess, den man als Selbstorganisation bezeichnet. Ähnliches gilt für das Wetter und längerfristig für das Klima, für Verkehrsströme in Städten, für Ökosysteme und für Märkte, für Fußballspiele und für Unternehmen. Man kann in deren Entwicklung bestimmte Randbedingungen und Impulse setzen, darüber hinaus aber ist oft kaum vorhersehbar und steuerbar, was passiert. Man muss sich auf die stattfindenden Entwicklungen einstellen, auf der Höhe des Geschehens agieren und eventuell nachjustieren.

Komplexität könnte man also durch folgende Merkmale charakterisieren:

- Begrenzte Vorhersehbarkeit
- Vielfalt und Variabilität in der Struktur und Dynamik von Systemen (also z.B. keine geradlinigen oder einfachen oszillierenden Entwicklungen), was mit verschiedenen Komplexitätsmaßen (z.b. der fraktalen Dimensionalität) erfassbar ist
- Emergenz neuer Qualitäten, z.b. Auftreten von Schwellen, Übergang zu neuen Ordnungen und Funktionen, womit komplexe Systeme immer auch für Überraschungen sorgen
- Multistabilität, was bedeutet, dass unterschiedliche Dynamiken und Muster verfügbar sind, zwischen denen das System bei minimalen Veränderungen innerer und äußerer Bedingungen „switchen" kann
- Die Beurteilung der Komplexität der Struktur oder Dynamik eines Systems (z.B. eines Musikstücks) hängt von der Wahrnehmungs- und Verarbeitungskapazität

und der Eigenkomplexität des Beobachters oder des damit interagierenden Systems ab (*law of requisite variety* nach Ashby, 1956/1985, 1965), womit Komplexität immer relativ ist und ein System-zu-System-Verhältnis bezeichnet

15.2 Komplexität in der Psychotherapie

Am Beispiel der Psychotherapie ist das Phänomen der Komplexität gut zu erkennen. Es tritt auf verschiedenen Systemebenen auf, etwa in der Dynamik und Konnektivität des Gehirns, in psychischen Prozessen des Lernens und der Emotionsverarbeitung, in der Therapeut-Klient-Beziehung, also der Mikro-Abstimmung der therapeutischen Kommunikation, usw. Die Beschreibung solcher Systeme kann auf verschiedenen Auflösungsebenen erfolgen, im Gehirn etwa auf der Ebene der Plastizität von Neuronen und Neuronennetzen (Vorgänge der Genexpression und der Veränderung von Synapsen und Dendritenbäumen) oder der funktionellen Neuroanatomie (darstellbar über wiederholte fMRT-Messungen im Therapieprozess, Schiepek et al., 2013), im Erleben und Verhalten z.B. durch Transkriptanalysen oder eine videobasierte Rekonstruktion von Sitzungen (Beirle & Schiepek, 2002) sowie durch tägliche Selbsteinschätzungen mit geeigneten Fragebögen (Haken & Schiepek, 2010), auf der Ebene der Therapeut-Klient-Interaktion durch Verfahren des Bewegungsmonitorings (Ramseyer & Tschacher, 2008) oder der video- und transkriptbasierten Sequenziellen Plananalyse (Haken & Schiepek, 2010, Kap. 6; Strunk & Schiepek, 2014). In Studien, die eine detaillierte Erfassung von Veränderungsprozessen vornehmen und geeignete Zeitreihendaten generieren, kann selbstorganisierte Komplexität im Sinne von deterministischem Chaos und von Ordnungsübergängen zwischen dynamischen Mustern (Attraktoren) positiv nachgewiesen werden (Schiepek et al., 2016a, 2017). Diese Nachweise motivierten uns vor über zehn Jahren, ein internet-basiertes Monitoringsystem für die Praxis zu entwickeln, das die Komplexität therapeutischer Prozesse auf der Höhe des Geschehens darstellbar macht (Tabelle 15.1). Es beruht auf täglichen Selbsteinschätzungen der Patienten, auf der Möglichkeit, Zeitreihen in geeigneter Weise darzustellen und mit nichtlinearen Methoden zu analysieren, und natürlich auf einer engen und partnerschaftlichen Kooperation mit den Patienten (Schiepek et al., 2015, 2016b).

Funktionen des Internet-basierten Synergetischen Navigationssystems

- Patientenverwaltung und Patientendokumentation
- Datenerfassung (Ratings, Tagebücher)
- Fragebogeneditor
- Outcome-Erfassung mit unterschiedlichen wählbaren Fragebögen
- Graphische Darstellung von Zeitreihen und Histogrammen (einzeln, überlagert, gemittelt)
- Ampelfunktion (z.B. nach den generischen Prinzipien)
- Zeitreihenanalyse
  - Faktorendarstellung
  - Dynamische Komplexität
  - Komplexitäts-Resonanz-Diagramme
  - Recurrence Plots
  - Synchronisation: Dynamische Korrelationsmatrizen
  - Permutationsentropie

Tabelle 15.1: Funktionen des SNS

Komplexität manifestiert sich in der Psychotherapie vor allem in der Chaotizität der Veränderungsdynamik, wobei das Chaos (Formen komplexer Ordnung) selbst Resultat von Selbstorganisationsprozessen ist. Dies bedeutet:

- Es besteht keine mittel- oder langfristige Vorhersehbarkeit der Entwicklung
- Es bestehen keine sicheren Input-Output-Relationen, was heißt, dass ein und dieselbe Intervention unter anderen Bedingungen oder zu einem anderen Zeitpunkt andere Effekte haben kann
- Man muss mit einer sensiblen Abhängigkeit der Dynamik von Ausgangsbedingungen, externen Mikroeinflüssen und systeminternen Fluktuationen rechnen
- Es liegt eine große Individualität und Vielfalt der Verläufe vor
- Das Superpositionsprinzip ist nicht gültig, was bedeutet, dass eine Mittelung von Verläufen keinen Sinn macht. Eine Annahme von Norm- oder Standard-"Tracks" ist nicht sinnvoll

Fokussiert man auf Komplexität oder zumindest auf bestimmte Aspekte der Komplexität, so ist dies eine bewusste Entscheidung. Sie beruht auf der empirisch wie theoretisch begründeten Annahme, dass die Komplexität therapeutischer Prozesse essenzielle Information enthält, die für das Verständnis, die Gestaltung und die Unterstützung von Therapieprozessen entscheidend ist. Was man für einen adäquaten Umgang mit Komplexität hält, ist also von Vorannahmen, von einer vorab erfolgten Konzeptualisierung des Gegenstandes abhängig, z.b. der Frage, worauf man fokussiert (z.b. auf das Erleben und die therapierelevanten Kognitionen und Emotionen des Patienten), mit welchen Messmethoden man arbeitet, welche Messfrequenz (Abtastrate) man wählt und welche Analyseverfahren man nutzt, um die Komplexität von Veränderungsprozessen abzubilden. Von solchen Entscheidungen wird abhängen, was man sehen kann.

Ein entscheidendes Kriterium ist die Abtastrate von Prozessen. In Therapieprozessen haben wir gute Erfahrungen mit täglichen Messungen (Selbsteinschätzungen) gemacht. Es handelt sich dabei um eine äquidistante, regelmäßige und relativ hochfrequente Prozesserfassung, die verschiedenste zeitreihenanalytische Verfahren zulässt, um Aussagen über Synchronisation, Frequenzmuster, (In-)Stabilität und Phasenübergänge (Musterwechsel) zu machen. Zugleich erweist sich für die Patienten ein Zeitaufwand von etwa 5-10 Minuten pro Tag für das Ausfüllen eines Fragebogens im Internet als praktikabel und nicht überfordernd. Die meisten Patienten erleben dies als wertvolle Zeit einer ruhigen, fokussierten Tagesreflexion und nutzen dabei auch die Möglichkeit eines elektronischen Therapietagebuchs. Analyseverfahren wie die dynamische Komplexität (Schiepek & Strunk, 2010), die dynamische Interkorrelation der Itemverläufe (Schiepek et al., 2015, 2016d) oder die Permutationsentropie (Bandt & Pompe, 2002) werden in einem Gleitfenster berechnet, womit man ohne große Verzögerung Zustandsänderungen, Instabilitäten und dynamische Übergänge erkennt und rückgemeldet bekommt. Abbildung 15.2 illustriert, welchen Informationsverlust man hinnehmen muss, wenn man einen Therapieverlauf mit täglicher Abtastrate seltener und unregelmäßig erfasst, wobei Abbildung 15.2a das Original mit einem gut erkennbaren Musterwechsel in der Dynamik zeigt, 15.2b, c und d dagegen wöchentliche Erfassungen mit leichter Variation der Wochentage, und e, f Beispiele mit noch selteneren und unregelmäßigen Messungen. Die Aussagen über den Prozess werden nicht nur gröber, sondern systematisch verzerrt, und trotzdem entsprechen solche unregelmäßigen Erfassungen (im Rahmen einzelner Therapiesitzungen) im Moment der üblichen Praxis des so genannten Therapiemonitorings.

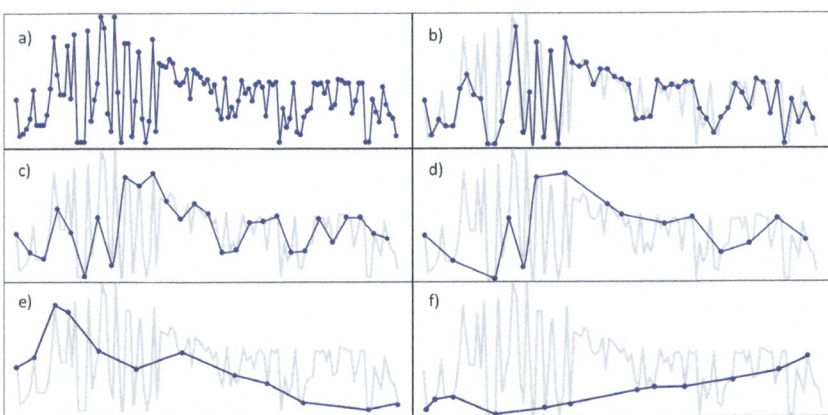

Abb. 15.2: (a) Zeitreihe „Selbstwertgefühl" einer Patientin im Verlauf einer stationären Psychotherapie (Diagnose: Borderline-Störung). Tägliche Selbsteinschätzungen mit Hilfe des Synergetischen Navigationssystems (SNS). (b), (c) und (d) zeigen ausgedünnte Zeitreihen. Dargestellt ist nur jeder 7. Wert mit leicht unterschiedlichen Zufallsschwankungen um den 7-Tages-Takt. (e) und (f) zeigen noch etwas seltenere und unregelmäßigere Selbsteinschätzungen (mit wöchentlichen bis vierzehntägigen Abständen). Dies entspricht in etwa den Frequenzen der bisher üblichen Praxis.

Ein weiteres Artefakt des Informations- und Komplexitätsverlusts besteht in der Annahme, Psychotherapien würden einem Standardverlauf (engl. *track*) folgen. Solche Tracks inklusive ihrer Konfidenzintervalle werden durch eine Mittelung von Verläufen gewonnen. Aus der Perspektive eines hochfrequenten Prozessmonitorings und der Theorie nichtlinearer Systeme handelt es sich jedoch dabei um ein Artefakt unangemessener Komplexitätsreduktion, resultierend aus einer Kombination linearen Denkens und unzulänglicher Verlaufsdaten. Die Konsequenz bestünde darin, die gleichen Ansprüche, die man an die Standardisierung von Messinstrumenten stellt, auch an die Abtastrate von Systemprozessen anzulegen. Nur wenn ein äquidistantes, in Relation zur Eigendynamik der Phänomene angemessenes und kontinuierliches Prozessmonitoring durchgeführt wird (Abtast-Theorem), können wir valide Aussagen über Themen wie *sudden changes*, kritische Instabilitäten im Verlauf, Rupture-Repair-Sequenzen in der therapeutischen Beziehung, frühzeitige Abbrüche, Nachhaltigkeit therapeutischer Effekte, usw. machen.

Ein adäquater Umgang mit (dynamischer) Komplexität setzt Hilfsmittel bei der Prozessmessung und Prozessanalyse voraus – wie eben das Synergetische Navigationssystem (SNS). Nichtlineare Dynamiken, Ordnungsübergänge, kritische Instabilitäten, zeitvariate Synchronisationsmuster usw. sind mit bloßem Auge nicht erkennbar. Sie sind im Hier und Jetzt nicht spürbar, weil sie sich in der Zeit abspielen und haben keine Qualia-Qualitäten wie etwa der Emotionsausdruck oder das Verhalten eines Gegenübers, auf die unser Spiegelneuronen-System zuverlässig anspricht. Die Arbeitsgruppe um Lambert (z.B. Lambert et al., 2002) konnte zeigen, dass sich auch erfahrene Therapeuten schwer tun, Verschlechterungen und sich anbahnende Therapieabbrüche (*drop outs*) ihrer Klienten ohne spezielle Feedback-Methoden rechtzeitig zu erkennen. Inzwischen konnte die Machbarkeit, Technikakzeptanz und Ausfüll-Compliance eines hochfrequenten Prozessmonitorings klar belegt werden (Schiepek et al., 2016b).

Was eine angemessene Form der Komplexitätsreduktion ist, hängt wie die Beurteilung eines methodischen Vorgehens in der Wissenschaft oder des Einsatzes von Entscheidungsroutinen in der Praxis von den Forschungsfragestellungen und Zielsetzungen ab. Geht es z.B. um die Legitimation eines Therapieansatzes, kann ein klassisches experimentelles Design (*Randomized Controlled Trial*) die Methode der Wahl sein. Ähnlich ist es bei Entscheidungsprozessen: Wenn es nur um einmalige Entscheidungen mit dem Ziel der Gewinnmaximierung geht, legen Personen andere Kriterien an und folgen anderen Entscheidungsstrategien als wenn es um längerfristige Beziehungen und zeitliche Entwicklungen geht (vgl. z.B. Studien zum iterativen *Prisoner's Dilemma Game* [Axelrod, 1984; Sigmund, 1993; Nowak & Sigmund, 1993] oder zur interpersonellen Konzeptbildung [King-Casas et al., 2005]). Ein Beispiel wären anhaltende Geschäftskooperationen, innerhalb derer die „Spielzüge" unterschiedlichen Kriterien folgen (z.B. langfristige Gewinnmaximierung, „Dankbarkeit" für Geleistetes, „Erziehung" des anderen, gegenseitige Absicherung und Stützung in Krisenzeiten), oder aber Konkurrenz- oder kombinierte Kooperations-Konkurrenz-Beziehungen zwischen mehreren Interaktionspartnern. Dann wird es komplex.

15.3   Konsequenzen für Praxis und Ausbildung

In psychologischen Anwendungsfeldern werden verschiedene Möglichkeiten der Komplexitätsreduktion genutzt. Eine Möglichkeit beruht auf der Annahme, dass in der Komplexität von Vernetzungsstrukturen und Zeitsignalen die Information steckt, die wir zur Steuerung von Veränderungsprozessen brauchen. Mit einer Internet-basierten Technologie (dem SNS) können in Therapie, Beratung und Organisationsentwicklung Daten erfasst und analysiert werden, womit ein Einblick in laufende Veränderungsprozesse auf der Höhe des Geschehens möglich wird. In der

Psychotherapie können wir die Erfassung von Systemen und deren Entwicklung inzwischen hochgradig individualisieren. Ein wesentlicher Zugang besteht in der Anwendung folgender methodischer Schritte:

• Ressourceninterview mit dem Klienten
• Idiographische Systemmodellierung (Schiepek et al., 2015; für ein Beispiel s. Abb. 15.3)
• Anlage eines individualisierten Prozessfragebogens zusammen mit dem Klienten auf Basis der relevanten Variablen im idiographischen Systemmodell
• Prozesserfassung mit Hilfe des individuellen Prozessfragebogens im SNS
• Therapiefeedback und Therapiesteuerung mit Hilfe der Zeitreihendaten und der Zeitreihenanalysen, wie mit dem SNS möglich
• Nutzung der generischen Prinzipien als Entscheidungsheuristiken für das weitere Vorgehen in der Therapie (Haken & Schiepek, 2010; Schiepek et al., 2015)

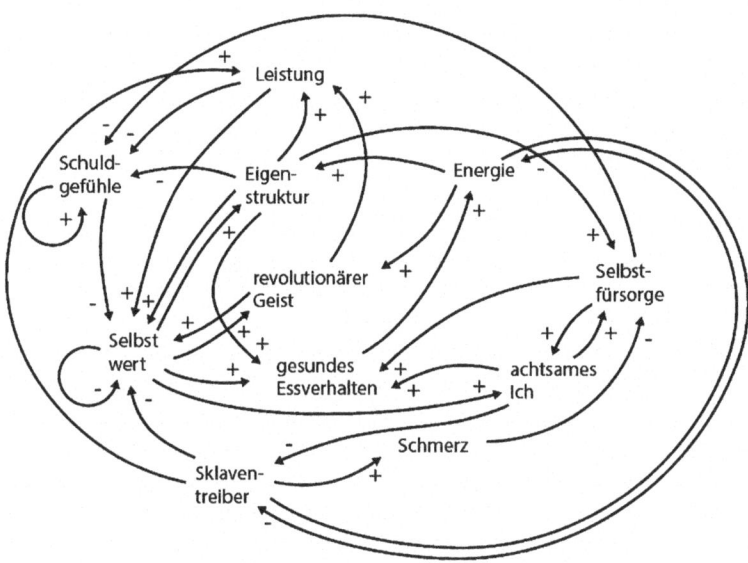

Abb. 15.3: Idiographisches Systemmodell einer Patientin mit chronischen Essstörungen. Es handelt sich um ein reduziertes Modell, in dem vor allem Ressourcen, Kompetenzen und stützende Ich-Anteile dargestellt sind (Ausnahme: „Schuldgefühle", „Schmerzen").

Damit sind für praktische Zwecke eine Reihe der bekannten Probleme im Umgang
mit komplexen Systemen zumindest sinnvoll eingrenzbar, etwa die Kontraintuiti-
vität solcher Systemdynamiken (Dörner, 1989) oder die Vielfalt der Optionen, die
unter Zeitdruck (z.B. eines therapeutischen Gesprächs) einzugrenzen sind (Rufer,
2012). Das Auftreten unerwarteter Folge- und Nebenwirkungen kann zumindest
in allen Variablen, die in einem Systemmodell berücksichtigt sind, erfasst werden.
Sprunghaftigkeit im Verhalten und das Fehlen linearer oder proportionaler Input-Output-Relationen sowie Zeitverzögerungen zwischen Ereignis und Ereignis-
folgen können anhand der Zeitreihen direkt nachvollzogen werden. An die Stelle
von immer wieder gescheiterten Langzeitvorhersagen tritt ein Frühwarnsystem für
auftretende kritische Instabilitäten, Musterwechsel (z.B. der Synchronisation von
psychologischen Variablen) oder Therapiekrisen (Fartacek et al., 2016). Eben weil
es nur begrenzte Möglichkeiten der linearen Steuerbarkeit und Intervenierbarkeit
gibt, d.h. lineare Input-Output-Relationen oder direkte Sollwertvorgaben (wie bei
einem Thermostat) nicht möglich sind, und jedes therapeutische Vorgehen auf Re-
sonanz und Synchronisation des Therapeuten mit der Eigendynamik des Klienten-
systems angewiesen ist, erscheint ein konsequentes Prozessmonitoring notwendig.

   Natürlich sind damit nicht alle Probleme im Umgang mit komplexen Systemen
gelöst, vor allem bleiben einige systemtheoretische und wissenschaftsphilosophische
Fragen offen. Prinzipiell sind zum Beispiel Strukturmodellhaftigkeit und Verhaltens-
modellhaftigkeit nicht deckungsgleich, d.h. ein bestimmtes Systemverhalten kann
prinzipiell von verschiedenen Modellen generiert werden, und ein Modell kann (je
nach Parametereinstellung und Randbedingungen) ganz unterschiedliches Verhalten
erzeugen (Dörner, 1984). Es gibt mithin eine mehr-mehr-deutige Beziehung zwischen Struktur und Dynamik. Ein anderer Punkt betrifft die Emergenz neuer Qualitäten: Welche Eigenschaften können in einem System auftreten und seine Entwicklung bestimmen, die auf der Ebene der Teile und ihrer Relationen nicht vorhanden
waren? Handelt es sich um schwache/reduktive Emergenz, was bedeutet, dass die Eigenschaften aus den Gesetzmäßigkeiten auf der Ebene der Teile erklärbar sind, oder
um starke Emergenz (vgl. Mainzer, 2007b; Stephan, 1999; Haken & Schiepek, 2010)?

   Die Konsequenz für die Ausbildung von professionell handelnden Menschen
in der Psychotherapie oder der Organisationspsychologie besteht darin, auf den
Umgang mit komplexen Systemen vorzubereiten. Das Kompetenzspektrum hierfür wurde als *Systemkompetenz* beschrieben (Haken & Schiepek, 2010). Es enthält
ein Vielzahl von mentalen (kognitiven wie emotionalen) und handlungsbezoge-
nen Teilkompetenzen sowie theoretischem, methodischem und praktischem Know
How, um mit der Komplexität dynamischer Systeme umgehen zu können. Für Be-
rufe, bei denen es auf das Handeln in und mit komplexen Systemen ankommt,
z.B. in der Psychotherapie, in Beratung und Coaching oder im Management kann

hierin eine Schlüsselkompetenz für erfolgreiche Professionalität gesehen werden (Mainzer, 2007a, 2008, 2010). Ein angemessener Umgang mit Komplexität (unter anderem durch die Nutzung geeigneter technologischer Hilfsmittel) ist möglich und stellt eben jene Informationen bereit, die wir beim Navigieren und Steuern durch die Dynamik selbstorganisierender Systeme brauchen. Studiengänge und Ausbildungen sollten Systemkompetenz daher an zentraler Stelle positionieren.

15.4 Berechenbarkeit und mathematische Modellierung psychischer Prozesse

Ein zentrales Hilfsmittel für das Verständnis von Komplexität und Dynamik ist seit Newton und Leibniz die formale Modellierung. Mathematische Modellierungen sind in der Psychologie aber ungewöhnlich, in der Psychotherapieforschung noch ungewöhnlicher, und man mag berechtigterweise nach ihrem Sinn fragen. Anwendungen der Mathematischen Psychologie, welche die Formalisierung von Theorien und die systematische Organisation von Daten zum Zweck hat (Falmagne, 2005; Heath, 2000), sind allerdings auch in der Klinischen Psychologie und der Psychiatrie nicht neu, z.B. haben Schiepek et al. (1992) bereits vor 25 Jahren die von Ciompi und Müller (1976) beschriebenen Verlaufsmuster der Schizophrenie in einem 5-Variablen-Modell simuliert.

In den theoretischen Neurowissenschaften (*systems neuroscience*) sind mathematische Modellierungen unverzichtbar: „Offensichtlich hat die Evolution das Gehirn mit Mechanismen zur Selbstorganisation ausgestattet, die in der Lage sind, auch ohne eine zentrale Instanz globale Ordnungszustände herzustellen. (…) Wir werden zur Analyse und Beschreibung dieser Systemzustände mathematisches Rüstzeug und den Einsatz sehr leistungsfähiger Rechner benötigen. Und wir werden das gleiche Problem haben, mit dem die moderne Physik konfrontiert ist. Die Modelle werden unanschaulich sein und vermutlich auch unserer Intuition von der Verfasstheit unserer Gehirne widersprechen" (Singer, 2007). Innovativen Methoden der nichtinvasiven Neurostimulation (z.B. dem Coordinated Reset bei chronischem tonalen Tinnitus, Tass et al., 2012) oder der Konnektivitätsanalyse neuronaler Netze (z.B. Functional Connectivity Dynamics, Hansen et al., 2015; nonlinear DCM, Stephan et al., 2008) liegen Simulationen neuronaler Netzwerke und ihrer Dynamik zugrunde (z.B. Leon et al., 2013; Messé et al., 2015; Tass, 2003). Aus Simulationen pathologischer Netzwerkdynamiken bei psychischen Störungen (z.B. MDD, Ramirez-Mahaluf et al., 2015) lassen sich – so die Hoffnung – in Zukunft spezifische therapeutische Verfahren ableiten und im Modell prüfen. Zum experiumentum in vitro und in vivo kommt das „experimentum in silico", also das Experiment im Computermodell (Mainzer, 2010, 2014).

Ein spezieller Bedarf ergibt sich nun aber tatsächlich auch in der Psychotherapieforschung. Erstens liegen hier Befunde zu unterschiedlichen Wirkfaktoren vor, die isoliert nebeneinander stehen (Duncan et al., 2010). Für die einzelnen Wirkfaktoren werden unterschiedliche Impactwerte für den Therapieeffekt (prozentualer Beitrag zur Outcomevarianz) geschätzt, eine Vorstellung über deren nichtlineares und rekursives Zusammenwirken gibt es allenfalls in Form von qualitativen Modellen (das bekannteste dürfte das „Generic Model" von Orlinsky et al. (1994, 2004) sein). Eine Formalisierung der die Variablen verbindenden Funktionen liegt bislang noch nicht vor. Zweitens ist Psychotherapie ein Prozess, der inzwischen engmaschig erfasst und abgebildet werden kann (Schiepek et al., 2016b). Wir verfügen also über umfassende Verlaufsdaten, die einer Erklärung bedürfen (Abb. 15.4a). Das Explanandum der Psychotherapieforschung ist der Prozess.

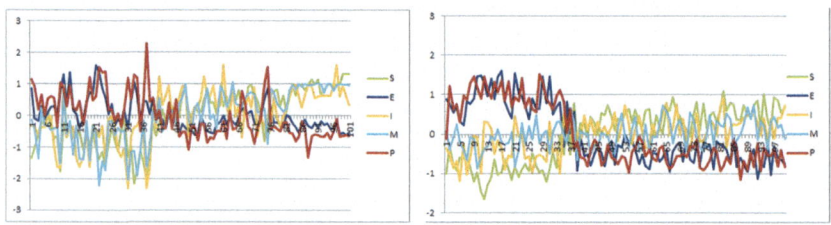

Abb. 15.4: (a) Links: Verlauf eines psychotherapeutischen Prozesses. Die Darstellung beruht auf täglichen Einschätzungen (100 Messpunkte) mit Hilfe des Therapie-Prozessbogens, erfasst mit Hilfe des Internet-basierten Synergetischen Navigationssystems (SNS). Problem- und Symptomausprägung (P), therapeutische Fortschritte (S), Veränderungsmotivation (M), Emotionsintensität (E), und Einsicht bzw. Entwicklung neuer Perspektiven (I). (b) Rechts: Simulierter Therapieverlauf, beruhend auf den gleichen Variablen.

Nun braucht man nur eins und eins zusammenzählen: Netzwerkmodelle (z.B. zwischen Wirkfaktoren) müssen in dynamischer Weise formalisiert werden (z.B. in gekoppelten Differenzen- oder Differentialgleichungen), um Prozesse und dynamische Muster zu erzeugen. Ein aktuelles Modell (Schiepek et al., 2016c, 2017) bezieht sich auf fünf Variablen bzw. Wirkfaktoren: Problem- und Symptomausprägung (P), therapeutische Fortschritte (S), Veränderungsmotivation (M),

Emotionsintensität (E), Einsicht bzw. Entwicklung neuer Perspektiven (I). Die Auswahl dieser Variablen ist dadurch motiviert, dass Klientenvariablen eine zentrale Rolle im Konzert der therapeutischen Wirkfaktoren spielen (Bohart & Tallman, 2010; Orlinsky et al., 2004), und dass genau diese fünf Variablen in einem standardisierten Therapieprozessbogen, der seit Jahren zur Verlaufserfassung (tägliche Selbsteinschätzungen von Klienten/-innen mit Hilfe des SNS) verwendet wird, als Faktoren vorliegen. Die Prozessdaten können damit zur Validierung des Modells herangezogen werden.

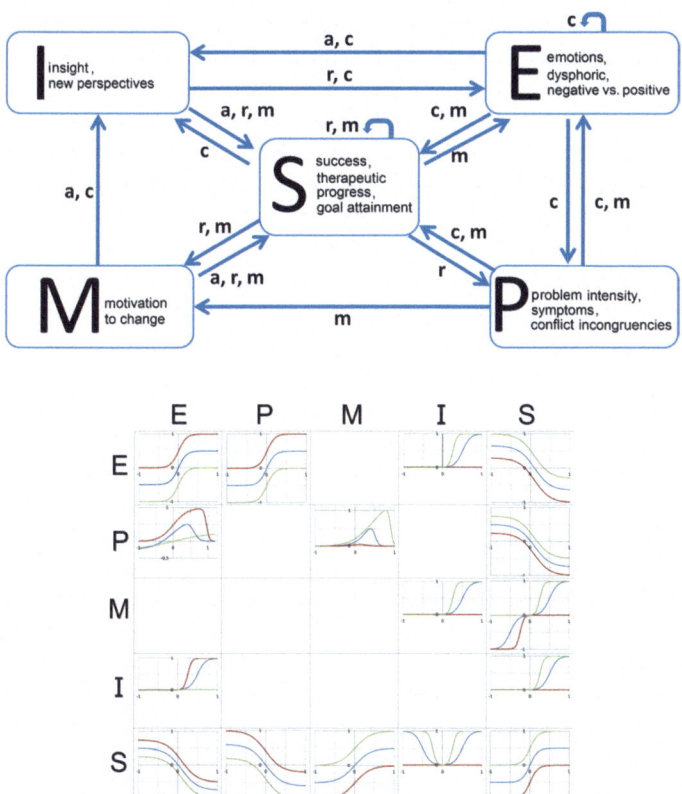

Abb. 15.5: (a) Links: Die Struktur des Wirkfaktorenmodells: P, S M, E, I sind die Variablen, a, c, r, m die Parameter des Modells. (b) Rechts: Graphische Veranschaulichung der Funktionen, welche die Wirkung der Variablen aufeinander vermittelt. Die genaue Form der Funktionen wird durch die Ausprägung der Parameter bestimmt.

Das Modell (Abb. 15.5a,b) formuliert Annahmen über die Wirkung der einzelnen Variablen aufeinander, die in nichtlinearen Funktionen dargestellt sind. Die vier Parameter des Modells sind als Dispositionsgrößen zu verstehen, welche die Form der Funktionen und damit die Wechselwirkung zwischen den Variablen beeinflussen (Schiepek et al., 2016d, 2017). Inhaltlich sind die Parameter wie folgt zu interpretieren: Fähigkeit zu einer vertrauensvollen Therapiebeziehung und Qualität der Arbeitsbeziehung (a), kognitive Kompetenzen, z.b. Mentalisierungsfähigkeit und Emotionsregulation (c), Verhaltensressourcen und Skills (r), sowie Selbstwirksamkeit und Belohnungserwartung (m). Das Modell enthält fünf nichtlineare gekoppelte Differenzengleichungen, wobei jede die Veränderung einer Variablen über die Zeit in Abhängigkeit von anderen Variablen und gegebenenfalls von sich selbst beschreibt.

Die Modellierung reproduziert wesentliche Merkmale psychotherapeutischer Verläufe, z.B. chaotische Dynamik, parameterabhängige Phasenübergänge, plausible Relationen zwischen den Variablen, Multistabilität, geringe oder keine Wirkung von Interventionen in stabilen Phasen und bei konstanten Parameterwerten, dagegen deutliche Wirkung in kritischen Phasen, sowie eine verlaufsabhängige Modulation der Parameter (interpretierbar als Persönlichkeitsänderung). Die Dynamik des Systems hängt nicht nur von den Startwerten und vom Input in das System ab, sondern wesentlich auch von den Parametern. Diese definieren in der inhaltlichen Interpretation des Modells die Kompetenzen und Vorbelastungen eines Patienten (traits). Umgekehrt können aber konkrete Erfahrungen in einer Therapie und im Leben (Dynamik der Variablen, states) eine Veränderung der traits bewirken. Ohne Parameteränderung keine nachhaltigen Therapieeffekte. Die Veränderung der traits ist im Modell in eigenen weiteren Gleichungen definiert, welche die Ausprägung der Parameter in Abhängigkeit von den Variablen beschreiben. Es liegt also eine Kreiskausalität zwischen „traits" und „states" vor.

Unter bestimmten Bedingungen und bei externer Stimulation (*dynamic noise*) verlaufen empirische (reale) und simulierte Therapieprozesse sehr ähnlich (Abb. 15.4b). Über diese Ähnlichkeit hinaus leisten formale Modellierungen noch weiteres:

Modelle und damit auch Theorien sind grundsätzlich selektiv, abstrahierend und perspektivenabhängig, also nie „ganzheitlich" (Stachowiak, 1978). Formalisierungen zwingen aber noch mehr als verbale oder nur als Graphiken veranschaulichte Modelle dazu, diese Selektionen und Abstraktionen explizit zu machen. Man kann nicht im Vagen bleiben, sondern muss jeden Schritt transparent machen und mathematisch ausformulieren. Es wird erkennbar, welche Zusatzannahmen erforderlich sind, um bestimmte Effekte zu erzielen.

Diese Präzision macht aber auch angreifbar, weil durchschaubar. Das Vorgehen macht Probleme der Modellbildung deutlich, die man bei nur verbalen Beschreibungen kaum entdeckt hätte. Rhetorische Nebelbomben sind praktisch unmöglich.

Mathematische Modelle erbringen Emergenzleistungen, die ansonsten kaum möglich wären. Eine davon ist die Generierung von Prozessen, die im Grunde von allen Theorien des Lernens, der Entwicklung und der Veränderung erwartet werden müsste. Nur in Prozessen finden sich Phänomene wie Chaos, selbstorganisierte Kritikalität, emergente Schwellen, Stabilität und (kritische) Instabilität, Phasenübergänge, differentielle Reagibilität auf Interventionen oder die Modulation von Prozessmustern durch Rauschen (vgl. Hütt, 2000). Der Mehrwert von Simulationen liegt in den (emergenten) dynamischen Eigenschaften, die nicht schon als Annahmen hineingepackt wurden.

Nichtlineare Modelle der Psychotherapie erweitern auch unsere Vorstellung von Interventionen über einen simplen Input-Output-Mechanismus hinaus. Unter bestimmten Bedingungen (Multistabilität) geht es darum, mit explorativen Strategien das Systemverhalten in einen anderen (aber potenziell vorhandenen) Attraktor zu „kicken", unter anderen Bedingungen darum, über das Verhalten der Variablen die Kontrollparameter des Systems zu beeinflussen, unter wieder anderen Bedingungen das System einfach unspezifisch anzuregen (*dynamic noise*), denn allein schon dadurch sind Ordnungsübergänge möglich.

Über Computersimulationen lassen sich Therapiespiele generieren, die für die Therapieausbildung von Nutzen sein können, weil sie zeigen, wie kontraintuitiv und unvorhersehbar sich „Menschen" verhalten können. Das Gegenteil von „gut" ist hier oft „gut gemeint".

Modelle lassen sich in Computerexperimenten testen, die aus praktischen oder ethischen Gründen in vivo nicht möglich wären („experimentum in silico", Mainzer, 2010, 2014). Modelle haben auch die Aufgabe, für empirische Studien Hypothesen zu generieren und zu kreativen Designs bei Validierungsstudien anzuregen.

Ob mathematische Modelle im Einzelfall die präziseren Vorhersagen liefern, sei dahingestellt – da müsste man die Ausgangs- und Randbedingungen des jeweiligen Einzelfalls schon sehr genau kennen. Im Falle nichtlinearer und damit oft auch chaotischer Systeme kann die Wikipedia-Aussage (Stichwort „Mathematische Psychologie") „Mathematische Formalisierungen haben den Vorteil, dass sie genauere Vorhersagen erlauben" jedenfalls angezweifelt werden.

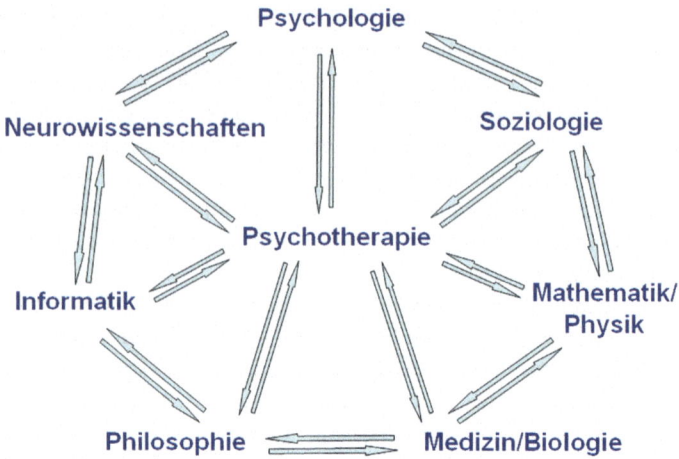

Abb. 15.6: „Einheit in der Vielheit" (Gottfried Wilhelm Leibniz). Psychotherapie als innovative Schnittstellendisziplin.

Immerhin sollte deutlich werden, dass wir uns mit mathematischen Modellierungen und Simulationen nicht nur auf eine interessante Spaß-Spielwiese begeben, sondern wo immer es um die Interaktion von Wirkfaktoren und um Veränderungsprozesse geht, auf dieses Instrumentarium angewiesen sind.

Psychotherapie wird mit der Einbettung in Synergetik und *Complexity Science* (Mainzer, 2007a,b, 2010, 2014; Mainzer & Chua, 2013) inter- und transdisziplinär erweitert zu einer Schnittstellendisziplin, für die nicht nur die Neurowissenschaften, sondern auch Informatik, Mathematik und Physik eine substanzielle Rolle spielen (Abb. 15.6). Die Entwicklungen laufen hier mit rasantem Tempo in Richtung Innovation und Technologieentwicklung, so dass wir gespannt sein können, was die Zukunft bringt. In einer Hinsicht hat sich die Praxis allerdings schon heute verändert, nämlich in der Nutzung von Internet-basierten Verfahren des Prozessmonitorings und Prozessfeedbacks. In innovativen Technologien wie dem SNS sind Methoden der nichtlinearen Zeitreihenanalyse implementiert, mit deren Hilfe individuelle Entwicklungsprozesse auf Musterveränderungen hin abgescannt werden können. Die Ergebnisse werden regelmäßig zusammen mit dem Patienten und in Kliniken auch mit dem Behandlungsteam besprochen (*continuous cooperative process control*). Insofern solche Technologien und Praxiskonzepte zunehmend Verbreitung finden, werden „flächendeckend" Daten zu Prozess und Outcome von Psychotherapien erhoben und fließen in umfassende Datenpools für Forschungszwecke ein (völlig anonymisiert und unter Berücksichtigung des Daten- und Patientenschutzes, versteht sich). Damit gehen Forschung und Praxis einen neuen Konnex ein (ein „Junktim", das schon Sigmund Freud eingefordert hatte), die Systemkompetenz des *Scientist-Practitioners* wird gefragt, und die Psychologie betritt zumindest im Bereich der Psychotherapie das Gebiet der *Big Data* (Mainzer, 2014).

Literatur

Ashby, W.R. (1965). *Design for a Brain.* London: Chapman & Hall.
Ashby, W.R. (1985). *Einführung in die Kybernetik.* Frankfurt am Main: Suhrkamp. (Originalausgabe: *An Introduction to Cybernetics.* 1956).
Axelrod, R. (1984). *The Evolution of Cooperation.* New York: Basic Books.
Bandt, C. & Pompe, B. (2002). Permutation Entropy: A natural complexity measure for time series. *Physical Revue Letters, 88,* 174102 1-4.
Beirle, G. & Schiepek, G. (2002). Psychotherapie als Veränderung von Übergangsmustern zwischen „States of Mind". Einzelfallanalyse einer systemisch-ressourcenorientierten Kurzzeittherapie. *Psychotherapie, Psychosomatik und Medizinische Psychologie, 52,* 214-225.
Bohart, A. C., & Tallman, K. (2010). Clients: The neglected common factor in psychotherapy. In B. Duncan, S. Miller, B. Wampold, & M. Hubble (2010) (Eds.), *The Heart and Soul of Change* (2nd. ed., pp. 83–111). Washington, DC: American Psychological Association.
Ciompi, L. & Müller, C. (1976). *Lebensweg und Alter der Schizophrenen.* Berlin: Springer.
Dörner, D. (1984). Modellbildung und Simulation. In E. Roth (Hrsg.), *Sozialwissenschaftliche Methoden* (S. 337-350). München: Oldenbourg.
Dörner, D. (1989). *Die Logik des Misslingens.* Reinbek: Rowohlt.
Duncan, B., Miller, S., Wampold, B., & Hubble, M. (2010) (Eds.). *The Heart and Soul of Change* (2nd Ed.) (pp. 49–82). Washington, DC: American Psychological Association.
Falmagne, J.C. (2005). Mathematical psychology - A perspective. *Journal of Mathematical Psychology, 49,* 436-439.
Fartacek, C., Schiepek, G., Kunrath, S., Fartacek, R., & Plöderl, M. (2016). Real-time monitoring of nonlinear suicidal dynamics: methodology and a demonstrative case report. *Frontiers in Psychology for Clinical Settings, 7: 130. doi: 10.3389/fpsyg.2016.00130*
Haken, H. (2004). *Synergetics. Introduction and Advanced Topics.* Berlin: Springer.
Haken, H. & Schiepek, G. (2010). *Synergetik in der Psychologie.* Göttingen: Hogrefe (2nd. Ed.).
Hansen, E.C.A., Battaglia, D., Spiegler, A., Deco, G., & Jirsa, V.K. (2015). Functional connectivity dynamics: Modeling the switching behavior of the resting state. *Neuroimage, 105,* 525-535.
Heath, R. A. (2000). *Nonlinear Dynamics: Techniques and Applications in Psychology.* Mahwah, NJ: Erlbaum.

Hütt, M.T. (2001). *Datenanalyse in der Biologie. Eine Einführung in die Methoden der nichtlinearen Dynamik, fraktalen Geometire und Informationstheorie.* Berlin Heidelberg: Springer.

King-Casas, B., Tomlin, D., Anen, C., Camerer, C.F., Quartz, S.R. & Montague, P.R. (2005). Getting to know you: reputation and trust in a two-person economic exchange. *Science, 308,* 78-83.

Lambert, M.J., Whipple, J.L., Vermeersch, D.A., Smart, D.W., Hawkins, E.J., Nielsen, S.L. & Goates, M. (2002). Enhancing psychotherapy outcomes via providing feedback on client progress: a replication. *Clinical Psychology and Psychotherapy, 9,* 91-103.

Leon, P.S., Knock, S.A., Woodman, M.M., Domide, L., Mersmann, J. McIntosh, A.R., & Jirsa, V. (2013). The Virtual Brain: A simulator of primate brain network dynamics. *Frontiers in Neuroinformatics, 7*:10. doi: 10.3389/fninf.2013.00010

Luhmann, N. (1984). *Soziale Systeme.* Frankfurt am Main: Suhrkamp.

Mainzer, K. (2007a, 5th ed.). *Thinking in Complexity. The Complex Dynamics of Matter, Mind, and Mankind.* Berlin, Heidelberg, New York: Springer.

Mainzer, K. (2007b). *Der kreative Zufall. Wie das Neue in die Welt kommt.* München: C.H. Beck.

Mainzer, K. (2008). *Komplexität.* München: W. Fink. Mainzer, K. (2010).

Mainzer, K. (2010). *Leben als Maschine? Von der Systembiologie zur Robotik und Künstlichen Intelligenz.* Paderborn: Mentis.

Mainzer, K. (2014). *Die Berechnung der Welt. Von der Weltformel zu Big Data.* München: C.H.Beck.

Mainzer, K. & Chua, L. (2013). *Local Activity Principle. The Cause of Complexity and Symmetry Breaking.* London: Imperial College Press.

Messé, A., Hütt, M.-T., König, P., & Hilgetag, C.C. (2015). A closer look at the apparent correlation of structural and functional connectivity in excitable neural networks. *Scientific Reports, 5*: 7870. doi: 10.1038/srep07870

Nowak, M. & Sigmund, K. (1993). A strategy of win-stay, lose-shift that outperforms tit-for-tat in the Prisoner's Dilemma game. *Nature, 364,* 56-58.

Orlinsky, D.E., Grawe, K. & Parks, B. (1994). Process and outcome in psychotherapy - noch einmal. In A.E. Bergin & S.L. Garfield (Eds.), *Handbook of Psychotherapy and Behavior Change* (4th Ed.) (pp. 270-376). New York: Wiley.

Orlinsky, D. E., Ronnestad, M. H., & Willutzki, U. (2004). Fifty years of psychotherapy process-outcome research: Continuity and change. In M. J. Lambert (Ed.), *Bergin and Garfield's handbook of psychotherapy and behavior change* (pp. 307–390). New York, NY: Wiley.

Ramirez-Mahaluf, J.P., Roxin, A., Mayberg, H.S., & Compte, A. (2015). A computational model of Major Depression: the role of glutamate dysfunction on cingulo-frontal network dynamics. *Cerebral Cortex, 2015,* 1-20. doi: 10.1093/cercor/bhv249

Ramseyer, F. & Tschacher, W. (2008). Synchronisation in dyadic psychotherapy sessions. In: S. Vrobel, O.E. Rössler, T. Marks-Tarlow (Eds.) Simulataneity: Temporal Structures and Observer Perspectives (pp. 329-347). Singapore: World Scientific.

Rufer, M. (2012). *Erfasse komplex, handle einfach: Systemische Psychotherapie als Praxis der Selbstorganisation.* Göttingen: Vandenhoeck & Ruprecht.

Schiepek, G., Schoppek, W. & Tretter, F. (1992). Synergetics in psychiatry: Simulation of evolutionary patterns of schizophrenia on the basis of nonlinear difference equations. In

W. Tschacher, G. Schiepek & E.J. Brunner (Eds.), *Self-Organization and Clinical Psychology* (pp. 163-194). Berlin: Springer.

Schiepek, G., Tominschek, I., Heinzel, S., Aigner, M., Dold, M., Unger, A., Lenz, G., Windischberger, C., Moser, E., Plöderl, M., Lutz, J., Meindl, T., Zaudig, M., Pogarell, O. & Karch, S. (2013). Discontinuous patterns of brain activation in the psychotherapy process of obsessive compulsive disorder: converging results from repeated fMRI and daily self-reports. *PloS ONE, 8(8)*, e71863

Schiepek, G., Eckert, H., Aas, B., Wallot, S. & Wallot, A. (2015). *Integrative Psychotherapy. A Feedback-Driven Dynamic Systems Approach*. Boston, MA: Hogrefe International Publishing.

Schiepek, G., Heinzel, S., Karch, S., Plöderl, M. & Strunk, G. (2016a). Synergetics in psychology: patterns and pattern transitions in human change processes. In A. Pelster & G. Wunner (Eds.), *Selforganization in Complex Systems: The Past, Present, and Future of Synergetics. Springer Series Understanding Complex Systems* (pp. 181-208). Berlin Heidelberg: Springer.

Schiepek, G., Aichhorn, W., Gruber, M., Strunk, G., Bachler, E., & Aas, B. (2016b). Real-time monitoring of psychotherapeutic processes: concept and compliance. *Frontiers in Psychology for Clinical Settings*, 7:604. doi: 10.3389/fpsyg.2016.00604

Schiepek, G., Aas, B. & Viol, K. (2016c). The mathematics of psychotherapy – a nonlinear model of change dynamics. *Nonlinear Dynamics in Psychology and the Life Sciences*, 20, 369-399.

Schiepek, G., Stöger-Schmidinger, B., Aichhorn, W., Schöller, H., & Aas, B. (2016d). Systemic case formulation, individualized process monitoring, and state dynamics in a case of dissociative identity disorder. *Frontiers in Psychology for Clinical Settings*, 7:1545. doi: 10.3389/fpsyg.2016.01545.

Schiepek, G., Viol, K., Aichhorn, W., Hütt, M.-T., Sungler, K., Pincus, D., & Schöller, H. (2017). Psychotherapy is chaotic—(not only) in a computational world. *Frontiers in Psychology for Clinical Settings*, 8:379. doi: 10.3389/fpsyg.2017.00379.

Sigmund, K. (1993). *Games of Life*. Oxford: Oxford University Press.

Singer, W. (2013). Auf der Suche nach dem Kern des Ich. *DIE ZEIT, 13. 9.2007*.

Stachowiak, H. (1978). Erkenntnis in Modellen. In: H. Lenk & H. Ropohl (Hrsg.), *Systemtheorie als Wissenschaftsprogramm* (S. 50-64). Königstein: Athenäum.

Stephan, A. (1999). *Emergenz. Von der Unvorhersehbarkeit zur Selbstorganisation*. Dresden: Dresden University Press.

Stephan, K.E., Kasper, L., Harrison, L.M., Daunizeau, J., den Ouden, H.E.M., Breakspear, M. & Friston, K.J. (2008). Nonlinear dynamic causal models for fMRI. *Neuroimage, 42*, 649–662.

Strunk, G. & Schiepek, G. (2006). *Systemische Psychologie*. München: Elsevier.

Strunk, G. & Schiepek, G. (2014). *Therapeutisches Chaos*. Göttingen: Hogrefe.

Tass, P. A. (2003). A model of desynchronizing deep brain stimulation with a demand-controlled coordinated reset of neural subpopulations. *Biological Cybernetics, 89*, 81–88.

Tass, P. A., & Hauptmann, C. (2007). Therapeutic modulation of synaptic connectivity with desynchronizing brain stimulation. *International Journal of Psychophysiologiy, 64*, 53-61.

Tass, P. A., Adamchic, I., Freund, H. J., von Stackelberg, T., & Hauptmann, C. (2012). Counteracting tinnitus by acoustic coordinated reset neuromodulation. *Restorative Neurology and Neuroscience, 30*, 137–159.

16. Creative Algorithms and the Construction of Meaning

Ruth Hagengruber[1]

Abstract

This paper investigates what we acknowledge to be meaningful and new knowledge. It refers to examples from history, illustrating how human knowledge and knowledge produced by machines were often opposed to each other. Taking into account that knowledge is confirmed by reciprocal processes, it argues that by acknowledging and integrating the results of machine processes into our daily lives, this knowledge also becomes "meaningful" for humans. The acceptance of machine-produced knowledge depends on the cultural network of knowledge confirmation. The strict difference between the two kinds of knowledge, that which is produced by human beings and that which is produced by machines, is vanishing.

---

1   Ruth Hagengruber is a professor of philosophy and has headed the Philosophy Department at the University of Paderborn since 2007. There she continued her studies on philosophy and information science and founded the Teaching and Research Area EcoTechGender. Economics, technology and gender are defined as the challenging and decisive factors of the future. She is an honorary member of the International Association for Computing and Philosophy (I-ACAP) and a member of the Advisory Board of the Munich Center for Technology in Society at the TU München. In 2015 she received the Wiener-Schmidt-Award from the Deutsche Gesellschaft für Kybernetik, Informations- und Systemtheorie. Main publications are: Hagengruber, Ruth, Riss, Uwe. (Eds.). 2014. Philosophy, Computing and Information Science. London: Pickering & Chatto; Hagengruber, Ruth, Ess, Charles. (Eds.). 2011. The Computational Turn: Past, Presents, Futures? Münster: MV-Wissenschaft. www.upb.de/hagengruber

16.1   Appropriating knowledge

Philosophical ontology determines entities and their relations. It defines classes of entities and hierarchies of relations. Intelligent machines rely on such knowledge bases. Algorithms explore them. How do human beings identify what knowledge is? Is this a qualitatively different operation from that of the digital machine? Intelligent machines are also able to supply knowledge, used by both human beings and machines. This procedure to identify knowledge can be regarded as similar, since both human beings and machines identify objects and even predict action.

According to Luciano Floridi questioning the unique prolific and creative manner of human beings is a provocation to the humanist self-concept (2014). Moreover, the anthropocentric claim supposes humans to be rational beings. But how do creativity and rationality go together? When Alan Turing investigated the question of whether machines can think, he constructed his famous Turing test, based on the insight that the claim for rationality is relative within a framework of time and culture, and dependent on claims to power. Answering to what he called the *Theological objection*: "Thinking is a function of man's immortal soul. Hence no animal or machine can think," he responds: "I am unable to accept any part of this." To substantiate his doubts, he refers to cultural differences. The arbitrary character of the orthodox view becomes clearer if we consider how it might appear to a member of some other religious community. "How do Christians regard the Moslem view that women have no souls?" he asks (1950, 443). Here, the outstanding attribute of human rational capacity interacts with claims of dominance. It has been employed for dominance over women and animals, denying them the capability to think. And it is clear that Turing assumes that this kind of anthropocentric dominance has also been applied to machine-produced thinking.

The most famous philosopher to hold the view that animals and machines are equal in their inability to think was René Descartes (1596-1650). Human beings, animals, and machines differ in their capacity to process knowledge. He compared animals to machines, by identifying "automated" modes of knowledge creation in them both. Not only Turing objected to that view. Many centuries before him, Gottfried Wilhelm Leibniz (1646-1716), philosopher, engineer, inventor of a calculating machine, and strong opponent of various Cartesian ideas, also disagreed with Descartes. Leibniz, and later Turing, denied an ontological gap between animals and people. Leibniz also refused to accept that human beings were created from two different substances, as Descartes had proposed to strengthen his claim that spiritual and corporeal substances were completely different. He objected to the Cartesian idea that only human beings were endowed with intellectual capacities and animals deprived of them, because in Leibniz' philosophical approach a world of complete interconnection was presented. The world was imagined to be interconnected by reason. Differences were not substantially well-founded (Leibniz 1923, Mainzer 1994). Leibniz interpreted the world as a rational unity, made up of tiny monads, which he called the most sophisticated "automata". Each monad was a miracle of complexity, reflecting the whole world and all its parts from its own individual perspective. According to Leibniz, there was nothing absolutely "new" and nothing absolutely different, as everything was in relation to each other. Thus, Leibniz began to invent a system to enable us to understand and read the world and also to come up with new concepts. Knowledge could be discovered by reducing "composite notions" to a simple alphabet, leading to inventions of every kind when combined according to strict rules (Hecht 1992).

Although Kant owes a lot to Leibniz, his view was beyond any comprehension of the world as a sophisticated automaton. According to him, knowledge was not obtained by the investigation of complex rules in order to detect the laws governing the world. Instead, he reintroduced the anthropocentric approach. According to Kant, human creative power exceeds the natural order. It even invents the rules, and nature is obliged to follow. The Kantian creative act of invention was, by definition, an arbitrary action of "genius". The world was not investigated in order to learn its rules, much more, it was the human genius that gave the rules to the world. Kant defined creative power as an expression of a volitional mind, opposing the idea of constructible and analytically determined knowledge, as is reflected exemplarily in Leibniz.

In the early 19th century, the field of technical engineering advanced and the old ideas were reconsidered. Thus, Ada Lovelace, in her famous interpretation of Charles Babbage's *Analytical Engine,* confirmed that this analytical machine was able to process all kinds of knowledge. The *Analytical Engine* was not constrained to automated processes, but able to provide "analytical development," as it iterated and amplified its own knowledge by processing "cycles of cycles" of variables. Although Lovelace's interpretation confirmed that the machine was no longer limited to the reproduction of knowledge, she qualified this with her statement, that the machine could only do "what we know how to order it to do, and thus never does anything really new"(Lovelace 730). According to Turing, we should be careful about what we consider to be new. In his view, the *Lovelace theorem* was open to harsh misunderstandings and subjected "philosophers and mathematicians particularly to a fallacy". It was wrong to interpret it to mean "there is nothing new under the sun." As a result, Turing tried to provide a "better variant of the objection" and wrote "a machine can never take us by surprise." The problem emerges, he says, from the widespread belief "that as soon as a fact is presented to a mind, all consequences of that fact spring into the mind simultaneously with it!" (Turing 1950, 447). But of course, this is not the case. So the discussion of the "new" is a discussion about the analytical depth of knowledge. Being the master of the principle that rules a procedure does not imply being the master of the procedure itself, and even less, of the outcome of it. Knowing the method does not imply being comprehensively conscious of or mastering all the results that are subsequently produced or provided.

Therefore, Turing investigated thinking and the quest for new knowledge in a different way. He did not regard creativity as an ingenious and spectacular occurrence but argued in favor of a learning process. Differing from Leibniz, he did not claim one type of substance for all "thinking beings". In fact, Turing claimed that the substance was irrelevant to the issue: The human being had no advantage because of her physical and biological difference, being a "continuous machine," as compared to the poor capacities of a "discrete-state machine", as he writes.

The famous Turing test, usually interpreted as a test of machine intelligence, where people are asked to identify whether a human being or a machine is responding to their questions, presents a highly sophisticated scenario. Interestingly, many aspects of this were consequently overlooked by many of its interpreters. Turing creates a situation of everyday knowledge, everyday deceit, and everyday conventional views which is situational for all aspects of human knowledge. He explains a scenario where the interrogator must identify the gender of the agents, who are on the other side of a wall and not perceivable to the interrogator, by means of questions and answers. "The object of the game for the interrogator is to determine which of the other two is the man and which is the woman". The second step is to confront a machine with the problem. It takes the part of the deceiving Agent A in this game: "Will the interrogator be wrong as often when the game is played like this as when the game is played with a man and a woman?" Turing asks (Turing 1950, 433).

Turing hereby proves his insight that knowledge is deeply anchored in culture. It was not a machine's task to "think", whatever that could mean. Turing invented a knowledge-gaining process determined by the cultural habits of acceptance, negation, and knowledge deception. To know and to think are not individual "activities", for example one singular brain activity. Knowledge is determined by a whole cultural network of knowledge confirmation, negation and deception. It is a context-driven process of agreement and rejection.

Turing's sophisticated arrangement based on complex facts of common knowledge and the construction of the "test" situation for machines is able to expose illusions of "truth". Knowledge is a complex net of information and the machine's task is to reproduce the kind of behavior which is expected from people. "The reader will have anticipated that I have no very convincing arguments of a positive nature to support my views," Turing states. But it is not Turing's intention to prove an act of "thought" when he speaks about the "thinking capacities" of machines. In regard to his response to the Lovelace argument, he confirms her statement that the rules of operation of the machine are pre-determined. However, "The teacher is ignorant of what is going on inside of the learning machine." Like children, the machine undergoes a process, arising from a child machine of well-tried design or programing and becoming—as a child does—an agent of full competence. "Most of the programs which we can put into the machine result in its doing something we cannot make sense of." Fallibility and the process-related nature of knowledge are a genuine and necessary part of the machine's development, just as is the case with learning children (Turing 1950).

16.2   The paradox of creativity

The rules of the program of a *learning* machine, as Turing called it, are determined, but the outcome undergoes changes during the process. Turing calls this a paradox, as something *new* is produced by a rigid rule and new knowledge can only be gained on the basis of what has already been disclosed. When Turing speaks of the paradoxical idea of the *learning* machine, he emphasizes the fact of the time-invariant rules by means of which new rules during the learning process are created, these being, however, of a "less pretentious kind", claiming only "an ephemeral validity". From the rigid rules more flexible ones emerge, the old enables the new (1950, 459).

The paradoxical situation is, however, adaptable to knowledge acquisition as it is processed in human circumstances, whether or not the stable basis of the rule giver to the community is accepted and the certain kind of knowledge as being a part of the knowledge community is given. What the rules allow is what the community accepts. By accepting "new" knowledge, which, of course, results from the procedures within a community, the community itself enlarges its own knowledge base. Since the new knowledge is actually new, it must be integrated into the corpus of knowledge already at hand, and perhaps, to a certain extent, it can be considered less valid in comparison to the knowledge corpus that legitimates it. The paradox is that something new can only be defined in dependency on what is already known. This process is similar to that of a *learning* machine. While for the community the new knowledge is defined in dependence on the known, the legitimate base that governs the *learning* machine is the basic rule. While in the world of humans knowledge is defined as new in relation to the known, for the machine the variations of the *learned* knowledge depend on the rules, but relevant aspects of the rule interpretation are also hereby broadened.

The acceptance of a creative act and its integration into the knowledge base is a process that works reciprocally. The digital machines driven by today's *learning* algorithms provide us with "new" bits of reality. These machines provide new segments of reality which have become—more or less successfully—integrated into our knowledge world. The knowledge we draw from Google, Siri, and all the algorithms that govern our knowledge is constructed by machines, but nonetheless accepted as relevant in the "real" world. The construction of new "meaning" is in full process.

Margaret Boden has investigated the question of whether creative acts can be distinguished from outstanding learning achievements, and she finds that this is not probable. She started an experiment with children to determine the categorical shift that occurs when "new" knowledge is created. Her experiment showed that creative human ideas emerge from transpositions of knowledge into new contexts. She presented conceptual transformations done by 4 to11-year-olds to prove the gradual evolution of the creative process. She characterized this development, starting from an arbitrary multitude of conceptual transpositions provided by the younger children and continuing to the more complex and also more adaptable transformations of the older ones. To be creative, various skills are necessary. The process of learning is, to a certain extent, a process of adaption, transformation and reintegration of concepts within a conceptual space (Boden 1990, 54-88).

Communities are knowledge communities that share a certain kind of knowledge organization and hierarchy. The community identifies itself by what it shares and what it ignores. Therefore, the transposition of any element into another knowledge context can signify a creative invention. The new is dependent on the confirmation process by the already disclosed, hence it is a reciprocal process, where the new is only new if it is accepted by the "old" knowledge.

Examples taken from the history of art and science illustrate this effectively. Picasso used African sculptures as a source of inspiration and he and his art were consequently regarded as brilliant. He delivered a transposition of ideas and for this he gained the attention of those who were important enough to generate acceptance. Others before him who did the same or similar things were laughed at and were not successful in becoming a part of the prevailing creative understanding. It is well- known that Picasso performed various kinds of adaptations at the expense of his colleagues. The African sculptures he used were not new, but their powerful transposition into a new context were his creative element. This came at a time when the community was willing to advance and to renew its concept of art. Exceptional knowledge is thus dependent on the response. The most brilliant results are not confirmed as such if their mapping within an "expert" system of their epoch is either not available or unacceptable. Many artists have experienced this, and this is true for many creative people in general, who were often only accepted as innovators many years later or within a different knowledge community. Thus, the paradoxical nature of what is new has been corroborated. To earn the attribute of "new," the knowledge at our disposal is necessary in order to confirm that the new idea is different from what has previously existed. Additionally, the creative act must be judged to be a valuable contribution to established knowledge.

16.3   In defense of creativity

Karl Popper and Hans Reichenbach investigated creative performance, asking how scientific knowledge could be discovered. They believed, in accordance with the Kantian claim, that scientific discoveries were due to an act of creativity which is not rationally approachable. Popper denied that scientific discoveries were approachable by logical means, but he accepted that the explanation and validation of a scientific discovery were done by logical analysis. "There is no such thing as a logical method of having new ideas, or a logical reconstruction of this process," Popper claimed (1959, 33) Hereby, Popper construed a conceptual distinction between the two processes, the process of discovery and the process of corroboration. He established the view of a twofold act, and moreover, of an insuperable difference between these two acts, the one determined as an inventive act, the other as an act of reproductive acceptance. This view was confirmed by Hans Reichenbach. „The act of discovery escapes logical analysis (…) logic is only concerned with the context of justification" (1958, 231). "The transition from data to theory requires creative imagination," confirmed Hempel (1966, 15; 1985). In a clear position against the claims of artificial intelligence, these authors rooted the origins of science and new knowledge in the irrational sphere of the ingenious being. "They agreed that physical laws explain data, but they obscured the initial connection between data and laws" (Hanson 1958, 71). The core ideas can, therefore, be traced back to the heritage of Kant, who questioned any possible apprenticeship of a new idea in his *Critique of Judgment*. Much more, he countered that the scientific framework itself was created by brilliant arbitrary ideas, instead of being a formative structure of rules by which new ideas could be discovered or generated. Like the mentioned authors, Kant also rejected the idea that the function of the rule itself enabled new insights (Critique of Judgment § 50).

The idea that scientific knowledge is due to uncontrollable interactions is also claimed by Thomas Kuhn. Kuhn differs from those mentioned above, as he insists on the reciprocal dependency of scientific inventions and their historic backgrounds.

He confirmed the requisite intertwining of a scientific theory and the knowledge community from which it arose. But instead of investigating the structure of these interdependencies in order to understand what supports the acceptance of a scientific innovation, like Popper, Reichenbach and many others, he idealized the act of new ideas emphatically as "imaginative posits, invented in one piece" (1970).

It is not by chance that artificial intelligence researchers have positioned themselves as epistemological alternatives to these theorists. Herbert A. Simon, Alan Newell and others working in the field of artificial intelligence reject the view that scientific discoveries can be triggered by irrational or random events. They emphasize that scientific findings are bound to the context from which they emerge. The discovery, if it can be assessed as such, is bound to the context which allows the validation of what is new. Strengthened by this insight, they have constructed their theories on the basis of the necessary and reciprocal relationship between scientific discovery and the background knowledge from which it arose. Their well-defined heuristic procedures were finally able to reconstruct specific scientific discoveries. The BACON One program successfully "discovered" Boyle's law among others. This program was based on simple algorithmic instructions. First: "Look for variables (or combinations of variables) with a constant value". Second: "Look for linear relations among variables." Third: "If two variables increase simultaneously, consider their ratio." Fourth: "If one variable increases while another decreases, consider their product" (Langley, Simon, Bradshaw and Zytkow 1987).

Laws of Galileo and Kepler were reconstructed. The program *Automated Mathematician* successfully employed an algorithm to demonstrate the generation of primes according to De Morgan's law and to prove Goldbach's Conjecture (Hayes 1989, Simon 1977). This "ingenious" outcome was discovered within the framework of rules, logically driven by heuristic procedures and a selection of data. The objections to the findings of BACON clashed with this procedure. The critics objected that the "invention" was not "new", but strategically prepared, as it was processed according to strict rules within a certain amount of data. But the critics of the automated procedure have overlooked exactly this intention: There was no difference in the act of discovery and validation. Based on these insights, Newell, Shaw and Simon came closer to how "creative" thinking might be explained (Newell, Shaw, Simon 1967). Researchers at Brunel University developed the *WISARD pattern recognition machine*, which succeeded in including and processing unknown patterns. A further strategy, allowing the weighting of the elements in the process, refined the results, as hereby elements of the pattern could be more easily selected (Mainzer, Balke 2005).

Today, heuristic procedures have increased and even dominate in the knowledge we gain of the world we live in. "Algorithms are little more than a series of step-by-step instructions ... however ... their inner workings and impact on our lives are anything but," stated Dormehl, asserting that algorithms solve all our problems but also create many more (2014). Self-adapting and self-constructing programs generate "new" knowledge based on a wide variety of data bases and according to rules. Today a huge and unmanageable data volume has become manageable via algorithms, which organize "packets", using parallel processing computers and achieving tremendous results. They are "so to say ... able to identify the needle in a haystack" (Mainzer 2014).

These machines sketch our cars and our houses, our way to the moon and beyond. Through their data collection and processing, they are able to define health and illness. They operate in the human body or in the earth's crust. We accept Siri's answers as often more reliable than the answer from some human being we have asked for directions. Our community life is shaped by the social networks invented in the digital world. All this intelligence has become part of our knowledge community. It provides us with "new" insights and "new" objects. All this deserves to be meaningful, in the case that we accept its "ideas" as answers to our questions. We construct the "meaning" of our lives when we accept what is proposed to us by the intelligent machines. We act with machines, we reflect on the knowledge provided by them and together with them. We have become one intellectual community. The concepts produced by our digital machines provide a stable part of our world interpretation. We live in a world which is, to a great extent, still structured by artificial concepts which we are free to accept and give "meaning" to.

What is commonly called creative thinking is actually a selective procedure, accomplished by delving through a multitude of data, according to an inner working of the algorithms that widely shape the knowledge of today (Hagengruber 2005). Intelligant Algorithms affect our lives in many ways, how we think and whom we know and what we know. Our knowledge is still part of the artificially provided knowledge: The feedback loops between human knowledge and machine-produced knowledge are constitutive for what is "meaningful" to us. The knowledge bases have mixed and the knowledge procedures enhance each other reciprocally. The knowledge provided by means of technology reflects on our own, and vice versa. This is true with regard for the knowledge we discover, as it is true of the knowledge we validate. The knowledge community is no longer only constrained to human beings. We share our world of meaningful entities with our digital machines.

References

Boden, M. A. 1990. *The Creative Mind. Myths and Mechanisms.* London: Routledge.

Dormehl, L. 2014. *The Formula.* How Algorthms solve all our problems and create more. London: Virgin Books.

Floridi, L. 2014. The Fourth Revolution in our Self-Understanding. Ruth Hagengruber and Uwe Riss (Eds.) *Philosophy, Computing and Information Science.* London: Pickering and Chatto, 19-29.

Hagengruber, R.. 2005. Mapping the Multitude. Categories in a Process Ontology. In *Professional Knowledge Management Experiences and Visions.* 3rd Conference WM 2005. Berlin u.a.: Springer.

Hagengruber, Ruth and Riss, Uwe (Eds.) 2014. *Philosophy, Computing and Information Science.* London: Pickering and Chatto.

Hanson, N. R. 1958. *Patterns of Discovery.* Cambridge: Cambridge University Press.

Hayes, J. R. 1989. Cognitive Processes in Creativity, In *Handbook of Creativity,* hrsg. J. A. Glover, R. R. Ronning und C. R. Reynolds, 135-145. New York: Springer.

Hecht, H. 1992. *Gottfried Wilhelm Leibniz. Mathematik und Naturwissenschaften im Paradigma der Metaphysik.* Stuttgart-Leipzig: B.G. Teubner.

Hempel, C. G. 1966. *Philosophy of Natural Science.* New Jersey: Engelwood Cliffs.

Hempel, C. G. 1985. Thoughts on the Limitations of Discovery by Computer. In *Logic of Discovery and Diagnosis in Medicine,* hrsg. K. F. Schaffner, 115-122. Berkeley: University of California Press.

Kuhn, Th. S. 1970. Logic of Discovery or Psychology of Research. In *Criticism and the Growth of Knowledge,* I. Lakatos und A. Musgrave, 1-23. Cambridge: Cambridge University Press.

Langley, P., Simon, H.A., Bradshaw G.L. und Zytkow J.M. 1987. *Scientific Discovery: An Account of the Creative Processes.* Cambridge, Massachusetts: MIT Press.

Leibniz, G. W. 1923-. *Sämtliche Schriften und Briefe.* Hg. v. der Deutschen Akademie der Wissenschaften zu Berlin, Berlin: Akademie-Verlag, II, 1, 160.

Lovelace, A. 1843. Sketch of the analytical engine; with notes from the translator Ada Lovelace. in: Taylor, R.: *Scientific Memoires* Vol. 3. Selected from the Transactions of Foreign Academies of Science and Learned Societies and from Foreign Journals. London: Richard and John E Taylor.

Mainzer, Klaus. 1994. *Computer – Neue Flügel des Geistes? Die Evolution computergestützter Technik, Wissenschaft, Kultur und Philosophie.* Berlin u.a.: De Gruyter.

Mainzer, Klaus. 2014. *Die Berechnung der Welt. Von der Weltformel zu Big Data.* München: C.H.Beck.

Mainzer, K., Balke, W.-T. 2005. Knowledge Representation and the Embodied Mind: Towards A Philospohy and Technology of Personalized Informatics. In *Professional Knowledge Management Experiences and Visions.* 3rd Conference WM 2005, hrsg. C.D. Althoff, A. Dengel, R. Bergmann, M. Nick und T. Roth Berghofer, 586-597. Berlin u.a.: Springer.

Newell, A. Shaw, J.C. und Simon, H.A. 1967.The Process of Creative Thinking. In *Contemporary Approaches to Creative Thinking,* hrsg. H. E. Grubner, G. Terell, M. Wertheimer, 63-119. New York: Atherton.

Popper, K. R. 1959. *The Logic of Scientific Discovery.* London: Hutchinson.

Reichenbach, H. 1958. *The Rise of Scientific Philosphy*. Berkeley: University of California Press.

Simon, H. A. 1977. *Models of Discovery*. Dordrecht: D. Reidel.

Turing, A.M. 1950. Computing Machinery and Intelligence. *Mind* 49: 433-460.

17. Zur Frage der Berechenbarkeit der Welt bei Immanuel Kant

Tobias Jung[1, 2]

Abstract

Seit dem Aufkommen der neuzeitlichen Naturwissenschaften wurden immer weitere Bereiche der Welt berechenbar, bis schließlich mit dem Laplaceschen Dämon der Gedanke von der vollständigen „Berechenbarkeit der Welt" formuliert wurde. Kant, der in jungen Jahren selbst einen bedeutenden Beitrag zu dieser Entwicklung geleistet hatte, durchschaute die Konsequenzen, die sich aus dieser vollständigen Berechenbarkeit der Welt für die Bestimmung des Menschen zwingend ergaben. In seiner kritischen Philosophie zeigte er Grenzen der Erkenntnis auf, die den Bereich, welcher der Berechenbarkeit unterworfen ist, auf den Bereich möglicher Erfahrung einschränkten. Die Grenzen der Erkenntnis erweisen sich dabei als Bedingungen der Möglichkeit des Menschen.

Since the emergence of modern science, more and more areas of the world have turned out computable until eventually the notion of Laplace's demon implied the idea of the world's full computability. Kant, who in his early days himself made an important contribution to this development, understood the profound consequenc-

1   Tobias Jung, geboren 1972, studierte von 1994 bis 2000 Physik in München und Cambridge. Bis 2004 promovierte er in Augsburg bei Prof. Mainzer in Philosophie und hatte dort bis 2011 einen Lehrauftrag inne. Seit 2005 arbeitet er als Gymnasiallehrer für Mathematik und Physik und ist seit 2011 an den Lehrstuhl für Philosophie und Wissenschaftstheorie der TUM abgeordnet. Interessensschwerpunkte sind u. a. die Philosophie der MINT-Fächer sowie Platon und Kant.

2   Wolfgang Pietsch möchte ich für seine sorgfältige Lektüre und zahlreiche wertvolle Kommentare und Hinweise zu einer früheren Version des Manuskripts danken.

es on man's determination which resulted from this complete computability. In his
critical philosophy he pointed out the boundaries of knowledge which restrict the
computable part of the world to the part of possible experience. The boundaries of
knowledge prove to be the conditions of the possibility of man.

17.1   Kant als Höhe- und zugleich Wendepunkt der Entwicklung der Berechenbarkeit der Welt – ein erster Überblick

Verfolgt man die Entwicklung der Astronomie und Physik vom Beginn des 17. Jahrhunderts bis zum Beginn des 19. Jahrhunderts über Johannes Kepler (1571–1630), Galileo Galilei (1564–1642) und Isaac Newton (1643–1727) bis hin zu Pierre Simon Laplace (1749–1827), so tritt in ihr ein gewaltiger Fortschritt der „Berechenbarkeit der Welt" zu Tage, der schließlich im Laplaceschen Dämon kulminiert. Die Universalität der Naturgesetze leistete, gerade weil sie keine Ausnahmen gestattete, einen bedeutenden Beitrag zur Aufklärung, insbesondere hinsichtlich der Beschränkung des Einflusses der Religion, die sich „durch ihre *Heiligkeit* [...] gemeiniglich derselben [nämlich der Kritik, d. h. der Aufklärung] entziehen" will (Kant[3], KrV, A XI).[4]
Immanuel Kant (1724–1804), der sich in jungen Jahren selbst an dieser Entwicklung ganz wesentlich beteiligt hatte, durchschaute die Konsequenzen, die sich aus der vollständigen Berechenbarkeit der Welt für die Bestimmung des Menschen zwingend ergaben. Er sah sich von daher genötigt, aber auch berechtigt, Grenzen dieser Berechenbarkeit der Welt aufzuzeigen. Somit erweist sich Kant, wenn man die Entwicklung der Berechenbarkeit der Welt als Leitlinie nimmt[5], als „Höhe- und zugleich Wendepunkt der europäischen Aufklärung" (Höffe 2001, S. 189). Seine kritische Philosophie ist eine „Aufklärung der Aufklärung" (Picht 1967, S. 219), denn sie begnügt sich nicht damit, immer weitere Bereiche von den Fesseln der

---

3   Die Schriften Kants werden nach der Akademieausgabe (AA) mit Bandnummer und
      Seitenzahl zitiert. Die „Kritik der reinen Vernunft" (KrV) wird wie üblich nach den
      Seitenzahlen der Auflagen A und B zitiert, die „Kritik der Urteilskraft" (KU) nach den
      Seitenzahlen der Auflage B.
4   Vgl. zur Bedeutung der Naturwissenschaften für die Sichtweise Gottes auch Jung (2014).
5   In dieser Hinsicht kann der vorliegende Beitrag als Ergänzung zu den jüngsten
      Darstellungen der „Berechenbarkeit (oder Berechnung) der Welt" in Küppers (2012)
      und Mainzer (2014) angesehen werden, in denen die Auseinandersetzung mit Kant
      weitgehend ausgespart wurde.

kirchlichen oder staatlichen Macht zu befreien, sondern versucht die Bedingung ihrer eigenen Möglichkeit in den Blick zu bekommen und „aufzuklären".

Die Auseinandersetzung mit der Naturwissenschaft seiner Zeit durchzieht das gesamte Werk Kants, angefangen von seinem Erstlingswerk aus dem Jahre 1746 bis zum „Opus postumum", an dem er bis zum endgültigen Nachlassen seiner Kräfte kurz vor seinem Tode am 12. Februar 1804 gearbeitet hat. Der frühe Kant kann vielleicht sogar eher als Naturwissenschaftler denn als Philosoph gelten (vgl. Picht 1980a, S. 22, und Picht 1998, S. 237; dagegen: vgl. Richter 1997). Die Naturwissenschaft, namentlich die Physik Newtons, machte einen tiefen Eindruck auf Kant. In seiner „Allgemeinen Naturgeschichte und Theorie des Himmels" versuchte er die Anfangsbedingungen für die Entstehung des Planetensystems, die Newton noch Gott überlassen zu müssen glaubte, selbst naturwissenschaftlich zu erklären. In diesen astronomischen Fragestellungen ist zweifellos eine der Wurzeln zu sehen, aus der Kants spätere Untersuchung der Bedingungen der Möglichkeit theoretischer Erkenntnis in der „Kritik der reinen Vernunft" hervorging und aus der sich die grundlegende Frage seiner Philosophie entwickelte, wie Moralität des Menschen angesichts seiner durchgehenden naturgesetzlichen Bestimmung überhaupt möglich sein kann. Ist die Newtonsche Physik universell gültig und damit die Welt vollständig berechenbar, dann ergeben sich, wie sich zeigen lässt, schwerwiegende Konsequenzen für die drei großen metaphysischen Fragen nach der Unsterblichkeit der Seele, der Freiheit des menschlichen Willens und der Existenz Gottes. Ohne Freiheit des menschlichen Willens ist nach Kant jedoch nicht nur die Moralität des Menschen unmöglich, sondern sogar der theoretischen Erkenntnis selbst wäre der Boden entzogen. Die Möglichkeit des Menschen als Wissenschaftler wie als moralisches Wesen – beides muss sich ja nicht unbedingt ausschließen – bedingt Grenzen hinsichtlich der Erkennbarkeit der Welt und das heißt auch hinsichtlich ihrer Berechenbarkeit. Dies wies Kant in der „Antinomienlehre" innerhalb der „Transzendentalen Dialektik" der „Kritik der reinen Vernunft" nach. Die Vermittlung der zunächst getrennten Bereiche, in denen die Naturgesetze beziehungsweise die Freiheit gelten, erfolgt in der Geschichte, denn die Geschichte ist der Bereich der Erscheinung der Freiheit in der Natur (vgl. Kant, AA VIII 17; vgl. Picht 1980a, S. 25). Hierbei griff Kant den Umstand der statistischen Berechenbarkeit von menschlichen Handlungen auf, die möglich ist, wenn eine Vielzahl solcher Handlungen, die unabhängig voneinander vollzogen werden, vorausgesetzt werden kann (vgl. Kant, AA VIII 17 f.).

Im Rahmen des vorliegenden Aufsatzes sollen wesentliche Stationen der Auseinandersetzung Kants mit der Berechenbarkeit der Welt, ihren Grenzen und Implikationen skizziert werden. Dazu werden in den Abschnitten 2 bis 5 am Beispiel des astronomischen Weltbildes und seines Wandels einige Hinweise gegeben, wie von

Kepler über Newton bis hin zum jungen Kant immer weitere Bereiche im Weltall einer Berechnung zugänglich wurden. Im sechsten Abschnitt wird dargestellt, wie diese Entwicklung bei Laplace ihren Höhepunkt und gewissermaßen Abschluss erreichte. Kants kritische Philosophie kann, wie im siebten Abschnitt ausgeführt ist, als unmittelbare Reaktion auf die von Laplace explizierte vollständige Berechenbarkeit des Weltalls verstanden werden. Zu den in Kants Entwurf dargelegten Grenzen der Erkenntnis werden im achten Abschnitt exemplarisch für die kosmologischen Antinomien einige Hinweise gegeben. Im neunten Abschnitt folgen Bemerkungen zur Bedeutung der Berechenbarkeit für Kants Geschichtsphilosophie am Beispiel der Schrift „Idee zu einer allgemeinen Geschichte in weltbürgerlicher Absicht". Abschließend werden im zehnten Abschnitt die wesentlichen Ergebnisse zusammengefasst.

17.2   Das Jahrhundert der fortschreitenden Berechenbarkeit der Welt: Von Keplers Himmelsmechanik und Galileis irdischer Physik zu den universellen Naturgesetzen Newtons

Kepler war es in immer neuen Anläufen von „Versuch und Irrtum" (vgl. Lombardi 2000, S. 46) schließlich geglückt, auf der Grundlage der Beobachtungsdaten, die der dänische Astronom Tycho Brahe (1546–1601) von der Bahn des Planeten Mars gesammelt hatte (vgl. Bialas 2004, S. 31), drei Gesetze zur mathematischen Beschreibung der Bewegung der Planeten um die Sonne abzuleiten und damit die *Himmelsmechanik* zu formulieren (vgl. Carrier und Mittelstraß 1989). Galilei hatte die Bewegung von Körpern in der Nähe der Erdoberfläche, also unter dem Einfluss des homogenen Gravitationsfelds der Erde, untersucht (vgl. Drake 2003, S. 370–372, und Hermann 1981, S. 7–18). Er fand insbesondere, dass ein frei nach unten fallender Körper eine Strecke Δs zurücklegt, die direkt proportional zum Quadrat der Fallzeit Δt ist: $\Delta s = \frac{1}{2} g (\Delta t)^2$. Damit hatte Galilei einen wesentlichen Beitrag zur *irdischen Physik* geleistet.

Dem englischen Naturforscher Newton gelang mit seinem erstmals im Jahre 1687 veröffentlichten Werk „Philosophiae naturalis principia mathematica" („Mathematische Grundlagen der Naturphilosophie [d. h. der Physik]", vgl. in deutscher Übersetzung Newton 1963, Newton 1988 und Newton 1999) die Formulierung von drei Axiomen der Bewegung, aus denen unter Hinzunahme seines Gravitationsgesetzes sowohl die *Himmelsmechanik* Keplers als auch Galileis Fallgesetz der *irdischen Physik* deduziert werden konnten.[6] Mit dieser Unifizierung war die von Aristoteles (384 v. Chr. – 322 v. Chr.) eingeführte Unterscheidung zwischen einer sub- und einer supralunaren Sphäre, die eine Trennung von „Himmel und Erde" hinsichtlich der Materie, der ausgezeichneten („natürlichen") Bewegungen und der zeitlichen Stabilität bedeutete, endgültig aufgehoben worden (zur Naturphilosophie von Aristoteles vgl. Craemer-Ruegenberg 1980 und Seeck 1975). Die drei Newtonschen Axiome der Bewegung und das Gravitationsgesetz konnten als universelle Gesetzmäßigkeiten, denen alles in der Natur unterworfen ist, gelten, sie brachten die nomologische Einheit der Welt zum Ausdruck.

Die kosmologischen Konsequenzen aus der Anwendung seiner physikalischen Grundlagen auf das Weltall zog Newton vor allem in seinem Briefwechsel mit Richard Bentley (1662–1742) in den Jahren 1692/1693 (für den Briefwechsel vgl. Newton 2004, S. 94–105; vgl. von Weizsäcker 2006, S. 21). Newton konstatierte, dass das Weltall unendlich sein muss, weil eine endliche Materieansammlung, wenn man von einem möglichen Gleichgewicht unter Hinzunahme einer Zentrifugalkraft infolge einer Rotationsbewegung absieht[7], aufgrund der Gravitationskraft in sich zusammenstürzen müsste (dies teilte Newton in seinem Brief an Bentley vom 10. Dezember 1692 mit; vgl. Newton 2004, S. 94 f.). Dagegen konnte er sich vorstellen, dass ein unendliches Weltall mit einer anfänglich homogenen Materieverteilung gegenüber einem vollständigen Gravitationskollaps stabil wäre. Ausgehend von einem solchen Anfangszustand wären dann aufgrund lokaler Gravitationsinstabilitäten die Sterne, die Sonne und die Planeten entstanden (vgl. den Brief von Newton an Bentley vom 10. Dezember 1692). Dabei zweifelte Newton aber daran, dass ein solcher Entstehungsprozess ausschließlich auf naturgesetzlicher Grundlage begründet werden könnte. Dies führte er später in seinen „Opticks" („Optik") weiter aus (Newton, *Opticks, Query 31*, hier zitiert nach der deutschen Übersetzung in Newton 2001, S. 144):

„Mit Hilfe dieser Principien scheinen nun alle materiellen Dinge [...] bei der Schöpfung nach dem Plane eines intelligenten Wesens verschiedentlich angeordnet zu sein; denn ihm, der sie schuf, ziemte es auch, sie zu ordnen. Und wenn er dies gethan hat, so ist es unphilosophisch, nach einem anderen Ursprunge der Welt zu

---

6   Die Ableitung der Kepler-Gesetze aus den drei Axiomen seiner Mechanik und dem Gravitationsgesetz nahm Newton *more geometrico* vor, während heute üblicherweise die analytische Formulierung der Newtonschen Physik, wie sie im 18. Jahrhundert entwickelt wurde, herangezogen wird. Für eine analytische Ableitung vgl. Nolting (2011), S. 252–258, Rebhan (2006), S. 93–99, und Scheck (2007), S. 13–17.

7   Diese Möglichkeit erwog Newton in seinem dritten Brief an Bentley, der vermutlich am 11. Februar 1693 verfasst wurde (vgl. Newton 2004, S. 101).

suchen oder zu behaupten, sie sei durch die blossen Naturgesetze aus einem Chaos
entstanden, wenn sie auch, einmal gebildet, nach diesem Gesetze lange Zeit fortbe-
stehen kann. [...] [Es] konnte doch niemals ein blinder Zufall bewirken, dass alle
die Planeten nach einer und derselben Richtung in concentrischen Kreisen gehen,
einige unbeträchtliche Unregelmässigkeiten ausgenommen, die von der gegenseiti-
gen Wirkung der Kometen und Planeten auf einander herrühren und wohl so lange
anwachsen werden, bis das ganze System einer Umbildung bedarf."

Hinsichtlich des Planetensystems sah Newton zwei Aspekte, die auf ausschließli-
cher Grundlage der Naturgesetze nicht erklärt werden konnten und für die er sich
auf Gott berufen zu müssen glaubte, nämlich erstens die Entstehung und zweitens
die Stabilität des Planetensystems auf hinreichend großen Zeitskalen. Die Beschreibung der zeitlichen Entwicklung eines Systems auf Grundlage der Newtonschen
Mechanik gelingt, wenn die Newtonschen Gesetze gültig sind, die wirkenden
Kräfte angegeben werden können und für einen beliebigen Zeitpunkt t_0 der Zustand des Systems bekannt ist.[8] Für das Planetensystem konnte die Gültigkeit der
Newtonschen Axiome vorausgesetzt werden, als Kräfte waren die zwischen je zwei
massebehafteten Körpern wirkenden Gravitationskräfte anzusetzen, aber hinsichtlich der Anfangsbedingungen meinte Newton von einer gewissen Feinabstimmung
ausgehen zu müssen. Seiner Ansicht nach konnte also nicht irgendeine beliebige
Anfangsbedingung unter Wirkung der Newtonschen Gesetze schließlich zur Entstehung des Planetensystems führen, sondern nur aus einer bestimmten anfängli-
chen Ordnung der Teilchen konnte sich die spätere Ordnung des Planetensystems
ergeben. Folglich musste Gott die geeigneten Anfangsbedingungen gesetzt haben.
Hinsichtlich der Stabilität des Planetensystems war klar geworden, dass zwar das
Zwei-Körper-Problem der Bewegung von Sonne und Planet zu geschlossenen Bahnen führen konnte, sich aber bei Berücksichtigung der störenden Einflüsse der anderen Planeten infolge der von ihnen jeweils ausgeübten Gravitationskraft eine zeitliche Veränderung der Kepler-Ellipsen ergeben müsste. Über eine genügend große
Zeitspanne hinweg waren folglich die Bahnen der Planeten instabil, und das ganze
Planetensystem einer Zerstörung ausgesetzt. Die Stabilität des Planetensystems, die
Newton einfach voraussetzte[9], war seiner Ansicht nach nur zu gewährleisten, wenn

---

8   Dazu ist die Angabe der Orte $x_{k,0} = x_k\,(t_0)$ und Impulse $p_{k,\,0} = p_k\,(t_0)$ für alle N im
    System befindlichen Teilchen erforderlich ($k \in \{1, 2, ..., N\}$). Diese Angaben bezeichnet
    man zusammenfassend als *Anfangsbedingungen*.

9   In den letzten zwei Jahrzehnten wuchs in einer Reihe von Arbeiten die Erkenntnis, dass
    sich erstens die Bahnen von Planeten beziehungsweise Exoplaneten verändern kön-
    nen (*Migration von Planeten*) und zweitens Planetensysteme über hinreichend lange

Gott dann und wann „eine[...] Umbildung" vornahm. Um die an sich defiziente Natur (vgl. Kutschmann 1989, vor allem S. 177) zu bewahren, musste Gott also mit Wundern eingreifen. Diese Sichtweise von Gott als einem Setzer von Anfangs-bedingungen und einem Wundertäter an der von ihm schwächlich geschaffenen Natur stieß bei Gottfried Wilhelm Leibniz (1646–1716) auf harsche Kritik[10] und führte zu einer Auseinandersetzung, die in einem Briefwechsel mit Samuel Clarke (1675–1729), einem engen Vertrauten Newtons, in den Jahren 1715/1716 ausge-tragen wurde und letztlich durch Leibniz' Tod am 14. November 1716 ihr Ende fand.[11] Im ersten Brief an Clarke vom November 1715 schrieb Leibniz (vgl. Leibniz, „Extrait d'une lettre écrite au mois de Novembre 1715", in Leibniz 2008, S. 352, hier zitiert nach der deutschen Übersetzung in Clarke 1990, S. 10)[12]:

„Herr Newton und seine Anhänger haben zudem eine sehr sonderbare Ansicht von Gottes Werk. Nach ihnen muß Gott von Zeit zu Zeit seine Uhr aufziehen. Andernfalls bliebe sie stehen. Er hat nicht genug Einsicht gehabt, um ihr eine immerwährende Bewegung zu geben. Nach ihrer Ansicht ist diese Maschine Gottes sogar derart unvollkommen, daß er sie von Zeit zu Zeit durch einen au-ßergewöhnlichen Eingriff reinigen und sogar flicken muß, wie ein Uhrmacher sein Werk [...]. Nach meiner Meinung bleibt darin immer dieselbe Kraft und Stärke erhalten und geht nur von Materie auf Materie über, gemäß den Geset-zen der Natur und der schönen prästabilierten Ordnung."

Während die nach Newton nötigen Eingriffe Gottes die Universalität der von ihm selbst vorgeschlagenen Naturgesetze einschränken, vertrat Leibniz in bestimmter Hinsicht die Auffassung einer durchgängigen Berechenbarkeit des Weltalls, wenn „dieselbe Kraft und Stärke erhalten" bleibt. Diese Sichtweise schließt sich in na-türlicher Weise an Leibniz' Hoffnung an, jegliches philosophisches Problem durch Rückführung auf einen geeigneten Algorithmus lösen zu können (Leibniz, Ent-

---

Zeiträume instabil sein können (vgl. Batygin und Laughlin 2008, Kley 2000, Laskar 1996 und Soter 2007).

10   Zu Leibniz vgl. Mainzer (1997).

11   Diese Auseinandersetzung ist als *Leibniz-Clarke-Debatte* bekannt (vgl. Alexander 2005, Clarke 1990 und Vailati 1997).

12   Leibniz orientiert sich mit der Analogie zwischen Gott und einem Uhrmacher bezie-hungsweise zwischen dem Weltall als göttlicher Schöpfung und einer vom Uhrmacher verfertigten Uhr am Stand der Technik seiner Zeit. In ähnlicher Weise wird heute ein Verständnis des Weltalls versucht, indem man eine Analogie zur Informatik und der Computertechnik herstellt (vgl. zum Beispiel Lloyd und Ng 2005 und Wolfram 2002).

würfe zur *Scientia generalis*, Kapitel XIV, abgedruckt in Leibniz 2008, S. 200, meine Übersetzung; vgl. Busche 1997, S. 134, Fußnote 321):

> „Hierauf, wenn Streitfragen auftreten, ist keine Unterredung zwischen zwei Philosophen mehr nötig, sondern zwischen zwei Rechenexperten. Es genügt nämlich, die Schreibfedern in die Hände zu nehmen, sich an die Rechner zu setzen und sich gegenseitig [...] zu sagen: Lasst uns rechnen [*calculemus*]!"

17.3 „Der bestirnte Himmel über mir" – Kant und ein Komet über Königsberg im Jahre 1744

Kants frühe Schriften behandeln naturwissenschaftliche Fragestellungen. Sie zeigen, dass er während seiner Studienzeit an der *Albertina* in Königsberg in den Jahren 1740 bis 1746 mit der Newtonschen Physik in Kontakt gekommen war, wenngleich die ihm vermittelten Kenntnisse hinter dem andernorts erreichten Niveau der damaligen Forschung zurückblieben (vgl. Kühn 2003, S. 107). Die Bedeutung von Kants beiden hauptsächlichen Lehrern im naturwissenschaftlichen Bereich, nämlich Johann Gottfried Teske (1704–1772) und Martin Knutzen (1713–1754), für seine weitere Entwicklung ist schwer einzuschätzen. Dagegen lässt sich der Einfluss, den ein astronomisches Ereignis und seine Begleitumstände während Kants Studienzeit auf ihn hatten, kaum bezweifeln (vgl. Waschkies 1987). Knutzen hatte im Jahre 1737 das Wiedererscheinen eines Kometen für das Jahr 1744 prognostiziert und damit Aufsehen erregt (vgl. Waschkies 1987, S. 79 f.). Tatsächlich war im Januar 1744 ein Komet am Himmel über Königsberg zu beobachten (es handelte sich um den Kometen C/1743X1; vgl. Kronk 1999, S. 408–411). Auch wenn sich im Nachhinein zeigte, dass Knutzens Überlegungen falsch waren, weil er verschiedene Kometen als ein und denselben identifizierte und auf dieser Basis eine – nicht zutreffende – Umlaufperiode des fraglichen Kometen um die Sonne abgeleitet hatte, so war die vermeintlich erfolgreiche Prognose doch geeignet, das Vertrauen in die Newtonsche Physik zu stärken.[13] An diesem astronomischen Beispiel lässt sich exemplarisch die aufklärerische Wirkung der neuzeitlichen Physik erkennen. Lange

---

13 Dass dieses Vertrauen tatsächlich gerechtfertigt war, zeigte sich im Jahre 1758, als die von dem Astronomen Edmond Halley (1656–1742) vorausgesagte Wiederkehr des Kometen von 1682 pünktlich eintrat (vgl. Heidarzadeh 2008, S. 133–135, und Schechner 1997, S. 156–177).

Zeit hatten Kometen als Unglücksboten und Künder von bevorstehenden göttlichen Strafen gegolten. Lässt sich aber ihre Bahn am Himmel berechnen und weist zumindest ein Teil der Kometen periodische Bahnen um die Sonne auf, dann wird einem solchen Aberglauben die Grundlage entzogen. Umgekehrt zeigt sich, dass ein weiteres Phänomen auf Grundlage der Newtonschen Physik verstanden werden kann und mit seiner Erklärung die Welt wieder ein Stück berechenbarer geworden ist.

17.4 „Gedanken von der wahren Schätzung der lebendigen Kräfte" – Ein Fehlgriff des jungen Kant

Mit seinem Erstlingswerk „Gedanken von der wahren Schätzung der lebendigen Kräfte", das 1746 verfasst und 1749 publiziert worden ist, knüpfte Kant thematisch an Leibniz' oben zitierte Aussage an, dass im Weltall „immer dieselbe Kraft und Stärke erhalten" bleibt. Zwischen Leibnizianern und Cartesianern war nämlich der berühmte *Vis-viva*-Streit hinsichtlich der richtigen Berechnung der „Kraft" entbrannt, der letztlich mit der Frage zusammenhängt, welche physikalische Größe als eigentliche *Erhaltungsgröße* gelten kann (zum Zusammenhang von Erhaltungsgrößen und Symmetrien vgl. Mainzer 1988, S. 240–607). Während Leibniz und seine Anhänger davon ausgingen, dass die Kraft das Produkt aus Masse eines Körpers und seiner Geschwindigkeit zum Quadrat war, vertraten die Anhänger von René Descartes (1596–1650) die Ansicht, die Kraft lasse sich als Produkt aus Masse und Geschwindigkeit berechnen. Im Rückblick zeigt sich einerseits, dass beide Ansätze in der Physik ihre Berechtigung haben, und andererseits, dass beide „Formeln" gerade nicht auf das führen, was wir heute als Kraft bezeichnen. Das Produkt aus Masse m eines Körpers und seiner Geschwindigkeit v zum Quadrat nennen wir *kinetische Energie*:[14] $E_{kin} = \frac{1}{2} \cdot m \cdot v^2$. Das Produkt aus Masse m und Geschwindigkeit v eines Körpers heißt sein *Impuls*: $p = m \cdot v$. Beide physikalische Größen sind in einem abgeschlossenen System im Rahmen der klassischen Mechanik Erhaltungsgrößen.[15] Kant wollte damit in „eine[r] der größten Spaltungen, die jetzt unter den Geometrern von Europa herrscht" (Kant, AA,

---

14 Der vermeintlich zusätzliche Faktor ½ im Vergleich zur Angabe von Leibniz fällt weg, wenn man, wie zu Leibniz' Zeit üblich, Verhältnisse betrachtet.
15 Die kinetische Energie ergibt sich als Wegintegral, der Impuls als Zeitintegral aus dem 2. Newtonschen Axiom.

I, 16), vermitteln. Die Relevanz der Klärung dieser Fragestellung für die Physik hat Kant klar gesehen (Kant, AA, X, 1)[16]:

> „Die wichtige Sache der wahren Kräftenschätz[ung] darauf in der Naturlehre so vieles ankommt erfordert es wenigstens daß die Bemühung der Deutschen die in Absicht auf diesen Punckt eingeschlafen zu seyn scheinet zu einer endlichen Entscheidung derselben aufgeweckt werde."

Sein Lösungsvorschlag blieb allerdings hinter dem bereits außerhalb von Königsberg erreichten Stand der Wissenschaft zurück (vgl. Kühn 2003, S. 111 f.)[17], was ihm ein im Juli 1751 verfasstes Spottgedicht von Gotthold Ephraim Lessing (1729–1781) einbrachte (Lessing 1996, S. 47):

> „K* unternimmt ein schwer Geschäfte,
> Der Welt zum Unterricht.
> Er schätzet die lebendgen Kräfte,
> Nur seine [Kräfte] schätzt er nicht."

Kant ließ sich durch den Misserfolg seiner ersten Publikation aber nicht entmutigen, vielmehr zeigte sich im Laufe seines Lebens die Berechtigung der Sätze, die er

---

16 Bei Newton sind infolge seines Postulats göttlicher Eingriffe Erhaltungsgrößen noch nicht als Grundlage einer naturwissenschaftlichen Beschreibung der Welt präsent, wohingegen Leibniz die Rolle von Größen, die sich mit der Zeit nicht ändern, bereits gegen Newton erkannt hatte. Mit dem *Vis-viva*-Streit ergab sich eine Verschiebung von der Frage, ob es Erhaltungsgrößen gibt, die man in gewisser Weise als Teil der Leibniz-Clarke-Debatte ansehen kann, hin zu der Frage, welche physikalische Größe denn in materialer Hinsicht eine Erhaltungsgröße ist. In seiner „Kritik der reinen Vernunft" nimmt Kant von dieser materialen Frage Abstand (vgl. Kant, KrV, B 228 f.) – er deutet lediglich an, dass im Zuge der Chemie von Antoine Laurent de Lavoisier (1743–1794) die *Masse* als Erhaltungsgröße gelten kann –, um die formale Frage nach Erhaltungsgrößen als Bedingung der Möglichkeit von Erfahrung zu stellen. So formuliert er die erste Analogie der Erfahrung (Kant, KrV, B 225): *„Bei allem Wechsel der Erscheinungen beharrt die Substanz, und das Quantum derselben wird in der Natur weder vermehrt noch vermindert."* Vgl. hierzu die bahnbrechende Arbeit von Carl Friedrich von Weizsäcker (1912–2007) (von Weizsäcker 1995), die aber nach seiner eigenen Einschätzung durch die Dissertation seines Schülers Peter Plaaß (1935–1965) (vgl. Plaaß 1965) „inhaltlich in vielen Punkten überholt" (von Weizsäcker 1995, S. 383) ist.

17 Vgl. ferner Dugas (1988), S. 235–238, Falkenburg (2000), S. 26 f., Friedman (1994), S. 3 f., Laudan (1968) und Wolff (1978), S. 304–312.

am Anfang seiner ersten veröffentlichten Schrift so selbstbewusst formuliert hatte (Kant, AA, I, 10):

> „Ich habe mir die Bahn schon vorgezeichnet, die ich halten will. Ich werde meinen Lauf antreten, und nichts soll mich hindern ihn fortzusetzen."

17.5 „Allgemeine Naturgeschichte und Theorie des Himmels" – Kants chaotischer Anfang des Planetensystems

Mit seiner 1755 anonym publizierten Schrift „Allgemeine Naturgeschichte und Theorie des Himmels" griff Kant unmittelbar die von Newton negativ beschiedene Frage auf, ob die Entstehung des Planetensystems auf Grundlage der Newtonschen Physik erklärt werden kann. Dies zeigt schon die Fortführung des Titels seines Werkes: „oder Versuch von der Verfassung und dem mechanischen Ursprunge des ganzen Weltgebäudes nach Newtonischen Grundsätzen abgehandelt". Kant zielte darauf, „die Bildung der Weltkörper selber und den Ursprung ihrer Bewegungen aus dem ersten Zustande der Natur durch mechanische Gesetze herzuleiten" (Kant, AA, I, 221). Er war entschlossen, eine der beiden Lücken, die Newton gelassen hatte, auf Grundlage dieser Newtonschen Physik zu schließen (Kant, AA, I, 230):

> *„Gebet mir Materie, ich will eine Welt daraus bauen!"*

Kant ging von einem ungeordneten Anfangszustand aus, den er als „ein vollkommenes Chaos" (Kant, AA, I, 225) bezeichnete, und legte qualitativ auf Grundlage der Newtonschen Gesetze unter Annahme zweier Kräfte, der anziehenden Gravitationskraft und einer abstoßenden Kraft, dar, wie sich das Planetensystem aus dem Chaos gebildet haben konnte. Im Unterschied zur Gravitationskraft war diese abstoßende Kraft noch nicht mathematisch formuliert worden, Kant sah sie „bei der feinsten Auflösung der Materie, wie z. E. [also z. B.] bei den Dünsten" (Kant, AA, I, 235) am Werk. Den Hintergrund könnte beispielsweise die Beobachtung bilden, dass sich ein Geruch ausgehend von seiner Quelle innerhalb kurzer Zeit in einem Raum ausbreitet. Im 19. Jahrhundert wurde dieser Befund im Rahmen der kinetischen Gastheorie mathematisch formuliert. Die von Kant herangezogene Kraft kann also als Kraft infolge des Drucks in einem Gas gesehen werden. Kant war sich bewusst, dass, wenn man beispielsweise von einer Rotationsstabilisierung absieht, sich nur aus dem Wechselspiel von anziehenden und abstoßenden Kräften ein Gleichgewichtszustand einstellen kann und somit stabile Strukturen auftreten können (Kant, AA, I, 476, hier zitiert nach der deutschen Übersetzung in Kant 1998, S. 519 und S. 521):

> „[F]erner [ist man], setzt man bloß eine zurückstoßende Kraft, nicht in der Lage [...], die Verbindung der Elemente, um Körper zusammensetzen, zu verstehen, sondern eher die Zerstreuung, setzt man aber bloß eine anziehende, zwar die Verbindung, nicht jedoch die bestimmte Ausdehnung und den Raum[.]"

Wenngleich Kant aus seiner Sicht die Entstehung und Entwicklung des Planetensystems rein physikalisch erklärt hatte, war er doch ähnlich wie Newton noch davon überzeugt, dass die Wohlordnung der Natur ein Hinweis auf Gott ist (vgl. zum Beispiel Kant, AA, I, 228).[18]

17.6   Die vollständige Berechenbarkeit der Welt bei Laplace

Etwa 40 Jahre nach Kants Versuch in der „Allgemeinen Naturgeschichte und Theorie des Himmels", die Entstehung des Planetensystems auf Grundlage der Newtonschen Physik zu erklären, kam der französische Mathematiker, Astronom und Physiker Laplace in seiner 1796 publizierten Schrift „Exposition du système du monde" („Darstellung des Weltsystems", vgl. für eine deutsche Übersetzung Laplace 2008a und Laplace 2008b) unabhängig von Kant (vgl. Henrich 2010, S. 38) zu in gewisser Hinsicht ähnlichen Resultaten (vgl. Henrich 2010, S. 38–41, und Mainzer 2014, S. 48–56).[19] Gemeinsam ist Laplace und Kant insbesondere die Aufgabe, die sie sich stellten, nämlich ein naturgesetzliches Verständnis des Beginns des Planetensystems zu erreichen. Im Unterschied zu Kant, der von einem ungeordneten Zustand ausging, stand Laplace aber in der Tradition der Wirbeltheorie von Descartes

---

18   Auch in seiner Schrift „Der einzig mögliche Beweisgrund zu einer Demonstration des Daseins Gottes" aus dem Jahre 1763 trat Kant in gewisser Weise für den physikotheologischen Gottesbeweis ein.

19   Zusammenfassend wird von der *Kant-Laplace-Theorie* der Entstehung des Planetensystems gesprochen (vgl. Schmidt 1925). Hieran knüpfte von Weizsäcker mit seinen physikalischen Arbeiten zur Entstehung des Planetensystems an (vgl. von Weizsäcker 1943).

(vgl. zum Beispiel Descartes, *Principia Philosophiae*, 1644, Pars Tertia, XXX, in Descartes 2005, S. 202). Er nahm als Anfangszustand eine den Raum des Planetensystems ausfüllende rotierende Flüssigkeit an, aus der durch Gravitationskollaps die Sonne hervorging (vgl. Henrich 2010, S. 40). Am Rand der entstehenden Sonne konnten sich Ringe ablösen, wenn die Zentrifugalkraft infolge der Rotation mindestens gleich der Gravitationskraft zwischen Sonne und jeweiligem Flüssigkeitselement des Rings war. Aufgrund von Gravitationsinstabilitäten kondensierte dann die Materie auf derartigen Ringen jeweils zu einem Planeten. Während Kant mit seinem Versuch einer physikalischen Ableitung der Genese des Planetensystems aber nur einen der beiden oben genannten Punkte, die Newton eben als nicht mehr naturwissenschaftlich erklärbar ansah, aufgriff und die vermeintliche Instabilität des Planetensystems aufgrund von gravitativen Störungen einer Planetenbahn durch die anderen Planeten außen vor ließ, bemerkte Laplace hinsichtlich der von Newton bezweifelten Stabilität (zitiert nach der deutschen Übersetzung in Laplace 2008b, S. 221):

„Wenn schließlich die Vermutungen, die ich gerade über den Ursprung des Planetensystems vorgeschlagen habe, begründet sind, ist die Stabilität dieses Systems auch eine Folge der allgemeinen Bewegungsgesetze."

Wenngleich sich diese Aussage von Laplace letztlich *inhaltlich* nicht halten lässt, so weist sie doch *methodisch* auf die durchgehende Berechenbarkeit der Welt hin. Laplace nimmt die Universalität der Newtonschen Gesetze im Unterschied zu Newton selbst ernst. Dies zeigt sich insbesondere an der Schlussfolgerung, die er aus der allgemeinen, das heißt keine Ausnahme zulassenden, Gültigkeit dieser Gesetze in seiner Schrift „Essai philosophique sur les probabilités" („Philosophischer Versuch über die Wahrscheinlichkeit") aus dem Jahre 1814 zog (zitiert nach der deutschen Übersetzung in Laplace 2003, S. 1 f.):

„Wir müssen also den gegenwärtigen Zustand des Weltalls als die Wirkung seines früheren und als die Ursache des folgenden Zustands betrachten. Eine Intelligenz, welche für einen gegebenen Augenblick alle in der Natur wirkenden Kräfte sowie die gegenseitige Lage der sie zusammensetzenden Elemente kennte, und überdies umfassend genug wäre, um diese gegebenen Größen der Analysis zu unterwerfen, würde in derselben Formel die Bewegungen der größten Weltkörper wie des leichtesten Atoms umschließen; nichts würde ihr ungewiß sein und Zukunft wie Vergangenheit würden ihr offen vor Augen liegen."

Die genannte Intelligenz ist als *Laplacescher Dämon* bekannt.[20] Der Gedanke, dass sich auf Grundlage der Newtonschen Gesetze, der Kenntnis aller wirkenden Kräfte und gegebenen Anfangsbedingungen, also den Orten und Impulsen aller Teilchen zu einem beliebigen Zeitpunkt t_0, die zeitliche Entwicklung des Systems für alle Zeitpunkte $t \in T \subseteq \mathbb{R}$ exakt berechnen lässt, ist Ausdruck des vollständigen *Determinismus der Newtonschen Physik*.

17.7 Die Berechenbarkeit der Welt als Ausgangspunkt für Kants kritische Philosophie

Laplace hatte die letzte Konsequenz des vollständigen Determinismus aus der fortschreitenden und alles umfassenden Berechenbarkeit der Welt, die Kant für sich schon früher gezogen haben muss, klar ausgesprochen. Aus ihr lassen sich für das Gebiet der Metaphysik, in die Kant, wie er schon 1766 in der Schrift „Träume eines Geistersehers" sagte, „das Schicksal [...] verliebt zu sein" (Kant, AA, II, 367) hatte, drei gewichtige Folgerungen ziehen. Die traditionelle Metaphysik, die Kant durch die Philosophie von Leibniz und Christian Wolff (1679–1754) kennengelernt hatte, war eingeteilt in *allgemeine Metaphysik* (*metaphysica generalis*) und *spezielle Metaphysik* (*metaphysica specialis*). Die allgemeine Metaphysik war die *Ontologie*, die sich mit dem Sein des Seienden und seinen Prinzipien befasste, also danach fragte, was allem, was *ist*, gemeinsam ist. Die spezielle Metaphysik untergliederte sich in drei Teilbereiche, die sich mit bestimmten Bereichen des Seienden auseinandersetzten. Diese drei Teilbereiche waren die *rationale Psychologie* (*psychologia rationalis*), die *rationale Kosmologie* (*cosmologia rationalis*) und die *rationale Theologie* (*theologia rationalis*), also *Seelenlehre*, *Weltlehre* und *Gotteslehre*. Die mit diesen

---

20 Es stellt sich natürlich die Frage nach der Verortung dieser Intelligenz in Laplaces Gedankenexperiment. Handelte es sich um eine „reine Intelligenz", die nicht auf eine materielle Grundlage angewiesen ist, dann ergäbe sich ein Widerspruch zur Universalität der Naturgesetze und dem durchgängigen Materialismus. Setzte die Intelligenz einen materiellen Träger voraus, so müsste sie außerhalb des Weltalls stehen, um das Weltall in der von Laplace beschriebenen Weise berechnen zu können. Dies steht im Widerspruch zu einer sinnvollen Definition des Grenzbegriffs *Weltall* als dem umfassendsten physikalischen System (vgl. Jung 2005). Ist aber die Intelligenz mit ihrem materiellen Träger im Weltall verortet, so ergibt sich ein Problem der Selbstreferentialität, denn sie müsste ihren eigenen Anfangszustand bestimmen und die eigene zeitliche Entwicklung mitberechnen.

Teilbereichen verbundenen Fragen waren die Fragen nach der Unsterblichkeit der Seele, der Freiheit des Willens und der Existenz Gottes.

Wenn, wie Kant in der „Allgemeinen Naturgeschichte und Theorie des Himmels" selbst sagte, „[d]ie Materie [...] der Urstoff aller Dinge ist" (Kant, AA, I, 228), dann lässt sich hieraus ein konsequenter Materialismus aufbauen. Die Folgerung, die sich in Anwendung auf den Menschen ergibt, hat der französische Arzt und Philosoph Julien Offray de La Mettrie (1709–1751) mit dem Titel seines 1748 erschienenen Werks gegeben: „L'homme machine" – „Die Maschine Mensch" (vgl. de La Mettrie 1990). Für die menschliche Seele kann das nur bedeuten, dass auch sie rein materiell konstituiert und damit in den Kreislauf des Werdens und Vergehens einbezogen ist. Folglich ist die *Unsterblichkeit der Seele* in Frage gestellt, die rationale Psychologie als eigenständiger Bereich der speziellen Metaphysik weicht einer empirischen Physiologie.

Wenn der Mensch auf die Verfasstheit und Bewegung der ihn konstituierenden Materie vollständig reduziert werden kann, dann ist angesichts des Laplaceschen Dämons sein Denken und Handeln durch die physikalischen Gesetze determiniert. Folglich ist fraglich, wie es eine *Freiheit des menschlichen Willens* überhaupt noch geben kann. Wenn der Mensch aber keinen freien Willen hat, führt dies nach Kant unmittelbar zu zwei schwerwiegenden Konsequenzen. Erstens kann es dann gar keine Wissenschaft geben, denn die naturgesetzliche Determination der Gedanken hebt zugleich die Möglichkeit auf, ihre Wahrheit zu erkennen. Zweitens entfällt mit dem Verschwinden der menschlichen Willensfreiheit die Verantwortung des Menschen für seine Handlungen. Wenn die Handlungen eines Menschen naturgesetzlich determiniert sind, dann trägt er weder für die Bestimmung des Willens, der dieser Handlung zugrunde liegt, noch für die Folgen dieser Handlung irgendeine Verantwortung. Eine menschliche Handlung ist dann nicht anders zu beurteilen als ein Stein, der aufgrund des Gravitationsgesetzes zu Boden fällt. Die Bestrafung eines Menschen beispielsweise für einen Mord wäre dann gleichbedeutend mit der Bestrafung eines Steins für die gleichförmig beschleunigte Bewegung im homogenen Gravitationsfeld der Erde. Ohne Freiheit des menschlichen Willens ist der Moral und dem Recht jeglicher Boden entzogen.

Wenn auf der Grundlage der Naturgesetze bei Kenntnis der Kräfte und des Zustandes eines in Frage stehenden Systems zu einem beliebigen Zeitpunkt die zeitliche Entwicklung des Systems für alle Zeitpunkte berechnet werden kann, dann liegt die Schlussfolgerung nahe, die auch Laplace aus seinen Untersuchungen der Entstehung des Planetensystems gezogen hat. Vom französischen Machthaber Napoleon (1769–1821) wurde er um 1811 gefragt, welche Rolle denn Gott in seinem astronomischen Weltbild spiele. Daraufhin soll Laplace geantwortet haben (übliche Übersetzung)[21]:

„Majestät, ich habe diese Hypothese nicht nötig."

Auf Grundlage einer konsequenten naturgesetzlichen Erklärung der Welt stellt sich in kompromissloser Schärfe die Frage nach der *Existenz Gottes*. Diese Spannung zwischen der vollständigen Berechenbarkeit der Welt auf der einen Seite und den großen Fragen der Metaphysik auf der anderen Seite bildete den Ausgangspunkt für Kants kritische Philosophie, die mit der „Kritik der reinen Vernunft" einsetzt. Gleich zu Beginn seiner „Vorrede" zur ersten Auflage aus dem Jahre 1781 greift Kant dies auf (Kant, KrV, A VII):

„Die menschliche Vernunft hat das besondere Schicksal in einer Gattung ihrer Erkenntnisse: daß sie durch Fragen belästigt wird, die sie nicht abweisen kann, denn sie sind ihr durch die Natur der Vernunft selbst aufgegeben, die sie aber auch nicht beantworten kann, denn sie übersteigen alles Vermögen der menschlichen Vernunft."

Kant steht in der philosophischen Tradition, die sich bis auf die Anfänge der Philosophie in der griechischen Antike zurückführen lässt, wenn er dem Menschen ein Erkenntnisvermögen, das er „Vernunft" nennt, zuschreibt (vgl. Schnädelbach 2007). Aristoteles hatte in seiner „Politik" im Anschluss an seinen Lehrer Platon (428/427 v. Chr. – 348/347 v. Chr.) den Menschen als ζῷον λόγον ἔχον, das heißt als „Lebewesen, das einen Logos hat"[22] (vgl. Aristoteles, *Politik*, 1253 a 10 f.), bestimmt. Die scholastische Philosophie bewahrte das antike Erbe dieser Bestimmung des Menschen im *animal rationale*, dem „vernünftigen Lebewesen". Kant knüpfte an diese Bestimmung an, wenn er auf den Einwand, dass der Mensch ja nicht immer rational handle, reagiert und ihn einschränkend als *animal rationabile* (vgl. Kant, AA, VII, 321; vgl. Schnädelbach 2007, S. 7 f.), also als *zur Vernunft fähiges* oder *vernunftbegabtes Lebewesen*, bezeichnet. Auf einem wie auch immer gearteten Erkenntnisvermögen beruht die Möglichkeit der Philosophie. Wer dem Menschen

---

21   Das französische Original lautet in einer der zahlreichen kursierenden Versionen (hier zitiert nach Dijksterhuis 1956, S. 548): „Sire, je n'avais pas besoin de cette hypothèse-là[.]" Eine andere Version gibt der französische Astronom Hervé Faye (1814–1902) in Faye (1884), S. 110:

22   Das schwer übersetzbare Wort λόγος wird hier üblicherweise mit „Vernunft" oder „Sprache" wiedergegeben.

das Vermögen der Vernunft absprechen möchte, kann sein Vorhaben zumindest nicht argumentativ in Angriff nehmen. Denn jede Argumentation macht von den Formen des Denkens Gebrauch, deren Bedingung der Möglichkeit Kant gerade auf Grundlage der Annahme, dass der Mensch ein Vernunftwesen ist, aufzuzeigen versucht. Die Fragen, die sich der menschlichen Vernunft aufdrängen, sind die großen Fragen der Metaphysik, die Fragen nach der Unsterblichkeit der Seele, der Freiheit des Willens und der Existenz Gottes. Diese Fragen wurzeln in der Natur der Vernunft und damit in der Natur des Menschen.[23] Daher sind sie nach Kant unauslöschlich und würden bleiben, „wenn gleich die übrigen [Wissenschaften] insgesammt in dem Schlunde einer alles vertilgenden Barbarei gänzlich verschlungen werden sollten" (Kant, KrV, B XIV). Dennoch werden wir diese Fragen nicht beantworten können, „denn sie übersteigen alles Vermögen der menschlichen Vernunft" (Kant, KrV, A VII). Kant beansprucht, im Rahmen der in der „Kritik der reinen Vernunft" vorgenommenen Selbstbestimmung der Vernunft hinsichtlich ihrer Quellen, ihres Umfanges und ihrer Grenzen (vgl. Kant, KrV, A XII) in dem „Transzendentale Dialektik" betitelten Teil zu zeigen, dass die Unsterblichkeit der Seele, die Freiheit des Willens und die Existenz Gottes nicht objektiv bewiesen werden können. Daher sprach Kants Freund Moses Mendelssohn (1729–1786), der selbst in seinen Schriften „Phaedon oder über die Unsterblichkeit der Seele in drey Gesprächen" von 1767 oder „Morgenstunden oder Vorlesungen über das Daseyn Gottes" von 1785 Antworten auf die großen Fragen der Metaphysik zu geben versucht hatte, vom „alles zermalmenden Kant[…]" (Mendelssohn 2009, S. 219).

---

23 Die Grundfrage von Kants Philosophie ist die Frage „Was ist der Mensch?" (Kant, AA, IX, 25). Hierin schließt Kant an Platon an, der eben diese Frage nach dem Wesen (φύσις) des Menschen ins Zentrum seines Philosophierens stellt (vgl. Platon, *Politeia*, Buch VII, 514 a, also auf dem Höhepunkt des Werkes zu Beginn des „Höhlengleichnisses", vgl. ferner Platon, *Theaitetos*, 174 b, sowie Platon, *Alkibiades I*, 129 e) und versucht, sie im Horizont der φύσις des Alls zu beantworten. Deshalb ist Platons Philosophie als praktische Naturphilosophie zu verstehen (vgl. Königkrämer 2009). Kant entfaltet die Grundfrage nach dem Wesen des Menschen in die drei Fragen „Was kann ich wissen?", „Was soll ich thun?" und „Was darf ich hoffen?" (Kant, AA, IX, 25, sowie Kant, KrV, B 833). Weil Kant den Menschen als vernunftbegabtes Lebewesen bestimmt und die metaphysischen Fragen nach der Unsterblichkeit der Seele, der Freiheit des Willens und der Existenz Gottes dem Menschen durch die Natur seiner Vernunft aufgegeben sind, ist die Beantwortung der Fragen „*Wie ist Metaphysik als Wissenschaft möglich?*" (Kant, KrV, B 22) und „*Wie ist Metaphysik als Naturanlage möglich?*" (Kant, KrV, B 22) unauflöslich mit der Frage nach dem Wesen des Menschen verbunden. Die „Kritik der reinen Vernunft" gehört also nach Kant zur Anthropologie (vgl. Kant, AA, IX, 25).

17.8   Die kosmologischen Antinomien in der „Kritik der reinen Vernunft" – Grenzen der Berechenbarkeit der Welt

Die von Kant offen gelegte Grenzziehung der Vernunft erweist, dass objektive Erkenntnis auf den Bereich möglicher Erfahrung beschränkt ist. Gegenstände, die keiner möglichen Erfahrung entsprechen beziehungsweise jede mögliche Erfahrung transzendieren, können nicht objektiv erkannt werden. Drei solche Gegenstände, die *transempirisch* sind und sich damit der objektiven Erkenntnis entziehen, sind die Untersuchungsgegenstände der Metaphysik, nämlich die *Seele*, die *Welt* und *Gott*. Kant weist in der „Kritik der reinen Vernunft" erstens die Fehlschlüsse (*Paralogismen*) in der rationalen Psychologie hinsichtlich der Seele auf. Er zeigt zweitens, wie sich die Vernunft in Widersprüche (*Antinomien*) verwickelt, wenn es um die Welt geht. Drittens destruiert er die gängigen *Gottesbeweise*. Dies ist die negative Seite der „Transzendentalen Dialektik": die menschliche Vernunft als Vermögen, wahre Urteile zu wahren Schlüssen zu verbinden (vgl. Picht 1998, S. 46–49), reicht nicht hin, um etwas Gesichertes über die Seele, die Welt und Gott auszumachen.

Für die „Berechenbarkeit der Welt" und die Reichweite der möglichen Erkenntnis der Natur folgen insbesondere aus den „kosmologischen Antinomien" Grenzen, die von der modernen Naturwissenschaft größtenteils ausgeblendet oder übergangen werden.[24] Kant stellt in vier Hinsichten dar, wie eine Reihe von Bedingungen zum Unbedingten hin fortgesetzt werden kann. Es gibt nach ihm jeweils zwei Möglichkeiten, zum Unbedingten zu gelangen: entweder wird ein erstes Unbedingtes angenommen, also ein Bedingendes, das selbst aber nicht bedingt ist, gewissermaßen eine *dogmatische Setzung*, oder die Reihe der Bedingungen wird in einem *unendlichen Regress* immer weiter geführt, sodass kein einzelnes Glied in der Kette der Bedingungen unbedingt ist, sondern nur die Reihe als Ganzes als unbedingt angesehen werden kann.

---

24   Für eine detaillierte und weitergehende Auseinandersetzung mit Kants Antinomienlehre vgl. insbesondere Falkenburg (2000). Eine formale Rekonstruktion gibt Malzkorn (1999).

Die „erste kosmologische Antinomie" betrifft eine Reihe in räumlicher und zeitlicher Hinsicht (vgl. die Analyse in Beisbart 2010)[25]. Kant stellt zwei Positionen gegeneinander (Kant, KrV, B 454 f.):

„Thesis. Die Welt hat einen Anfang in der Zeit und ist dem Raum nach auch in Grenzen eingeschlossen."

„Antithesis. Die Welt hat keinen Anfang und keine Grenzen im Raume, sondern ist sowohl in Ansehung der Zeit als des Raums unendlich."

Durch den Verweis auf das Faktum, dass es „die moderne physikalische Kosmologie [als] eine an der Erfahrung orientierte Disziplin der Physik" (Mittelstaedt und Strohmeyer 1990, S. 145) gibt, in der „die Welt im Ganzen zum Gegenstand der Erfahrung wird" (Mittelstaedt und Strohmeyer 1990, S. 145)[26], scheint Kants „Kritik der reinen Vernunft" empirisch falsifiziert zu sein.[27] Dementsprechend wird von Kosmologen im Rahmen des kosmologischen Standardmodells aufgrund empirischer Evidenz beispielsweise die Frage nach dem zeitlichen Anfang der Welt als entschieden angesehen (Liddle 2009, S. 83):

„Der Streit um die Frage, ob es einen Urknall gab oder ob das Universum ewig in einem stabilen Zustand existiert, wurde im Jahr 1965 mit der Entdeckung der kosmischen Hintergrundstrahlung zu Gunsten der ersten Option entschieden."

In dieser Sichtweise stellt der *Urknall* vor etwa 13,7 Milliarden Jahren eine definitive Antwort auf die Kants „erster kosmologischer Antinomie" zugrunde liegende Frage dar. Er ist das „absolut Unbedingte" (Kant, KrV, B 445) für die ganze zeitliche Reihe aller Erscheinungen, er ist der erste Zeitpunkt, dem kein Zeitpunkt vorhergeht.

---

25 Die Bedeutung der „ersten kosmologischen Antinomie" für die heutige physikalische Kosmologie wird u. a. erörtert in Beisbart (2012) sowie in Mittelstaedt und Strohmeyer (1990).

26 Ob überhaupt „die Welt im Ganzen zum Gegenstand der Erfahrung" werden kann, scheint allein schon dadurch fraglich, dass von empirischer Seite immer nur eine endliche Datenmenge gegeben ist, von der nur unter Heranziehung zusätzlicher Annahmen wie zum Beispiel dem kosmologischen Prinzip ein Modell von der Welt rekonstruiert werden kann.

27 Die in Mittelstaedt und Strohmeyer (1990) vorgeschlagene Lösung, den Begriff von möglichen Gegenständen der Erfahrung um „Quasi-Gegenstände" zu erweitern und für diese dann die Möglichkeit objektiver Erkenntnis nachzuweisen, soll hier nicht weiter diskutiert werden.

Die weitere kosmologische Forschung dürfte dann die Frage nach einer möglichen Bedingung dieses ersten Unbedingten nicht mehr stellen. Aber die Kosmologen lassen es nicht dabei bewenden, sondern sie suchen nach dem Grund des Urknalls und formulieren Theorien, „was vor dem Urknall" (Titel von Clegg 2012)[28] war. Die „zweite kosmologische Antinomie" thematisiert die Frage, ob es Elementarteilchen als letzte Bausteine der Welt gibt oder nicht[29] (Kant, KrV, B 462 f.):

„Thesis. Eine jede zusammengesetzte Substanz in der Welt besteht aus einfachen Theilen, und es existieret überall nichts als das Einfache, oder das, was aus diesem zusammengesetzt ist."

„Antithesis. Kein zusammengesetztes Ding in der Welt besteht aus einfachen Theilen, und es existiert überall nichts Einfaches in derselben."

Im Rahmen der heutigen Elementarteilchenphysik wird von der „Entdeckung des Unteilbaren" (Titel von Resag 2014), den „ersten zwei richtigen Elementarteilchen" (Bleck-Neuhaus 2013, S. 14, in der Überschrift von Abschnitt 1.1.2.), den „wirklich fundamentalen Bausteinen des Universums" (Lincoln 2013, S. 53) oder der Vollständigkeit des Baukastens der Teilchenphysik (vgl. Eidemüller 2012, S. 18) berichtet. Diese Äußerungen suggerieren, dass die im Standardmodell der Teilchenphysik, auf das jeweils Bezug genommen wird, aufgeführten Teilchen tatsächlich Elementarteilchen, also nicht mehr weiter teilbare Entitäten sind, aus denen sich die Materie konstituiert. Andererseits bleiben die Physiker trotz der von ihnen verwendeten Sprechweise „Elementarteilchen" nicht beim „Standardmodell der Elementarteilchenphysik" stehen, sondern werfen vielmehr die Frage auf, ob sich nicht auch innere Strukturen von Quarks und Leptonen auffinden lassen (vgl. Lincoln 2013, S. 48). Bereits seit vier Jahrzehnten gibt es Vorschläge für die „wirklich fun-

---

28 Der Untertitel lautet: „Eine Reise hinter den Anfang der Zeit". Vgl. auch den Untertitel von Vaas (2010): „Wie es zum Urknall kam". Vgl. ferner Titel und Untertitel von Bojowald (2009): „Zurück vor den Urknall. Die ganze Geschichte des Universums". Diesen populärwissenschaftlichen Schriften liegen mathematisch formulierte Theorien zugrunde, in denen insbesondere versucht wird, die Allgemeine Relativitätstheorie mit der Quantentheorie zu vereinigen.

29 Vgl. die eingehenden Untersuchungen in Falkenburg (1994), Falkenburg (1995) und Falkenburg (2002); vgl. auch Malzkorn (1998).

damentalen Elementarteilchen" (Lincoln 2013, S. 53), *Preonen*[30] genannt, die als Bausteine der Leptonen und Quarks gesehen werden.

Die „dritte kosmologische Antinomie" ist die „Freiheitsantinomie", in der es um die Frage nach Determinismus und Freiheit des menschlichen Willens geht (Kant, KrV, B 472 f.):

> „Thesis. Die Kausalität nach Gesetzen der Natur ist nicht die einzige, aus welcher die Erscheinungen der Welt insgesammt abgeleitet werden können. Es ist noch eine Kausalität durch Freiheit zu Erklärung derselben anzunehmen nothwendig."

> „Antithesis. Es ist keine Freiheit, sondern alles in der Welt geschieht lediglich nach Gesetzen der Natur."

Im Rahmen der heutigen Hirnforschung wird die Behauptung vertreten, der freie Wille sei eine uns vom Gehirn vorgespiegelte Illusion und alle unsere Handlungen seien durch die neuronalen Prozesse im Gehirn vollständig bestimmt.[31] Hieran knüpft sich eine weitreichende Debatte um den Fragenkomplex von Freiheit und Determinismus an[32], in die nicht zuletzt auch die Quantentheorie einbezogen wurde. Nach der „Kopenhagener Interpretation" zeigt sich in dieser Theorie ein Indeterminismus[33], auf dessen Grundlage verschiedentlich versucht worden ist, die Freiheit des Willens aufzubauen. Allerdings lässt sich meinem Dafürhalten nach auf einer wie auch immer gearteten naturgesetzlichen Einschränkung des „Spielraums der Möglichkeiten", wie er sich auch für die Quantenmechanik mit ihrem Probabilismus ergibt, letztlich kein wirklich freier Wille begründen (vgl. Strohmeyer 2013). Daher wäre der quantenmechanische Indeterminismus nach der Kopenhagener Deutung inhaltlich eher der „Antithesis" zuzuordnen.

---

30 Diese Bezeichnung wurde 1974 von Jogesh Pati (geb. 1937) und Abdus Salam (1926–1996) eingeführt. Einen Vorschlag für eine entsprechende Theorie legten im Jahre 1979 unabhängig voneinander Haim Harari (geb. 1940) und Michael A. Shupe vor. Demnach sollte es zwei Preonen mit Ladung + ⅓e und 0 sowie ihre Antiteilchen geben. Ein Up-Quark beispielsweise würde durch drei Preonen konstituiert, zwei Preonen mit der Ladung + ⅓e und ein Preon mit der Ladung 0 (vgl. Lincoln 2013, S. 51).

31 Vgl. zum Beispiel Prinz 2013 und Singer 2013; vgl. ferner auch die Analyse in Falkenburg 2012.

32 Vgl. für einen ersten Überblick zur Debatte zum Beispiel Keil 2009; vgl. außerdem Geyer 2013 und Keil 2013.

33 Vgl. hierzu Jung und Nickel (2013).

Die „vierte kosmologische Antinomie" betrifft die Frage, ob sich die Existenz Gottes aus der Existenz der Welt folgern lässt oder nicht (Kant, KrV, B 480 f.):

„Thesis. Zu der Welt gehört etwas, das entweder als ihr Theil, oder ihre Ursache ein schlechthin notwendiges Wesen ist."

„Antithesis. Es existirt überall kein schlechthinnotwendiges Wesen weder in der Welt, noch außer der Welt als ihre Ursache."

Unter anderem die kosmologische *Feinabstimmung der Naturkonstanten* (vgl. Kanitscheider 1996, S. 113–127) wird für die auch heute wieder vertretene These, dass es einen „intelligenten Urheber" gibt, ins Feld geführt. Dieser Auffassung von einem *intelligent design* stehen rein naturwissenschaftliche Erklärungsversuche gegenüber, wobei manche Wissenschaftler die empirischen Befunde gerade als Beweis gegen einen göttlichen Urheber werten (vgl. beispielsweise Dawkins 2013; vgl. ferner Hemminger 2009 und Kessler 2012).

Die vier grundlegenden Fragen, die den kosmologischen Antinomien Kants zugrunde liegen, werden auf Grundlage der heutigen Naturwissenschaft oder mit Bezug zu ihr wieder so kontrovers diskutiert, dass man sich im 21. Jahrhundert erneut auf dem „Kampfplatz [...] endlose[r][...] Streitigkeiten" (Kant, KrV, AVIII), der für Kant der Ausgangspunkt seiner Vernunftkritik war (vgl. Kant, KrV, A VII), wiederzufinden vermeint. Nach Kant war sein *„kritische[r] Weg allein noch offen"* (Kant, KrV, B 884), also die einzige Möglichkeit, „die menschliche Vernunft in dem, was ihre Wißbegierde jederzeit, bisher aber vergeblich beschäftigt hat, zur völligen Befriedigung zu bringen" (Kant, KrV, B 884), das heißt einen „ewigen Frieden" der Vernunft mit sich selbst zu stiften. Dies kann nur gelingen, wenn naturwissenschaftliche Forschung im Bereich dieser größten kosmologischen Fragen in dem ständigen Bewusstsein betrieben wird, dass diese Fragen uns aufgegeben sind (vgl. Höffe 2007, S. 154 f.) und wir nach Antworten zu suchen durch die Natur unserer Vernunft genötigt sind, dass wir aber keine endgültigen Antworten finden können. Die Freiheit des menschlichen Willens, so die negative Seite der „Transzendentalen Dialektik", kann nach Kant nicht objektiv erkannt werden. Aber – und dies ist das positive Ergebnis der Grenzziehung – es kann auch nicht das Gegenteil erkannt werden, dass der menschliche Wille nicht frei ist.[34] Zwar kann weder die Wahrheit des Satzes „Der menschliche Wille ist frei" noch des Satzes „Der menschliche Wille

---

34 Aus Kants Sicht beruht die naturwissenschaftliche Demonstration, dass die menschliche Willensfreiheit eine Illusion ist, auf einem transzendentalen Schein.

ist nicht frei" *objektiv erkannt* werden, aber die Freiheit des menschlichen Willens kann widerspruchsfrei *gedacht* werden. Deshalb sagt Kant (Kant, KrV, B XXX):

> „Ich mußte also das *Wissen* aufheben, um zum *Glauben* Platz zu bekommen, und der Dogmatism der Metaphysik, d. i. das Vorurteil, in ihr ohne Kritik der reinen Vernunft fortzukommen, ist die wahre Quelle alles der Moralität widerstreitenden Unglaubens[.]"

In der „Kritik der reinen Vernunft" erweist Kant die äußere Möglichkeit der Freiheit (vgl. Picht 1980a, S. 25). In der „Kritik der praktischen Vernunft" stellt er die Prinzipien und die *innere Möglichkeit der Freiheit* „in metaphysischer Absicht" (Kant, AA, VIII, 17) dar. Damit drängt sich Kant die Aufgabe auf, wie die Wahrheit (*verum*), auf der die Erkenntnis beruht, mit dem Guten (*bonum*), welches das Fundament der Moralität bildet[35], in einer Einheit (*unum*) zusammenhängen kann[36] (Kant, KrV, B 868):

> „Die Gesetzgebung der menschlichen Vernunft (Philosophie) hat nun zwei Gegenstände, Natur und Freiheit, und enthält also sowohl das Naturgesetz, als auch das Sittengesetz, anfangs in zwei besonderen, zuletzt aber in einem einzigen philosophischen System. Die Philosophie der Natur geht auf alles, was da *ist*; die der Sitten nur auf das, was da *sein soll*."

Die Bedingung der Möglichkeit der Überbrückung der „unübersehbare[n] Kluft zwischen dem Gebiete des Naturbegriffs [...] und dem Gebiete des Freiheitsbegriffs" (Kant, KU, B XIX) zeigt Kant in der „Kritik der Urteilskraft" auf. Dieser Übergang zwischen dem Bereich der Natur und dem Bereich der Freiheit manifestiert sich in der Geschichte.

---

35 Vgl. zum Beispiel Kant, AA, IV, 393.
36 Vgl. Picht 1980b, S. 58 f.

17.9   Kants Geschichtsphilosophie

Deshalb eröffnet Kant seine Schrift „Idee zu einer allgemeinen Geschichte in weltbürgerlicher Absicht" mit den Worten (Kant, AA, VIII, 17):

> „Was man sich auch in metaphysischer Absicht für einen Begriff von der *Frei-
> heit des Willens* machen mag: so sind doch die *Erscheinungen* desselben, die
> menschlichen Handlungen, eben so wohl als jede andere Naturbegebenheit,
> nach allgemeinen Naturgesetzen bestimmt."

In der Geschichte *erscheint* die Freiheit in den menschlichen Handlungen. Wenn
man eine einzelne menschliche Handlung betrachtet, so kann sie ganz oder teilweise auf der Freiheit des menschlichen Willens beruhen. Die Freiheit bedeutet
dann die Möglichkeit, „eine Reihe von Begebenheiten *ganz von selbst* anzufangen"
(Kant, KrV, B 562; vgl. auch Kant, KrV, B 831), anders gesagt äußert sich die Freiheit in einer unbedingten Setzung des Beginns einer Handlung, deren Folgen aber
naturgesetzlich sind. Ist Geschichte die Summe aller menschlichen Handlungen, so
ergibt sich, weil sie *ex ante* nicht „berechenbar" ist, ein „planloses *Aggregat* menschlicher Handlungen" (Kant, AA, VIII, 29).[37] Wenngleich die einzelne menschliche
Handlung nicht berechenbar ist, so lassen sich nach Kant doch Aussagen über die
„Mittelwerte" einer hinreichend großen Anzahl gleichartiger Handlungen machen,
sofern sie jeweils als unabhängig angenommen werden können (Kant, AA, VIII,
17):

> „So scheinen die Ehen, die daher kommenden Geburten und das Sterben, da
> der freie Wille der Menschen auf sie großen Einfluß hat, keiner Regel unterwor-
> fen zu sein, nach welcher man die Zahl derselben zum voraus durch Rechnung
> bestimmen könne; und doch beweisen die jährlichen Tafeln derselben in gro-
> ßen Ländern, daß sie eben so wohl nach beständigen Naturgesetzen gesche-

---

37   Um die Bedingung der Möglichkeit einer Geschichts*wissenschaft* zu zeigen, wäre zu
     untersuchen, wie sich dieses „bloße[...] Aggregat" (Kant, KrV, B 860) auf Grundlage
     einer „Idee" (Kant, KrV, B 860; vgl. auch den Titel der Schrift „Idee zu einer allge-
     meinen Geschichte in weltbürgerlicher Absicht") zu einer systematischen Einheit
     zusammenschließt. Weil Geschichte mit Erscheinungen zu tun hat und der „Inbegriff
     der Erscheinungen" (Kant, KrV, B 446; vgl. auch Kant, AA, IV, 318) Natur genannt wird,
     ist die Mathematisierbarkeit als Spezifikum der naturwissenschaftlichen Seite einer
     Geschichtswissenschaft zu berücksichtigen (Kant, AA, IV, 470): „Ich behaupte aber, daß
     in jeder besonderen Naturlehre nur so viel *eigentliche* Wissenschaft angetroffen werden
     könne, als darin *Mathematik* anzutreffen ist."

hen, als die so unbeständigen Witterungen, deren Eräugniß man einzeln nicht vorher bestimmen kann, die aber im Ganzen nicht ermangeln den Wachsthum der Pflanzen, den Lauf der Ströme und andere Naturanstalten in einem gleichförmigen, ununterbrochenen Gange zu erhalten."

Kant unterstellt für die menschlichen Handlungen im Großen eine statistische Berechenbarkeit, wie sie auch in anderen Bereichen, die auf einer ungeordneten Bewegung beruhen (vgl. Picht 1980a, S. 31), fraglos gelten. Die statistischen Tendenzen wertet er als empirisches Indiz für „einen regelmäßigen Gang" (Kant, AA, VIII, 17) der Geschichte, die zur empirischen Ableitung der kreisförmigen Bewegung unseres Planetensystems um das Zentrum der Galaxis parallelisiert werden kann (vgl. Kant, AA, VIII, 27; vgl. Brandt 2011, S. 92–94). Dieser regelmäßige Gang der Geschichte könnte nach Kant am Leitfaden einer „Naturabsicht" (Kant, AA, VIII, 17) oder „Vorsehung" (Kant, AA, VIII, 30) verstanden werden, der gemäß „sich diejenigen Naturanlagen, die auf den Gebrauch seiner [d. h. des Menschen] Vernunft abgezielt sind, [...] in der Gattung [...] vollständig entwickeln" (Kant, AA, VIII, 18). Dieses teleologische Moment ist die Bedingung der Möglichkeit der statistischen Berechenbarkeit der Geschichte. Weil die Geschichte nach Gesetzen der Statistik berechenbar und damit nicht vollständig, sondern nur im Mittel bestimmt ist, besteht für die Menschheit die Möglichkeit, den „Naturplan" (Kant, AA, VIII, 30) zu beschleunigen (vgl. Kant, AA, VIII, 24 und 27; vgl. Brandt 2011, S. 94). Aber auch wenn die Menschen ihren eigenen, zumindest teilweise vernunftwidrigen Absichten folgen, wird sich nach Kant dennoch auf hinreichend großen Zeitskalen aufgrund der ihnen verborgenen Naturabsicht die Weltgeschichte als Geschichte der fortschreitenden Auswickelung der Vernunft zeigen (vgl. Kant, AA, VIII, 17 f.).

17.10   Die Grenzen der Berechenbarkeit und ihre Bedeutung für den Menschen

Der junge Kant knüpfte an die Naturwissenschaft seiner Zeit an. Die Unifizierung von irdischer Physik und Himmelsphysik bei Newton hatte nahegelegt, alles naturgesetzlich zu verstehen. In seiner „Allgemeinen Naturgeschichte und Theorie des Himmels" nahm Kant die Gesetze der Newtonschen Physik ernst und versuchte sie auch auf die Entstehung und Entwicklung des Planetensystems anzuwenden. Damit ging er einen Schritt, den Newton selbst mit Hinweis auf das Wirken Gottes ausgespart hatte. Eine Weiterführung dieser Gedanken führt, wie sich dann bei

Laplace zeigte, zur vollständigen Berechenbarkeit der Natur, das heißt auf das deterministische Weltbild der Newtonschen Physik. Infolgedessen stehen die Seele als etwas Nichtmaterielles, die Freiheit des Willens des Menschen und die Existenz Gottes in Frage – und mit ihnen auch die Moralität des Menschen. Vor diesem Hintergrund unternahm es Kant, in seiner kritischen Philosophie zu zeigen, dass die Naturgesetze auf den Bereich der Erscheinungen beschränkt sind. Die Fragen nach der Unsterblichkeit der Seele, der Freiheit des Willens und der Existenz Gottes liegen außerhalb dieses Bereiches möglicher Erfahrung. Durch seine Grenzziehung gelang es Kant, die Freiheit des Willens als denkmöglich zu erweisen und ihr Bestehen neben dem Bereich der Naturgesetze zu sichern. Für die Naturwissenschaften bedeutet diese Grenzziehung, dass die großen Fragen nach der zeitlichen und räumlichen Ausdehnung der Welt, der Teilbarkeit der Materie, dem Verhältnis von Freiheit und Kausalität sowie der Notwendigkeit beziehungsweise Verzichtbarkeit der Existenz Gottes als Urheber der Welt letztlich nicht beantwortbar sind, sondern der Forschung als leitende Fragen aufgegeben bleiben. Dass die beiden Bereiche der Gesetzgebung der menschlichen Natur, nämlich die Natur und die Freiheit, nicht unvermittelt nebeneinander stehen, zeigt sich in der Geschichte. Der Mensch ist Naturwesen und damit an die Naturgesetze gebunden, aber er kann sich auch als Wesen denken, das einen freien Willen hat und seine Handlungen an selbst gegebenen Gesetzen orientieren kann. Dort, wo also die Berechenbarkeit aufhört, fangen die Möglichkeiten des Menschen an. In diesem Sinne sind die Grenzen der Berechenbarkeit die Bedingungen der Möglichkeit des Menschen.

Literatur

Alexander, Hubert Griggs (Hrsg.). 2005. The Leibniz-Clarke-Correspondence. Together with Extracts from Newton's "Principia" and "Opticks". Manchester: Manchester University Press

Batygin, Konstantin und Gregory Laughlin. 2008. On the Dynamical Stability of the Solar System. The Astrophysical Journal 683: 1207–1216

Beisbart, Claus. 2010. Kants mathematische Antinomie (I): Anfang und räumliche Grenzen der Welt. In Kants Grundlegung einer kritischen Metaphysik. Einführung in die „Kritik der reinen Vernunft", hrsg. von Norbert Fischer, 243–263. Hamburg: Felix Meiner

Beisbart, Claus. 2012. Können wir wissen, wie das Universum beschaffen ist? Echte und vermeintliche Erkenntnisprobleme der Kosmologie. In Philosophie der Physik, hrsg. von Michael Esfeld, 247–266. Frankfurt am Main: Suhrkamp

Bialas, Volker. 2004. Johannes Kepler. München: C. H. Beck

Bleck-Neuhaus, Jörn. 2013. Elementare Teilchen. Von den Atomen über das Standard-Modell bis zum Higgs-Boson. Berlin: Springer Spektrum

Bojowald, Martin. 2009. Zurück vor den Urknall. Die ganze Geschichte des Universums. Frankfurt am Main: S. Fischer

Brandt, Reinhard. 2011. Die einheitliche Naturgeschichte der Menschheit („Idee", Achter Satz). In Immanuel Kant: Schriften zur Geschichtsphilosophie, hrsg. von Otfried Höffe, 91–101. Berlin: Akademie Verlag

Busche, Hubertus. 1997. Leibniz' Weg ins perspektivische Universum: Eine Harmonie im Zeitalter der Berechnung. Hamburg: Felix Meiner

Carrier, Martin und Jürgen Mittelstraß. Johannes Kepler (1571–1630). In Klassiker der Naturphilosophie. Von den Vorsokratikern bis zur Kopenhagener Schule, hrsg. von Gernot Böhme, 137–157. München: C. H. Beck

Clarke, Samuel. 1990. Der Briefwechsel mit G. W. Leibniz 1715/1716. Übersetzt und mit einer Einführung, Erläuterungen und einem Anhang hrsg. von Ed Dellian. Hamburg: Felix Meiner

Clegg, Brian. 2012. Vor dem Urknall. Eine Reise hinter den Anfang der Zeit. Reinbek bei Hamburg: Rowohlt

Craemer-Ruegenberg, Ingrid. 1980. Die Naturphilosophie des Aristoteles. Freiburg im Breisgau: Karl Alber

Dawkins, Richard. 2013. Der blinde Uhrmacher. Warum die Erkenntnisse der Evolutionstheorie beweisen, dass das Universum nicht durch Design entstanden ist. Aus dem Englischen von Karin de Sousa Ferreira. München: Deutscher Taschenbuch Verlag

Descartes, René. 2005: Die Prinzipien der Philosophie. Lateinisch – Deutsch. Übersetzt und hrsg. von Christian Wohlers. Hamburg: Felix Meiner

Dijksterhuis, Eduard Jan. 1956. Die Mechanisierung des Weltbildes. Ins Deutsche übertragen von Helga Habicht. Berlin: Springer

Drake, Stillman. 2003. Galileo at Work. His Scientific Biography. Mineola (New York/USA): Dover

Dugas, René. 1988. A History of Mechanics. Translated into English by John R. Maddox. New York: Dover

Eidemüller, Dirk. 2012. Alle Dinge sind drei. Der Baukasten der Teilchenphysik ist wohl vollständig. Süddeutsche Zeitung 289: 18

Falkenburg, Brigitte. 1994. Teilchenmetaphysik. Zur Realitätsauffassung in Wissenschaftsphilosophie und Mikrophysik. Mannheim: Bibliographisches Institut

Falkenburg, Brigitte. 1995. Kants zweite Antinomie und die Physik. Kant-Studien 86 (1): 4–25

Falkenburg, Brigitte. 2000. Kants Kosmologie. Frankfurt am Main: Vittorio Klostermann

Falkenburg, Brigitte. 2002. Metamorphosen des Teilchenkonzepts. Praxis der Naturwissenschaften – Physik in der Schule 51 (4): 14–21

Falkenburg, Brigitte. 2012. Mythos Determinismus. Wieviel erklärt uns die Hirnforschung? Heidelberg: Springer

Faye, Hervé. 1884. Sur l'origine du monde, théories cosmogoniques des anciens et des modernes. Paris: Gauthier-Villars

Friedman, Michael. 1994. Kant and the Exact Sciences. Cambridge (Massachusetts/USA): Harvard University Press

Geyer, Carl-Friedrich (Hrsg.). 2013. Hirnforschung und Willensfreiheit. Zur Deutung der neuesten Experimente. Frankfurt am Main: Suhrkamp

Heidarzadeh, Tofigh. 2008. A History of Physical Theories of Comets, from Aristotle to Whipple. Ohne Ort: Springer

Hemminger, Hansjörg. 2009. Und Gott schuf Darwins Welt. Schöpfung und Evolution, Kreationismus und Intelligentes Design. Gießen: Brunnen

Henrich, Jörn. 2010. Die Fixierung des modernen Wissenschaftsideals durch Laplace. Berlin: Akademie Verlag

Hermann, Armin. 1981. Weltreich der Physik. Von Galilei bis Heisenberg. Esslingen am Neckar: Bechtle

Höffe, Otfried. 2001. Kleine Geschichte der Philosophie. München: C. H. Beck

Höffe, Otfried. 2007. Immanuel Kant. München: C. H. Beck

Jung, Tobias. 2005. Universum und Multiversum – „E pluribus unum"? Philosophia naturalis 42 (1): 77–101

Jung, Tobias. 2014. Das Herausdrängen Gottes aus dem astronomischen Weltbild. Versuch einer Skizze wichtiger Stationen von Johannes Kepler bis Stephen Hawking. Beiträge zur Astronomiegeschichte 12: 67–192

Jung, Tobias und Lukas Nickel. 2013. Messung und Unschärfe in der klassischen Physik. Philosophia naturalis 50 (2): 251–273

Kanitscheider, Bernulf. 1996. Im Innern der Natur. Philosophie und Physik. Darmstadt: Wissenschaftliche Buchgesellschaft

Kant, Immanuel. 1998. Werke in sechs Bänden, Band I: Vorkritische Schriften bis 1768. Hrsg. von Wilhelm Weischedel. Mit Übersetzungen von Monika Bock und Norbert Hinske. Darmstadt: Wissenschaftliche Buchgesellschaft

Keil, Geert. 2009. Willensfreiheit und Determinismus. Stuttgart: Philipp Reclam jun.

Keil, Geert. 2013. Willensfreiheit. Berlin/Boston: Walter de Gruyter

Kessler, Hans. 2012. Evolution und Schöpfung in neuer Sicht. Kevelaer: Butzon & Bercker

Kühn, Manfred. 2003. Kant. Eine Biographie. Aus dem Englischen von Martin Pfeiffer. München: C. H. Beck

Kley, Willy. 2000. On the Migration of a System of Protoplanets. Monthly Notices of the Royal Astronomical Society 313 (4): L47–L51

Königkrämer, Lutz. 2009. Der historische Hintergrund von Platons politischer Naturphilosophie. Dissertation an der Philosophischen Fakultät der Westfälischen Wilhelms-Universität zu Münster

Küppers, Bernd-Olaf. 2012. Die Berechenbarkeit der Welt. Grenzfragen der exakten Wissenschaften. Stuttgart: S. Hirzel

Kronk, Gary W. 1999. Cometography. A Catalogue of Comets, Volume 1: Ancient – 1799. Cambridge: Cambridge University Press

Kutschmann, Werner. 1989. Isaac Newton (1643–1727). In: Klassiker der Naturphilosophie. Von den Vorsokratikern bis zur Kopenhagener Schule, hrsg. Gernot Böhme, 171–186. München: C. H. Beck

de La Mettrie, Julien Offray. 1990. L'homme machine. Die Maschine Mensch. Französisch – deutsch. Übersetzt und hrsg. von Claudia Becker. Hamburg: Felix Meiner

Laplace, Pierre Simon. 2003. Philosophischer Versuch über die Wahrscheinlichkeit. Hrsg. von Richard von Mises. Thun: Harri Deutsch

Laplace, Pierre Simon. 2008a. Darstellung des Weltsystems, Band I. Hrsg. und aus dem Französischen übersetzt von Manfred Jacobi und Franz Kerschbaum. Thun: Harri Deutsch

Laplace, Pierre Simon. 2008b. Darstellung des Weltsystems, Band II. Hrsg. und aus dem Französischen übersetzt von Manfred Jacobi und Franz Kerschbaum. Thun: Harri Deutsch

Laskar, Jaques. 1996. Large Scale Chaos and Marginal Stability in the Solar System. Celestial Mechanics and Dynamical Astronomy 64 (1–2): 115–162

Laudan, Larry L. 1968. The Vis Viva Controversy, a Post-Mortem. Isis 59 (2):131–143

Leibniz, Gottfried Wilhelm. 2008. Die philosophischen Schriften, Band 7. Hrsg. von Carl Immanuel Gerhardt. Hildesheim/Zürich/New York: Georg Olms

Lessing, Gotthold Ephraim. 1996: Werke, Band I: Gedichte, Fabeln, Lustspiele. Darmstadt: Wissenschaftliche Buchgesellschaft

Liddle, Andrew. 2009. Einführung in die moderne Kosmologie. Übersetzung der zweiten englischen Auflage [von Sybille Otterstein] – überarbeitet und erweitert. Weinheim: Wiley-VCH

Lincoln, Don. 2013. Das Innenleben der Quarks. Spektrum der Wissenschaft 36 (12): 46–53

Lloyd, Seth und Y. Jack Ng. 2005. Ist das Universum ein Computer? Spektrum der Wissenschaft 28 (1): 32–41

Lombardi, Anna Maria. 2000. Johannes Kepler. Einsichten in die himmlische Harmonie. Übersetzung aus dem Italienischen von Michael Spang. Heidelberg: Spektrum der Wissenschaft

Mainzer, Klaus. 1988. Symmetrien der Natur. Ein Handbuch zur Natur- und Wissenschaftsphilosophie. Berlin/New York: Walter de Gruyter

Mainzer, Klaus. 1997. Gottfried Wilhelm Leibniz (1646–1716). In Die großen Physiker, Band 1: Von Aristoteles bis Kelvin, hrsg. von Karl von Meÿenn, 212–228. München: C. H. Beck

Mainzer, Klaus. 2014. Die Berechnung der Welt. Von der Weltformel zu Big Data. München: C. H. Beck

Malzkorn, Wolfgang. 1998. Kant über die Teilbarkeit der Materie. Kant-Studien 89 (4): 385–409

Malzkorn, Wolfgang. 1999. Kants Kosmologie-Kritik. Eine formale Analyse der Antinomienlehre. Berlin: Walter de Gruyter

Mendelssohn, Moses. 2009. Ausgewählte Werke. Studienausgabe. Band II: Schriften zu Aufklärung und Judentum 1770–1786. Hrsg. und eingeleitet von Christoph Schulte, Andreas Kennecke und Grażyna Jurewicz. Darmstadt: Wissenschaftliche Buchgesellschaft

Mittelstaedt, Peter und Ingeborg Strohmeyer. 1990. Die kosmologischen Antinomien in der Kritik der reinen Vernunft und die moderne Kosmologie. Kant-Studien 81 (2):145–169

Newton, Isaac. 1963. Mathematische Prinzipien der Naturlehre. Mit Bemerkungen und Erläuterungen hrsg. von Joseph Philipp Wolfers. Darmstadt: Wissenschaftliche Buchgesellschaft

Newton, Isaac. 1988. Mathematische Grundlagen der Naturphilosophie. Ausgewählt, übersetzt, eingeleitet und hrsg. von Ed Dellian. Hamburg: Felix Meiner

Newton, Isaac. 1999. Die mathematischen Prinzipien der Physik. Übersetzung und hrsg. von Volkmar Schüller. Berlin/New York: Walter de Gruyter

Newton, Isaac. 2001. Optik oder Abhandlung über Spiegelungen, Brechungen, Beugungen und Farben des Lichts. Aus dem Englischen übersetzt und hrsg. von William Abendroth. Frankfurt am Main: Harri Deutsch

Newton, Isaac. 2004. Philosophical writings. Hrsg. von Andrew Janiak. Cambridge: Cambridge University Press

Nolting, Wolfgang. 2011. Grundkurs: Theoretische Physik 1. Klassische Mechanik. Berlin: Springer

Picht, Georg. 1967. Was heißt aufgeklärtes Denken? Zeitschrift für Evangelische Ethik 11 (4): 218–230

Picht, Georg. 1980a. Kants transzendentale Grundlegung des Völkerrechts. In Hier und Jetzt. Philosophieren nach Auschwitz und Hiroshima, Band I, hrsg. von Georg Picht, 21–56. Stuttgart: Klett-Cotta

Picht, Georg. 1980b. Philosophie und Völkerrecht. Die anthropologischen Voraussetzungen des Rechts. In Hier und Jetzt. Philosophieren nach Auschwitz und Hiroshima, Band I, hrsg. von Georg Picht, 57–115. Stuttgart: Klett-Cotta

Picht, Georg. 1998. Kants Religionsphilosophie. Mit einer Einführung von Enno Rudolph. Stuttgart: Klett-Cotta

Plaaß, Peter. 1965. Kants Theorie der Naturwissenschaft. Eine Untersuchung zur Vorrede von Kants „Metaphysischen Anfangsgründen der Naturwissenschaft". Göttingen: Vandenhoeck & Ruprecht

Prinz, Wolfgang. 2013. Der Mensch ist nicht frei. Ein Gespräch. In Hirnforschung und Willensfreiheit. Zur Deutung der neuesten Experimente, hrsg. von Christian Geyer, 20–26. Frankfurt am Main: Suhrkamp

Rebhan, Eckhard. 2006. Theoretische Physik: Mechanik. Heidelberg: Spektrum

Resag, Jörg. 2014. Die Entdeckung des Unteilbaren. Quanten, Quarks und die Entdeckung des Higgs-Teilchens. Berlin: Springer Spektrum

Richter, Peter H. 1997. Kants Theorie des Himmels von 1755. Sterne und Weltraum 36 (7): 640–644

Schechner, Sara J. 1997. Comets, Popular Culture, and the Birth of Modern Cosmology. Princeton (New Jersey/USA): Princeton University Press

Scheck, Florian. 2007. Theoretische Physik 1: Mechanik. Von den Newton'schen Gesetzen zum deterministischen Chaos. Berlin: Springer

Schmidt, Heinrich (Hrsg.). 1925: Die Kant-Laplace'sche Theorie. Ideen zur Weltentstehung von Immanuel Kant und Pierre Laplace. Leipzig: Alfred Kröner

Schnädelbach, Herbert. 2007. Vernunft. Leipzig: Reclam

Seeck, Gustav Adolf (Hrsg.). 1975. Die Naturphilosophie des Aristoteles. Darmstadt: Wissenschaftliche Buchgesellschaft

Singer, Wolf. 2013. Verschaltungen legen uns fest: Wir sollten aufhören, von Freiheit zu sprechen. In Hirnforschung und Willensfreiheit. Zur Deutung der neuesten Experimente, hrsg. von Christian Geyer, 30–65. Frankfurt am Main: Suhrkamp

Soter, Steven. 2007. Are Planetary System Filled to Capacity? Computer simulations suggest that the answer may be yes. But observations of extrasolar systems will provide the ultimate test. American Scientist 95 (5): 414–421

Strohmeyer, Ingeborg. 2013. Kausalität und Freiheit. Eine Untersuchung des quantenmechanischen Indeterminismus im Lichte der kantischen Freiheitsantinomie. Kant-Studien 104 (1): 63–99

Vaas, Rüdiger. 2010. Hawkings neues Universum. Wie es zum Urknall kam. München: Piper

Vailati, Ezio. 1997. Leibniz & Clarke. A Study of Their Correspondence. New York: Oxford University Press

Waschkies, Hans-Joachim. 1987. Physik und Physikotheologie des jungen Kant. Die Vorge-
schichte seiner Allgemeinen Naturgeschichte und Theorie des Himmels. Amsterdam: B.
R. Grüner
von Weizsäcker, Carl Friedrich. 1943. Über die Entstehung des Planetensystems. Zeitschrift
für Astrophysik 22: 319–355
von Weizsäcker, Carl Friedrich. 1995. Kants „Erste Analogie der Erfahrung" und die Erhal-
tungssätze der Physik. In: Die Einheit der Natur. Studien. Hrsg. von Carl Friedrich von
Weizsäcker, 383–404. München: Deutscher Taschenbuch Verlag
von Weizsäcker, Carl Friedrich. 2006. Die Tragweite der Wissenschaft. Stuttgart: S. Hirzel
Wolff, Michael. 1978. Geschichte der Impetustheorie. Untersuchungen zum Ursprung der
klassischen Mechanik. Frankfurt am Main: Suhrkamp
Wolfram, Stephen. 2002. A New Kind of Science. Champaign: Wolfram Media

III
Komplexität und Information

18. Complexity Studies: Interdisciplinarity in Action

Helena Knyazeva[1]

Abstract

Modern complexity studies and their interdisciplinary character are under consideration in this article. The conceptual framework of complexity studies includes such methodologically significant notions as nonlinearity, self-organization, creative chaos, co-evolution and blow-up regimes. It is argued that interdisciplinarity corresponding to a holistic worldview becomes a powerful trend in modern science. At the same time, evolutionary holism constitutes a philosophical basis for complexity studies. The possible future developments of complexity studies as a fundamental interdisciplinary paradigm are discussed as well. It is shown that interdisciplinary research will define the character of science in the medium-term future.

"Interdisciplinary dialogues are needed to find transdisciplinary problems and new portfolios of technologies." K. Mainzer (2011)

---

1   Helena Knyazeva is Dr.habil. in Philosophy, Professor of the School of Philosophy at the National Research University Higher School of Economics. Member of Bertalanffy Center for the Study of Systems Science and of Scientific Council "Multiversidad Mundo Real Edgar Morin" in Mexico. Academician of the International Academy for Systems and Cybernetic Sciences. Her fields of research are epistemology and philosophy of science. She has published 10 monographs in Russian and more than 30 research articles in international professional journals.

18.1   Principles of Holistic Thinking

All systems of human knowledge are based on certain principles. The latter can be considered as initial attitudes to any research or as starting points of human reasoning and thinking, including modern complex system studies and evolutionary thinking. Aristotle said: "The first principles have to be accepted, all the rest has to be justified." As a rule, there are some initial statements, or principles, in any theory, and then, via a top-down approach, an entire system of theoretical knowledge is constructed. The principles render scientific theories essentially open. Kurt Gödel discovered this fundamental fact for mathematical theorems. He proved that, for most sets of axioms, there are true theorems that cannot be deduced. These theorems are, in other words, "random truths".

Other questions arise, when we enter the field of methodology and begin to deal with methodological principles. Generally speaking, any well-elaborated theory can be used as an instrument for getting new knowledge; thereby, the theory becomes a method of research. Rene Descartes, being a genuine liberator of pure reason, introduced the very idea of method, i.e. an order which we should follow in developing our thoughts (Descartes 1953, p. 14-16). If one manages to formulate the methodological principles in an intelligible and plain form, they can serve as strategies of scientific—or, in general, intellectual—search. Although this article does not pursue the goal to deal either with the nature of principles, or of human knowledge, or with specific features of methodology, understanding these creates the necessary prerequisites for further consideration.

The modern principles of thinking should be holistic and at the same time evolutionary, complex, humane and ethically oriented—a demand that has been time and again emphasized by Klaus Mainzer (1997, 2007a, 2007b, 2008, 2011, 2014) as a brilliant and influential thinker in the field of complexity studies.

Inasmuch as modern holism is inextricably linked with the ideas of evolutionism, it becomes evolutionary holism. Holism can be considered from an ontological viewpoint as integrity of structures, of forms in nature and society. From an epistemological viewpoint, holism signifies an integral, all-in-one vision of complex phenomena of cognition and creativity, and specifically: perception and thinking, logic and intuition, analysis and synthesis, etc. (Pink 2005). From the methodological perspective, holism means bridge-building between natural sciences and the humanities, science and engineering (the emergence of techno-science), science and culture, and even convergence of the highest values of humanity: truth, virtue, and beauty (Knyazeva 2009). Interdisciplinarity has become a powerful trend of modern science. This is a manifestation of the methodological meaning of holism.

Ideas of holism can be found already in the philosophy of Plato, who treated the cosmos as a single whole that determines itself, using only its own resources. According to E. Laszlo, holism is a new fundamental paradigm which sets down methods and strategies of scientific research:

"Holism today is present not only as a philosophy within the general framework of science, to use Smut's phrase, but as a new fundamental paradigm: a basic hallmark of scientific theories themselves. This brings the new sciences significantly closer to the hopeful subcultures in society, a convergence that is all the more significant as it blazes a trail towards a civilization in which holistic thinking embraces all things in nature and society." (Laszlo 2012, p. 80)

Synergetics developed by Hermann Haken is a well accepted and wide spread name for the whole field of complexity studies in Russia. The version of synergetic theory elaborated by me together with Sergei P. Kurdyumov in the 1990s and in the beginning of the 2000s, fits into the mainstream of evolutionary holism as a worldview and as a methodology. Co-evolution turns out to be a key idea here. The content of this idea is not limited to ecology as a branch of biological knowledge. It is much broader and covers methods of integration and of nonlinear synthesis of complex systems.

Biologists talk about co-evolution of living organisms in biocenosis in terms of a specific biological community; environmentalists consider co-evolution of man and nature from the perspective of maintaining the ecological balance in the environment. In a more general and non-trivial sense, one can talk about co-evolution of complex systems developing at different paces and being at different stages of development. In this respect, co-evolution is the joint and mutually concerted, sustainable development of complex systems and—in the case of successfully coordinated systems—their delicate set of interdependences in the same tempo-world. The idea of co-evolution with respect to complex self-organizing systems was proposed by Kurdyumov and developed in a number of our joint works (Knyazeva & Kurdyumov 2001, 2008).

Thus, my view is based on research results of the Moscow school of synergetics led by Kurdyumov, who was the Director of the Keldysh Institute of Applied Mathematics of the Russian Academy of Sciences from 1989 to 1999. Synergetics has been under development here as a theory of very fast processes, blow-up regimes, localized structure formation and their evolution in complex (open and nonlinear) dissipative media. I prefer to use the term "medium" instead of "system" to underline that an emerging structure is a localized process in a continuous dissipative media. Most of the methodological conclusions presented here are founded on

382   Helena Knyazeva

reliable results of mathematical modeling and computer simulations of nonlinear evolutionary processes in such complex media.

18.2   Complexity Studies: On the Way towards Interdisciplinarity

The complexity studies based on the theory of self-organization of complex systems, which are often called, after Haken (1977), synergetics in Russia, play a central role in the indispensable reform of thinking in the modern world. The most substantial innovative discoveries are made when scientists overstep the boundaries of separate disciplinary fields and turn to think in an interdisciplinary way. Mainzer is absolutely right when accentuating certain characteristics of this innovative trend in modern science:

> "In recent time, innovations emerge from problem-oriented research overcoming traditional boundaries of disciplines (e.g., material research, energy, environment, health, ageing society). If problem-oriented research is beyond former divisions of faculties, it is sometimes called 'transdisciplinary'. Interdisciplinary dialogues are needed to find transdisciplinary problems and new portfolios of technologies." (2011, p. 278)

Thus, the first key feature of complexity studies is their *interdisciplinarity or transdisciplinarity*. They are oriented to reveal common patterns of self-organization of complex systems of any kind, independent of the concrete nature of their elements. Recently, complexity studies have moved beyond the domains of mathematical physics, laser physics, physics of plasma and physical chemistry, i.e. the fields of research in which the basic models have been elaborated. Models of complex systems are fruitfully applied to the understanding of human beings, human culture and society, human psychology and cognition, creative work and education.

Complexity studies have discovered "all-pervading complexity" manifesting itself at different levels of reality. In this sense, complexity studies follow the same rational particular to transdisciplinary research. To put it in another way, the modern theory of complexity is highly transdisciplinary research. Theoreticians in transdisciplinarity proclaim that nature itself, in its internal complexity and ever-changing, diverse character, but also in its deep internal unity, requires crossing borders between disciplines; and transdisciplinarity seeks to discover the complexity which underlies all these processes and phenomena. And it is transdisciplinarity which is peculiar for research strategies of the theory of complexity.

The holistic character of complexity studies coincides with the holistic nature of transdisciplinary approaches. Besides, holism in complexity studies has an evolutionary character, so that the processes of evolution of inanimate nature, of living nature, the origin of human beings, the development of humankind and the emergence of networks of collective intelligence are all considered different stages of single universal process of evolution ("Big History"). In the past, natural philosophy sought to understand nature in its integrity. It relied on the concepts of natural science of the corresponding historical epoch, and often established an identity between phenomena observed in the micro- and macrocosmos. This intellectual tradition continues today in terms of the theory of complexity. It is the holistic tendency inherent in this theory that defines the face of modern science. And, presumably, this tendency will become more pronounced; and the ability of scientists to think in non-linear and holistic ways will be highly valued.

Complexity studies lead to a new *constructive dialogue* between specialists in different disciplinary fields. They work towards the *synthesis* of the natural sciences and the humanities, of the Eastern and the Western worldviews, of the new science of complexity and the old traditions of culture, art and philosophy, of theory and practice of decision making, management and entrepreneurship.

Complexity studies can provide us with *strategies of research*. Once the general patterns of self-organization and of nonlinear synthesis of complex systems are revealed, one can forecast processes of structure formation and their evolution in certain domains of the natural or the human world. This theoretical vision may suggest a next step in research, namely *savoir faire* for researchers. As Mainzer once said, "the complex systems approach is not a metaphysical process ontology. Synergetic principles (among others) provide a heuristic scheme to construct models of nonlinear complex systems in the natural sciences and in the humanities" (2007b, p. 15).

Our thinking should be complex enough to cope with the complexity of the world around us. The principles of complex nonlinear thinking bear in themselves the imprint of the nature of principles as such. They are elements of open-minded and open-ended systems of knowledge. They are connected not only to pure, rationalized knowledge, but to human intuitions and convictions as well. Therefore, to cultivate complex thinking means to learn the *art of thinking*. Equally, to acquire the ability to efficiently act in complex surroundings means to learn the *art of activity* and of management. This understanding is consonant with an apophthegm of Paul Valéry: "Il y a *science* des choses simples et *art* des choses compliquées… L'intérêt de la science gît dans l'*art* de faire la science" ("There is the *science* of simple things and the *art* of complex things… The interest of science consists in the *art* of doing science"—*my translation, H.K.*) (Valéry 1943, p.52).

18.3   Complexity Studies: Their Conceptual Frames

18.3.1 Nonlinearity

"Nonlinearity" is a fundamental conceptual knot of the paradigm of complexity studies. This paradigm may also properly be called the paradigm of nonlinearity. Complex thinking is, first of all, nonlinear thinking. Therefore, it is important to understand various implications of the notion of nonlinearity, including its most general, philosophical sense.

Considered from the mathematical point of view, nonlinearity signifies a certain type of mathematical equation which contains unknown quantities in powers more than 1 or coefficients which depend on properties of a medium (system). Nonlinear equations can have several qualitatively different solutions. From this follows the physical sense of nonlinearity. A certain set of solutions of a nonlinear equation corresponds to a multitude of evolutionary paths of a nonlinear system which is described by the equation. In most problems of the theory of self-organization, different paths of evolution are connected with bifurcations that result from changing the characteristics of a medium (system). A certain control parameter changes in differential equations; when the parameter reaches a definite critical quantity, the thermodynamic branch (i.e. branch which corresponds to a stable stationary solution of nonlinear equations) becomes unstable, and at least two possible paths of development appear.

Ilya Prigogine and Isabel Stengers describe the notion of bifurcation in their model of the Brusselator (i.e. a theoretical model of a certain type of autocatalytic reaction developed by Prigogine and his colleagues in the Free University of Brussels) as follows:

> "At equilibrium or near-equilibrium, there is only one steady state that will depend on the values of control parameters. We shall call λ the control parameter, which, for example, may be the concentration of substance B in the Brusselator… We now follow the change in the state of the system as the value of B increases. In this way the system is pushed farther and farther away from equilibrium. At some point we reach the threshold of stability of the 'thermodynamic

branch'. Then we reach what is generally called a 'bifurcation point'... At bifurcation point, the thermodynamic branch becomes unstable with respect to fluctuations. For the value λ_c of the control parameter λ, the system may be in three different steady states: C, E, D" (I. Prigogine and I. Stengers 1984, p.160-161).

The process of branching of evolutionary paths is surprising, but it is well known among mathematicians. One of the peculiarities of a nonlinear world consists in the fact that, in a certain range of characteristics of a system and of parameters of nonlinear equations, there are no qualitative changes in the picture of a process. In spite of the quantitative variation of parameters, the attraction of one and the same attractor remains, the process reduces to one and the same structure, and the evolutionary regime of the system doesn't change. But if a definite threshold, or a critical quantity of a control parameter, is attained, the evolutionary regime of the system changes qualitatively. The system passes the field of another attractor. The picture of integral curves on the phase plane transforms qualitatively.

The transformation described above is quite easily explained. The change of parameters of nonlinear equations above critical quantities creates the possibility of passing to another state, i.e. in reality, to enter into another world. And if a medium, be it a medium of physical interactions, chemical reactions, or a medium in which living beings dwell, changes in a qualitative way, it is quite natural to expect the emergence of new possibilities, namely, new structures, new paths of evolution, and new bifurcations.

The research done by Kurdyumov with his collaborators (Kurdyumov 1990, Samarskii et al. 1995) went in another direction. Along with the solution of problems, in which parameters of a medium undergo changes, Kurdyumov considered problems of a different kind. These are problems in which only the character of the initial influence upon one and the same medium varies. What is a significant change of character of the initial influence? The change of spatial configuration, topology of an influence (for example, its symmetry or color symmetry) is of great importance, in contrast to the change of the intensity of the influence. Because of such changes, different structures appear in a medium. Other researchers have gone to study the problem elaborating models of finite-state cellular automata in the game "Life", etc, to explain the nature of complexity in living nature.

It is somewhat paradoxical that different structure-attractors as asymptotes of different evolutionary paths can appear in one and the same system. Moreover, based on the study of different stages of evolutionary processes in an open nonlinear system, one may expect a qualitative change of the picture of processes, including reorganization—an increase of complexity or degradation—of the medium. The openness of a system signifies that there is an exchange of matter, energy and/

or information between the system and its environment. This change of the picture of processes does not occur by modifying the parameters of a system, but as a result of its own development and self-organization.

The *idea of nonlinearity* has a profound philosophical sense. Kurdyumov and I have disclosed the content of this idea of nonlinearity using a few more accessible notions, namely:

- the *idea of multiplicity of evolutionary paths*, i.e. the availability of alternative paths of evolution (it is well worth underlining here the fact that a large number of evolutionary paths is characteristic even for one and the same, invariable, open and nonlinear medium, or system);
- the idea of choice between these alternative paths of evolution;
- the *idea of tempo of evolution*, i.e. the speed of evolutionary processes in an open nonlinear system;
- the idea of irreversibility of evolution.

The specific features of the phenomenon of nonlinearity are as follows.

First, owing to nonlinearity, an important principle of "amplification", or "the strengthening of fluctuations" holds in these systems. Under certain conditions, nonlinearity can strengthen fluctuations. An insignificant difference in initial conditions may transform into an appreciable one with macroscopic consequences.

Second, special types of open nonlinear systems show another remarkable property—the existence of thresholds of sensitivity. Below a threshold, everything diminishes and disappears. Such events are eventually forgotten, and don't leave any traces in nature, science or culture. On the contrary, everything increases excessively above a threshold.

Third, nonlinearity gives rise to a particular quantum effect: the discreteness of evolutionary paths of nonlinear systems. It means that, in a given nonlinear system, it is not any arbitrary (either conceivable or desirable) evolutionary path that is possible, but only a discrete spectrum of evolutionary paths is available and feasible for implementation in the system. The above-mentioned existence of thresholds of sensitivity for special types of open nonlinear systems is an indicator of this quantum nature as well.

Fourth, nonlinearity signifies the possibility of unexpected, or "emergent", changes of direction in the course of a certain process. This entails some consequences for those human activities connected with predictions of the future. Predictions as extrapolations from an available state of affairs are still quite widespread in at least those sciences which deal with explorations over a short-term horizon. Because of the nonlinearity of evolutionary processes in the world, such predictions are in principle unreliable and insufficient. The development occurs through accidental choices of a path around bifurcation points, and the change is, as a rule, never repeated (such is the nature of things).

Results of our research show that the picture of processes at initial or intermediate stages may be completely opposite to their picture at a developed, asymptotic stage. For example, the processes, which initially spread and faded, may flame up in time and become localized near the center of a structure. Such bifurcations are determined not by changes of parameters of a medium, but by the course of processes of self-organization in the medium. Of course, some forced or spontaneous changes of an open nonlinear medium itself can occur as well. If a medium becomes different, this entails quite naturally qualitative changes in the course of its evolution. At a deeper level, a transformation of the field of possible evolutionary paths of the medium takes place in such a case.

And finally, nonlinearity implies the possibility of super-rapid development at certain evolutionary stages. These very fast evolutionary processes together with their singularities in open nonlinear media, so-called blow-up regimes, have been studied extensively at the Moscow synergetic school.

18.3.2 New facets of complexity

Obviously, "complexity" is another key notion of complexity studies. The notion is closely related with the concepts of "self-organization", "openness", "nonlinearity" and "chaos." According to the second law of classical thermodynamics, in observed integrated systems evolution leads to the simplification of organization, to the degradation of structures and formations in the world, and precisely to the increase of entropy, i.e. chaotic, unordered elements. Taken to the extreme, these notions result in Clausius' well-known hypothesis of heat death of the universe. Synergetics, which is based on nonequilibrium thermodynamics, examines mostly the opposite process: the path to increasing complexity, and the origin of complexity and its growth, i.e. the processes of morphogenesis. The processes of simplification of organization and possible transitions to chaos are considered by synergetics only as necessary evolutionary stages of the emergence of complexity and the functioning of complex systems.

The model of morphogenesis was a research subject of Alan Turing as far back as 1952 (Turing 1952). How does a complex structure appear? Why do self-organizing structures in the world have definite (mostly spiral or lattice) forms, such as the symmetrically organized hexagonal lattices constructed by honey bees? How is a change of forms that is an increasing complexity at all possible? How does the process of morphogenesis occur? In other words, how is a chain reaction of increasing complication possible? How does the assembly of integrated systems from parts developing at different speeds occur? Specialists in complexity studies have only discovered the tip of the iceberg in their search for answers to these questions.

How can we define complexity? It is hardly possible to give a precise definition of complexity. To strive for exact definitions of general concepts is rather often (even in the exact sciences!) an unrewarding task. Complexity is a multi-faceted phenomenon.

It is quite clear that complexity is not only a complicated composition of elements within a system. That is, complexity is not simply connected with a large number of interacting elements or components and intricate interactions between them. It is rather a characteristic of the behavior of open nonlinear systems, in particular their manner of structure formation, i.e., spatial and temporal patterning.

As J.A. Scott Kelso has written, there are two problems, namely: the problem of complexity of substance, and the problem of pattern in the exploration of complexity. Kelso has correctly identified the second problem as more important in the theory of self-organization:

"Any principle of pattern formation has to handle two problems. The first is how a given pattern is constructed from a very large number of material components. We might call this *the problem of complexity of substance...* Biological structures like ourselves, for example, are multifunctional: we can use the same set of anatomical components for different behavioral functions as in eating and speaking, or different components to perform the same functions (try writing your name with a pencil attached to your big toe). We might call this second problem *the problem of pattern complexity.*" (Kelso 1995, p.5)

Trying to penetrate into the very nature of complexity, be it chemical, biological or social complexity, Gregoir Nicolis and Ilya Prigogine explore general components of complex behavior. These are "nonequilibrium, feedback, transition phenomena, and evolution." (Nicolis and Prigogine 1989, p.40) Complex behavior, rather than complex systems, is the proper subject of research. "While we cannot yet attempt a clear-cut definition of complexity, we begin to perceive some of its essential ingredients: the emergence of bifurcations in far-from-equilibrium conditions, and in the presence of suitable nonlinearities; the generation of broken symmetries beyond bifurcation; and the formation and maintenance of correlations of macroscopic range." (Nicolis and Prigogine 1989, p.78)

Complexity is connected with the hierarchical principles of system organization. Besides, complex hierarchical formations should be regarded from an evolutionary perspective. Instabilities and nonequilibrium phase transitions, which include oscillations, spatial structure formation, and chaos, are the subject of analysis here. Haken (1977) and Nicolis (1986) develop this very approach.

In his book *The Quark and the Jaguar* Murray Gell-Mann shows that, no matter how paradoxical it seems, the world of quarks has much in common with the world of a jaguar wandering in the darkness (Gell-Mann 1995). These two poles of the universe—the simple and the complex—are closely connected with each other. He proposes a new term "plectics" which from his point of view expresses quite well the relations between simplicity and complexity in all their diverse manifestations. The term "plectics" derives from Greek and is related semantically with the "art of weaving", "composition", and "complication". Thus, a turn "from complexity to perplexity" occurs in modern discussions of complexity. Complexity is poised *at the edge of chaos*.

Complexity is the unity of a great number of diverse elements. According to Edgar Morin, who discusses the problem in the appropriate philosophical context, complexity is "unitas multiplex", i.e., "the unity of diversity and in diversity":

"Complexity appears therefore in the heart of the One simultaneously as relativity, connectedness, diversity, duality, ambivalence, ambiguity, uncertainty, and in the unity of these complementary, competitive and antagonistic notions. The system is a complex entity that is more, less and other than itself. It is simultaneously open and closed. There is no organization without disorganization. There is no functioning without dysfunction." (Morin 1977, p.147)

Complexity and tempos of evolution. In my joint work with Kurdyumov (Knyazeva and Kurdyumov 2001, 2008), we developed the view that complexity of structures and of their behavior is determined, first of all, by their tempos of evolution. The tempo, or the rate of evolution of open nonlinear systems, is a key characteristic in exploring complexity. This thesis can be explained by a few more concrete ideas, such as the following: 1) There are very fast, avalanche-like processes, so-called blow-up regimes, which are of great importance, which display effects of localization, i.e., structure formation, and the emergence and development of extremely complicated structures; 2) Periodical alternation of various evolutionary regimes may take place. The change of the tempo of evolution as well as of the general character of the processes is one possible basis for self-maintenance of complex structures in the world; 3) This is tempo of evolution that serves as an indicator that separate structures developing with different speeds become parts (substructures) of a larger complex structure; 4) The synchronization of tempos of evolution of different complex structures is a way for co-evolution, and sustainable development, to take place in the world.

Narrow evolutionary channels towards systems of increasing complexity. The world is built in such a way that it admits complexity. This very fact is one of the main miracles of the world. One popular viewpoint is that complexity can be justified by the anthropic   principle, i.e. the world must be complex because the existence of an observer already presupposes complexity. The complexity of the observed universe is made possible by an extremely narrow range of primary elementary processes and by corresponding values of fundamental constants. If, for example, the cross sections of elementary processes in the epoch shortly after the Big Bang had been a little bit larger, the whole universe would have consumed itself during a short period of time. This view takes the anthropic principle as principle of existence of complexity in the world. For complex systems to exist and to develop, very specific elementary processes at the micro-level had to occur in a highly peculiar way right from the outset.

One of the most important results of studies of mathematical models of open nonlinear media is the discovery of the phenomenon of *inertia of heat*, i.e. localization of processes and structure formation in dissipative media, such as how the plasma medium organizes itself in the form of non-stationary structures developing in blow-up regimes (Samarskii et al. 1995). There are some reasons to extend the anthropic principle to cover the conditions for the emergence and growth of complexity in the processes of self-organization.

The wording of the hypothesis might be as follows: complex spectra of structure-attractors, which differ from each other in their sizes and forms, exist only for a narrow, unique class of mathematical models with exponential nonlinear functions. It is astonishing that all complex formations in the world are, in fact, built in an extremely selective way, i.e. that the evolutionary channel to complexity is very narrow. The evolutionary motion upwards towards more and more complex formations and structures corresponds to the realization of less and less probable events.

Thus, relatively simple mathematical models contain complexity, i.e. complex spectra of structure-attractors. Kurdyumov and I have shown that complex spectra of non-stationary structures, i.e., structures developing in blow-up regimes, can emerge and maintain themselves in a meta-stable way in a restricted class of open nonlinear media. The path of increasing complexity is a path towards media with bigger nonlinearities and new properties, with more and more complex spectra of forms and structures. There are reasons to consider the world as a hierarchy of media with different nonlinearities.

Reduction of complexity. One of the main methodologically fruitful ideas is the idea of a radical reduction of complexity. Complexity can be reduced essentially to simpler representations of evolutionary processes. Attractors of evolution of complex systems are described much more easily than zigzagging and ramified paths to meta-stable final states.

Synergetics allows us to overcome certain psychological barriers, in particular the fear of systems organized in a very complicated manner. When studying such systems, it is unjustified to excessively complicate the corresponding models and to introduce many parameters of evolution. A super complex and multidimensional system, which has a chaotic, unorganized motion at the level of its elements, may be described by a few fundamental ideas and parameters or even, if possible, by mathematical equations that determine general trends of the respective evolutionary processes.

The approach developed by Haken (1977) is of great significance here. To find out the small number of order parameters, which are generated by elements of a complex system and which enslave the behavior of the elements, means to simplify an observed picture of complex behavior in a radical way. This is really a key to our understanding of complex systems, i.e. a way to cope with complexity.

"According to the slaving principle of synergetics, the order parameters determine the behavior of all the individual parts of the system, for instance the motion of the individual volume elements. The number of components may be very large, whereas, in general, the number of order parameters is rather small or may even be only one. In this way, the slaving principle implies an enormous information compression. Instead of characterizing the system by its many individual components and their activities, it is sufficient to characterize it by the order parameters." (Haken 1997, p.154)

Synergetic models of another kind have been developed by Kurdyumov and his collaborators. Radical reduction of complexity takes place when an evolutionary process reaches an asymptotic stage. The asymptotes, or structure-attractors of evolution, are much simpler than the complicated, unstable course of intermediate processes in nonlinear complex systems.

"It proves that at an asymptotic stage the processes in the system with a lot of parameters (infinite-dimensional systems) are satisfactorily described by very simplified finite-dimensional models. If we consider a series expansion in harmonics of the solution to a complex nonlinear system, it proves that at an asymptotic stage only a few harmonics make considerable contribution to the solution. As a result, at an asymptotic stage only a few harmonics out of an infinite variety remain essential." (Kurdyumov 1990, p.313)

The structure-attractors determine evolutionary trends, or "purposes", of ongoing processes. The evolutionary processes at an asymptotic stage become simpler to an enormous degree. A strength of this approach is that new opportunities for forecasting arise. When building scenarios of the future, one may proceed: 1) from "end states" being like "purposes" of processes (structure-attractors of evolution), 2) from "the whole", i.e., from general tendencies of evolutionary processes in complex systems (media); and, therefore, 3) from some desirable "an ideal" that corresponds to specific and particular trends of evolutionary processes in complex systems.

18.3.3. The art of complex thinking

The theory of complexity has discovered certain common patterns in the behavior of different complex systems/media. Thus, science makes progress towards a new nonlinear evolutionary worldview as well as towards a statement of the principles of complex nonlinear thinking. The most essential elements of such thinking are the understanding of:

1. The constructive role of chaos in evolution, and the connection of chaos on the micro-level with the evolution of structures on the macro-level;
2. The non-uniqueness of purposes (structure-attractors) and the elements of pre-determination in the field of multiple evolutionary paths;
3. The laws of very fast, avalanche-like processes, blow-up regimes in complex systems;
4. The existence of changing rhythms and regimes of evolutionary processes, and nonlinear dynamics of evolutionary processes;
5. The patterns of constructing complex totalities from simpler elements (or subsystems), that is, the elaboration of a new kind of an evolutionary holism.

The style of thinking in classical science is subjected to radical criticism from the standpoint of the theory of complexity. This style is characterized mostly by patterns of linear thinking. Some of these patterns of thinking are not yet eliminated. Therefore, it is worthwhile to consider in detail the crucial paradigm shift from linear to nonlinear thinking.

Creative chaos. Even at present, chaos is popularly portrayed as something fearful and not fundamentally different from the gaping abyss of classical mythology. Chaos seems to be an exclusively destructive element of the world. It seems that chaos leads nowhere.

Furthermore, the classical view of "chance" is still widely prevalent. In the classical conception, chance (randomness) is considered as a secondary and subsidiary factor, which is not of principal importance. Chances are forgotten and fade away; no traces of them remain in the course of events in nature, science, culture or society. The world we are living in is viewed as independent from small fluctuations at underlying levels of reality as well as from insignificant cosmic influences.

By contrast, complexity studies reveal the creative role of chaos (randomness, fluctuations) in evolutionary processes, which occur in nonlinear complex systems (Mainzer 2007a). There must be a certain degree of chaos and destruction in the world. Chaos and fluctuations on the micro-level play an essential role in determining the actual trends, or "aims", of processes at the macro-level. Chaos manifests itself as a "force", as a mechanism underlying the development towards one of several evolutionary structure-attractors. Macro-organization evolves from chaos on the micro-level. Dissipative processes, being the macroscopic revelation of micro-chaos, act like a sculptor who chisels and shapes a statue from a block of marble.

Order and chaos, organization and disorganization, construction and destruction, all seem to be well-balanced in the world. Thus, it is senseless to struggle against chaos, or to strive to completely eliminate the negative, destructive elements from the world. Chaos is a necessary condition for self-organization.

Besides this, chaos (exchange processes of different kind) serves as a basis for integration of relatively simple evolutionary structures into more complex ones. It is a mechanism for coordinating their tempos of evolution. Chaos and fluctuations on the micro-level can also be a way of "evolutionary switching", that is, of allowing a periodical transition from one evolutionary regime to another.

Thus, the theory of complexity reveals a sympathetic, creative face of chaos. Chaos is ultimately a field producing sparks of social and cultural innovations. Because chaos opens up the possibility for the appearance of something completely new, an element of chaos is desirable. The theory of complexity allows us to understand destruction as a creative principle. "A passion for destruction is creative," wrote the Russian philosopher Mikhail Bakunin in his article "Reaction in Germany" in 1842. Liberation from old conditions can mean the turn of evolutionary processes and can transform them into something different and even opposite. Hence the new can emerge from the old.

Discrete sets of evolutionary paths. From the classical point of view, development is understood as a linear and progressive course, without any alternatives. The past is only of historical interest. Even if there are some alternatives, they represent only accidental deviations from a main stream. They are subordinated to the main stream, which is itself determined by general laws of the universe. All these alternatives flow, in the end, into the main course of historical events. The world is a kind of net of hard links of cause and effect. The causal links have linear character. When following these causal links, the course of development can be infinitely calculated without limit in the past as well as in the future. The development is both pre-dictable and retro-dictable. The present is determined by the past, the future is determined by the past and the present.

One of the most essential and paradoxical consequences of the theory of complexity is a notion of pre-determination, which is conceived in a new nonlinear sense. Evolutionary processes, even as they emerge out of chaos, are to some extent predetermined. The well-known notions of final and formal causes coined by Aristotle might be at issue here. They could be interpreted in a new and quite materialistic way in light of the theory of complexity.

A whole system of concepts, notions and ideas connected with nonlinearity and "aims" of evolution, i.e. the construction of a kind of evolutionary teleology, might be developed in the framework of complexity studies. First of all, these are the notions of localization of processes (structure formation) in open dissipative media, spectra of structure-attractors as stable formations, which evolutionary processes in such systems lead to, and ways of resonant excitation of evolutionary structure-attractors.

In relatively simple mathematical and computational models, a result of fundamental importance has been obtained: a continuous nonlinear medium potentially contains in itself different kinds of localization processes, that is, different kinds of structures. The medium acts both as a carrier of different forms of future organization and as a field of different evolutionary paths. In other words, any open nonlinear medium contains in itself a discrete set of possible evolutionary paths which is in keeping with a solution of the corresponding nonlinear equations (Knyazeva and Kurdyumov 2001, 2008).

The same notion follows from the original synergetic model of order parameters and slaving principle elaborated by Haken. There is a principle of circular causality which describes the relationship between the order parameters and the parts that are enslaved by the former: the individual parts of a system generate the order parameters that in turn determine the behavior of the individual parts. It can also be expressed in quite another, more anthropomorphic, form, namely: the order parameters represent a consensus finding among the individual parts of a system. Thus, the several order parameters and the several possibilities that exist for determining individual states of elements of systems reflect the fact that in complex systems only a small number of definite structures that are "self-consistent" with respect to the elements is possible. Or to put it differently, even if some configurations can be generated artificially from the outside, only a few of them are really viable (Knyazeva and Haken 1997, Haken and Knyazeva 2000).

The future states of complex systems escape our control and prediction. The future is open, not unequivocal. But at the same time, there is a definite spectrum of "purposes", or "aims", of development available in any given open nonlinear system. If we choose an arbitrary path of evolution, we have to be aware that this particular path may not be feasible for a given system. Only a definite set of evolutionary pathways is possible; only certain kinds of structures can emerge.

These spectra of evolutionary structure-attractors look much like spectra of purposes of evolution. There is a "tacit knowledge" on the part of a system itself. The spectra are determined exclusively by the inner properties of nonlinear complex systems. The future turns out to be open in the form of spectra of pre-determined possibilities. In spite of the existence of a set of possible evolutionary paths, many structure-attractors remain hidden, and many possibilities will not be actualized. Many inner purposes cannot be achieved within the given parameters of a medium. It almost looks as if many things exist in a hidden world.

Nevertheless, the evolutionary structure-attractors, as possible future forms of organization, determine the course of historical events. The future pre-determines the present. The future is in some sense available in the present. How can this be so? How can the future influence the present state of affairs? The attractors as future states are pre-determined (they are determined by the properties of a given open nonlinear medium). Patterns precede processes. They can be interpreted as a memory of the future, a "remembrance of future activities." All attempts that go beyond one of the basins of attraction are in vain. Everything which is not in accordance with the structure-attractors will be eliminated. For example, a human can fight unconsciously against those forces, i.e., those of his attitudes and plans which look like structure-attractors of his life and "pull him" from the future, but all these attempts are doomed to failure.

Hyperbolic growth. Classical science conceived instabilities as "regrettable troubles" which have to be overcome. Instabilities and small deviations from an equilibrium are something negative and destructive; they are indicators of confusion. However, the theory of complexity shows that the most astonishing events may occur exactly far from equilibrium, i.e. in extremely unstable states. Localized complex self-organizing structures may appear in such states.

The processes of rapid development were under investigation already in classical science. For example, Thomas Malthus proceeded from the supposition that the population on the Earth grows in geometrical progression, whereas the resources grow only in arithmetical progression. In the past it was usually, and still is occasionally believed that such processes of explosive development obey an exponential law. This notion is actually one of the misconceptions of classical science, because, as a rule, the processes in complex systems can occur faster, in blow-up regimes.

The theory of complexity determines conditions for self-organization far from equilibrium and discovers laws of very fast, avalanche-like growth during which meta-stable localized complex structures can emerge. Such processes of nonlinear, self-stimulating growth take place all over the space of a complex system, i.e., in all of its local domains.

Most evolutionary processes—like, for instance, population growth on Earth, growth of scientific information, or the sharp upswing of the economy,—occur in blow-up regimes. These processes obey a hyperbolic rather than an exponential law of development, i.e. due to nonlinearity they proceed faster than exponentially. It is important to understand in this context how we can initiate such very fast processes in nonlinear complex systems and what the requirements are for avoiding a possible decay of complex structures in moments of their maximal development.

Rhythms of evolution. The theory of complexity explores not only mechanisms of fast development, but various types of blow-up regimes as well. The evolution of structures undergoes a periodic change of states from one regime of development to another. There cannot be sharp growth of a structure without the threat of its fall and destruction. Certain universal laws govern these rhythms. They are peculiar to living beings as well as to inanimate nature. Some cyclical changes of state also occur: upsurge—slump—stagnation—upsurge—slump, etc; only by obeying these "life rhythms", or oscillatory modes, can complex systems maintain their integrity and develop dynamically. This mechanism of "self-movement" or auto-oscillation reminds one of oriental images—in particular the image of Yin-Yang. Yin is complete potentiality and aspiration: the subconscious, the non-verbalized and the hidden. Yang is the realized: the verbalized and the revealed.

According to models elaborated by the Moscow synergetic school, a change between the two different, complementary regimes, HS- and LS-modes with peaking, takes place in open media (systems) with strong nonlinearity. This could be considered as a mathematical analogy of the Yin-Yang alternation, i.e. changing between two complementary world elements. The HS-mode is a wave that is "infinitely running out" when there is no localization, i.e. all structures and heterogeneities are being washed away. The LS-mode with peaking is a "converging wave of heat (or burning)", a mode of localization and intensive growth of processes in a more and more narrow area near the maximum.

Evolutionary whole. The main principle of holism that "the whole is more than the sum of its parts" can be traced back to ancient philosophical studies. One of the earliest formulations of it can be found in Taoism, in the philosophy of Lao Tzu. However, a complete and profound sense of the principle has been revealed only by such theories as gestalt-psychology, systems theory, theory of complexity, and synergetics.

The research approach from the whole to the parts is quite unusual and un-conventional for classical science. It moves in the course of analysis mostly from separate parts to the whole. From the standpoint of the theory of complexity, it is order parameters which determine the behavior of the parts (subsystems) of complex systems. They allow the enormously complex description of the system under consideration, to be reduced.

The classical principle of superposition is not valid in the nonlinear world: the sum of partial solutions is generally not a solution of the equation. The whole is not equal to the sum of its parts. Generally speaking, it is neither more nor less than the sum of parts. Rather, it is qualitatively different in comparison to the parts which are integrated in it. Besides, an emerging whole alters the parts. The co-evolution of different systems means a transformation of all the subsystems by mechanisms of cooperation and exchange between the subsystems.

New principles of organization of an evolutionary whole from parts, or the formation of complex structures from simple ones, are revealed by the theory of complexity. Holism acquires an evolutionary character in this theory. A complex structure is an integration of structures of "different ages", that is: structures at different stages of evolution. The principles which govern the integration of such structures of "different ages" are gradually being revealed. The integration of simple structures into a complex one occurs by the establishment of a common tempo of development in all unified parts (fragments, simple structures). Structures of "different ages" start to co-exist in one and the same tempo-world.

Inasmuch as the appearance of non-classical science was connected, first of all, with the birth of quantum mechanics and the theory of relativity in the first third of the 20th century, the rapid development of the theory of complexity may be said to represent the next scientific revolution, i.e. the rise of a post-nonclassical scientific paradigm. This new paradigm includes patterns of nonlinear evolutionary thinking as well as new strategies of research.

One may find many non-traditional and highly astonishing notions and ideas in the theory of complexity. Turning the magic crystal of knowledge, the theory teaches us to view the world differently.

To sum up, one can express the quintessence of the theory of complexity as a source of nonlinear, holistic, complex thinking by the following theses:

1. The theory investigates evolutionary mechanisms and laws of self-organization and co-evolution of complex systems of any kind.
2. It is interdisciplinary in character.

3. Models of the behavior of complex systems possess a heuristic value for many disciplinary fields, because they are based on an understanding of the inner mechanisms of complex systems behavior in general.
4. It serves as a bridge between the natural sciences and the humanities.

18.3.4 Art of soft, nonlinear management

The classical approach to the management of human activities is based on the linear notion of the functioning of complex systems. According to this notion, the result of the governing influence is single-valued and linear. It is directly proportional to the applied efforts. It can be described by a simple scheme: *governing influence* → *desired result.*

It seems that the more energy you put in, the more recoil you get. By contrast, the theory of complexity reveals five principles of non-linear management of complex systems. This theory teaches us the art of soft management.

1. The future is open and unpredictable, but not arbitrary. There are spectra of possible future states, certain discrete sets of structure-attractors of complex evolutionary processes (Knyazeva 1999). Initial conditions do not determine the vector of human activity; the initial conditions will be forgotten when one of the structure-attractors is reached. Latent (unconscious) and revealed (conscious) attitudes determine present activities. The present state of affairs is constructed by and from the future. The soft lines of the future presuppose ways of special, soft management.
2. The art of soft management consists in ways of self-management and self-control of complex systems. How to manage a complex system without forceful management is the main problem. How can we push a system in a favorable, preferred evolutionary path by a small, resonant influence? How can we provide a system with self-maintaining and sustainable development? As a matter of fact, the notions of the theory of complexity are in accordance with the behavioral rules of Eastern people, first of all, with the principle of non-violence. "The Taoist school held that the rulers should go with natural law (the 'Tao'), governing as little as possible" (Min 1993, p. 226). Francis Bacon gave us a similar advice: "Nature cannot be ordered about, except by obeying her".
3. Soft management is management by "clever" and appropriate influences. Weak, but proper—i.e. resonant—efforts are of great efficiency. They have to correspond with the inner trends of the development of a complex behavioral system.

Correct resonant actions can lead to the revelation of tremendous inner forces and possibilities of a human being or a human community. Thus, the theory of complexity rediscovers the well-known philosophical principle: small events cause large results.

4. Certain human actions are doomed to be unsuccessful. They fail because they are not in step with the inner trends of the complex system. There are *evolutionary prohibition rules*, which are imposed on some kinds of human action. Taking into account this general consideration, we can explain quite naturally the historical (the past and probably future) failures of voluntaristic management of scientific, technical, and social progress. A management is ineffective, if it attempts to construct those structures, which are not adequate to the inner trends of an evolving system, figuratively speaking, which "violate" reality. Man has either to look for ways of changing the features of the given open nonlinear system or to give up the attempts "to force" the system to develop in an inappropriate way.

5. There must be a certain topology of action. The managing influence need not be particularly forceful, but only topologically organized in the right way. Its topological configuration—symmetric "architecture"—is important, not the intensity of the influence; these weak, but topologically correctly organized—resonant—perturbations upon complex systems are extremely effective. The resonant action is a spatially distributed action. It is a sting in the right place and at the right time.

There is an intrinsic connection between space and time in the evolutionary structure-attractors of a complex system. On the one hand, it seems that time has a topological structure. We can imagine a tree of time, a tree of historical events. On the other hand, the spatial configuration of a structure-attractor contains in itself information of both the past and the future of that structure. We could try to extract this information by simply analyzing the available spatial configurations of a given evolutionary structure. The future and the past manifest themselves by showing their "faces" in and through the present configuration of the structure-attractor. So, we can speak about the special status of the "here and now" (Knyazeva 2011).

The theory of complexity determines how it is possible to reduce the amount of time and required effort to generate, by a resonant influence, a desirable and, what is no less important, feasible complex structure. This feature of complex organizations was understood, intuitively, already thousands of years ago, by the father of Taoism Lao Tzu. He said that the weak defeats the strong, the soft defeats the hard, the quiet defeats the loud.

18.4   The Future of Complexity Studies

Complexity studies have to be self-critical towards their own future development (Knyazeva 2003). Synergetics is one of the branches in them. Synergetics in the sense of Haken was born in the late 1960s, almost 50 years ago and has developed very fruitfully during the last decades. But what is to happen further? What is the destiny of synergetics and of the whole field of complexity studies?

According to a point of view that I share, synergetics is a modern and close analogue of cybernetics which passed its highest point of development in the fifties and sixties; systems theory was under active development one decade later. As mentioned above, the development of cybernetics was accompanied by an enthusiastic mood of the scholars involved. But what happened then? Cybernetics was replaced first with general systems theory and later with synergetics. Now, what is the future of synergetics? Although during the last decades synergetics has proved to be a highly promising field of scientific research, we should admit possible limitations of the theory.

In my opinion, cybernetics as well as synergetics can be considered as certain "constellations" of developments, conclusions and arguments collected from some basic scientific disciplines, such as mathematics, physics, engineering science and technology, biology. Both of them are interdisciplinary theories. Cybernetics with its principles of feedback, purposive behavior and information processing is closer to technology and the study of the construction of automatic devices and apparatuses. Synergetics is immersed deeper, because it tries to reveal the profound evolutionary mechanisms of complex systems in general.

Such a "constellation" is determined by the logic of historical development of science. This "constellation" may take on certain forms; it may transform itself and may even diffuse for a while, thereby enriching the basic scientific disciplines. The "constellation" may be assembled, may fall apart and then be re-assembled later in a different way, whereas the basic disciplines remain in their frameworks.

Proceeding from the analysis of the historical development of cybernetics, systems theories and synergetics, one may assume that in the second part of the 20th century, a new, quite unknown type of scientific research started to take shape. It is not simply interdisciplinary, or transdisciplinary, research, though it is considered already good style to be engaged in an interdisciplinary study. Rather, it is the kind of "ferment" or "catalyst" which does not replace the basic disciplines, but stimulates the development of knowledge within their frames. One pioneering "ferment" approach may disappear giving way to another one, which is a new igniting stimulus.

This constitutes a new phenomenon in the development of science. The focus of interdisciplinary investigation may change in time, other problems and means of investigation may turn to be at the center of attention. But the essence of such an interdisciplinary structure based on the sciences of complexity in general remains the same.

As an enthusiastic scholar in the field of synergetics, I hope for the successful and prolonged prospective of synergetics as a theory of self-organization and of the co-evolution of complex systems. Perhaps, the culminating point in the development of synergetics has not yet been reached. But only history will put all things in their due places. The future is charming because of its uncertainty and openness.

Acknowledgement

I would like to express my gratitude to Wolfgang Pietsch from the Technical University of Munich and to John Hannon from the Academic Writing Center of the National Research University Higher School in Moscow for their great help in editing this article.

The article was prepared partly by the support of the Russian Foundation of Basic Research (project No. 13-06-00813).

References

Descartes, R. 1953. Œuvres et lettres. Paris: Gallimard.
Gell-Mann, M. 1995. *The Quark and the Jaguar. Adventures in the Simple and the Complex.* London: Abacus.
Haken, H. 1977. *Synergetics.* Berlin: Springer.
Haken, H. 1997. Synergetics of the Brain. In *Matter Matters? On the Material Basis of the Cognitive Activity of Mind.*, Ed. by Arhem, P., Liljenström, H. and Svedin, U., 145-176. Berlin: Springer.
Haken, H. and Knyazeva, H. 2000. Arbitrariness in Nature: Synergetics and Evolutionary Laws of Prohibition. *Journal for General Philosophy of Science* 31 (1): 57-73.
Kelso, J.A.S. 1995. *Dynamic Patterns. The Self-Organization of Brain and Behavior.* Cambridge (MA): The MIT Press.
Knyazeva, H. 1999. Synergetics and the Images of Future. *Futures* 31 (3/4): 281-290.

Knyazeva, H. 2003. Self-Reflective Synergetics. *Systems Research and Behavioral Science* 20 (1): 53-64.

Knyazeva, H. 2009. Nonlinear Cobweb of Cognition. *Foundations of Science* 14 (3): 167-179.

Knyazeva, H. 2011. The Russian Cosmism and the Modern Theory of Complexity: The Comparative Analysis. *Analecta Husserliana* (Astronomy and Civilization in the New Enlightenment) 107: 229-235.

Knyazeva, H. and Haken, H. 1997. Perché l'Impossible è Impossible. *Pluriverso:* 2 (4): 62-66.

Knyazeva, H. and Kurdyumov, S.P. 2001. Nonlinear Synthesis and Co-evolution of Complex Systems. *World Futures* 57: 239-261.

Knyazeva, H.N. and Kurdyumov, S.P. 2008. Synergetics: New Universalism or Natural Philosophy of the Age of Post-Nonclassical Science. *Dialogue and Universalism* 11-12: 39-61.

Kurdyumov, S.P. 1990. Evolution and Self-organization Laws in Complex Systems. *International Journal of Modern Physics C* 1: 299-327.

Laszlo, E. 2012. *The Chaos Point. The World at the Crossroads.* London: Piatkus.

Mainzer, K. 1997. *Gehirn, Computer, Komplexität.* Berlin: Springer.

Mainzer, K. 2007a. *Der kreative Zufall. Wie das Neue in die Welt kommt.* München: C.H. Beck.

Mainzer, K. 2007b. *Thinking in Complexity. The Computational Dynamics of Matter, Mind, and Mankind.* 5. Auflage. Berlin: Springer.

Mainzer, K. 2008. *Komplexität.* Paderborn: Wilhelm Fink Verlag.

Mainzer, K. 2011. Interdisciplinarity and Innovation Dynamics. On Convergence of Research, Technology, Economy and Society. *Poiesis & Praxis* 7: 275-289.

Mainzer, K. 2014. *Die Berechnung der Welt. Von der Weltformel zu Big Data.* München: C.H. Beck.

Min, J. 1993. Transformations in the Chinese Cognitive Map. In *The Evolution of Cognitive Maps: New Paradigms for the Twenty-first Century,* Ed. by E. Laszlo. New York: Gordon and Breach Publishers.

Morin, E. 1977. *La Methode. La Nature de la Nature.* T. 1. Paris: Editions du Seuil

Nicolis, G. and Prigogine, I. 1989. *Exploring Complexity. An Introduction.* New York: W.H. Freeman.

Nicolis, J. 1986. *Dynamics of Hierarchical Systems: An Evolutionary Approach.* Berlin: Springer.

Pink, D.H. 2005. *A Whole New Mind: Moving from the Informational Age to Conceptual Age.* New York: Riverhead Book.

Prigogine, I. and Stengers, I. 1984. *Order out of Chaos. Man's New Dialogue with Nature.* New York: Bantam Books.

Samarskii, A.A., Galaktionov, V.A., Kurdyumov S. P., and Mikhailov A. P. 1995. *Blow-up in Problems for Quasilinear Parabolic Equations.* Berlin: Walter de Gruyter-Verlag.

Turing, A. 1952. The Chemical Basis of Morphogenesis. *Philos.Trvans.Roy.Soc.* 237: 37-72.

Valéry, P. 1943. *Tel Quel.* II. Paris: Gallimard.

19. Science in an Unstable World. On Pierre Duhem's Challenge to the Methodology of Exact Sciences

Jan C. Schmidt[1]

Dedicated to Klaus Mainzer, pioneer in the philosophy of complex systems, and interdisciplinary scholar bridging scientific cultures.

Abstract

Complex systems theory—encompassing nonlinear dynamics, chaos theory, synergetics, dissipative structures, fractal geometry, catastrophe theory, and the like—is a young and fascinating field of scientific inquiry that spans many established disciplines (Mainzer 1996). Long before its full development from the end of the 1960s, Pierre Duhem, at the beginning of the 20th century, anticipated that unstable and complex systems pose challenging problems for scientific methodology. Citing groundbreaking works of Poincaré and Hadamard, Duhem relentlessly questioned widely acknowledged, implicit assumptions in the exact sciences, such as mathematical deducibility/predictability and empirical testability. Duhem was aware that the common denominator of these challenges is instability—a fact that is well known in today's complex systems theory. But Duhem did not proceed from that point; he restricted science to the domain of

---

1   Jan Cornelius Schmidt is a physicist and a philosopher. He received a Ph.D. in Theoretical Physics and a Habilitation in Philosophy. He was a Professor's Assistant at the Institute of Physics, University of Mainz (1996-1999), as well as at the Institute of Philosophy, Darmstadt University of Technology (1999-2006). Schmidt was Associate Professor for Philosophy of Technology at Georgia Tech (2006-2008). Since 2008 he has been Professor of Philosophy of Science and Technology at Darmstadt University of Applied Sciences. He was a Visiting Professor at the University of Jena (2011-2012) and invited Guest Professor at the University of Klagenfurt (2015) and at the University of Natural Resources and Life Sciences (Boku), Vienna (2016).

stable systems—which is in sharp contrast to recent complex systems theory. Thus, Duhem can be regarded as an interesting watershed figure on the way towards modern philosophy of complex systems.

19.1 Introduction

The French physicist, philosopher, and historian Pierre Duhem is not solely a central figure in the philosophy of science of the 20th century and, besides Henri Poincaré, one of the fathers of modern conventionalism, i.e., the idea that certain fundamental elements within scientific theories have to be understood as conventions. A point seldom acknowledged is that he can also be regarded as a forerunner of philosophical reflection on complex systems. He was one of the first scientists to anticipate methodological issues of science in an unstable and complex world.

Complex systems theory—encompassing nonlinear dynamics, chaos theory, synergetics, dissipative structures, fractal geometry, catastrophe theory, and the like—is a young and fascinating field of scientific inquiry that spans many established disciplines (Mainzer 1996; Schmidt 2011). Long before its full development from the end of the 1960s, Duhem anticipated that unstable and complex systems pose challenging problems for scientific methodology. Citing groundbreaking works of Poincaré and Hadamard, Duhem relentlessly questioned widely acknowledged, implicit assumptions in the exact sciences, such as mathematical deducibility and predictability on the one hand, and experimental reproducibility of empirical objects on the other, and thus the possibility of testability. Duhem was aware that the common denominator of these challenges is instability—a fact that is well known in today's complex systems theory.[2]

Although many philosophers and historians have done extensive work on Duhem's view of science—in particular on: his theory holism based on his view of the impossibility of crucial experiments (Bacon's as well as Popper's *experimentum crucis*); the theory-ladenness of observation; the underdetermination thesis; and theoretical vs. experimental language (cp. Quine 1951; Schäfer 1978; Crowe 1990; Martin 1991)—Duhem's outstanding reflection on the structure of theories and, more specifically, on theory-centered deductions has not been widely perceived. Duhem was precursory in distinguishing between two types of deductions: those that are useful to sciences and those that are not. This distinction is essential to Duhem's view of physics, in particular in contrast to mathematics. Because deduc-

2   For example, see Mainzer (1996), Batterman (2002), Kuhlmann (2007), and Schmidt (2008a; 2008b; 2011).

tions are central elements of the structure of physical theories for uncovering the consequences of a theory for the empirical-experimental level, deductions also determine whether a theory can be regarded *a priori* as empirically useful and, in particular: experimentally testable and, thus, physically relevant.[3]

Duhem's challenge to physical methodology was induced, as we can formulate from a present-day perspective, by those complex systems that are nonlinear and can cause unstable behavior and, thus, provide diverging deductions. For Duhem, the general point of departure was that physics needs to cope with uncertainties, experimental errors, and empirical vagueness; it has to enable approximations and approximate reasoning (cp. Batterman 2002). To achieve this, the mathematical structure of a physical theory has to provide non-diverging (stable) deductions; only this kind of deduction is *empirically useful*. However, although Duhem was aware of the existence of unstable systems in the physical world, he restricted physical sciences to the domain of stability. He believed that stability has to be built into the structure of deductions: Deductions, he maintained, are required to be non-diverging in order to enable approximate conclusions, reasoning, and predictions and, thus, to ensure the possibility of matching theoretical facts with the empirical-experimental world.[4] Considering the development of complex systems theory, this view turns out to be mistaken.

If one views Duhem through the lens of the recent state of the art in the sciences, he can be regarded as an interesting watershed figure on the way towards modern philosophy of complex systems.[5] In the following I will examine Duhem's concept of science not from a general perspective but from the perspective of stability/instability, and while doing so address Duhem's understanding of deductions and, hence, of the mathematical structure of physical theories (Sec. 2-6). Based on this analysis, I will expose Duhem's approach to today's complex systems theory (cp. Mainzer 1996/2005/2008; Schmidt 2008a/2011) (Sec. 7). In approaches such as Duhem's we can identify a "dogma of stability" (Guckenheimer/Holmes 1983, 256). Duhem mainly perceived a threat to physical methodology, and therefore he stuck to the dogmatic stability requirement; he did not realize that this threat is a challenge that could also be the initial point to advance science in a novel direction.

---

3   Batterman (2002) argues in his book "The Devil is in the Detail" from the same angle, but without reference to Duhem.

4   Duhem speaks explicitly about "divergence" (Duhem 1991, 139).

5   The main idea behind this paper was inspired by three short remarks—one by Ilya Prigogine and Isabelle Stengers (1990, 316) in "Dialogue with Nature", one by Jules Vuillemin (1991, xxviii) in his introduction to the English edition of Duhem's "Aim and Structure of Physical Theory", and one by David Ruelle (1994, 64f) in his "Chance and Chaos". Even the historian of mathematics David Aubin (1998) mentions this point merely in passing and, thus, misses the main issue that is involved here.

In the last 40 years complex systems theorists have replaced the dogma of stability by other requirements (Sec. 8).

In sum, Duhem urges us to rethink our thinking. He can be regarded, as will be shown, as a somewhat ambivalent precursor to a philosophy of complex systems theory—in particular to the work of Klaus Mainzer that has earned worldwide attention and is profoundly acknowledged in both the sciences and the philosophy of science.

19.2   Duhem's Aim

Right from its publication, Duhem's main work *Aim and Structure of Physical Theory*[6] was considered a pacemaker towards modern philosophy of science. Besides Poincaré, Duhem has been credited with advocating a conventionalist picture of exact sciences that is based on a holistic understanding of a physical theory.[7] Duhem formulated his famous underdetermination thesis that crucial experiments aiming to sort out false theories or false theory elements are impossible: Theories are not determined by empirical facts; scientists have to make decisions and, therefore, science needs to be seen as an inherently social enterprise.[8]

More specifically, contrary to the view held at the Vienna Circle, the problem of inductivism is obvious to Duhem:

"[W]e can make infinity of different formulas or distinct physical laws correspond to the same group of facts." (Duhem 1991, 169) "[...] it is impossible to construct a theory by a purely inductive method." (ibid., 219). "[A] set of experimental laws does not [...] suffice to suggest to the physicist what hypotheses he should choose in order to give a theoretical representation of these laws." (ibid., 255)

Duhem's critique of inductivism also constitutes the basis of his underdetermination thesis, however his thesis goes beyond that: Underdetermination is, indeed, not just

---

6   The title *Aim and Structure* can be regarded, as Vuillemin (1991, xxvi) points out, as a substitution for the most common terms such as *goals* and *means*. By choosing such a title Duhem is also stressing that he regards science as a collective human action.

7   Indeed, there are big differences between Poincaré's and Duhem's account of conventionalism. This will not be reflected on in detail in this paper.

8   Duhem states that scientists need discernment or "good sense" ("bon sens") in order to accomplish this selection. Social constructivists frequently refer to Duhem's underdetermination thesis.

another form of the problem of induction (Humeian problem). According to Duhem, theories are not determined—and also not rejected—by experimental evidence alone; a freedom of choice remains.[9] If an anomaly emerges, it is unclear which kind of hypothesis or theory element should be rejected. "An experiment in physics can never condemn an isolated hypothesis but only a whole theoretical group" (ibid. 183). A Baconian *experimentum crucis* is not feasible for eliminating a particular hypothesis or for refuting a specific theory element—which also poses problems for any falsificationist philosophy of science. Thus, Duhem argued against both inductivism and many varieties of deductivism (Schäfer 1978; Wilholt 2012). He maintained that theories are always encompassed by a group of other theories; theories have to be regarded as a kind of seamless web.[10] Generally, Duhem's holism posed, and still poses, challenges to any modern philosophy of science and to any present-day epistemology.

In line with a view most prominently developed by the Vienna Circle—besides his critique of inductivism—, Duhem firmly upheld the idea of separating physics from metaphysics and supported an instrumentalist, non-explanatory account of theory. He characterized the aim of a theory in a similar way to Ernst Mach's "economy of thought" and was frequently accused of being a positivist.[11] According to Duhem, the aim of a theory is to provide a "natural classification" of empirical facts and their phenomenological regularities ("laws"). A theory is best understood as a synthetically consistent, common umbrella in order to economically subsume empirical laws,[12] whereas a law is seen as a symbolic and approximate relation combining symbolic terms in a quantitative mathematical way to describe empirical phenomena (cp. ibid., 165f/168).

"A physical theory is not an explanation. It is a system of mathematical propositions [...]. Instead of a great number of laws offering themselves as independent of one another, each having to be learnt and remembered on its own account,

9 This requires again "good sense" of the physicist.
10 Later on, Quine gave more substance to Duhem's idea of underdetermination and developed the well-known *Duhem-Quine thesis*. In general, the underdetermination thesis can be considered as one main fundament of obviously different positions, such as conventionalism, instrumentalism, and also constructivism.
11 Theories aim to describe and "save phenomena". They rarely do more than that; they do not provide an explanation and have no explanatory power; explanation is not a concern of science, but rather of metaphysics. Theories relate to the phenomena (only) in a "symbolic", "approximate", and "provisory" way. But this does not mean that theories are arbitrary. In the process of being perfected during the course of its further development throughout history, physical theory progressively takes on the character of a "natural classification" of phenomena. By this, Duhem implicitly stresses that we should attribute to a physical theory a "deeper" significance than that of a mere methodological classification of facts already known.
12 Duhem talks of "reduction of physical laws to theories" (ibid., 21).

physical theory substitutes a [...] number of propositions [by ...] fundamental hypotheses. [...] Such condensing of a multitude of laws into a small number of principles affords enormous relief to the human mind [...]. The system of hypotheses and deducible consequence [...] constitutes a physical theory in our definition." (Duhem 1991, 19/21/55)

The aim of a theory is to reduce or condense the description length of a phenomenon and to provide a condensed "natural classification" of experimental laws. This classification[13] is based on physical magnitudes that are to be regarded as the result of symbolic abstraction by which we bring a number or a set of numbers into relation with an empirical phenomenon.[14]

However, there is no final or fixed one-to-one correspondence between the symbolic-theoretical and empirical-physical level. Every theory remains "provisional and relative because it is approximate" (cp. ibid. 172). The obvious evidence in favor of the provisional and relative character is given by the historical development of science.

"The provisional character [...] is made plain every time we read the history of this science. [...] Any physical law, being approximate, is at the mercy of the progress which, by increasing the precision of experiments, will make the degree of approximation of this law insufficient: the law is essentially provisional." (Duhem 1991, 172f.)

Thus, finding a unique relation between a law on the one hand and an experimental phenomenon on the other hand is impossible, not solely because of the conventions involved, but also because of the different degrees of approximation that can be achieved and among which there is to choose. Laws approximate experimental facts in a quantitative sense—it is a process of assigning a symbol and, in particular, a magnitude or value. The degree of approximation might change when experimental technology advances, and new apparatuses and new experimental setups are developed.[15] So, part of the progress of physics is induced by the development of increasingly better experimental precision.

---

13 Although Duhem is clearly against metaphysics he believes that an ontological structure of reality is given: "[I]n the physicist's theory there is something like a transparent reflection of an ontological order" (Duhem 1991, 298).

14 Temperature, for instance, is a symbol of primary quality.

15 Duhem (ibid., 172) stresses that the "degree of approximation is not something fixed; it increases gradually as instruments are perfected [...]. As experimental methods gradually improve, we lessen the indetermination of the abstract symbol brought into correspondence with the concrete fact by physical experiment." Therefore, indetermination is not an invariant.

19.3   Structure of Theories: Deductions, Bundles, Translations

Generally speaking, theories organize knowledge and condense the empirical data of a phenomenon to a shorter description length.[16] Central to theories is that they enable deductions: Deductions are methodologically indispensable elements to uncover the implications of a theory for the empirical level; they are part of the theory and built-in in the mathematical structure of the theory (cp. Duhem 1991, 55).[17]

A point that has scarcely been reflected on by philosophers of science is Duhem's strong claim that the methodology of physical sciences requires specific *physically useful or utilizable* deductions—that is, deductions that enable the physicist to link the theoretical with the empirical level, and to compare both.[18] In a very general sense, Duhem characterizes deduction as a symbolic-nomological conclusion scheme—based on (a) hypothesis or general laws, and on (b) facts ("circumstances") representing the empirical level, such as data, e.g., about initial and boundary conditions.[19]

"Deduction is an intermediary process; its object is to teach us that, on the strength of the fundamental hypotheses of the theory, the coming together of such and such circumstances will entail such and such consequences; if such and such facts are produced, another fact will be produced." (Duhem 1991, 132)

---

16   Duhem regards a theory as the result of a twofold process (cp. ibid., 55). (1) *Abstraction:* Laws, and thus theories, subsume several practical ("concrete, diverse, complicated, particular"; ibid., 55) facts; they form an umbrella of different empirical propositions. Laws are abstractions of concrete practical facts. Duhem calls this subsumption an "abstraction". (2) *Generalization:* A group of laws is itself substituted by "a very small number of extremely *general* judgements" (ibid., 55): the theory, by referring to hypotheses.—This twofold translation process is regarded as a double "reduction": the "reduction of facts to laws and the reduction of laws to theories" (ibid., 55). This will "truly constitute intellectual economies" (ibid.). In any reduction, several simplifications and approximations are inevitable.

17   Duhem frequently uses terms such as deduction, conclusion, and prediction synonymously.

18   Duhem explicitly deals with deductions in chapter three of part II of his book *Aim and Structure of Physical Theory* ("Mathematical Deduction and Physical Theory," 132ff.).

19   All this takes place on the theoretical level: From a "theoretical fact" follows, by deduction, another theoretical fact. This thesis was later referred to as the structural identity thesis of deduction, explanation, and prediction.

Two kinds of facts need to be considered: those that can be empirically and experimentally observed, and those that we find in the realm of a theory. Duhem calls the former "practical" or "experimental facts", and the latter "theoretical facts" or "mathematical data" (ibid.). Theoretical facts are precisely given;[20] in the case of practical facts we no longer see "anything of the precision". For instance, "The body is no longer a geometrical solid; it is a concrete block. However sharp its edges, none is a geometrical intersection of two surfaces" (ibid., 134).[21]

When we attribute, for instance, an error interval to a practical fact, measured in an experiment, we also characterize it by quantitative symbols: we construct various theoretical facts. Duhem assumes a non-*one-to-one* relation between both kinds of facts, which he calls "translation". Two types of language are involved here: the "experimental language" of practical facts and the "language of numbers" of theoretical facts (ibid., 133).[22]

"Between the concrete [= practical] facts, as the physicist observes them, and the numerical symbols by which the [practical] facts are represented in the calculations of the theorist [e.g., the theoretical facts], there is an extremely great difference" (ibid., 133).[23]

---

20 For example, the temperature distributed in a certain manner over a certain body or, in general, the body studied, is geometrically defined in a precise manner and its sides are true lines without thickness. Duhem talks of a "translation": "Opposite this theoretical fact let us place the practical fact translated by it" (ibid., 134).

21 Theoretical facts are characterized as symbolic; practical facts do not possess a unique numerical or symbolic structure. They are concrete objects in the real world. From a quantitative (exact, pure mathematical, theoretical) perspective the practical fact is somewhat hidden.

22 The Vienna Circle later distinguished "Theoriesprache" (theoretical/theory language) from "Beobachtersprache" (observational language).

23 This is not to be viewed from a classic representationalist perspective; according to Duhem it is impossible to represent *one* practical fact by *one* theoretical fact, and vice versa. "Translation," Duhem argues, "is treacherous: *traduttore, traditore* (to translate is to betray)" (Duhem 1991, 133).

From a mathematical point of view, theoretical facts are precise, whereas practical facts do not exhibit such precision; they do not exist in a unique symbolic or clear numerical way. They are always

> "misty, fringed, and shadowy. It is impossible to describe the practical fact without attenuating by the use of the word 'approximately' or 'nearly' whatever is determined too well by each proposition." (ibid., 134)

Theoretical facts can be stated, for example, as lines having a precise length: 1 cm or 0.999 cm or 1.002 cm. Obviously, these are different theoretical facts, consisting of different symbolic quantities: Due to empirical uncertainties and measurement errors, *one* practical fact does not correspond to only *one* theoretical fact, but to an infinity of different quantitative judgments. There is non-uniqueness in the translation between both.[24] Each practical fact has to be described by a "bundle of theoretical facts," as Duhem calls this empirically induced vagueness and the involved uncertainties (ibid., 136).[25] "Bundle" is a key term in Duhem's concept of physical methodology: Empirical errors need to be taken into account by a physicist; quantitative error estimation, error theory, statistical analysis, and approximation theory are common tools of any empirical scientist.

The bundle concept is interlaced with translations and deductions—and their interplay. Three steps are involved here: *First*, as already mentioned, we need to consider a "translation" between the practical-experimental level and the theoretical facts: Practical facts, including the "experimental conditions, given in a concrete manner, are translated by a bundle of theoretical facts" (ibid., 136).[26] For instance, the pressure condition P existing at a particular ice block may vary between 9.95 and 10.05 atm.[27]—*Second*, an initial bundle of theoretical facts is now given by the values between 9.95 and 10.05 atm. They constitute the initial conditions that are

---

24 *Different* theoretical facts can serve as a translation of *one* practical fact. This can also be regarded as a variation of one of the background premises of the well-known thesis of underdetermination (ibid., 169).

25 In many standard cases this point is represented by specific error curves or uncertainty distributions, e.g., Gauss's curve.

26 Duhem is very concerned about the translation process that is linked to the accuracy of measurement: "This evaluation may also depend on the sensitivity of the means of measurement used to translate into numbers the practically given conditions of experiment" (ibid., 137). The choice of measure functions, metrical conventions, methods of measurement, and instrument-specific errors cannot be neglected.

27 This translation should not be understood as an empirically based inductive methodology.

part of any kind of deduction; the deduction, which is central to the "development of the theory", maps that initial bundle to a second one: "The mathematical development of the theory correlates this first bundle of theoretical facts with a second, intended to stand for the result of the experiment" (ibid., 136). Deduction is, evidently, an operation within the domain of a theory; it is central to the development of a theory (ibid., 136/7).[28]—*Third*, there is a second translation, or more specifically: a *re*translation, from the level of theoretical facts to practical facts in order to reach the experimental level.

> "[T]his new translation [is] intended to translate theoretical into practical facts, the inverse of the one with which we first concerned ourselves." (ibid., 136)[29]

This *re*translation is especially important. After having deduced theoretical facts (second bundle) from other theoretical facts (initial bundle) and after having retranslated these deduced facts to the empirical level, the physicist confronts these theory-deduced and translated facts with his empirical observation or experimental result (cp. ibid., 28). Duhem stresses the necessity of comparing the deduced consequences of a theory with experimental data, which he refers to as "testing a theory" (ibid., 21; cp. 180/219).[30] The deductions and translations drawn from theories

> "can be submitted to test by facts [obtained by the experiment]. If [...] there is among these consequences one which is sharply in disagreement with the facts

---

28   Deduction can always be considered as a determination of *one* exact proposition from another exact initial one. Expressed in classical words: The *explanans* comprises the initial conditions of the (deterministic) law or hypothesis; the explanandum follows via deduction. Duhem seems to have had in mind this kind of theory-based or hypothesis-based operation. But, it also holds that: "Deduction will not forecast the experimental result in the form of a unique theoretical fact but in the form of an infinity of different theoretical facts" (ibid., 135).

29   Duhem seems slightly unsure about whether to call the whole process (translation, deduction, retranslation) "prediction" or not. Also, Duhem uses the words "calculation", "prediction", and "forecast" interchangeably (ibid., 135). For example, he sometimes refers to deduction as "the calculation of the theorist" (ibid., 135; cp. ibid. 139). Duhem therefore advocates what was later called the structural identity thesis of deduction, description/explanation, and prediction (cp. Salmon 1989).

30   Or, in other terms, "the comparison of the theory with experiment" (ibid., 21). In general, four different operations are involved in the process: "(1) the definition and the measurement of physical magnitudes; (2) the selection of hypotheses; (3) the mathematical development of the theory [involves the three steps (translation, deduction, retranslation) just mentioned]; (4) the comparison of the theory with experiment" (ibid., 21).

[of the experiment …], [the theory] will have to be more or less modified, or perhaps completely rejected." (ibid., 28)[31]

An interesting point here is that Duhem's concept of science does not prohibit speaking about "truth".[32] "Truth" is attributed to "good" and "evident" theories that provide a "provisional" "natural classification" of phenomena. A theory is said to be "true" *if and only if* it describes *approximately* a group of practical facts (or experimental laws) in a "natural" and approximate manner and entails acceptable deductions and predictions. "Reality" is the sole criterion for a good approximation; Duhem did not believe in any form of idealism but rather in the existence of a structural reality external to man (cp. Psillos 1999, 149f).[33] Nevertheless, this does not mean that reality itself provides precise and unique suggestions that would enable us to select theories by simply following a (kind of algorithmic or mechanistic) rule: "Physics is not [proceeding like] a machine"—as symbolic constructions, choices, judgments, and "bon sens" ("good sense") play a key role (Duhem 1991, 187).

---

31   According to Duhem, the confrontation of a theory with reality is somewhat *indirect*; it is via (the detour of empirical-phenomenological) laws. A theory does not refer to reality directly: this is Duhem's well-known non-referential thesis. It only "represents […] a group of experimental laws" (ibid., 20). Laws are closer to the phenomena and to the practical facts than abstract hypotheses and symbolic theories. An example of such a law is Kepler's law, whereas the theory in this case is Newton's Classical Mechanics or Einstein's General Relativity; the non-referentialist thesis also figures prominently in Cartwright's writing (Cartwright 1983).

32   See also Vuillemin (1991, xxxii f.).

33   According to Duhem, "agreement with experiment is the sole criterion of *truth* for a physical theory" (ibid., 21; see also: ibid., 167f). But he also believed that an experiment is not simply an observation of a phenomenon. It is based on abstract and symbolic judgment, on hypotheses and theories; the experiment and any observation is theory-laden (cp. ibid., 28f./144f.). "If this agreement between the conclusions of theory and the facts of experiment were not to manifest a satisfactory approximation, the theory might well be logically constructed, but it should nonetheless be rejected because it would be contradicted by observation, because it would be physically false" (ibid., 206). It is interesting to see that, Duhem therefore does not argue here, as Mach and as he himself does in other parts of his book, in favor of the criterion of simplicity or economy of thought as the main truth criterion.

19.4   Duhem's Discovery of Diverging Bundles through Unstable Deductions

Indeed, physics is an empirical-experimental science; it differs from mathematics. Duhem was pioneering in making a distinction between accuracy and vagueness, between precision and approximation, between a single fact and a whole bundle. By this distinction he was able to anticipate what has today become known as sensitive dependence on initial or boundary conditions—that are central characteristics of dynamical and structural instability (cp. Mainzer 1996; Schmidt 2011).

As mentioned above, Duhem argues that it is not a single deduction, but deductions with respect to bundles, encompassing very many theoretical facts, which constitute the methodological center of physics.[34] In most standard cases of physical systems, Duhem states, it holds that

"the bundle of infinitely numerous theoretical facts by which mathematical deduction assigns to our experiment the result that should be produced will not furnish us after the [re-] translation with *several* different practical facts, but only with *a single* practical fact. It may happen, for instance, that two numerical values found for the letter T never differ by even a hundredth of a degree, and that the limits of sensitivity of our thermometer is a hundredth of a degree, so that all these different theoretical values of T correspond practically to one and the same reading on the scale of the thermometer." (Duhem 1991, 136)

In this case, as Duhem asserts by referring to the measurement resolution as well as to the mathematical structure of the deduction, "deduction will have made possible the comparison of the consequences of the theory with the [practical, empirical] facts" (Duhem 1991, 136).

However, the issue is that there are other cases in which a single practical fact will *not* be found, but several practical facts may coexist. These cases can only be handled to some extent. According to Duhem, scientists usually make the optimistic and somewhat naïve assumption that they can improve the translation-deduction-retranslation procedure and thereby reduce the narrowness of the bundle by improving the experimental setup and the measurement precision in order to obtain a convergence—in particular: a *single* practical fact. In other words, reducing measurement uncertainties seems to be merely a technical-experimental problem,

---

34   While deductions belong to the realm of theories, they are the basic elements that enable theories to be related to the concrete physical world.

simply a matter of the state-of-the-art and the progress of experimental technology. Physicists core presuppose that they can

"increase [...] the precision of the methods of measurement that were used to translate the practically given conditions of an experiment into theoretical facts; in that way, [they ...] have tightened more and more the bundle of theoretical facts which this *translation* correlated with a *single* practical fact. At the same time [they assume to] have also tightened the bundle of theoretical facts with which our mathematical *deduction* represents the result predicted for the experiment [= re-translation]; it has become narrow enough for our method of measurement to correlate it with a *single* practical fact, and at that moment our mathematical deduction has become useful." (Duhem 1991, 138)

At first glance,

"[i]t seems as if it should always be so. [...] Whatever narrowness is needed for the bundle of theoretical facts which we wish to obtain as a result, [...] deduction will always be able to guarantee it that narrowness, provided that we tighten sufficiently the bundle of theoretical facts representing the data given." (Duhem 1991, 138)[35]

The core question that Duhem raises here is: What can happen to a bundle on applying mathematical deduction? As he shows, the narrowness of bundle mapping—by the deduction from the first (initial) bundle to a second one—cannot be taken for granted; a preservation of narrowness is not guaranteed by itself. Duhem brands the assumption of a preservation of narrowness as "nothing but a deception": "we can cite cases where it is in plain contradiction with the truth" (ibid., 139). Owing to the structure of a theory, bundles of the theoretical facts can diverge on applying mathematical deduction.

"Although the first [initial] bundle [representing the initial practical fact] is infinitely narrow, the blades forming the second bundle [can] diverge and separate out without being able to reduce their mutual deviations below a certain limit. Such a mathematical deduction is and always will remain

---

35   The notion of "narrowness" refers to the inner structure of the bundle, more specifically to the neighborhood of theoretical facts within a bundle that could be specified in a set- and measure-theoretical way. Complex systems theorists also speak here about "likeness" and "similarity" in the sense of facts (or points) in the neighborhood.

useless to the physicist; however precise and minute are the instruments by which the experimental conditions will be translated into numbers, this deduction will still correlate an infinity of different practical results with practically determined experimental conditions, and will not permit us to predict [and to conclude] what should happen in the given circumstances." (Duhem 1991, 139)[36]

In other words, Duhem argues that we need to consider cases in which the narrowness is not preserved on applying mathematical deduction: *similar* theoretical facts within the first bundle, located within a neighborhood (narrowness), can diverge through deduction. They will not remain *similar or like* facts within the second, or deduced, bundle.[37] In line with Duhem's terminology and, in addition, by referring to recent complex systems theory, we call this kind of deduction "diverging" or "unstable" because it is based on instabilities generated by the mathematical structure of hypotheses or general laws.[38] Let us now look at how Duhem supports his thesis that diverging (unstable) deductions exist and pose challenges to empirical sciences.

---

36   Duhem asserts that "a great many problems well defined for the mathematician lose all their meaning for the physicist" (ibid., 141). In addition, with regard to the terminology Duhem uses here, it is worth stressing that, within the framework of the deductive-nomological scheme of explanation, prediction is often formally parallelized with conclusion (structural identity thesis). As widely known, this logical equivalence was later formalized by P. Oppenheim and C.G. Hempel (cp. Hempel 1965).

37   Duhem expands his thoughts on what he calls "facts" to general propositions and statements. Generally speaking, a "deduction is of no use to the physicist so long as it is limited to asserting that a given *rigorously true* proposition has for its consequence the *rigorous accuracy* of some other proposition" (Duhem 1991, 143). Duhem stresses convincingly that if deductions diverge, there is no possibility of relating the deduced propositions to experimentally gained practical facts in a quantitative manner; empirical tests are rendered impossible.

38   Duhem explicitly says that a bundle can "diverge" through deduction (cp. Duhem 1991, 139), but he does not mention "stable" or "unstable" deductions. Such terminology is, however, used by Vuillemin (1991, xxviii), who argues convincingly that Duhem "limits his reflections to extrapolation concerning the stability of systems."

19.5 Poincaré, Hadamard, Maxwell: Physics vs. Mathematics

Diverging (unstable) deductions are not just an abstruse invention of an armchair philosopher of science. From the development of complex systems theory over the last 50 years, the divergence is well known as the "butterfly effect" or, synonymously, as "sensitive dependence on initial and on boundary conditions", "dynamic instability" (deterministic chaos) and "structural instability" (bifurcations, phase transitions) (see Fig. 19.1; cp. Mainzer 1996/2014; Schmidt 2011/2015). Duhem was a pioneer in anticipating instabilities as possible aspects of the structure of theories.[39] To give further substance to his thesis of the existence of diverging deductions, Duhem considers the work of his contemporaries such as Henri Poincaré and Jacques Hadamard, and especially their analysis of the mathematical structures of theories and, hence, of deductions.

Based on their works, Duhem presents examples of deductions that cannot be utilized by a physicist at all. These examples can be drawn from the lessons of classical physics.[40] A *single* event, a unique state or, equivalently, a single "theoretical fact"—in Duhem's terminology—can be regarded as an *exactly* given point[41] that is governed by a differential equation. A differential equation is seen as the skeleton of any hypothesis or general law; therefore, it constitutes the center of a theory and, further, is a basic element of any kind of deduction. From one theoretical fact (initial point) others can be deduced. Duhem talks explicitly about a "trajectory" (the time evolution of the system), and ties this term to deduction:[42] "[M]athematical deduction can determine the trajectory of this point" (ibid., 141).

---

39   According to the mathematical physicist David Ruelle, this kind of methodological reflection on unstable deductions by Duhem is pioneering (Ruelle 1994, 64f).

40   Duhem says that we can "borrow" these cases "from one of the simplest problems that the least complicated of physical theories, mechanics, has to deal with" (Duhem 1991, 139).

41   From a cognate perspective we can also consider here a *precise* boundary or initial condition.

42   In this regard, Duhem mentions differential equations. If one were to question "differential calculus [...], theoretical deduction would be stopped in its tracks from the start" (ibid., 208). Differential equations are a central element of deductions, as is, for operation and performance of the deduction, the theory of differential calculus.

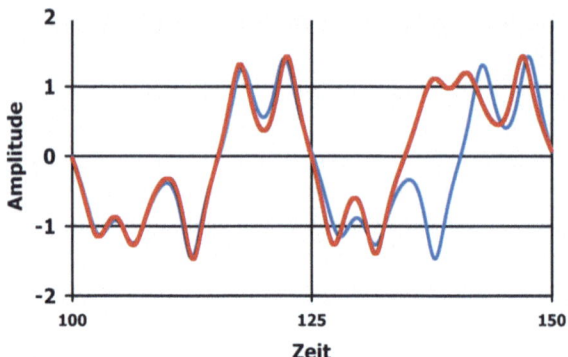

Fig. 19.1: Dynamical instability: This kind of instability is also known as the "butterfly effect". This system is sensitive to initial conditions and shows dynamical instability
("deterministic chaos"). Two distinct, but very close initial conditions show very
different trajectories during the time evolution; bundles diverge. Duhem identifies
the divergences by referring to bundles.

To explicate the existence of diverging (unstable) bundles, Duhem sets forth, together with some of Poincaré's work, Jacques Hadamard's (1898) construction of
certain kinds of trajectories (geodesics) on surfaces with negative curvature.[43] Topologically, the surface resembles the forehead of a bull with horns and ears which
are expanded mathematically to infinity. This surface also generates instabilities:
Given two distinct initial points[44] arbitrarily close to one another in a neighborhood ("bundle"), the trajectories—that Duhem interprets as being central to the
deductions—starting from these points can diverge during time evolution, at a rate
characteristic of the system.[45] That is to say: *one* trajectory may be attracted by the
one horn and proceed to infinity, while *the other* may go to the *other* horn and may
be periodic. Hadamard's construction shows a property that later became known

---

43 He takes an example from classical mechanics with trajectories on a specific surface of a
    differentiable manifold. Hadamard looks at a specific example: trajectories on a surface
    of negative curvature with multiple connections.
44 This means, in another terminology, two theoretical facts or two states in the phase
    space.
45 Different types of trajectories can be distinguished: (a) some of them are periodic or
    almost periodic, (b) some go to infinity at the one or the other horn, (c) some alternate
    between the two horns.

as "sensitive dependence on initial condition". Like Poincaré before, Hadamard anticipates this property:

"Any change, no matter how small, brought to the initial direction of a geodesic [= trajectory] [...] is enough to bring about absolutely any variation to the final outlook of the curve."[46]

Physicists are somewhat uncomfortable with this situation. Induced by instabilities, or by sensitive dependencies, the bundle can diverge on applying a mathematical deduction or, equivalently, during the time evolution of the system: preservation of the neighborhood of "theoretical facts" that are distinct but very close to each other does not occur.[47] The divergence of the bundle leads to very different theoretical facts. Duhem states that any

"deduction can [from a mathematical perspective] determine the [single] trajectory of this point and tell whether this path goes to infinity or not. But, for the physicist, this [kind of diverging] deduction is forever unutilizable." (Duhem 1991, 141)[48]

Duhem identifies as a central element of the "deduction" the calculation of a trajectory or the time evolution of the dynamical system.

In order to further elaborate on this finding, Duhem presents another case: the famous problem of the three (or n) bodies of the solar system, for instance, the sun-earth-moon, to which Poincaré contributed work in the 1890s (Poincaré 1892/1914). Duhem calls it "the problem of stability" and situates his considerations in the context of stability and instability (cp. Duhem 1991, 142ff). Again, the physicist is somewhat uncomfortable with this situation because bundles diverge.

"The practical data [taken from the experiment] that he [= the physicist] furnishes to the mathematician are equivalent for the latter to an infinity of theoretical data [= theoretical facts], neighboring on one another but yet distinct

---

46   My translation of Hadamard (1898).
47   If the initial conditions are not given by mathematical precision but practically ("practical facts"), a certain trajectory will not be determined without any quantitative uncertainty. This is equivalent to the situation in which—as Duhem argues—any practical fact is not given with accuracy but with some amount of empirical uncertainty, which invoked the bundle concept of theoretical facts.
48   It is worth underlining once more that Duhem uses the terminology of classical mechanics and also that of today's complex systems theory, such as "trajectory".

[and, thus, forming a bundle]. Perhaps among these data, there are some that would eternally maintain all heavenly bodies at a finite distance from one another, whereas others would throw some one of these bodies into the vastness of space. If such a circumstance analogous to the one offered by Hadamard's problem should turn up here, any mathematical deduction [… with regard] to the [question of the] stability of the solar system would be for the physicist a deduction that he could never use." (Duhem 1991, 142 f.)

Although diverging (unstable) deductions match *single* theoretical facts to other *single* theoretical facts, the neighborhoods within the bundles are not preserved. The mapping of *single* theoretical facts is a necessary but insufficient condition for a physically relevant methodology.

Duhem is clear on the point that diverging deductions cannot be considered as mere artefacts of mathematics. Moreover they constitute the structure of physical theories referring to the real physical world.

"One cannot go through the numerous and difficult deductions of celestial mechanics and mathematical physics without suspecting that many of these deductions are condemned to eternal sterility." (Duhem 1991, 143)

Therefore, the existence of diverging deductions as central elements of the mathematical structure of theories is nothing but a consequence of unstable systems in the physical world.

Certainly, there were also precursors to Duhem in the 19th century, such as James C. Maxwell. Maxwell stands out for his major contribution to electrodynamics—he is less known for his works on classic mechanics (Maxwell 1991). Duhem does not refer to Maxwell directly, and it is not sure whether Duhem was aware of Maxwell's very brief, but far-reaching and thought-provoking remarks on instability. Even as early as in the 1870s, Maxwell himself was very clear on the issues discussed above. He was one of the first physicists to identify a "general maxim of physics", as he called it, which requires stability in order to constitute a methodology utilizable in physics—that is, a methodology in which the non-divergence of bundles and, thus, the likeness of facts (initial theoretical facts: "cause"; second bundle of theoretical facts: "effects") is guaranteed.

"There is […] a maxim [or assumption of stability] which asserts that, *like* causes produce *like* effects'. [But] [t]his is only true when small variations in the initial circumstances produce only small variations in the final state of the system." (Maxwell 1991, 13)[49]

In a similar vein as Duhem, Maxwell views this maxim as being necessary in order to be able to formulate laws and obtain theories.

"[O]nly in so far as stability subsists [...] principles of natural laws can be formulated." (ibid., 13f)[50]

Like Duhem, Maxwell is explicitly aware that stability is not always present and, that because of this lack, the methodology of physics is bound to be threatened. According to Maxwell, weather dynamics can serve as a paradigm for an unstable system.

"Insofar as the weather may be due to an unlimited assemblage of local instabilities, it may not be amenable to a finite scheme of law at all." (ibid., 14)

But Maxwell did not subject this methodological problem, induced by instabilities, to further considerations. He did not articulate deeper reflections on the matter or, more fundamentally, come up with a solution.

Taking Maxwell's observation further, Duhem was among the first to recognize here a fundamental challenge that needs to be tackled.[51] Duhem's profound thinking was confirmed within the last four decades of the 20th century by the most recent development in complex systems theory—although hardly any complexity theorist or philosopher of complex systems theory makes reference to Duhem in this regard.[52]

In accord with John Guckenheimer and Philip Holmes (1983, 259) more than 50 years after Duhem's time, we have to concede that, since many physical phenomena are unstable and induce diverging trajectories, "details of their dynamics, which do not persist in perturbations, may not correspond to testable [...] properties." To the extent that "test" in this context denotes a measure-dependent quan-

---

49 From a different angle, similar formulations regarding the relation and the interaction between *res cogitans* and *res extensa* can be found in Descartes' works (cp. Leiber 1996).
50 Maxwell contradicts Laplace and argues against ontological determinism: "It thus puts a limitation on any postulate of universal physical determinacy such as Laplace was credited with" (Maxwell 1991, 13f).
51 This point will be elaborated on in the next paragraph.
52 One exception is Rueger/Sharp (1996).

titative agreement of facts obtained by the theory-based procedure of translation, deduction, and retranslation ("theoretical facts") with the experimentally measured practical facts, such *point by point testing* is impossible: theoretically deduced and experimentally observed facts are not correlatable point by point: An unstable phenomenon giving rise to a certain data sequence *on one side*, and an unstable theory, with which the physicist aims to approximate the phenomenon's unstable behavior, *on the other side*, differ quantitatively.

> "[N]onlinear theory itself shows that 'point by point testing' is inadequate [...]. If we [try to] test a theory in this [traditional] way, we will not find a precise quantitative fit, and this is to be expected if the theory is 'true' of the system. Because of [...] sensitivity to initial conditions in chaotic regimes, no individual trajectory can be compared with experiment or with a computed orbit." (Rueger/Sharp 1996, 103)

In line with this statement, Robert Batterman (2002) argues that, whenever instabilities are present, due to the involved sensitivities the "devil is in the detail" and "approximate reasoning" is limited. According to Brigitte Falkenburg (2013), unstable and complex systems do not equip us with the possibility to find "mechanical explanations" that are based on deductions.[53] Is there a way out?

19.6   Duhem's Solution: Impose what is not given!

According to Duhem, methodology can, in general, be regarded as a normative enterprise which is indispensable to the foundation of sciences. This is the guiding idea of any type of conventionalist or, going further, constructivist picture of science. Following Duhem's line of thought, we can now ask what should be done if not all types of deductions are non-diverging—and if, at the same time, this challenges scientific methodology? According to a general deductivist's approach, such as Duhem's,

> "Deduction is sovereign mistress, and everything has to be ordered by the rules she imposes." (Duhem 1991, 266)

---

53   This also holds for quantum-mechanical systems.

That is to say, if something regarded as being indispensable is not given it has to be demanded. Facing the issue of diverging deductions, Duhem argues that the scientist is normatively

"bound to impose rigorous conditions [...] on mathematical deduction": "To be useful to the physicist, it must still be proved that the second proposition remains approximately exact when the first is only approximately true." (Duhem 1991, 143)[54]

Science requires non-diverging (stable) deductions in order to allow small variations within the experiment and also within the differential equation—this is Duhem's strong normative claim.[55] Assuming, at first sight, the plausibility of this claim, Duhem maintains that deductions define what physics is and will always be. Consequently, an a priori choice of the mathematical structure of theories is indispensable to enable the possibility of empirical sciences. From this point Duhem then extends his non-divergence (stability) requirement to a priori choices of hypotheses that constitute the center of theories. Duhem asks, "What are the conditions imposed by logic on the choice of hypotheses on which a physical theory is to be based?" (Duhem 1991, 219) His answer is:

"Hypotheses shall be chosen in such a manner" that it is possible to compare "the entire system of theoretical representation on the one hand [= theoretical facts], and the entire system of observed data on the other [= practical facts; experimental laws]. As such they are to be compared to each other and their resemblance judged." (Duhem 1991, 220) [56]

Hypotheses (as a part of theories and, thus, as central elements of deductions) need to be chosen in a way that enables the neighborhood of bundles to be conserved. This means that the bundle—in other words: the whole system—has to be taken into account: In order to enable the possibility of approximations, the neighbor-

---

54 Duhem's thinking moves from "facts" to "propositions" in a more general way: A single fact is identified by a "*rigorously true* proposition" (cp. Duhem 1991, 143).

55 Deductions—if they are to be considered physically relevant—*must* preserve the narrowness or likeness of the neighborhoods within the bundles when applying the deduction. The second bundle *must* be structurally similar to the first.

56 From today's perspective, this requirement is identical to what is called "structural stability", that is, the stability of general laws/equations under perturbations (cp. Schmidt 2011).

hood of a proposition encompassing the "whole group of hypotheses" has to be taken into account.[57]

Such a requirement entails consequences not solely for the mathematical structure of theories, but also for the physical objects under consideration. One has to remember that, according to conventionalist's approaches such as Duhem's, the presupposed mathematical structure of theories always plays a central role in the constitution of physical phenomena: Inasmuch as only stable theories (with non-diverging deductions) are assumed to represent a *physical* phenomenon, a phenomenon is seen as a *physical* phenomenon if, and only if, it is stable; methodology constitutes and constructs reality. In this vein, Jules Vuillemin points out that "Duhem limits his reflections to extrapolations concerning the stability of systems" (Vuillemin 1991, xxviii). Unstable phenomena—implying unstable theories with diverging deductions—are not regarded as being located in empirical science.[58]

In the history of sciences, the stability assumption was very common as a guiding maxim of physical methodology. In line with Duhem, the Russian mathematician and physicist Alexandr Andronov, for instance, reflects in the 1930s on mathematical a priori conditions that need to be guaranteed in order to enable the possibility of obtaining physical knowledge: "Which properties must a dynamical system (models, theories) possess, in order to cope with experimental data and to be applicable in physical contexts?" (Andronov et al. 1965, 403ff) Andronov requires the objects studied by physicists to "be stable both in relation to small variations of the coordinates and velocities, and in relation to small variations of the mathematical model itself" (Andronov et al. 1965, introduction). Because of empirical uncertainties, experimental vagueness, and measurement errors, an object and a model representing the object have to be "robust" or "structurally stable," Andronov argues.[59]

---

57   A subsequent, far-reaching thesis would be to say that the Duhemian type of theory holism is also a consequence of the stability requirement. The theory holism would, then, turn out to be a strong normative construct: Duhem's (and Quine's) theory holism would be nothing but a consequence of his stability assumption. Indeed, there is further work to be done on this point.

58   Further, Prigogine stresses that Duhem was pioneering in his reflection on instabilities, but—according to Prigogine—Duhem goes too far in his general critique of the ever-remaining uselessness of unstable deductions. In Prigogine's view, Duhem assesses instabilities in a negative sense only: as a threat to classical-modern physical methodology. "At this point we are decisively of a different opinion" (Prigogine/Stengers 1990, 316; my translation, J.C.S.).

59   In fact, Andronov, besides Birkhoff and Poincaré, was the father of dynamical systems theory. He undertook detailed work on structural stability and coined this term. Later,

19.7   On a Razor's Edge: Instabilities everywhere

In the past, physicists have conceptualized the physical world as well as scientific methodology in terms of stability and non-diverging (stable) deductions. In doing so they have obviously neglected to consider a broad variety of phenomena of the physical world. From the 1970s on, the restriction of physics to stability, as advocated by Duhem, Maxwell, Andronov, and others,[60] turned out to be nothing but a "stability dogma" (Guckenheimer/Holmes 1983, 259). In fact, unstable objects and phenomena are common; they do exist in the *physical* world—and, thus, theories describing them should also allow for instability. It is interesting that Duhem seems, at least to some extent, to have anticipated this. Very precisely he underlines the existence of "numerous and difficult deductions of celestial mechanics and mathematical physics" (Duhem 1991, 143).

A prominent critique of the stability assumption was raised in the 1980s by John Guckenheimer and Philip Holmes (1983). They argue that the normative stability presupposition—with regard to both physical systems and mathematical theories—has in fact hindered scientific progress. Stability was mistakenly imposed "as an a priori restriction on 'good' models [and theories] of physical phenomena" (ibid., 38/259). The traditional framing of the world in terms of stability is nothing but a mere "conventional conviction" without any justification (ibid.). With reference to unstable processes in real physical systems being analyzed by current-day scientists (cp. Mainzer 1996), Guckenheimer and Holmes clearly state that "[t]he logic which supports the stability dogma is faulty" (Guckenheimer/Holmes 1983, 259). Many

Wiggins points out, referring to Andronov, that the "mathematical models we devise to make sense of the world around and within us can only be approximations. Therefore, it seems reasonable that if they are to accurately reflect reality, the models themselves must be somewhat insensitive to perturbations. The attempts to give mathematical substance to these rather vague ideas have led to the concept of structural stability" (Wiggins 1990, 94). This debate is very prominent in physics. Arrowsmith and Place stress that "in physical applications we require our mathematical models to be robust. By this we mean that *all* their properties should not change significantly when the model is subjected to small perturbation" (Arrowsmith/Place 1990, 119).

60   In general, the stability assumption can be traced back to ancient Greece and to Plato; also Newton's and Einstein's understanding of nature was based on the stability assumption.

theories and models are dynamically and structurally unstable, since they correctly refer to unstable phenomena in the physical world,

"and we are confident that these [models...] are realistic models [...] of corresponding physical systems." (ibid., 259)

Duhem might not have accepted this statement. He believed that theories based on instabilities are impaired by deductive applicability and utilizability. Today, however, according to complexity theorists we have to admit that

"most dynamical systems that arise in classical mechanics are not structurally stable. [...] These systems cannot be approximated" in the classic way of asymptotic convergence of mathematical curves (Devaney 1987, 53).

Thus, instability is not—this is one interesting point we can draw from complex systems theory today—the end of exact sciences, as, contrary to the position developed here, John Horgan (1996) argues; rather, instability challenges established views of the sciences and calls for a change. In fact, exact sciences are coping successfully with unstable phenomena that induce diverging (unstable) bundles of deductions. Misleadingly, Duhem has argued that such kind of deduction will always remain "useless to the physicist" as it "will not permit us to predict" what will happen (Duhem 1991, 139). For current-day physicists and philosophers, Duhem's argumentation is inspiring and challenging because he is both right *and* wrong at the same time.

On the one hand, Duhem drew on the sciences at the end of the 19th century and beginning of the 20th century. Duhem rightly stressed the limits and problems of physical methodology when applying it to an unstable and complex world. However, he did not go further and seek other approaches; his understanding of the exact sciences was very narrow. Of course, he could not foresee that this kind of physics would turn out to not represent the entire exact sciences. Today, complex systems theory extends the methodology of sciences to include unstable phenomena and objects. As a result, Duhem's methodological stability requirement is, *on the other hand*, now rejected.[61] In this regard, Duhem can be seen as a watershed figure on the way towards recent philosophy of complex systems.

---

61   Further, instability is nowadays seen as *the* condition for the possibility of self-organizing systems, for growth and pattern formation (Schmidt 2015).

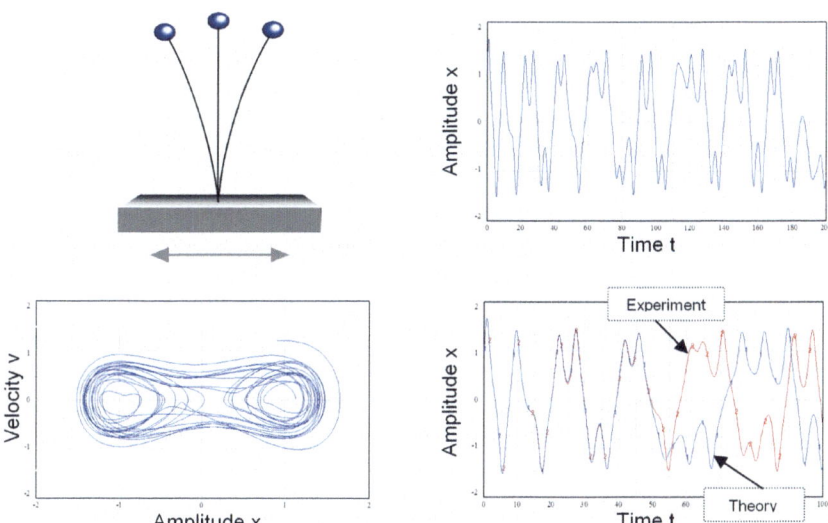

Fig. 19.2: Top left: Duffing Oscillator. Top right: Time evolution of the amplitude. Dynamical instability ("deterministic chaos") can be identified. The time series looks randomly. Bottom left: x-v plot of a chaotic attractor. Bottom right: Illustration of the
dynamical instability ("sensitive dependence on very close initial conditions").
Theoretical data and experimental data diverge because of instability. No approximation is possible.

Let us illustrate the rejection of the stability dogma today. A prominent example is
the Duffing oscillator (Fig. 19.2, top left) which has been extensively investigated
over the last 40 years. In the early part of the 20th century, Georg Duffing introduced
a periodically forced oscillator with nonlinear elasticity to describe the leaf spring
phenomenon or hardening spring effects observed in many physical and technical
fields (Duffing 1918; Parker/Chua 1989, 10f). The restoring force can be written as

$$F_{Duff}(x) = \alpha \cdot x - \gamma \cdot x^3$$

Duffing's model is given by the specific differential equation

$$dx/dt = y \ , \quad dy/dt = \alpha \cdot x - \gamma \cdot x^3 - \beta \cdot y + k \cdot \cos(\omega \cdot t) \ .$$

The term $k \cdot \cos(\omega \cdot t)$ represents an external driving force; x indicates the position
and y the velocity of the mass; the boundary conditions are α, β, γ, k; the initial
conditions are represented by $x = x_0$ and $y = y_0$.

For specific models of the model class—meaning for certain initial and boundary conditions—Duffing's model can be structurally as well as dynamically unstable. For instance, it generates dynamically unstable trajectories: chaotic time series and seemingly random data sequences are the result of this process (Fig. 19.2, top right). Therefore, the empirical data from the experiment—the "practical facts" in Duhem's terminology—cannot be quantitatively compared with the theoretically deduced data ("theoretical facts") point by point (Fig. 19.2, bottom right). Although Duffing's model has never been *quantitatively* proven point by point, physicists are confident that Duffing's equation is evident and that the model is a good model of a real oscillator (Wiggins 1990, 29/153; Parker/Chua 1989, 10f).

Their conviction—together with the impossibility of quantitative tests in a strict sense—is the most interesting point. We find similar convictions in the history of hydrodynamics, e.g., in the case of the Navier-Stokes equation (Darrigol 2005). For over 50 years the Navier-Stokes equation was not at all (mathematically and, hence,) empirically accessible. But physicists never doubted that the equation was a good model for real viscous fluids (Böhme et al. 1983). Since Prandtl's work, some linear approximations of the Navier-Stokes equation have become accessible. In particular, turbulent phenomena are not really understood to this day, although the theories governing these phenomena are classical theories of the 19th century. Hydrodynamics is certainly not a unique case. We find cognate methodological issues in General Relativity with its nonlinear (and, for some boundary conditions, unstable 10-dimensional) partial differential equations formulated by Albert Einstein. Physical Cosmology was often criticized for being nothing more than mere speculation and in this respect not a physically relevant theory.[62] Although such critique is one-sided, it nevertheless highlights the methodological problems, e.g. with regard to quantitative tests including the possibility of approximation.[63]

---

62   The so-called *Anthropic Principle* can be regarded as a selection attempt to eliminate some of the arbitrariness of instabilities.

63   However, unstable phenomena in our day-to-day environment—the "mesocosmic" objects—are somewhat similar to unstable macrocosmic phenomena of Physical Cosmology.

19.8 Other Requirements: Qualitative instead of Quantitative Criteria

To gain access to unstable objects, complex systems theorists had to question and relativize the stability conviction. In particular, physicists had to change their perspective and turn away from a quantitative-metrical (point-by-point) to a more qualitative-topological approach.[64] After instability had been disregarded and denounced for over hundreds of years, the stability dogma was replaced by an approach employing more qualitative features and relating to those physical properties of a theory or experiment which are relevant for the specific situation in question; a non-universalism and a contextualism have emerged.

"The definition of physical relevance will clearly depend upon the specific problem. This is quite different from the statement [...] of the stability dogma." (Guckenheimer/Holmes 1983, 259)

It is widely acknowledged in the methodological debate in complex systems theory that weaker requirements are (a) sufficient for physical methodology and, also, (b) that they are necessary to gain access to unstable phenomena. These requirements are based mainly on qualitative (topological) rather than on quantitative (metrical) properties.

"Solving the equation [in order to enable a deduction and a test of the theory] does not result in a series of values that agrees point by point with the observed series of values. Calculated and observed data differ quantitatively but show qualitative agreement." (Rueger/Sharp 1996, 103)

Qualitative properties play a key role in this novel account of methodology and do not confound a mathematical approach at all. Generally, the *qualitative* aspect is highly acknowledged in exact sciences today. For instance, Benoit Mandelbrot calls his "Fractal Geometry" a "qualitative theory of different kinds of similarities" (Mandelbrot 1991), and René Thom stresses that his "Catastrophe Theory" is a "qualitative science of morphogenesis" (Thom 1975). All of these works have

---

64 According to Rueger and Sharp (1996, 103), "sensitive dependence on initial conditions [= dynamical instability / chaos] [...] requires us to modify the notion of a comparison of data obtained from a real [= concrete physical] system with predictions [and, thus, with deductions] from the theory."

a common background: the qualitative theory of differential equations and differential topology which can be traced back to George David Birkhoff in the 1920s (Birkhoff 1927), as well as to the foundational work of Henri Poincaré (1892).

What does "qualitative" mean here? As used in this context, the term does not relate at all to subjective impressions or to qualia-like perceptions; it remains a mathematical concept but in a particular way. When addressing instabilities, complex systems theorists do not require theories and deductions to be quantitatively stable, i.e., to show a quantitative robustness of the structure of the theory, but rather to possess some qualitative properties or dynamical invariants, such as *complexity characteristics*. These qualitative, mostly differential topological properties refer to the geometrical shape, dynamical pattern, or "Gestalt" described in the phase space after or during the process of time evolution—and not solely to the bare equation (chaotic attractor; Fig. 19.2, bottom left).[65] The qualitative properties are not established on *single* trajectories and *single* measurable events but also on *possible* dynamics of the *whole* dynamical system. The entirety of all possible dynamics cannot be measured using classic quantitative metrical distance measures. Rather, it can be classified by a qualitative account of the shape of the whole attractor. Thus, complex systems theorists favor an approach to describe a phenomenon and to test a theory "that is holistic, historical, and qualitative, eschewing deductive systems and causal mechanisms and laws" (Kellert 1994, 114; cp. Mitchell 2002; Schmidt 2008a).

In the case study on the Duffing oscillator, the essential qualitative properties are the structure of the three equilibrium points of the spring (two of them are point attractors), and the dynamical pattern of the chaotic attractor; additional properties

---

65 Today, an increasing amount of research is carried out to define and explore such (qualitative, dynamical, topological invariant) properties ("characteristics," "parameters," "measures of complexity": cp. Wackerbauer et al. 1994). These properties are not suited to *all* physical problems in general. In some contexts, certain properties which are used in hypotheses and theories are meaningless, depending upon the goal of the research issue. But there will be others, allowing a classification, detailed description, partial explanation, and piecewise prediction of the dynamics of a specific physical system. In order to illustrate the amazing variety of these contextual properties, characteristics, and parameters, some of them are enumerated below without a mathematical definition: (a) entropies, (b) parameters of information theory, (c) Lyapunov exponents, (d) prediction time scaling exponents, (e) lengths of transients, (f) parameters of symbol dynamics, (g) fractal dimensions, (h) parameters of renorming, (i) topological characteristics, (j) quantity of stable and unstable periodic trajectories, (k) existence of chaos or hyperchaos, (l) parameters of basin boundaries, (m) types of bifurcations, (n) parameters of chaos control theory, (o) power and Fourier spectra, (p) phenomenological analysis of pattern, etc.

include the specific sensitivity exponents,[66] different kinds of fractal dimensions, and the types of bifurcations (cp. Moon 1992, 5f.). To gain access to the phenomena and to obtain an insight into the time evolution of the system, the phenomena have to be generated by numerical simulation and integration, and then represented by visualization techniques. It is worth noting that simulations are not restricted to the context of discovery but also encompass the context of justification—because of instability.[67]

Duhem and the advocates of non-diverging (stable) deductions would not have accepted such a more qualitative-oriented exact science. Their understanding of a physically relevant deduction—intrinsically related to predicting, concluding, and asymptotic reasoning—is based on the ideal of a limited quantitative point-by-point approach.

19.9   Outlook

Stability cannot be taken for granted, as Klaus Mainzer and others have shown: The advancement of exact sciences in the past 50 years into new interdisciplinary fields of complex phenomena has revealed that instability is a common and widespread phenomenon.[68]

Although it has been discussed in depth within the exact sciences, instability has, so far, not been adequately acknowledged by the broader community of philosophers of science.[69] Looking back on the history of philosophy of science, Pierre Duhem was notably one of the most profound thinkers in this regard. On

---

66   These "Lyapunov exponents" characterize the divergence of trajectories under the time evolution of the dynamical systems or the dynamical instability.

67   Thus, in addition to the well-established two domains of experimental and theoretical physics—beyond and in between the practical and theoretical facts—a third domain of physics emerges that addresses nonlinear complex systems with their unstable behavior, and it uses numerical techniques, computer simulations, and graphical representation. All this is not covered by the Duhemian hypothetico-deductive model and its cognates, as Poser (2001) has rightly stressed: The hypothetico-deductive model devaluates non-quantitative aspects and, in addition, also boundary and initial conditions and process phenomena. In this sense, explanation can be described as dynamical—and not just as nomological.

68   For example, see Mainzer (1996; 2008; 2014), Kuhlmann (2007) and Schmidt (2003; 2008b; 2011; 2015).

69   Here, Mainzer (1996f), Batterman (2002), and Rueger/Sharp (1996) are exceptions.

the one hand, Duhem was a pioneer with his early anticipation of the possibility of instability in physical systems and in theories, and in particular of diverging (unstable) deductions. On the other hand, Duhem believed, in line with the prevalent tradition, that exact sciences are threatened by instabilities and, based on such considerations, he argued in favor of stability, formulated a stability requirement, and pursued what later has become known as a *stability dogma*.

With reference to the seminal work of Mainzer and to complex systems theory, we can disclose the limitations of the restriction to stability that governed Duhem's picture of science. The discovery, perception, and acknowledgement of instability, complexity, and chaos in the world we live in engender, as Mainzer emphasizes, the emergence of a new picture of nature as well as of science, and urge us to rethink our thinking. The time has come to advance "from linear to nonlinear thinking" (Mainzer 1996, 1).

References

Achinstein, Peter (1998), Explanation vs. Prediction: Which Carries More Weight? In: Curd, M., Cover, J.A. (eds.) (1998), Philosophy of Science. The Central Issues; New York/London, 481-493.

Andronov, Alexander and Lev S. Pontryagin (1937), Systèmes Grossiers. In: Dokl. Akad. Nauk. (Doklady) SSSR 14: 247-251.

Andronov, Alexander A., et al. (1965), Theorie der Schwingungen. Teil I. Berlin: Akademie-Verlag.

Arrowsmith, David K., and C.M. Place (1990), An introduction to dynamical systems. Cambridge: Cambridge University Press.

Aubin, David (1998), A Cultural History of Catastrophes and Chaos. Around the Institute des Hautes Études Scientifiques, France 1958-1980, Ph. D. thesis (Princeton University).

Aubin, David, and A.D. Dalmedico (2002), Writing the History of Dynamical Systems and Chaos: Longue Durée and Revolution, Disciplines and Cultures. In: Historia Mathematica 29: 273-339.

Batterman, Robert (2002), The Devil in the Detail: Asymptotic Reasoning in Explanation, Reduction and Exergence. Oxford: University Press.

Birkhoff, Georg David (1927), Dynamical Systems. New York: John Wiley & Sons.

Böhme, Gernot, et al. (eds.) (1983), Finalization in Science. The Social Orientation of Scientific Progress. Dordrecht: Reidel.

Cartwright, Nancy (1983), How the laws of physics lie. Oxford: Clarendon Press.

Chaitin, Gregory J. (1974), Information-Theoretic Limitations of Formal Systems. In: Jour. Ass. Comp. Mach. 21: 403-424.

Crowe, Michael J. (1990), Duhem and the History and Philosophy of Mathematics. In: Synthese 83: 431-47.

Darrigol, Oliver (2005), Worlds of Flow. A History of Hydrodynamics from the Bernoullis to Prandtl. Oxford: Oxford University Press

Devaney, Robert (1987), An Introduction to Chaotic Dynamical Systems. Redwood City: Addison-Wesley.

Duffing, Georg (1918), Erzwungene Schwingungen bei veränderlicher Eigenfrequenz und ihre technische Bedeutung. Braunschweig: Vieweg.

Duhem, Pierre ([1906] 1991), The Aim and Structure of Physical Theory. Reprint. Translated by P. P. Wiener. Originally published as La théorie physique: son objet, et sa structure (Paris: Marcel Rivière & Cie). Princeton: Princeton University Press.

Falkenburg, Brigitte (2013), Mythos Determinismus. Berlin/New York: Springer.

Flügel, Martin (1996), Duhems Holismus. In: Philosophia Naturalis 33, 143–167.

Guckenheimer, John, and Philip Holmes (1983), Nonlinear oscillations, dynamical systems, and bifurcations of vector fields. New York: Springer.

Hadamard, Jacques (1898), Les surfaces à courbures opposées et leurs lignes géodésiques. In: Journal de Mathématiques pures et appliquées 4: 27-73.

Hempel, Carl G. (1965), Aspects of Scientific Explanation; New York: The Free Press.

Horgan, John (1996), The End of Science. Facing the Limits of Knowledge in the Twilight of the Scientific Age. Massachusetts: Addison-Wesley.

Kuhlmann, Meinard, 2007: Theorien komplexer Systeme: Nicht-fundamental und doch unverzichtbar?; in: Bartels, Andreas, Stöckler, Manfred (Hg.), 2007: Wissenschaftstheorie; Paderborn, 307–328.

Leiber, Theodor (1996), Kosmos, Kausalität und Chaos; Würzburg.

Mainzer, Klaus (1996), Thinking in complexity. The complex dynamics of matter, mind, and mankind. Heidelberg/Berlin: Springer.

Mainzer, Klaus (2005), Symmetry and Complexity. The Spirit and Beauty of Nonlinear Science. Singapore: World Scientific.

Mainzer, Klaus (2007), Der kreative Zufall. Wie das Neue in die Welt kommt. München: Beck.

Mainzer, Klaus (2008), Komplexität. Paderborn: Fink.

Mainzer, Klaus (2014), Die Berechnung der Welt. München: Beck.

Mandelbrot, Benoit ([1977] 1991), Die fraktale Geometrie der Natur. Originally published as The Fractal Geometry of Nature (New York: Freeman). Basel: Birkhäuser.

Martin, Nial (1991), Pierre Duhem. Philosophy and History in the Work of a Believing Physicist. Open Court: La Salle/USA.

Maxwell, James Clerk ([1876] 1991), Matter and motion. New York: Dover Publications.

Mitchell, Sandra (2002), Integrative Pluralism. In: Biology and Philosophy 17: 55-70.

Mitchell, Sandra (2009), Unsimple Truth: Science, Complexity and Policy. Chicago: University of Chicago Press.

Moon, Francis (1992), Chaotic and fractal dynamics. New York: Wiley.

Parker, Thomas S., and Leon O. Chua (1989), Practical Numerical Algorithms for Chaotic Systems. New York: Springer.

Poincaré, Hénri (1892), Les méthodes nouvelles de la méchanique céleste. Paris: Gauthier-Villars.

Poincaré, Hénri ([1908] 1914), Wissenschaft und Methode. Reprint. Originally published as Science et méthode (Paris: Flammarion). Leipzig: Teubner.

Poser, Hans (2001), Wissenschaftstheorie. Stuttgart: Reclam.

Prigogine, Ilya, and Isabelle Stengers ([1980] 1990), Dialog mit der Natur. München: Piper Verlag.

Psillos, Stathis (1999), Scientific Realism. How Science tracks Truth. London/New York: Routledge.

Quine, Willard V.O. (1951), From a Logical Point of View; in: Philosophical Review 60, 20-43.

Rueger, Alexander, Sharp, W. David (1996), Simple Theories of a Messy World: Truth and Explanatory Power in Nonlinear Dynamics. In: Brit. J. Phil. Sci. 47: 93-112.

Ruelle, David ([1991] 1994), Zufall und Chaos. Reprint. Originally published as Hasard et Chaos (Paris: Odile Jacob). Berlin: Springer.

Salmon, Wesley (1989), Four Decades of Scientific Explanation; in: Kitcher, Philipp, Salmon, Wesley (eds.), 1989: Scientific Explanation; Minnesota, 3-219.

Schäfer, Lothar (1978), Einleitung: Duhems Bedeutung für die Entwicklung der Wissenschaftstheorie und ihre gegenwärtigen Probleme. In: Duhem, Pierre ([1906] 1978), Ziel und Struktur der physikalischen Theorien. Originally published as La théorie physique: son objet, et sa structure (Paris: Marcel Rivière & Cie). Hamburg: Meiner, ix-xxxiv.

Schmidt, Jan C. (2001), Was umfaßt heute Physik? Aspekte einer nachmodernen Physik. In: Philosophia Naturalis 38: 271-297.

Schmidt, Jan C. (2003), Zwischen Berechenbarkeit und Nichtberechenbarkeit. Die Thematisierung der Berechenbarkeit in der aktuellen Physik komplexer Systeme. In: Journal for the General Philosophy of Science 34: 99-131.

Schmidt, Jan C. (2008a), Instabilität in Natur und Wissenschaft. Eine Wissenschaftsphilosophie der nachmodernen Physik. De Gruyter, Berlin.

Schmidt, Jan C. (2008b), From Symmetry to Complexity. On Instabilities and the Unity in Diversity in Nonlinear Science; in: International Journal for Bifurcation and Chaos 18 (4), 897–910.

Schmidt, Jan C. (2011), Challenged by Instability and Complexity. On the methodological discussion of mathematical models in nonlinear sciences and complexity theory. In: Hooker, Cliff (2011), Philosophy of Complex Systems. Elsevier, Amsterdam, 223-254.

Schmidt, Jan C. (2015), Das Andere der Natur. Neue Wege zur Naturphilosophie; Hirzel, Stuttgart.

Thom, René (1975), Structural Stability and Morphogenesis. Massachusetts: D.H. Fowler.

Vuillemin, Jules (1986), On Duhem's and Quine's Theses, in Hahn, L.E., Schilpp, P.A. (1986), The Philosophy of W.V.O. Quine. The Library of Living Philosophers, vol. XLVIII. LaSalle: Open Court, 594-618.

Vuillemin, Jules (1991), Introduction, in: Duhem, Pierre ([1906] 1991), The Aim and Structure of Physical Theory. Reprint. Translated by P. P. Wiener. Originally published as La théorie physique: son objet, et sa structure (Paris: Marcel Rivière & Cie). Princeton: Princeton University Press: xv-xxxiii.

Wiggins, Stephen (1990), Introduction to applied nonlinear dynamical systems and chaos. New York: Springer.

Wilholt, Torsten (2012), Conventionalism: Poincaré, Duhem, Reichenbach. In: Brown, J.R. (2012), Philosophy of Science: The Key Thinkers, London: Continuum Books, 32-52.

Wackerbauer, Renate, et al. (1994), A Comparative Classification of Complexity Measures. In: Chaos, Solutions & Fractals, 4(1): 133-174.

20. The Demise of Systems Thinking: A Tale of Two Sciences and One Techno-science of Complexity

Alfred Nordmann[1]

Abstract

Stuart Kauffman and Brian Goodwin treat organisms as complex systems. Their claim revitalizes the vitalism of Hans Driesch who situated biological processes within a general theory of order. Along with many others who emphasize the challenge of complexity all three maintain that "[w]e will have to rethink what science is itself" (Kauffman 2000, 22). But what is this challenge to canonical ways of doing science? Here, Kauffman and Goodwin part company. Though both fend off the accusation that the very idea of a "scientific vitalism" is conceptually incoherent, only Kauffman's scientific vitalism involves a wholly different science of complexity while Goodwin foregrounds a wholly new class of complex phenomena which, however, are subject to familiar modes of causal analysis.—Kauffman and Goodwin's disagreement about the requirements for a "science of complexity" reappears in a new guise in discussions of systems biology and synthetic biology and generally of complex systems.

---

1   Alfred Nordmann is Professor of Philosophy of Science and of Technoscience at Technische Universität Darmstadt. His research focuses on epistemological, metaphysical, and aesthetic aspects of technoscientific research in chemistry, nanotechnology or synthetic biology. Grounding an account of scientific knowledge production in the philosophy of technology, he explores the notion of working knowledge. Nordmann is the editor of the book series *History and Philosophy of Technoscience*, author of introductions to the philosophy of technology and the philosophy of Wittgenstein.

20.1   Introduction

Systems biology appeared on the scene at roughly the same time when nanotechnological researchers began dreaming of bottom-up rather than top-down engineering. The shift from the study of organs and their functions to that of whole systems appears to signify the embrace of a holistic paradigm and thus a new philosophical orientation of scientific research. Similarly, bottom-up engineering seeks to harness principles of self-organization in the design process and thus to establish a new engineering paradigm that does not impose control but utilizes the self-control in the non-linear complex dynamics of emerging systems. These and other contemporary developments seemed to testify to a new age of science and new insights that were grounded in 19th and 20th century *Naturphilosophie* with its critique of a mechanistic uni-directional causality (Mainzer 2007, Schmidt 2008).

Arguably, however, systems biology and dreams of bottom-up engineering do not testify to a new way of doing science and of understanding natural phenomena humbly in terms of non-linear complexity and integrated wholes. Instead, they speak to a kind of disinterest in "paradigms" or *Naturphilosophie* altogether. As systems biology feeds into synthetic biology, the systems in question are not awesomely complex in the sense of challenging us to understand them as integrated wholes. They are very complicated technical units of more or less robust functioning that need not be "understood" at all but that can be blackboxed or modularized and investigated for their behaviors and parameter-dependencies (for different perspectives on this see Nordmann 2014 and Schmidt 2014, see also Fox Keller 2007). Similarly, the dream of bottom-up engineering relies on effective design principles that allow researchers to breed desired material properties. Instead of a sublime new scientific way of thinking in complexity, we witness a kind of unthinking or technoscientific research that produces proofs of concept and explores in the mode of engineering not how things are but what they can do. The competition between paradigms for understanding phenomena and thus between different conceptions of science becomes absorbed into technoscientific research that does not rely on intellectual understanding in order to elaborate technical capabilities for managing large sets of data (Nordmann 2010).

Should we be surprised that "(complex) system" ceased serving as a category of understanding and became a technical or organismic unit of functioning? In other words, should we be surprised that in the face of increasing complexity science becomes less philosophical rather than more so? More pointedly perhaps than anyone else it was philosopher of biology William Wimsatt who detailed the need to adopt piecemeal strategies where causal processsses elude intellectual tractability (Wimsatt 2007). And though visionary "data scientist" Jim Gray misleadingly insinuates a kind of continuity when he speaks of a data-driven "fourth paradigm," he exhibits the prominence of an engineering mindset in contemporary research efforts that aim not so much to understand and reduce but to generate and manage complexity (Hey, Tansley, Tolle 2009).

Today's technoscientific accounts of nanotechnology or of systems and synthetic biology pit a metaphysically indifferent engineering mindset against, say, the traditional *naturphilosophische* tug of war between mechanistic and holistic conceptions of science. There is another way to discover, however, over that a "science of complexity" need not foster a new kind of understanding of biological systems as integrated wholes but that it might also treat of a new kind of entity—order-seeking dynamic systems—in the otherwise conventional ways of causal analysis. One can make this discovery by going back to the 1990s and thus the heyday of philosophical aspirations for a new kind of science.

20.2   Driesch's "Scientific Vitalism".

The following story of different attempts to establish a science of complexity begins with the biologist and philosopher Hans Driesch (1867-1941). His proposal of neo-vitalism was the first attempt to establish an explicitly scientific rather than metaphysical theory of complexity and the emergence of order. Driesch distinguished his neo-vitalism from "old vitalism" precisely on the grounds that it involves nothing obscure, neither teleology nor life force. Driesch arrives at his vitalist position by an abductive inference from the insuffficiency of received evolutionary accounts to the *prima facie* plausibility of any certifiably scientific hypothesis capable of filling the explanatory gap: "All *proofs* of Vitalism [...] can only make it clear that mechanical or singular causality is *not* sufficient for an explanation of what happens" (1914, 208). Physico-chemical laws cannot account for processes of life, the growth of organisms cannot be explained by Weismannian or Rouxian "machine theories" or in analogy to the growth of crystals, and Darwinism falls short with its proposal to explain "how by throwing stones one could build houses of a typical style" (1914, 137). In particular, Driesch advances two theses, the first establishing the insuffiency of extant theory and the second closing the explanatory gap abductively:

(1) Darwinian evolution by natural selection relegates to chance all construction of biological form (1914, 141). It thus leaves important phenomena unexplained, such as homologies between historically and cladistically distant species or the

regeneration of the salamander's lost tail. Most significantly, Darwinism or any "machine theory" of organismic development does not explain why the drastic removal of germ-cells during early cell-division does not inhibit the development of complete organisms (1914, 208f.).

(2) A form-giving principle ("entelechy") can provide the requisite explanation. This physically non-localizable principle, a feature of the spatial arrangement itself, regulates, controls, or directs the construction of biological form.

According to Driesch, these two propositions lead to a new conception of science where individualising and unifying causality complements mechanistic or linear causality. Instead of asking how some physico-chemical state specifically determines another in space (Driesch 1914, 198f.), this science asks how the transition from possibility or potentiality to determinate individuality is generally regulated, or what principle unifies the spatial distribution of parts in individually different organisms.[2] "Understanding nature" means to conceive nature as something ordered, and the theory of nature is part of a non-metaphysical theory of order and ordering (1914, 223, 233).

These, to be sure, are only the most abstract and generic features of Driesch's account. They are prominent for a tale of two sciences of complexity since it is due to these two theses that Driesch can be considered a precursor of Kauffman and Goodwin.

20.3   Philipp Frank's Critique of Scientific Vitalism.

Driesch's two theses hinge on the claim that there is a deficit as well as a scientifically credible, that is, metaphysically innocuous way to compensate the deficit through a neo-vitalist form-giving principle or entelechy. That this claim is far from trivial becomes evident from its critique by Vienna Circle philosopher of science Philipp Frank (1884-1966). Frank devotes a chapter to "Causality, Finalism and Vitalism" in his *The Law of Causality and its Limits* of 1932. The chapter provides an extended discussion of Driesch's proposed proofs for the autonomy of the phenomena of life

---

2   This individualising and unifying causality is said to be a non-spatial agent that acts into space (Driesch 1914, 204). Driesch speaks of "non-spatial agents of a controlling type" when there is no quantitative re-arrangement of energy or matter but an increase "in the number of relations among the things" (1914, 200).

(1998, 105-116).[3] Frank's aim is to show that "scientific vitalism" either signifies no departure at all from ordinary scientific practice or - if it is taken to signify such a departure - that it is an incoherent notion.

According to Frank's critique, the first of Driesch's two theses states an impossibility but does not deliver the requisite impossibility-proof. In particular, Driesch maintained that "it is impossible to explain the processes of life with the help of the laws of mechanics" (Frank 1998, 106). Driesch's evidence for this claim was, most prominently, the case of the sea urchin. In early stages of cell division, some cells can be removed and a complete sea urchin larva develops from the remainder. The removal of parts from a classically constructed machine would destroy it. Also, each part of a machine has an assigned function. In the case of a sea-urchin, a cell that might develop into some particular organ is removed and will now develop into a complete organism. Since mechanical laws cannot explain this phenomenon they need to be supplemented by "a magnitude E which is a constant similar to specific heat in physics" and which aims the organism towards a definite shape (Frank 1998, 106f., but see Driesch 1908, 202-205).

Frank appreciates the clarity of Driesch's argument, namely that it is formulated "so sharply and with so much reference to real facts that it is possible to point out exactly where the realm of genuine scientific proof is abandoned" (Frank 1998, 105f., 128). In particular, Frank asks whether Driesch provided any proof for the impossibility of explaining the sea-urchin phenomenon by the laws of mechanics or a physico-chemical theory of life:

"To prove the impossibility of such a theory that rests on the experimental biological facts that Driesch uses, at a minimum, proofs of the following kind should be attempted: a definite physico-chemical theory should be taken as the basis; for example, to take the simplest case, Newton's equations of mechanics with definite laws of force, and it would have to be shown that there could not be any initial states of point masses which, after removal of some masses, might develop in the same way as the original system. [...] Demonstrating this however seems very improbable if we consider that systems with very many mass-points are involved, and that the recurrence of similar groupings need not occur with mathematical precision, but only approximately." (Frank 1998, 107f.)[4]

---

3   Here, only one of these proofs will be considered and only one of Frank's "misgivings" about that proof.

4   Frank goes on to point out that this kind of demonstration will have to be provided for all physico-chemical theories, including those of the future.

To be sure, Driesch never claims to provide positive evidentiary proof for the existence of entelechies. He follows Kant and Mach when he maintains that only material constellations can be known, while entelechies can only be inferred by way of
indirect proof of the systematic incompleteness and limitation of our mechanical
knowledge of the dynamics of matter in space (Driesch 1913, 231; Frank 1998, 62).
However, by exposing the difficulty of motivating and justifying Driesch's abductive argument, Frank has shown that a need to embrace entelechies does not arise
automatically from the supposed limitations of the empirical sciences. It therefore
becomes questionable whether Driesch's second thesis and the notions entelechy,
purpose, or intent really do provide additional explanatory power where it was
lacking before.

Driesch's second thesis proposes a determination of the present by an intended
or implied future state—a well-ordered future state draws less well-ordered states
to reconfigure themselves. However, according to Frank, to the extent that one can
give pure physical meaning to such processes, the notions of purpose or intent drop
out and one is simply fixing a physico-chemical trajectory in a specific manner. For
this, Frank uses the analogy of shooting at a target with a gun.

> "The trajectory is either determined by position and velocity of the missile at a
> time A, or by the specification of two points A and B of the trajectory and the
> corresponding times. Here nobody will have the idea of saying: the problem of
> shooting a bullet from a gun can be explained in terms of causality or finalism.
> In the first instance, the trajectory is determined by its past, in the second by its
> future." (Frank 1998, 100)

Here, then, one is free to refer to the causal role played by the target or goal, but any
such teleological reference can be eliminated easily. Accordingly, one can remain
in the traditional framework of causal mechanisms and has provided no reason
to extend it along the lines of Driesch's second thesis. To be sure, Frank continues,
such reasons do exist in some cases and it is indeed possible, though perhaps not
palatable to go beyond the mechanistic framework and to assign scientific meaning
to notions of purpose or intent:

> "If a stone falls down from a building and arrives somewhere at time B, we
> do not speak of a purpose that should be inferred from the fact that the stone
> has fallen just where it is now found. If we want to regard the final state of an
> organism developing from an ovum as a purpose toward which an entelechy
> is striving, [...] the consideration of purpose [...] has a scientific meaning only

if we assume that a being exists who aims with this purpose in mind." (Frank 1998, 103)

If someone drops a stone from a building so as to kill someone, this intention to kill needs to enter an explanation of the fact that the stone has fallen just where it is now found. This involves the testable empirical claim that there was someone there who had that intention. In other words, if Driesch does not consider the second thesis eliminable, he should back this up by explicitly introducing the factual or metaphysical claim that there is a purpose or intent at play. Driesch's "scientific" vitalism, however, is premised on the conviction that he can infer entelechies without having to attribute purposiveness to nature, let alone a Creator. Indeed, it is the point of his metaphysically inoccuous vitalism that he wants entelechy and directedness without committing to untestable empirical claims about the presence of purpose or intent. Against this, Frank maintains that claims about the purposive directedness of an entelechy are empirically meaningful only if one assumes the existence of a being that needs to solve a specific problem. That being can only intend to get from A to B by integrating knowledge of initial values (point A and aiming direction at A) and of boundary values (point B in relation to A) into the likely trajectory (Frank 1998, 101). Though Frank's consideration shows that psychological and spiritual notions of purpose or intent do have empirical content, it spells doom for the proposed scientific vitalism and a science of self-organizing complexity. Seeking to be scientific, such theories cannot assume an intelligent designer or intentional ovum that aims at the organism's final state with a purpose in mind—but according to Frank they also cannot do without this assumption:

"If in a vitalist theory, the being that establishes the purpose is simply left out, then the word 'purpose' is completely out of gear; it does not express any scientific knowledge." (Frank 1998, 103)

If Frank is right, Driesch is caught in a dilemma. In order to secure its meaningfulness, he "is forced, in spite of his reluctance, to make the entelechy that is at work here much more of a psychological factor." Though he wants to limit himself to the "description of external experiences" he ends up endorsing spiritualism (Frank 1998, 115f.).[5] Whether one assumes a neo-vitalist stance or not thus depends "only

---

5   Frank quotes Driesch: "In fact, entelechy is affected by acts upon spatial causality as if it came out of an ultra-spatial dimension; *it does not act in space, it acts into space*, it is not in space, it only has points of manifestation in space. This analogy with some theoretical views that are advocated by so-called spiritualists to explain facts which are admitted by

on whether we believe that the spiritualist hypothesis is usable for the study of nature" (Frank 1998, 116). Frank generously concedes that such a spiritualist hypothesis "makes a kind of empathy possible in nature" but agrees with Bergson that this kind of empathetic knowledge can be confirmed only philosophically and not scientifically:

> "Within science there can be no struggle between mechanism and vitalism [...];
> what is basically disputed is whether there is, outside scientific knowledge, still
> another essentially different kind of knowledge." (Frank 1998, 117)

In sum, then, one either describes the trajectory from ovum to organism mechanically such that the notions of purpose or intent drop out, or one insists on a specifically vitalist as opposed to mechanist interpretation of these terms. In the latter case, however, the conception of a "scientific vitalism" becomes a contradiction in terms.[6]

Frank's two arguments against Driesch's two theses are independent in the following sense. If the requisite impossibility proof for (1) could be delivered, after all, "scientific vitalists" and proponents of the new science of complexity would still have to contend with the argument about (2). As a consequence of Frank's arguments, one can take therefore adopt one of three stances:

i.   Accepting Frank's argument against (2), the stated impossibility (1) to explain certain phenomena of life can be taken as indicative of the limits of science: "We may have to admit that science can't know everything."

ii. One can reject Frank's argument against (2) on the grounds that, in light of the impossibility (1), the very enterprise of science needs to be radically reconceived: The traditional conception of science needs to make room for an "essentially different kind of knowledge."

iii. Frank's argument against (2) can be accepted, but a scientific vitalism proposed that is not spiritualist and yet does not deny beings which establish a purpose. In agreement with Frank's claim that "the consideration of purpose [...] has a scientific meaning only if we assume that a being exists who aims with this

---

them to exist, is a very good *description*, indeed, of what happens in any natural system upon which entelechy is acting" (1908, 235, see also 182).

6   "In my opinion, there are only two possibilities: either we purify teleology of all anthropomorphic elements and thus blur each real difference with respect to the causal approach of physics after we removed all overtones of old animism from it; or we offer a real teleology, that is a world-plan which is concretely laid out at least in its main features, and which has been created by a real spiritual being" (Frank 1998, 119).

purpose in mind," one can seek to empirically characterize all organisms as purposive, problem-solving entities. The purpose or intentionality in question is attributed neither to nature nor to a Creator but to the causal dynamics of the, e.g., organic systems under consideration.

20.4   Scientific Vitalism Revitalised.

For the 1990s, some 60 years after Phillip Frank formulated his arguments against Driesch, these three stances serve well to describe the situation.[7]

Frank had conceptualized the trajectory from ovum to organism in analogy to the trajectory of a missile. The removal of some cells from the ovum does not always affect its trajectory; it can still arrive at the same organism. Frank was therefore willing to concede the autonomy of the phenomena of life only if it could be shown that there "could not be any initial states of point masses which, after removal of some masses, might develop in [approximately] the same way as the original system." His demand had not been met but it had been shown by non-linear complex dynamics that for certain physical systems even a small change in the initial states results in widely divergent trajectories. If, as in the sea-urchin this divergence is not observed, this requires special explanation—and thus one arrives at Driesch's first thesis. Even if it remained possible, as Frank pointed out, to find a mechanistic explanation in the case of the sea-urchin, this possibility could be discounted for the non-linear dynamics of a genuinely complex system. The acknowledgment of such systems makes room for a rejoinder to Frank's second argument. Given that organisms are highly sensitive complex systems that nevertheless behave like ordered systems, organisms must possess a tendency towards order that counteracts their complexity.

On the basis of considerations like these, Stuart Kauffman and Brian Goodwin both endorse closely similar theses to those advanced by Hans Driesch.

(1') Like Driesch, Kauffman and Goodwin single out a class of phenomena that cannot be explained in terms of Darwinian evolution by natural selection. Driesch describes these phenomena most generally as the problem of how "a *homogeneous* distribution of possibilities (*i.e.* cells which all have the same morphogenetic potencies) *is transformed into a heterogeneous distribution of*

---

7   The following discussion focuses on the adoption of the second and third of these modes of response by Stuart Kauffman and Brian Goodwin, respectively. It appears that Richard Lewontin, for example, adopts the first, most straightforwardly empiricist stance.

realities (*i.e.* specific parts of cells with specific physiological functions)" (1914, 209f., see 200f., 203). Goodwin describes the problem as the "bifurcation from spatial uniformity to pattern" (1994, 106) and Kauffman as the propagation of organization (2000, 104). According to Goodwin, a Darwinian account "makes biological form unintelligible" (Goodwin quoted in Lewin 1999, 41), and Kauffman notes that "much of the order in organisms, from the origin of life itself to the stunning order in the development of a newborn child from a fertilized egg, does not reflect selection alone" (2000, 1f., see 17). Like Driesch, both object to the implication of these limitations, namely that Darwinism renders the origins of life merely accidental (Driesch 1914, 138; Kauffman 2000, 9f.; Goodwin 1994, 104, 116).

(2') Again like Driesch, Goodwin and Kauffman consider the requisite biological theory "as a part of the theory of order" (Driesch 1914, 233). Driesch proposes "entelechy" as a non-mechanical agent that regulates the creation or release of biological form (1914, 203). In the language of complexity theorists, Driesch's entelechy corresponds to the constrained release of energy (Kauffman 2000, 4, 6, 102f.) or, more generally, to the control parameters that govern bifurcations in the logistic equation (Goodwin 1994, 40, 97, 110-114; Webster and Goodwin 1996, 121, 230, 250).

Kauffman and Goodwin differ, however, in that Kauffman rejects Frank's argument against (2) and (2'), whereas Goodwin accommodates it. That is, of the three stances in response to Frank, Kauffman pursues (ii) and advocates a whole new way of doing science, whereas Goodwin chooses option (iii) and speaks only of a new class of entities as an object of inquiry. The very fact of their disagreement raises the question as to whether the endorsement of (1') and (2') needs to be attended by a wholesale transformation of the scientific enterprise. Entelechies need not be conceived as paradigms or categories of understanding but might serve to define the order-seeking units of functioning that are subject to classical causal analysis.

20.5   Kauffman's New Vitalism.

As the title of theoretical biologist Stuart Kauffman's (1939-) best-known books indicates, the Santa Fe Institute complex systems researcher seeks to explain *Origins of Order* by considering *Self Organization and Selection in Evolution*. When he "seeks origins of order in the generic properties of complex systems" (1993, 644), Kauffman finds that even simple living systems are sufficiently complex to share in these generic system properties. He therefore emphasizes that an organism is "just another physical system" (2000, 5, 8, 109). Instead of highlighting the special features of biological organisms, Kauffman's "General Biology" considers all complex physical systems as autonomous agents that co-construct an evolving biosphere. These agents propagate organization by patterning the conditions that then enable them to contribute to the further development of these conditions. Together, Kauffman's autonomous agents write the story of the continuous production of novelty, a story that cannot be told in terms of algorithms alone since there are no algorithms powerful enough to create genuine novelty (2000, 124f.). This limit of traditional science forces Kauffman to go "beyond Newton, Einstein, and Bohr" (2000, 119, 123, 125):

> "If [...] we cannot prestate the configuration space, variables, laws, initial and boundary conditions of a biosphere, if we cannot foretell a biosphere, we can, nevertheless, tell the stories as it unfolds. Biospheres demand their Shakespeares as well as their Newtons. We will have to rethink what science is itself." (2000, 22)

Kauffman's general biology thus sacrifices predictability to make room for novelty. Indeed, the traditional demand for algorithms that lead from a finitely prestated configuration space to predictions of future states now appears as a constraint that is willfully imposed upon nature as something that will not be so constrained: "the biosphere is not hampered by our failure at categorization" (2000, 125, see 135). In order to make room for novelty and the idea of a living, evolving universe, one must also rethink the role played by conservation laws, most prominently in the account of chemical reactions. Modern chemistry was based, after all, on Lavoisier's conservation of matter and exclusion of novelty.

> "We may lay it down as an incontestable axiom, that, in all the operations of art and nature, nothing is created; an equal quantity of matter exists before and after the experiment." (1965, 130)

Kauffman replaces this picture by another one. Regardless of whether one can allow for an overall change in the quantity of matter, something novel is created as there is a change of universe in reference to which the total quantity of matter is defined: A chemical reaction moves us from the actual universe into its adjacent possible universe.

"Consider all the kinds of organic molecules on, within, or in the vicinity of the Earth, say, out to twice the radius of the moon. Call that set of organic molecules the 'actual.' [...] The adjacent possible consists of all those molecular species that are not members of the actual, but are *one reaction step away*. [...] The chemical potential of a single reaction with a single set of substrates and no products is perfectly definable [...] The substrates are present in the actual, and the products are not present in the actual, but only in the adjacent possible. It follows that every such reaction couple is displaced from its equilibrium in the direction of an excess of substrates compared to its products. The displacement constitutes a chemical potential driving the reaction toward equilibrium. The simple conclusion is that there is a real chemical potential from the actual to the adjacent possible." (2000, 142f.)

Kauffman's scientific vitalism endows all complex systems, including the biosphere and the universe as a whole, with intentionality and directionality: "the total system 'wants' to flow into the adjacent possible" and "[s]omething has obviously happened over the last 4.8 billion years. The biosphere has expanded, indeed, more or less persistently exploded, into the ever-expanding adjacent possible" (2000, 143). Biospheres and the universe construct themselves through this explosion into ever more diversified patterns of constraints as they tend chaotically into disequilibrium states and spontaneously regain order in an adjacent universe (2000, 4; 1993, 232).

Kauffman's general theory of order thus posits the behavior of complex systems as something that cannot be explained by the best traditional theories. He goes on to show how an account of this behavior reconstitutes "what science is itself."

20.6   Goodwin's Science of Complex Systems.

Just like Stuart Kauffman, Brian Goodwin (1931-2009) studied biological complexity through the lens of so-called mathematical or theoretical biology. In contrast to Kauffman, however, Goodwin took complex systems for his phenomena. Instead of offering a vitalist conception of biospheres or universes and instead of promoting a new way of "thinking in complexity," Goodwin envisions a science of complex systems that identifies mechanisms, seeks out algorithms and generalizations, and discovers causal dependencies in more or less familiar ways. However, this otherwise traditional enterprise is concerned with a different kind of entity, namely a complex system or, in Goodwin's terminology, a morphogenetic field—a term that owes considerably to Driesch's interest in the dynamics of the spatial organization of matter. On this account, organisms are powerful particulars that express their species-being dynamically in the course of development (Goodwin 1994, 116, 176-178; Webster and Goodwin 1996, 97, 100, 229f.). Quite literally, then, Goodwin deals with an entirely new kind of animal, one that has been rendered accessible through the algorithmic generation and computer simulation of complex systems. Indeed, one can say that these techniques provide Goodwin with the morphogenetic field, allowing him to investigate some of its behaviors.[8] But this construction of the domain of inquiry does not constitute a theory of morphogenetic fields, nor does it somehow explain them.

Implicitly at least, Goodwin himself reflects the distinction between his approach and Kauffman's. From Goodwin's point of view, Kauffman is concerned with generic properties of systems and not with biological organisms *sui generis*. Drawing on general properties of complex systems "to simulate a variety of biological processes," a view like Kauffman's maintains that

> "much (and perhaps most) of the order that we see in living nature is an expression of properties intrinsic to complex dynamic systems organized by simple rules of interaction among large numbers of elements. This order is generic, and what we see in evolution may be primarily an emergence of states generic to the dynamics of living systems." (Goodwin 1994, 186; see Webster and Goodwin 1996, 243f.)

In contrast to this way of viewing ordered organismic states as expressions of properties of complex dynamic systems, Goodwin emphasizes that "an organism or a work of art expresses a nature and a quality that has intrinsic value and meaning, with no purpose other than its own self-expression" (1994, 199). Organisms express only their own nature. Complexity, order, or integration are therefore not properties of organisms that require explanation but are simply what an organism is: "The emergent qualities that are expressed in biological form are directly linked to the nature of organisms as integrated wholes" (1994, 199). Biological form is therefore generic only for a particular taxon or species and should be referred to the morphologists' "unity of type" (1994, 143f.). Formal features that have no adaptive value are produced

> "not because natural selection has stabilised them and then cannot remove them, but because this is a natural form for this type of organism. [...] In this

---

8   The significance of this point will be highlighted in the conclusion of this paper.

sense the form can be explained in the same way that forms are explained in physics, as consequences of the way in which natural systems are organised in terms of their generative fields." (Webster and Goodwin 1996, 215)

For the natural process that is thus organized in a generative field Goodwin takes the term "hereditary inertia" in a strictly physical sense (Webster and Goodwin 1996, 197, 210). He thereby likens the organism's development of robust, intrinsic order to the movement of an inertial body through a field of forces. While the inertial movement itself requires no explanation but belongs to the nature of the organisms, the forces that impinge on this motion are genetic and environmental since both genes and the environment exert causal influence—sometimes through feedback loops—on the generative process (Goodwin 1994, 96f., 176f.; see Goodwin and Webster 1996, 246-250).

Kauffman wants to be Newton and Shakespeare in one and within a new way of doing science to tell stories of evolving biospheres. In contrast, Goodwin's ideal is to "transform biology from a purely historical science to one with a logical, dynamic foundation" by constructing "a theory of biological forms whose equivalent in physics is the periodic table of elements" (Goodwin 1994, 114, see 196 and Goodwin and Webster 1996, 215, 230, 250).

20.7   Conclusion.

Stuart Kauffman and Brian Goodwin agree to (1') and (2') and thus to Hans Driesch's general program of a new vitalism that acknowledges entelechies which cannot be accounted for in the terms of mechanistic science. But they disagree i) whether the envisioned biological science should represent the emergence of novelty or only the successions of states within a single, conserved universe; ii) whether it should take non-linearity as its paradigm or seek predictability and control by identifying mechanisms and causal dependencies; iii) whether it needs to become more overtly historical (telling stories) or more closely associated with the ahistorical physical sciences. So profound is this disagreement that it is difficult to fathom how Kauffman and Goodwin could be lumped together in popular presentations of the emergence of a new science of complexity (e.g., Lewin 1999, Waldrop 1993). At the same time, it does not come as a surprise that their disagreement is underwritten by ideological difference. Kauffman wrote as a neo-liberal who treats the production of unpredictable, yet generically inevitable biological novelty in analogy to unpredictable, yet generically inevitable innovation in a free-market economy (2000, 211-241). In contrast, when Goodwin seeks to exhibit the intrinsic value and meaning of organisms that have no other purpose than to express their nature, this is the view of a neo-marxist who offers a scientific foundation for a non-exploitative, unalienated relation to nature (1993, xvf., 200-237; 1994, 199).

In the present context, the very fact of the disagreement between Kauffman and Goodwin signifies that the recognition of "complex systems" changes our view of what kind of things there are in the world but does not necessitate a new kind of thinking or a different way of doing science. Indeed, if organisms are complex systems or integrated wholes (Driesch's "totalities"), they are therefore no aggregates of simple components or processes (Driesch's "sums"). This insight marks a limit to scientific understanding with its implicit commitment to compositionality, conservation, linearity, or uni-directional mechanical causality. However, the recognition of this limit need not herald an entirely different kind of science. By arguing along the lines of Goodwin that a science of complexity takes certain structures with their natures as its object, their entelechies become, in effect, black-boxed. Goodwin's science of complexity does not seek to explain how or what these are but only how they behave in response to causal influences.

This is where the story of Kauffman and Goodwin and their revivals of scientific vitalism comes full circle and joins up with the rather more contemporary considerations of technoscientific developments in the Introduction above. It is impossible to overestimate in this story the role of computing and the computer. It allowed for the creation, representation, systematic variation and therefore investigation of complex systems as self-contained, well-defined units with specific variable behaviors. By being implemented in a physical, electronic device these could be identified as fully material systems. At the same time, this implementation affords black-boxing: the system's behavior and sensitivity to parameter-variations can be studied without considering the physical processes that produce them. Whereas Stuart Kauffman took these implementations as illustrations or *models of* the theory and thus focused exclusively on their representational character, Brian Goodwin took these as model organisms and thus as *models for* similarly black-boxed biological organisms whose behavioral properties and parameter-dependencies he was studying (cf. Fox-Keller 2000).

This resonates with the technoscientific transition from systems biology to synthetic biology. Arguably, systems biology relies on the computer as an information processing device that can represent whole systems by integrating vast amounts of data. Synthetic biology takes this a step further by recognizing that these systems are materially implemented dynamic structures and thereby objects which are subject to variation and manipulation, as such blue-prints for biological design

(Nordmann 2014). In other words, synthetic biology is distinct from molecular or genetic engineering in that it is not biological structures or organisms that serve as models for other organisms but in that computer models serve that function. As a technoscience, synthetic biology does not share the mindset of the science of biology, but even so, its manner of dealing with and generating complexity also does not provide a paradigm for a new science of biology: Goodwin's scientific vitalism places complex systems in the black box of nature. Technoscientific synthetic biology places them in the black box of engineered devices.

References

Driesch, Hans (1908) *The Science and Philosophy of the Organism*, London: Black, 2 vols.
Driesch, Hans (1914) *The History and Theory of Vitalism*, London: Macmillan.
Fox Keller, Evelyn (2000) "Models Of and Models For: Theory and Practice in Contemporary Biology," *Philosophy of Science*, 67 (Proceedings), pp. S72-S86.
Fox-Keller (2007) "The disappearance of function from 'self-organizing systems'", in Fred Boogerd, Frank J. Bruggeman, Jan-Hendrik S. Hofmeyr, H.V. Westerhoff (eds.) *Systems Biology: Philosophical Foundations*, Amsterdam: Elsevier, pp. 303-317.
Frank, Philipp (1998) *The Law of Causality and its Limits*, Dordrecht: Kluwer.
Goodwin, Brian (1994) *How the Leopard Changed Its Spots: The Evolution of Complexity*, New York: Charles Scribner's.
Hey, Tony, Stewart Tansley, Kristin Tolle (eds.) (2009) *The Fourth Paradigm: Data-Intensive Scientific Discovery*, Redmond: Microsoft Research.
Kauffman, Stuart (1993) *The Origins of Order: Self-Organization and Selection in Evolution*, New York: Oxford University Press.
Kauffman, Stuart (2000) *Investigations*, Oxford: Oxford University Press.
Lavoisier, Antoine (1965) *Elements of Chemistry*, New York: Dover.
Lewin, Roger (1999) *Complexity: Life at the Edge of Chaos*, 2nd ed. Chicago: University of Chicago Press.
Mainzer, Klaus (2007) *Thinking in Complexity*, 5th ed. New York: Springer.
Nordmann, Alfred (2010) „Enhancing Material Nature," in Kamilla Lein Kjølberg and Fern Wickson (eds.) *Nano meets Macro: Social Perspectives on Nanoscale Sciences and Technologies*, Singapore: Pan Stanford, pp. 283-306.
Nordmann, Alfred (2014) „Synthetic Biology at the Limits of Science," in Bernd Giese, Christian Pade, Henning Wigger, Arnim von Gleich (eds.) *Synthetic Biology: Character and Impact*, Berlin: Springer, pp. 31-58.
Schmidt, Jan C. (2008) *Instabilität in Natur und Wissenschaft. Eine Wissenschaftsphilosophie der nachmodernen Physik*, Berlin: De Gruyter, 2008.

Schmidt, Jan C. (2014) "Synthetic Biology as Late-Modern Technology," in Bernd Giese, Christian Pade, Henning Wigger, Arnim von Gleich (eds.) *Synthetic Biology: Character and Impact*, Berlin: Springer, pp. 1-30.

Waldrop, Mitchell (1993) *Complexity: The Emerging Science at the Edge of Order and Chaos* (New York: Touchstone).

Webster, Gerry and Brian Goodwin (1996) *Form and Transformation: Generative and Relational Principles in Biology*, Cambridge: Cambridge University Press.

Wimsatt, William (2007) *Re-Engineering Philosophy for Limited Beings: Piecewise Approximations to Reality*, Cambridge: Harvard University Press.

21. Algorithmic Data Analytics, Small Data Matters and Correlation versus Causation[1]

Hector Zenil[2]

Abstract

This is a review of aspects of the theory of algorithmic information that may contribute to a framework for formulating questions related to complex, highly unpredictable systems. We start by contrasting Shannon entropy and Kolmogorov-Chaitin complexity, which epitomize correlation and causation respectively, and then surveying classical results from algorithmic complexity and algorithmic probability, highlighting their deep connection to the study of automata frequency distributions. We end by showing that though long-range algorithmic prediction models for economic and biological systems may require infinite computation, locally approximated short-range estimations are possible, thereby demonstrating how small data can deliver important insights into important features of complex "Big Data".

---

1   The chapter is based an invited talk delivered to UNAM-CEIICH via videoconference from The University of Sheffield in the U.K. for the Alan Turing colloquium "From computers to life" (http://www.complexitycalculator.com/TuringUNAM.pdf) in June, 2012.

2   Hector Zenil is a Principal Investigator and Assistant Professor affiliated to the Department of Computer Science, University of Oxford in the UK; and the Unit of Computational Medicine and SciLifeLab of the Karolinska Institute in Sweden. After a PhD in Theoretical Computer Science from the University of Lille 1 in France and a PhD in Philosophy and Epistemology awarded by the Sorbonne (Paris 1), he joined the Behavioural and Evolutionary Lab, University of Sheffield in the UK. He is also the head of the Algorithmic Nature Group and has been a visiting scholar and professor at MIT/NASA, Carnegie Mellon University and the National University of Singapore.

21.1   Introduction

Complex systems have been studied for some time now, but it was not until recently that sampling complex systems became possible, systems ranging from computational methods such as high-throughput biology, to systems with sufficient storage capacity to store and analyse market transactions. The advent of *Big Data* is therefore the result of the availability of computing power to process these complex systems, from social to economic to physical and biological systems. Much has been said on the importance of being able to deal with large amounts of data, amounts that require division into smaller segments if said data is to be analysed, and once understood, then exploited. As it can be seen in 1, generated digital information (mostly unstructured) is currently growing by 2 times the information stored but both stored and generated are outperforming computer power.

For example, common mathematical models of the dynamics of market prices (e.g., the Black-Scholes model) assume *a geometric Brownian motion*. In and of itself (the model can and is commonly tweaked) a geometric Brownian motion implies that price changes accumulate in (log-normal) Gaussian distributions prescribing constant volatility and a controlled, risk-free financial environment. These models work fine *on the average day* but are of limited use in turbulent times. In the long term, volatility is far from constant when price movement data is plotted; it is extreme price changes that bring on very rough behaviour (Mandelbrot and Hudson, 2005). As shown in (Zenil and Delahaye, 2011), the long-tail distribution observed in market prices may be due to behavioural decisions that resonate with algorithmic mechanisms, reflecting an algorithmic signature.[3]

The concepts and methods reviewed here are useful in cases where the complex systems' data to be divided is incomplete or noisy and cannot be analysed in full in the search for local regularities of interest whose frequency is an indication of an exploitable algorithmic signature, allowing the profiling of otherwise unmanageable amounts of data in the face of spurious correlations. This is a foundation of what I call *algorithmic data analytics*.

---

3   Furthermore, today more than 70% of trading operations in the stock market are conducted automatically, with no human intervention, cf. algorithmic trading and high-frequency trading (before human traders are even capable of processing the information they register).

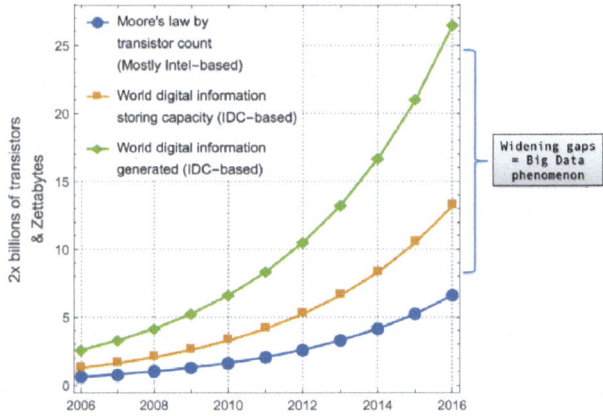

Fig. 21.1: Computational power versus generation and storage of digital data. The gap explains the 'Big Data' phenomenon and the plot shows that the gap is unlikely, if not impossible, to close and will rather widen in the future. The plot can only be fair if Moore's law is also multiplied by the world number of computers but their number per capita will never grow exponentially in average by number of CPUs or GPUs in the future and so this is a constant with no real effect. Most generated digital data is algorithmic even if unstructured, that means it is the result of a causal recursive (algorithmic) mechanism, e.g. genome sequencing or TV broadcasting. IDC source: IDC's Digital Universe in 2020, sponsored by EMC, Dec 2012.

21.1.1   Descriptive complexity measures

Central to information theory is the concept of Shannon's information entropy, which quantifies the average number of bits needed to store or communicate a message. Shannon's entropy determines that one cannot store (and therefore communicate) a symbol with n different symbols in less than $\log(n)$ bits. In this sense, Shannon's entropy determines a lower limit below which no message can be further compressed, not even in principle. Another application (or interpretation) of Shannon's information theory is as a measure for quantifying the *uncertainty* involved in predicting the value of a random variable.

For an ensemble $X(R, p(x_i))$, where R is the set of possible outcomes (the random variable), $n = |R|$ and $p(x_i)$ is the probability of an outcome in R. The Shannon information content or entropy of X is then given by

$$H(X) = -\sum_{i=1}^{n} p(x_i) \log_2 p(x_i)$$

Which implies that to calculate $H(X)$ one has to know or assume the mass distribution probability of ensemble X.

The Shannon entropy of a string such as 0101010101... at the level of single bits, is maximal, as there are the same number of 1s and 0s, but the string is clearly regular when two-bit blocks are taken as basic units, in which instance the string has minimal complexity because it contains only 1 symbol (01) from among the 4 possible ones (00,01,10,11). One way to overcome this problem is by taking into consideration all possible "granularities" (we call this *entropy rate*), from length 1 to n, where n is the length of the sequence. To proceed by means of entropy rate is computationally expensive as compared to proceeding in a linear fashion for fixed n, as it entails producing all possible overlapping $\binom{i}{n}$ substrings for all $i \in \{1,...,n\}$.

Most, if not all, classical methods and tools in data analytics are special cases, or have their roots in a version of this type of statistical function, yet as we will see later on, it is highly prone to produce false positives of data features, as the result of spurious correlations.

21.2   Computability theory for data analytics

A function f is said not to be computable (uncomputable or undecidable) if there is no computer program running on a *universal* Turing machine that is guaranteed to produce an output for its inputs or, in other words, if the machine computing f doesn't halt for a number of inputs. A universal Turing machine is a general-purpose machine that can behave like any specific purpose Turing machine. The theory of algorithmic information proves that no computable measure of complexity can test for all (Turing) computable regularity in a dataset. That is, there is no test that can be implemented as a Turing machine that takes the data as input and proceeds to retrieve a regularity that has been spotted (e.g., that every 5th place is occupied by a consecutive prime number). A computable regularity is a regularity for which a test can be set as a computer program running on a specific-purpose Turing machine testing for that regularity. Common statistical tests, for example, are computable because they are meant to be effective (on purpose), but no computable *uni-*

versal measure of complexity can test for every computable regularity. Hence only noncomputable measures of complexity are up to the task of detecting any possible (computable) regularity in a dataset, given enough computing time. Therefore:

$$\text{Universality} \Rightarrow \text{Uncomputability}$$

Which is to say that universality entails or implies uncomputability, for if a system is computable (or decidable), then it cannot be universal. One may wonder why more power is needed when dealing with finite strings in limited time (real data). Though finite in length, the number of possible finite strings is infinite, and detecting only a fraction of regularities leaves an infinite number of them undetected. This is therefore a real concern at the limits of what is and is not computable.

21.2.1   Finding the generating mechanism

The aim of data analytics (and for that matter of science itself) is to find the mechanisms that underlie phenomena, for purposes such as prediction. The optimal way to do this is to find the generating program, if any, of a piece of data that is of interest. To this end, the concept of algorithmic complexity (also known as Kolmogorov or Kolmogorov-Chaitin complexity) is key. The algorithmic complexity $K(x)$ of a piece of data x is the length of the shortest program p that produces x running on a universal Turing machine U (or a Turing-complete language, i.e., a language expressive enough to express any *computable object*). Formally,

$$K_U(x) = min\{|p|,\ U(p) = x\} \tag{1}$$

Data with low algorithmic complexity is compressible, while random "data" is not (Kolmogorov, 1968; Chaitin, 1969). For example, strings like 1111111111... or 010101010... have low algorithmic complexity because they can be described as n times 1 or m times 01. Regardless of their size the description only grows by about $log(k)$ (the integer size of n or m). The string 01101011001101101011001, however, has greater algorithmic complexity because it does not seem to allow a (much) shorter description than one that includes every bit of the string itself, so a shorter description may not exist.

For example, the repetitive string 01010101... can be produced *recursively* by the following program p:

1: N:= 0

2: PRINT N

3: N:= N+1 MOD 2

4: GOTO 2

The length of this p (in bits) is thus an upper bound of the complexity K of $010101\ldots 01$ (plus the size of the line to terminate the program at the desired string length). Producing a shorter version of a string is a test for non-randomness, but the lack of a short description (program) is not a guarantee of randomness.

As we will discuss, K is extremely powerful, if we bear in mind the caveat that it can only be approximated. Before explaining the other side of the coin, namely algorithmic probability (AP), let's study some properties of algorithmic complexity that are relevant (because AP encompasses all of them too) to an algorithmic approach to data analytics.

21.2.2 Uncomputability and the choice of language

Tailored measures of "complexity" have often been used for pattern detection in areas such as data analytics. Many of these "complexity" measures are based on Shannon's entropy. The chief advantage of these measures is that they are computable, but they are also very limited, and we are forced to focus on a small subset of properties that such a measure can test. A computable measure is implemented by an effective algorithm that, given an input, retrieves the expected output, for example, the Shannon entropy of a string. Attempts to apply more powerful *universal* (universal in the sense of being able to find any pattern in the data) measures of complexity traditionally face two challenges:

- Uncomputability, and
- Instability.

On the one hand, K would only make sense as a complexity measure for describing data if it showed signs of stability in the face of changes in the language used to describe an object. An invariance theorem guarantees that in the long term different algorithmic complexity evaluations (with different description languages) will converge to the same values as the size of data grows.

If U_1 and U_2 are two (universal) Turing machines and $K_{U_1}(x)$ and $K_{U_2}(x)$ the algorithmic complexity of a piece of data x when U_1 or U_2 are used respectively, there exists a constant c such that for all "binarisable" (discrete) data x (Kolmogorov, 1968; Chaitin, 1969; Solomonoff, 1964):

$$|K_{U_1}(x) - K_{U_2}(x)| < c_{U_1, U_2} \qquad (3)$$

One can think of $c_{U1, U2}$ as the length of the computer program that translates the programs producing x on U_2 in terms of U_1 and the other way around, usually called a compiler.

On the other hand, uncomputability has for a long time been seen as the main "drawback" of K (and related measures). No algorithm can tell whether a program p generating x is the shortest (due to the undecidability of the halting problem of Turing machines). But the uncomputability of $K(x)$ is also the source of its power, namely, the universality of this measure as based on a reference universal Turing machine. This is because measures based on non Turing-universal machines (eg finite automata) do not allow a general universal measure capable of characterizing every possible regularity in the data, including non-statistical ones, such as recursive or algorithmic (those that can compress the data by using generating mechanisms even if not obviously statistically present). While for a single piece of data only a finite number of possible patterns can be associated to that piece of data, upper bounded by its Shannon entropy (all possible combinations of the data symbols), the number of all possible patterns (statistical and algorithmic) in *any* piece of data (eg strings) is uncountably infinite by a Cantor diagonalization (cf. 4). Martin-Löef proves (Martin-Löf, 1966) that no computable measure can capture the typicality of all possible strings and therefore computable measures will miss specific important properties of these strings. He also proves there is a *universal test* that depends on the universality of the concept of effective method implemented by a universal Turing machine.

However, to be more specific, K is an upper semi-computable function, because it can be approximated *from above*. Lossless compression algorithms have traditionally been used to find short programs for all sorts of data structures, from strings to images and music files. This is because lossless compression algorithms are required to reproduce the original object, while the compressed version of the object together with the decompression program constitutes a program that when decompressed reconstructs the original data. The approximation to K is thus the size in bits of the compressed object together with the size in bits of the decompression program. By this means, one can, for example, find a short program for a piece of data x, shorter than the length $|x|$ of x. But even though one cannot declare an object x not to have a shorter program, one can declare x not random if a program generating x is (significantly) shorter than the length of x, constituting an upper bound on K (hence approximated from *above*).

21.2.3   Data from a system's unpredictability

Formalised by Peter Schnorr, the unpredictability approach to randomness involves actors and concepts familiar to economists and risk managers. It establishes that a gambler cannot make money following a computable betting system M against the digits of a random sequence x. In equational form,

$$\lim_{n \to \infty} M(s \uparrow n) = \infty \qquad (2)$$

Equivalently, a piece of data x is statistically atypical (also known as Martin-Löf random) if and only if there does not exist a computable enumerable martingale M that succeeds on x, that is, there exists no effective computable betting strategy. It then establishes a formal connection between concepts that were previously thought to be only intuitively related:

simple $\Leftrightarrow$ predictable

random $\Leftrightarrow$ unpredictable

That is, what is random is unpredictable and what is unpredictable is random; otherwise it is simple. For example, program A (cf. *Example of an evaluation of K)* trivially allows a shortcut to the value of an arbitrary digit through the following function $f(n)$ (if $n = 2m$ then $f(n) = 1$, $f(n) = 0$ otherwise) (Schnorr, 1971).

A series of *universality results* (both in the sense of *general* and in the sense of Turing *universal,* the latter concept being a version of the former) (Kirchherr et al., 1997) leads to the conclusion that the definition of random complexity is mathematically objective:

- Martin-Löf proves (Martin-Löf, 1966) that there is a *universal* (but uncomputable) statistical test that captures all computably enumerable statistical tests. His definition of randomness is therefore general enough to encompass all effective tests for randomness.
- Solomonoff (Solomonoff, 1964) and Levin (Levin, 1974) prove that the concept of *universal search* (cf. algorithmic probability) is the optimal learning strategy with no prior language. In fact Levin's distribution is also called *the Universal Distribution* (see Section 3).
- Schnorr (Schnorr, 1971) shows that a predictability approach based on martingales leads to another characterization of randomness, which in turn is equivalent to Martin-Löf randomness.
- Chaitin (Chaitin, 1975) proves that an uncompressible sequence is Martin-Löf random and that Martin-Löf randomness implies uncompressibility. That is, se-

quences that are complex in the Kolmogorov-Chaitin sense are also Martin-Löf random.

- The confluence of the above definitions (cf see explanation below).

One has to sift through the details of these measures to find their elegance and power. Kolmogorov, Chaitin, Levin, Schnorr, and Martin-Löf independently devised approaches to algorithmic randomness. It follows that the algorithmic characterizations of randomness converge. That means that the definitions are coextensive in the sense that the complex elements for one measure are also the complex elements for the others definitions, and the low complex elements are also low for the others. Furthermore, all the definitions assign exactly the same *complexity* values to the same objects, hence they are equivalent to each other. In summary:

$$\text{uncompressibility} \Leftrightarrow \text{unpredictability} \Leftrightarrow \text{typicality}$$

When this convergence happens in mathematics it is believed that a concept (in this case randomness) has been objectively captured.

This is a significant contribution to science which complex systems' models should take into account and build upon instead of designing new and ad hoc new measures of complexity, sometimes contradicting each other.

For algorithmic universal measures it does not matter whether the source is algorithmic and deterministic, it all boils down to the fact that if a phenomenon is unpredictable or uncompressible or statistically typical (there are no regularities in the data), then the model too will be unpredictable, uncompressible and statistically typical.

21.2.4 More structured data has more short descriptions

One preliminary observation is that even if one cannot tell when data is truly random, most data cannot have much shorter generating programs than themselves. For strings and programs written in binary a basic counting argument tells that most data must be random:

- There are exactly 2^n bit strings of length n, among all these strings
- there are only $2^0 + 2^1 + 2^2 + \ldots + 2^{(n-c)} = 2^{(n-c+1)} - 1$ bit strings of c fewer bits.
- In fact there is one that cannot be compressed even by a single bit.
- Therefore there are considerably fewer short programs than long programs.

Thus, one cannot pair off all n-length binary strings with binary programs of much shorter length (there simply aren't enough short programs to encode all longer strings).

21.2.5 Non-random data does not have disparate explanations

Next to the uncomputability of K, another major objection to K_U is its dependence on a universal Turing machine U. It may turn out that:

$K_{U_1}(x) \neq K_{U_2}(x)$ when evaluated using U_1 and U_2 respectively.

This dependency is particularly troublesome for small data, e.g., short strings, shorter, for example, than the length of T, the universal Turing machine on which K_T is evaluated (typically on the order of hundreds to thousands—a problem originally identified by Kolmogorov himself). As pointed out by Chaitin (Zenil, 2011):

> "The theory of algorithmic complexity is of course now widely accepted, but was initially rejected by many because of the fact that algorithmic complexity is on the one hand uncomputable and on the other hand dependent on the choice of universal Turing machine."

The latter downside is especially restrictive in real world applications because this dependency is particularly true of small data (e.g., short strings). Short strings are common in various disciplines that need measures capable of exploiting every detectable regularity. Biology is one instance, specifically in the area of DNA and protein sequences, and relating particularly to identification—questions such as which genes map which biological functions or what shape a protein will fold into.

21.2.6 Compressing small data to exploit big data

A problem common to implementations of lossless compression algorithms trying to compress data is that the compressed version of the data is the compressed data together with the program that decompresses it (i.e., the decompression instructions). If the data is too small the decompression program will dominate its overall complexity length, making it difficult to produce values for comparisons for data profiling purposes, for which sensitivity is key (e.g., the impact on the data of a small perturbation). If one wished to tell which of two pieces of small data are objectively more or less random and complex using a compression algorithm, one quickly finds out that there is no way to do so. The constant involved c_{U_1, U_2} comes from the Invariance theorem and it can be arbitrarily large, providing unstable evaluations of K(x), particularly for small data (e.g., amounting to properties of interest or trends in big data).

In (Zenil, 2011; Delahaye and Zenil, 2012; Soler-Toscano et al., 2014) we introduced a novel alternative to compression using the concept of algorithmic probability, which is proving to be useful for applications (Soler-Toscano et al., 2013; Zenil et al., 2015) and which we believe is relevant to complex systems where large amounts of data are generated.

21.3   Algorithmic hypothesis testing

There is a measure that defines the probability of data being produced by a random program running on a universal Turing machine (Solomonoff, 1964; Levin, 1974). This measure is key to understanding data in the best possible scientific way, that is, by exploring hypotheses in search of the most likely genesis of the data, so as to be able to make predictions. Formally,

$$m(x) = \sum_{p:U(p)=x} 1/2^{|p|} < 1 \tag{4}$$

i.e. it is the sum over all the programs that produce x running on a universal (prefix-free[4]) Turing machine U. The probability measure $m(x)$ is traditionally called Levin's semi-measure, Solomonoff-Levin's semi-measure or *algorithmic probability* and it defines what is known as the *Universal Distribution (semi* from *semi-decidable* and the fact that $m(x)$, unlike probability measures, doesn't add up to 1). $m(x)$ formalises the concept of Occam's razor (favouring simple—or shorter—hypotheses over complicated ones). algorithmic probability is deeply related to another fascinating uncomputable object that encodes an infinite amount of uncomputable and accessible information.

---

4   The group of valid programs forms a prefix-free set (no element is a prefix of any other, a property necessary to keep $0 < m(x) < 1$). For details see (Calude, 2013).

21.3.1   Real big data and Chaitin's Ω *number*

m is closely related to Chaitin's halting probability, also known as Chaitin's Ω number (Chaitin, 1969) defined by:

$$\Omega_U = \sum_{U(p) \text{ halts}} 1/2^{|p|} \qquad (5)$$

(the halting probability of a universal *prefix-free* Turing machine U)

Evidently, $m_U(x)$ provides an approximation of Ω_U plus the strings produced (like Ω_U, $m(x)$ is uncomputable, but just as is the case with Ω_U, $m(x)$ can sometimes be calculated up to certain *small* values).

While a Borel normal number locally contains all possible patterns, including arbitrarily long sequences of 0s or 1s, or any other digit, thereby containing all possible statistical correlations, all of them will be spurious because a Borel normal number has no true global statistical features (given the definition of Borel normality). A Chaitin Ω, on the other hand, epitomises *algorithmic Big Data*, where no statistical correlation is an algorithmic feature of the data. This is because a Chaitin Ω number is not only Borel normal but also algorithmically random, and therefore cannot be compressed. In fact, there are Chaitin Ω numbers for which none of the digits can be estimated and are therefore perfectly algorithmically random. Algorithmic Big Data is a subset of statistical Big Data and therefore reduces the likelihood of spurious hypotheses based on correlation results. The move towards algorithmic tools is therefore key to the practice of data science in the age of Big Data and complex systems.

21.3.2   Optimal data analytics and prediction

The concept of algorithmic complexity addresses the question of the randomness content of individual objects (independent of distributions). Connected to algorithmic complexity is the concept of algorithmic probability (AP), that addresses the challenge of hypothesis testing in the absence of full data and the problem of finding the most likely process generating the data (explanation). Algorithmic probability and algorithmic complexity are two faces of the same coin.

While the theory is very powerful but is hard to calculate estimations of complexity, it does make two strong strong assumptions that are pervasive in science and that does not mean they are necessarily right but seemingly reasonable (any other assumption seems less reasonable):

• The generating mechanism is algorithmic as opposed to uncomputable, e.g., the result of an 'oracle' (in the sense of Turing, i.e. a non-computable source), magic, or divine intervention.

• The likelihood distribution of programs is a function of their length (formalisation of Occam's razor)

The assumptions are reasonable because consonant with the original purpose of analysing the data with a view to identifying the generating mechanisms that would explain them, and while in assigning greater probability to shortest explanations consistent with the data (Occam's razor) they do favour the simplest computer programs, they also entail a Bayesian prior for predicting all the generating programs (all possible explanations accounting for the data regularities), hence complying with Epicurus' principle of multiple explanations. While one can frame criticisms of these principles, both principles are common and among the most highly regarded principles in science, thus transcending the practicalities of performing algorithmic data analytics. One of the chief advantages of algorithmic probability as an implementation of Occam's razor is that it is not prone to over-fitting. In other words, if the data has spurious regularities, only these and nothing else will be captured by an algorithmic probability estimation, i.e., by definition no additional explanations will be wasted on the description of the data.

21.3.3 Complexity and frequency

The intuition behind the Coding theorem (5) is a beautiful relation between program-size complexity and frequency of production. If you wished to produce the digits of a mathematical constant like π by throwing digits at random, you would have to produce every digit of its infinite irrational decimal expansion. If you seated a monkey at a typewriter (with, say, 50 keys), the probability of the monkey typing an initial segment of 2400 digits of π by chance is $(1/50^{2400})$.

If, instead, the monkey is seated at a computer, the chances of its producing a program generating the digits of π are on the order of $1/50^{158}$, because it would take the monkey only 158 characters to produce the first 2400 digits of π using, for example, C language.

So the probability of producing x or U by chance so that $U(p) = x$ is very different among all (uniformly distributed) data of the same length: $1/2^{|x|}$ and the probability of finding a binary program p producing x (upon halting) among binary programs running on a Turing machine U is $1/2^{|p|}$ (we know that such a program exists because U is a universal Turing machine).

The less random a piece of data the more likely it is to be produced by a short program. There is a greater probability of producing the program that produces the data, especially if the program producing x is short. Therefore, if an object is compressible, then the chances of producing the compressed program that produces said object are greater than they would be for a random object, i.e., $|p| << |x|$ such that $U(x) = x$.

The greatest contributor to the sum of programs $\Sigma_{U(p)=x}\ 1/2^{|p|}$ is the shortest program p, given that this is when the denominator $2^{|p|}$ reaches its smallest value and therefore $1/2^{|p|}$ reaches its greatest value. The shortest program p producing x is none other than $K(x)$, the algorithmic complexity of x.

For $m(x)$ to be a probability measure, the universal Turing machine U has to be a prefix-free Turing machine, that is, a machine that does not accept as a valid program one that has another valid program as its beginning. For example, program 2 starts with program 1, so if program 1 is a valid program then program 2 cannot be valid.

The set of valid programs is said to constitute a prefix-free set, that is, no element is a prefix of any other, a property necessary to keep $0 < m(x) < 1$. For more details see the discussion of Kraft's inequality, in (Calude, 2013).

21.3.4   Massive computation in exchange for an arbitrary choice

In order to truly approximate Kolmogorov-Chaitin complexity rather than, for example, entropy rate (which is what is achieved by currently popular lossless compression algorithms based on statistical regularities—repetitions—captured by a sliding window traversing a string), we devised a method based on algorithmic probability (Delahaye and Zenil, 2012; Soler-Toscano et al., 2014).

Let $(n,2)$ be the set of all Turing machines with n states and 2 symbols, and let $D(n)$ be the function that assigns to every finite binary string x the quotient:

$$\frac{\#\ \text{of times that a machine}\ (n,2)\ \text{produces}\ x}{\#\ \text{of machines in}\ (n,2)} \tag{6}$$

As defined in (Delahaye and Zenil, 2012; Soler-Toscano et al., 2014), $D(n)$ is the probability distribution of the data (strings) produced by all n-state 2-symbol Turing machines. Inspired by m, $D(n)$ is a finite approximation to the question of the algorithmic probability of a piece of data to be produced by a random Turing machine up to size n. Like $m(x)$, $D(n)$ is uncomputable (by reduction to Rado's Busy Beaver problem), as proven in (Delahaye and Zenil, 2012). Examples for $n = 1$, $n = 2$ (normalised by the # of machines that halt):

$$D(1) = 0 \rightarrow 0.5;\ 1 \rightarrow 0.5$$

$$D(2) = 0 \rightarrow 0.328;\ 1 \rightarrow 0.328;\ 00 \rightarrow .0834\ldots$$

By using the Coding theorem Eq. 7 one can evaluate $K(x)$ through $m(x)$, which reintroduces an additive constant. One may not get rid of the constant, but the choices related to $m(x)$ are less arbitrary than picking a universal Turing machine directly for $K(x)$, and we have proven that the procedure is not only theoretically sound but stable (Zenil et al., 2015), and in accordance with strict program-size complexity and compressibility (Soler-Toscano et al., 2013), the other traditional method for approximating $K(x)$, which fails for small data that an approximation of $m(x)$ handles well. The trade-off, however, is that approximations of $m(x)$ require extraordinary computational power. Yet we were able to calculate relatively large sets of Turing machines to produce $D(4)$ and $D(5)$ (Delahaye and Zenil, 2012; Soler-Toscano et al., 2014). D can be seen as averaging K over a large set of possible languages in order to reduce the possible impact of the constant from the Invariance theorem. We call CTM to the calculation of D and stand for the *Coding theorem method*.

It is common that an approach to deal with the involved constant in the Invariance theorem is to choose a small *reference* universal Turing machine on which all evaluations of K are made, but an arbitrary choice is still made. There is another approach to this challenge, and that is to exchange the choice of a reference machine for what may appear a less arbitrary choice of enumerating scheme. One can take the quasi-lexicographical enumeration as a natural choice because it sorts all possible Turing machines by size, hence consistent with the theory itself. Moreover, one can run all Turing machines of certain size at the same time and get a perfect sample of the whole space. The choice is still arbitrary in the sense that one could enumerate all possible Turing machines (which is itself a Turing universal procedure) in any other way, but the choice is *averaged* over a large number of small Turing machines.

21.4   Correlation versus causation

In (Calude and Longo, 2015) it has been shown that for increasing amounts of data, spurious correlations will increase in number and distinguishing between false and true positives becomes more challenging.

While Shannon entropy is a very common and, in some applications, a powerful tool, it can be described as a counting function that involves a logarithm. As such, it can count arbitrary elements. In a network, for example, it can reckon edge density, edge count, connected edge count, etc., from which it is clear that not all definitions converge, i.e., given one definition it is not always true that high or low entropy will remain high or low entropy for some other definition.

According to Shannon entropy, specifying the outcome of a fair coin flip (two equally likely outcomes) requires one bit at a time, because the results are *independent* and therefore each result conveys maximum entropy. This *independence* is generally in appearance only, and has to be construed as *statistical independence*.

While both definitions, entropy rate and Kolmogorov-Chaitin (algorithmic) complexity, are asymptotic in nature, they are essentially different, and connected in a rather trivial fashion. Low Shannon entropy implies an exploitable regularity that can be captured by recursion, i.e., a small computer program. However, maximal entropy rate (*Shannon randomness*) does not imply algorithmic randomness.

By means of the so-called *universal statistical test* (which tests for all effective—computable—statistical tests using a universal Turing machine), Martin-Löf proved that algorithmic randomness can test for every effective (computable) property of an infinite sequence. In practice, one does not apply such a test to infinite objects. However, the number of finite objects is classically numerable as infinite, thus leading to an infinite number of different features. In the context of networks, for example, every network can be a feature by itself, testing membership to a set with only itself in it, but in general one may be interested in properties such as clustering coefficients, motifs, geodesics, certain kinds of betweenness, etc., an infinite number of properties. This has significant implications for data science and big data.

We say that a function Δ is robust if for an object s, Δ_T is invariant to the description T, where T is a description that fully recovers s. In other words,

Definition: Let $\Delta_T(s)$ from $\Delta_S(s)$ be the same function applied to the same object s described using descriptions T and S. If $|\Delta T - \Delta S| > c$ then Δ is not a robust measure. Which is to say that a robust definition is a definition that is at most a constant away from each other, and thus grows in the same direction and rate for the same objects.

This definition is a generalization of the *Invariance theorem* in the theory of algorithmic information that does not require T or S to be computer programs to be run on universal Turing machines and therefore allow Δ to be also a computable function (i.e. one for which there is always a value for every evaluation of Δ).

By the *Invariance theorem*, Kolmogorov-Chaitin complexity and algorithmic probability are robust as they are invariant up to a constant to different object descriptions.

Observation 1: Entropy rate is more robust than Shannon entropy for an arbitrary choice of domain (tuple or n-gram size), but entropy rate is not invariant to description changes (e.g., 1-grams versus n-grams, or edges vs connected edges).

Observation 2: According to Martin-Löf's results, for every effective (computable) property X, there is a computable measure Y to test whether every system x has a property X, but there exists a property Z for some x that Y cannot test. The results also imply that for a measure to test every effective statistical property of any possible system the measure has to be uncomputable, and that any uncomputable measure of lowest degree implemented on a universal Turing machine will only differ by a constant from any other uncomputable measure of the same type. Hence, all uncomputable complexity measures are asymptotically the same.

In other words, for every effective feature, one can devise/conceive an effective measure to test it, but there is no computable measure (e.g., entropy rate) able to implement the universal statistical test, only a measure of algorithmic randomness (e.g., Kolmogorov-Chaitin complexity or algorithmic probability).

That no computable function can detect all possible (enumerable effective) computable regularities can be proven by contradiction using a diagonalization argument. By definition every computable regularity can be implemented as a computable test that retrieves 0 or 1 if the regularity is present in the input. Let q be an enumeration of all Turing machines, each implementing a computable test and let r be an enumeration of the computable regularities. We then take the negation of every (i, i) element (the diagonal) to build a (enumerable effective) computable regularity n that is not in r. Then we arrive at the contradiction that $n \notin r$ which was supposed to include all the (enumerable effective) computable regularities. We know that all Turing machines are enumerable therefore the assumption that r was enumerable has to be false. This means that no computable function can enumerate all possible regularities, not even only the (enumerable effective) computable ones. Constructive proofs are also possible by reducing the problem to the undecidability of the *halting problem* (encoding the halting probability as a computable test and then arriving at a contradiction).

The power of algorithmic complexity over Shannon entropy can be expressed in the following statements of logical implication:

- Presence of statistical regularities $\Rightarrow$ low Shannon entropy $\Rightarrow$ low Kolmogorov-Chaitin complexity
- Algorithmic compressibility $\not\Rightarrow$ statistical regularities
- Low Kolmogorov complexity $\not\Rightarrow$ low Shannon entropy.

To illustrate the above, let ω be the Thue-Morse sequence obtained by starting with 0 and recursively appending the Boolean complement of the sequence. The first few steps yield the initial segments 0, 01, 0110, 01101001, 0110100110010110, and so on.

x is of low Kolmogorov complexity because it can be compressed by a short recurrence relation: $t_0 = 0$(or 1), $t_{2n} = t_n$, and $t_{2n+1} = 1 - t_n$. For all positive integers n implemented in a program whose input is n and therefore only grows by log(i), with i the number of digits in the sequence produced while the sequence grows in linear fashion.

π is an example of a similar type. It is believed to be Borel normal and therefore of maximal entropy like most real numbers (Martin-Löf, 1966), yet π is of the lowest Kolmogorov-Chaitin complexity because small formulae can generate arbitrarily long initial segments of π.

It is therefore clear how, unlike other computable descriptive complexity measures such as Shannon entropy, algorithmic complexity measures are not assigned maximal complexity (e.g., entropy rate), but rather low complexity if they have a generating mechanism, thereby constituting a causal origin as opposed to a random one, despite their lack of statistical regularities.

In fact, Shannon entropy can be equated with the existence of statistical regularities in an object (e.g., a sequence) relative to other elements (the elements of interest to be counted) in a distribution. Shannon entropy can also be seen as a counting function, one that can count any element or property of interest but only one at a time (modulo properties that are related or derived), relative to the occurrence of the same property in other elements in a set (the distribution).

21.5   Divide-and-conquer data analysis

The greatest contributor to the sum of programs $1/2^{|p|}$ is the shortest program p, and the shortest program p producing x is none other than $K(x)$, the algorithmic complexity of x.

The algorithmic Coding theorem describes the reverse connection between $K(x)$ and $m(x)$ (Levin, 1974; Chaitin, 1969):

$$K(x) = \log_2(m(x)) + O(1) \tag{7}$$

where $O(1)$ is a linear term independent of x.

This tells us that if a piece of data x is produced by many programs, then there is also a short program that produces x (Cover and Thomas, 2012).

The Block decomposition method (BDM) that we introduced in (Zenil et al., 2014; Soler-Toscano et al., 2014), allows us to extend the power of the Coding theorem by using a combination of Shannon entropy and algorithmic probability. The idea is to divide a piece of data into segments for which we have algorithmic probability approximations, and then count the number of times that a local regularity occurs. We then apply the formula $\sum_{\forall x} C(x) + \log M(x)$, where $C(x)$ is the estimation of the complexity from CTM based on algorithmic probability and related to its Kolmogorov-Chaitin complexity, and $M(x)$ is the number of times such a pattern repeats. In other words, one does not count $K(x)\,M(x)$ times but only once, because the repetitions can be produced by the addition of another program that repeats the same pattern and requires no greater description length than the shortest program producing x and the logarithm of the number of repetitions. We have shown that by applying the method we can profile data in an accurate fashion (Zenil et al., 2015b).

21.5.1 The algorithmic Bayesian framework

Let's come back to attempting to approximate Kolmogorov complexity by chunking data. One can use BDM to estimate the complexity of a piece of data in order to attempt to reveal its possible generating mechanism (the possible cause of the regularity). The Bayesian application is as follows. Let x be the observed data, R a random process and D the possible generating mechanism (cause). The probability that x is generated by R can be estimated from Bayes' theorem:

$$P(R|x) = \frac{P(x|R)P(R)}{P(x|R)P(R) + P(x|D)P(D)},$$

where D stands for "not random" (or deterministic) and $P(x|D)$ the likelihood of x (of length l) being produced by a deterministic process D. This likelihood is trivial for a random process and amounts to $P(x|R) = 1/m^l$, where m, as above, is the size of the alphabet. The algorithmic approach, however, suggests that we plug in the complexity of x as a normative measure (equivalently, its algorithmic probability) as a natural formal definition of $P(x|D)$. $P(x|D)$ is then the distribution we get from running randomly chosen algorithms, with (by definition) simpler strings having greater higher $P(x|D)$ on the other hand, $P(x|R)$ is the ordinary probability distribution, with equal weight on any string.

To facilitate the use of this approach, we have made available an R package called ACSS (likelihood_d() returns the likelihood $P(x|D)$, given the actual length of x) and an Online Complexity Calculator http://www. complexitycalculator.org (the R package documentation is also in the Complexity Calculator website). This was done by taking $D(x)$—a function based on the distribution of programs produced by a large set of small random Turing machines—and normalising it with the sum of all $D(x_i)$ for all x_i with the same length l as x (note that this also entails performing the symmetric completions).

With the likelihood at hand, we can make full use of Bayes' theorem (the R package acss contains the corresponding functions). One can either obtain the likelihood for a random rather than a deterministic process via function likelihood_ratio() (the classical default for the prior is $P(R) = 0.5$). Or, if one is willing to make assumptions as to the prior probability with which a random rather than a deterministic process is the generating mechanism for the data x (i.e., $P(R) = 1 - P(D)$), one can obtain the posterior probability for a random process given x, $P(R|x)$, using prob_random().

The numerical values of the complexity of x are calculated by the Coding theorem method (CTM). CTM, however, is computationally very expensive (it grows faster than any computable function, equivalent to the Busy Beaver) and is ultimately uncomputable beyond small computer programs and therefore works at the short sequence granularity. We have shown, however, that CTM can profile objects by their family by simply looking at these short sequences (small data) as local regularities, just like other approaches (e.g., network motifs (Milo et al., 2002) shown to unveil underlying biological and physical mechanisms). Moreover, the BDM method described in Section 5 extends the power of CTM, its chief disadvantage being that it converges to the Shannon entropy rate the longer the sequence, if CTM (which, again, is computationally expensive but need only be run once) is not further executed. Nevertheless, we have produced applications that are proving to be useful in data analysis and data profiling in fields ranging from cognitive science (Gauvrit et al., 2015, 2014; Kempe et al., 2015; Chekaf et al., 2015) to network theory (Zenil et al., 2014, 2015a,b, 2016), to image classification (Zenil et al., 2015), molecular biology and finance (Zenil and Delahaye, 2011).

21.6   Further discussion and conclusion

Data is usually not generated at random but by a process. That is why, if asked, you would say you expected a '0' to follow 01010101, rather than a '1' (even though, according to probability theory, both have exactly the same classical probability of occurring among all strings of the same length).

In data generated by an algorithmic process, $m(x)$ would establish the probability of an occurrence of x in the data in the presence of any other additional information. m thus tells us that patterns which result from simple processes (short calculations) will be likely, while patterns produced by long uncompressible calculations are unlikely if the data is produced by a mechanistic cause (e.g., a type of customer preference for a product in a catalogue).

m is not simply a probability distribution establishing that there are some patterns that have a certain probability of occurring. Unlike traditional classical distributions, it also determines the specific order in which the elements are distributed, with those of low complexity more likely to be the result of a cause and effect process. On the other hand, those that are random looking and less interesting because not the result of an underlying behaviour or pattern will appear infrequently.

For example, market prices are considered non-random, even though models may be random, because prices follow laws of supply and demand. However, since regularities are erased by the same laws, making them look random, they would nevertheless appear to be algorithmic. Market patterns are artificially and quickly erased by economic activity itself, in a natural movement towards equilibrium. Shannon entropy will tend to over-fit the amount of apparent randomness in such a social phenomenon, despite the underlying algorithmic footprints that are exploitable in principle.

In the age of complex systems and Big Data, therefore, an algorithmic account is key to coping with complex systems and spurious correlations in Big Data. We strongly suggest that these algorithmic measures constitute a viable framework for analysing data from complex systems and to move away from the traditional practice of correlation and regression so pervasive, and often damaging if not misleading, in the natural and social sciences.

Acknowledgements

Part of this work was done while the author was visiting the Institute for Mathematical Sciences, National University of Singapore (NUS), as part of the *Algorithmic Randomness* program in 2014. The visit was supported by NUS and the Foundational Questions Institute (FQXi).

References

Calude, C. and G. Longo (2015). The deluge of spurious correlations in big data. *CDMTCS Research Report Series CDMTCS-488.*

Calude, C. S. (2013). *Information and randomness: an algorithmic perspective.* Springer Science & Business Media.

Chaitin, G. J. (1969). On the length of programs for computing finite binary sequences: Statistical considerations. *Journal of the ACM (JACM)* 16(1), 145–159.

Chaitin, G. J. (1975). A theory of program size formally identical to information theory. Journal of the ACM (JACM) 22(3), 329–340.

Chekaf, M., N. Gauvrit, A. Guida, and F. Mathy (2015). Chunking in working memory and its relationship to intelligence. In *Proceedings of the 37th annual meeting of the cognitive science society, Pasadena, California.*

Cover, T. M. and J. A. Thomas (2012). *Elements of information theory.* John Wiley & Sons.

Delahaye, J.-P. and H. Zenil (2012). Numerical evaluation of algorithmic complexity for short strings: A glance into the innermost structure of randomness. *Applied Mathematics and Computation* 219(1), 63–77.

Gauvrit, N., H. Singmann, F. Soler-Toscano, and H. Zenil (2015). Algorithmic complexity for psychology: a user-friendly implementation of the coding theorem method. *Behavior research methods,* 1–16.

Gauvrit, N., F. Soler-Toscano, and H. Zenil (2014). Natural scene statistics mediate the perception of image complexity. *Visual Cognition* 22(8), 1084– 1091.

Kempe, V., N. Gauvrit, and D. Forsyth (2015). Structure emerges faster during cultural transmission in children than in adults. *Cognition* 136, 247–254.

Kirchherr, W., M. Li, and P. Vitányi (1997). The miraculous universal distribution. *The Mathematical Intelligencer* 19(4), 7–15.

Kolmogorov, A. N. (1968). Three approaches to the quantitative definition of information. *International Journal of Computer Mathematics* 2(1-4), 157–168.

Levin, L. A. (1974). Laws of information conservation (nongrowth) and aspects of the foundation of probability theory. *Problemy Peredachi Informatsii* 10(3), 30–35.

Mandelbrot, B. B. and R. L. Hudson (2005). *The (mis) behavior of markets: a fractal view of risk, ruin, and reward.* Basic Books.

Martin-Löf, P. (1966). The definition of random sequences. *Information and control* 9(6), 602–619.

Milo, R., S. Shen-Orr, S. Itzkovitz, N. Kashtan, D. Chklovskii, and U. Alon (2002). Network motifs: simple building blocks of complex networks. *Science* 298(5594), 824–827.

Schnorr, C. P. (1971). *Zufzälligkeit und wahrscheinlichkeit: eine algorithmische begründung der wahrscheinlichkeitstheorie.* Springer.

Soler-Toscano, F., H. Zenil, J.-P. Delahaye, and N. Gauvrit (2013). Correspondence and independence of numerical evaluations of algorithmic information measures. *Computability* 2(2), 125–140.

Soler-Toscano, F., H. Zenil, J.-P. Delahaye, and N. Gauvrit (2014). Calculating kolmogorov complexity from the frequency output distributions of small turing machines. PLoS ONE 9(5).

Solomonoff, R. J. (1964). A formal theory of inductive inference: Parts i and ii. *Information and control.*

Zenil, H. (2011). *Une approche expérimentale à la théorie de la complexité algorithmique.* Ph. D. thesis, University of Lille 1. dissertation in fulfilment of the degree of Doctor in Computer Science (committee: J.-P. Delahaye, C.S. Calude, G. Chaitin, S. Grigorieff, P. Mathieu and H. Zwirn.

Zenil, H. and J.-P. Delahaye (2011). An algorithmic information theoretic approach to the behaviour of financial markets. *Journal of Economic Surveys* 25(3), 431–463.

Zenil, H., N. A. Kiani, and J. Tegnér (2015a). Numerical investigation of graph spectra and information interpretability of eigenvalues. In *Bioinformatics and Biomedical Engineering*, pp. 395–405. Springer.

Zenil, H., N. A. Kiani, and J. Tegnér (2015b). Quantifying loss of information in network-based dimensionality reduction techniques. *Journal of Complex Networks.* (online ahead of press).

Zenil, H., N. A. Kiani, and J. Tegnér (2016). Methods of information theory and algorithmic complexity for network biology. *Seminars in Cell and Developmental Biology.* (online ahead of press).

Zenil, H., F. Soler-Toscano, J.-P. Delahaye, and N. Gauvrit (2015). Two-dimensional kolmogorov complexity and validation of the coding theorem method by compressibility. *PeerJ Computer Science* 1(e23).

Zenil, H., F. Soler-Toscano, K. Dingle, and A. A. Louis (2014). Correlation of automorphism group size and topological properties with program-size complexity evaluations of graphs and complex networks. *Physica A: Statistical Mechanics and its Applications* 404, 341–358.

22. Der Begriff der Information: Was er leistet und was er nicht leistet

Holger Lyre[1]

Abstract

Der Informationsbegriff und seine vermeintliche fundamentale Rolle in der Wissenschaft, speziell der Physik, wird kritisch analysiert. Eine Hypostasierung von Information erweist sich als unplausibel, die Formalisierung führt auf das begriffliche Problem pluraler syntaktischer Informationsmaße. Weitere Probleme entstehen im Zusammenhang mit einem Subjektivismus in den Naturwissenschaften, der mit dem Informationsbegriff einhergeht, und den kategorialen Differenzen zwischen physikalischer Deskription und semantischer Zuschreibung.

The concept of information and its alleged fundamental role in science, especially in physics, is critically analyzed. A hypostatization of information proves to be implausible, the formalization leads to the conceptual problem of multiple syntactic information measures. Further problems arise in connection with a subjectivism in the sciences that comes along with the concept of information and the categorical differences between physical description and semantic attribution.

---

1   Holger Lyre ist Professor für Theoretische Philosophie/Philosophie des Geistes an der Universität Magdeburg mit Forschungsschwerpunkten in Philosophie der Physik, Wissenschaftstheorie der kognitiven Neurowissenschaften und Philosophie des Geistes. Wichtigste Buchveröffentlichungen: Quantentheorie der Information (1998), Informationstheorie. Eine philosophisch-naturwissenschaftliche Einführung (2002), Lokale Symmetrien und Wirklichkeit (2004), Kants 'Prolegomena': Ein kooperativer Kommentar (hg. mit O. Schliemann, 2012), Philosophie der Quantenphysik (mit C. Friebe, M. Kuhlmann, P. Näger, O. Passon & M. Stöckler, 2015).

22.1   Der Informationsbegriff in den Naturwissenschaften

Information ist allgegenwärtig. Informations- und Kommunikationstechnologien
weisen die höchsten Innovations- und Veränderungsraten auf – deutlich sichtbar
durch die Entwicklung immer leistungsfähigerer Computer, die umfassende und
alles durchdringende Medienausbreitung und die Entwicklung immer größerer
Datenspeicher, Datennetze und intelligenter Begleiter wie z.b. Smartphones. Information wird nicht selten als Rohstoff und Ware gehandelt. Wir machen uns Gedanken darüber, welche Rechte und Pflichten man ihr gegenüber hat. Zugang und
Schutz von Information stehen im Zentrum von Informationsrecht und Informa-
tionsethik. Unser Zeitalter lässt sich wahrhaft als *Informationszeitalter* charakterisieren.

Der Begriff der Information hat in den vergangenen Dekaden auch zunehmend
Einzug in die Naturwissenschaften gehalten, und in diesem Aufsatz soll die Frage
aufgegriffen werden, was er da leistet und was nicht. Einige gut sichtbare Bereiche,
in denen der Informationsbegriff eine Rolle spielt, seien hier zunächst nur genannt.
Bereits die Tatsache, dass eine eigene und bedeutsame Wissenschaftsdisziplin, die
Informatik, in der zweiten Hälfte des 20. Jahrhunderts entstanden ist, zeigt, wie
sehr das Konzept der Information die Wissenschaftslandschaft beeinflusst und
verändert hat. Im Großgebiet der Kognitionswissenschaften wird beständig die
Idee bemüht, dass es zum Wesen kognitiver Systeme und Agenten gehört, infor-
mationsverarbeitend zu sein – was immer das dann genauer besagt. Auch in der
Biologie hat sich informationstheoretischer Jargon durchgesetzt, speziell in Form
der gängigen Annahme, dass die Prozesse und Aktivitäten des molekulargeneti-
schen Levels so anzusehen sind, dass genetische Information prozessiert, codiert
und transformiert wird. Man kann die Sichtbarkeit des Informationsbegriffs in den
Naturwissenschaften vordergründig als einen Hinweis dafür lesen, dass Information in unserem modernen Verständnis zu einer Grundgröße der Natur avanciert
ist. Ob dies aber tatsächlich der Fall ist, sei hier auf den Prüfstand gestellt. Dabei
soll der Blick vornehmlich auf die Physik gerichtet werden. Denn gerade in der
Physik existieren gleich mehrere Anknüpfungspunkte, an denen traditionelle und
grundlegende physikalische Konzepte mit dem Informationsbegriff verknüpft oder
assoziiert werden. Hierzu ein kleiner Überblick (vgl. zu zahlreichen der hier genannten Punkte Lyre 2002):

- In der Statistischen Thermodynamik ist seit längerem bekannt, dass der Entro-
  pie-Begriff eine formale Ähnlichkeit mit der Shannonschen Definition von Information aufweist, genauer: Boltzmannsche Entropie ist bis auf eine Konstante

gleich Shannonscher Informations-Entropie (daher die Brillouinsche These, Information sei Negentropie). Hierauf wird später noch näher eingegangen.

• In den Grundlagen der Thermodynamik wird unter dem Stichwort „Maxwellscher Dämon" die Möglichkeit diskutiert, inwieweit ein Wesen, das in der Lage ist, Moleküle wahrzunehmen, den Zweiten Hauptsatz verletzen kann. Nicht selten wird dabei die Annahme vertreten, dass es möglich ist, den Maxwellschen Dämon zu vertreiben und damit den Zweiten Hauptsatz zu retten, wenn man eine informationstheoretische Überlegung über die Energiedissipation bei der Speicherlöschung zugrunde legt (siehe speziell die Arbeiten von Landauer und Bennett in Leff & Rex 2002). In jüngerer Zeit wird an dieser Annahme jedoch wieder Kritik angemeldet (Norton 2013).

• In dem stetig wachsenden Arbeitsgebiet der Quanteninformationstheorie bzw. Quanteninformatik zeichnet sich ab, dass durch die Behandlung von Information als genuin quantenmechanisch völlig neue Möglichkeiten erwachsen – sowohl grundlagentheoretisch als auch technologisch (vgl. Timpson 2013 für eine kritische Diskussion des Informationsbegriffs in der Quanteninformatik).

• Die Quantentheorie ist die mit Abstand erfolgreichste physikalische Theorie – so wie in einer profaneren Weise auch die Thermodynamik empirisch hochgradig bestätigt ist; denn die Anwendungserfolge beider Theorien umgeben uns tagtäglich. Gleichzeitig sticht die Quantentheorie philosophisch heraus, weil sie trotz ihres großen Erfolgs zugleich auch diejenige Theorie ist, um die sich seit ihrer Entstehung eine erbitterte Interpretationsdebatte rankt, die bis heute zu keinem versöhnlichen Ende geführt worden ist (vgl. Friebe et al. 2015 und Esfeld 2012, Teil II). Gerade in jüngerer Zeit treten dabei wieder Ansätze in den Vordergrund, die im Verbund mit dem Informationsbegriff zur Klasse der epistemischen Interpretationen der Quantentheorie zählen, und die, sofern sie an den Bayesianismus anknüpfen, auch als Quanten-Bayesianismus bezeichnet werden. Demnach bringt die quantenmechanische Zustandsfunktion im Kern nichts anderes als subjektiv verfügbare Beobachter-Information über ein Objekt zum Ausdruck (vgl. Timpson 2013, Kap. 9-10). Dies war auch ein Kerngedanke der traditionellen Kopenhagener Deutung.

• In Grundlagenfragen der Raumzeit und der Kosmologie betrachtet man mögliche tief liegende Zusammenhänge zwischen Information, Entropie und extremalen raumzeitlichen Zuständen. Nennenswert sind hierbei die Bekenstein-Hawking-Entropie schwarzer Löcher oder das „information loss paradox", das das scheinbare Zusammenbrechen der unitären Entwicklung der Wellenfunktion im Zusammenhang mit schwarzen Löchern thematisiert. Weiterhin existieren spekulative Überlegungen, wie sich aus der theoretischen Verbindung von Quantenmechanik und Thermodynamik schwarzer Löcher neue physikalische

Prinzipien extrahieren lassen. Die Bekenstein-Hawking-Entropie kann etwa als diejenige Informationsmenge angesehen werden, die einem physikalischen Objekt mittels des seiner Masse zugeordneten Ereignishorizonts zugeschrieben werden kann. Der physikalische Informationsgehalt eines Raumgebiets lässt sich dann auf dessen Oberfläche zurückführen, eine Idee, die man als holographisches Prinzip bezeichnet (vgl. Artikel von Beisbart und Hedrich in Esfeld 2012).

Die Präsenz und Relevanz des Informationsbegriffs in den Wissenschaften lässt vordergründig drei Erklärungen zu: Der Begriff der Information ist, erstens, ein trivialer, da undifferenzierter und unspezifischer Begriff. Seine triviale Vagheit bewirkt, dass er sich überall finden und anknüpfen lässt, substantiell ist damit aber nicht viel gewonnen. Es könnte aber, zweitens, auch das Gegenteil der Fall sein: vielleicht sind wir mit dem Informationsbegriff auf einen fundamentalen Begriff gestoßen, dessen grundlegender Charakter sich in den Wissenschaften mehr und mehr zeigt. Drittens könnte es sein, dass es sich bei den vielfachen Verwendungsweisen des Wortes Information um bloße Äquivokationen handelt, dass wir in verschiedenen Zusammenhängen zwar dasselbe Wort benutzen, aber eben nicht denselben Begriff. Dieser letzten Möglichkeit wollen wir hier nachspüren und dabei die Diskussion in vier große Fragedomänen aufteilen, die unter den folgenden vier Schlagworten stehen: 1. Hypostasierung von Information, 2. Pluralität der Begriffe, 3. Subjektivismus und 4. Semantik.

22.2   Teil 1. Hypostasierung von Information

Norbert Wiener hat es 1948 so formuliert: „Information ist Information, weder Materie noch Energie. Kein Materialismus, der dieses nicht berücksichtigt, kann den heutigen Tag überleben." Dem Wienerschen Diktum zufolge ist Information als neue physikalische Grundgröße gleichrangig neben die arrivierten Grundgrößen Materie und Energie zu stellen. Information wäre dann eine irreduzible Größe, die nicht auf Energie oder Materie zurückgeführt werden kann. Man kann insofern von einer Ontologisierung oder auch Hypostasierung von Information bei Wiener sprechen. Verfolgt man diesen bereits radikalen Gedanken weiter, so ergibt sich eine schwächere und eine stärkere Form. Eine schwächere These wäre es zu sagen, dass Information zwar ontologisch irreduzibel ist, aber in dieser Hinsicht gleichrangig neben anderen ontologisch irreduziblen Größen steht. Genau dies scheint Wiener in Hinblick auf Information, Materie und Energie sagen zu wollen. Die stärkere und mithin radikalste These würde demgegenüber besagen, dass sich sämtliche physikalischen Größen letztlich auf Information reduzieren lassen, also auch Materie und Energie. Dies wäre das ambitiöse Programm eines Informations-Reduktionismus oder, ontologisch gesprochen, eines Informations-Monismus.

Tatsächlich sind solche Programme in der Physik hin und wieder versucht worden bzw. werden bis zum heutigen Tage von verschiedentlichen Autoren in der einen oder anderen Weise insinuiert. Streng und überzeugend durchgeführt wurde der Informations-Monismus jedoch nie. Einige bekanntere Autoren seien genannt: John Archibald Wheeler (1989) verwendete den Slogan „It from Bit" und die Idee eines partizipatorischen Universums um auszudrücken, dass physikalische Realität auf Beobachterentscheidungen zurückzuführen ist. Carl Friedrich von Weizsäcker (1985) verfolgte das Programm einer fundamentalen Quantentheorie binärer Alternativen, elementarer Quantenbits, aus deren Symmetrien sich sogar die Struktur der Raumzeit begründen lassen sollte (vgl. Lyre 1998). Auch Anton Zeilinger (2005) knüpft hier entfernt an. Konrad Zuse (1969) hat vom „Rechnenden Raum" gesprochen, ähnliche Ideen über die physikalische Welt als zellulären Automaten verfolgten oder verfolgen Edward Fredkin oder Steven Wolfram, und schließlich sieht Seth Lloyd (2004), wie schon von Richard Feynman (1982) vorgedacht, die Welt als Quantencomputer. Doch keine dieser Ideen oder keines dieser Programme ist als Grundlagenprogramm der Physik in nennenswerter Weise durch- oder ausgeführt worden (in dem Sinne, dass ein mathematisch konkreter Zusammenhang zum Standardmodell der Elementarteilchen sichtbar würde), auch wenn es zahlreiche Populärliteratur hierzu gibt (z.B. Stonier 1990, Baeyer 2003, Gleick 2011).

Für das weitere ist es hilfreich, ein wenig philosophischen Jargon einzuführen. Sowohl bei Platon als auch bei Aristoteles findet sich der Begriff der Form als ontologischer Grundbegriff. Beide Autoren sind auch der Meinung, dass Form – zum Beispiel die sichtbare Gestalt eines konkreten Dings – einem raumzeitlichen Gegenstand nur insofern zukommt, als dass dieser Gegenstand aus Materie besteht. Der Materie, griechisch „hyle", wird eine Form, griechisch „eidos", aufgeprägt – ein konkreter raumzeitlicher Gegenstand ist dann ein Zusammengesetztes, griechisch „synholon", aus Form und Materie. Weizsäcker weist in diesem Zusammenhang darauf hin, dass der moderne Begriff Information ja genau diesen Ausdruck „Form" bereits enthält und schlägt vor, Information abstrakt als ein „Maß für Form" anzusehen (Weizsäcker 1985, S. 167).

Folgen wir den Klassikern, dann sollten wir bezüglich physikalischer Gegenstände davon ausgehen, dass Form immer eines materiellen Trägers bedarf. Diesen materiellen Träger wegzudenken, also Materie (oder aufgrund der Einsteinschen Äquivalenzformel genauer: Energie-Materie) selbst auf Form zu reduzieren, wäre das Programm eines konsequenten Informations-Monismus. Insofern dies in keinem modernen Programm realisiert ist, sollten wir bis auf weiteres davon ausgehen, dass es ein Irrtum wäre, Information zu ontologisieren oder zu hypostasieren. Stattdessen gibt es in der modernen Physik drei sehr basale Grundkonzepte, nämlich Energie-Materie, Wechselwirkung und Raum-Zeit, die sich nach heutigem Kenntnisstand nicht aufeinander reduzieren oder abbilden lassen.

In der heutigen Physik ist also Energie-Materie ontologisch ein „factum brutum", und dies ungeachtet der Tatsache, dass es Versuche gibt, das Konzept der Masse der Elementarteilchen genauer zu verstehen. Dabei spielt zum Beispiel nach dem Standardmodell der Higgs-Mechanismus eine Rolle. Gelegentlich wird sogar behauptet, das Higgs-Teilchen sei dasjenige Teilchen, das allen anderen Teilchen Masse verleiht. Man muss aber davor warnen, solche Formulierungen allzu wörtlich zu verstehen. Zahlreiche Arbeiten zu den konzeptionellen und philosophischen Grundlagen der Eichtheorien haben sich in jüngerer Zeit diesem Fragenkreis kritisch gewidmet (z.B. Friederich 2014, Smeenk 2006). Insbesondere sollte der Higgs-Mechanismus nicht mit einem Kausal-Mechanismus im landläufigen Sinne verwechselt werden, einem Mechanismus also, der sich durch einen oder mehrere dynamische Prozesse in Raum und Zeit manifestiert. Stattdessen bezieht sich der Ausdruck „Higgs-Mechanismus" auf einen eleganten mathematischen Weg, durch eichinvariantes Umtransformieren von Freiheitsgraden von einer Lagrange-Formulierung der elektroschwachen Theorie zu einer anderen zu gelangen, wobei gleichzeitig eine verborgene Eichsymmetrie ausgenutzt und respektiert wird (Lyre 2008). Ferner liefert der Higgs-Mechanismus keine Erklärung des Massenspektrums der Elementarteilchen – also eine Antwort auf die Frage, warum und in welchen Verhältnissen die Massen der Teilchen zueinander stehen. Ein vollständiges Verständnis vom Wesen der Energie-Materie sollte aber eine solche Erklärung mit umfassen. Energie-Materie ist also nach wie vor ein irreduzibler Grundbegriff der Physik.

22.3   Teil 2: Pluralität der Begriffe

Wir haben bereits gesehen, dass Information sich allgemein als Maß oder Quantifikation von Form oder Struktur auffassen lässt. Dabei handelt es sich um die Form oder Struktur eines materiellen Trägers. Ein nächster Schritt muss sein, das Maß der Information mathematisch exakt zu bestimmen und zu quantifizieren. Wie sich zeigt, bietet sich hierfür mehr als nur eine mathematische Quantifizierung, und das heißt letztlich, mehr als nur ein Maßbegriff, an. Stattdessen landen wir bei einer Pluralität der Informationsmaße.

Die erste, mathematisch präzise formulierte Informationstheorie ist mit dem Namen Shannons verbunden, der Grundgedanke geht aber bereits auf Ralph Hartley (1928) zurück. Seine Idee war es, den Informationsgehalt eines Ereignisses oder eines Zeichens an dessen Unwahrscheinlichkeit zu binden. Ein sicheres, also mit Wahrscheinlichkeit eins auftretendes Zeichen ist uninformativ, besitzt keinerlei Informationsgehalt, während ein sehr unwahrscheinliches Zeichen einen hohen Informationsgehalt besitzt. Folgende Formalisierung bietet sich an:

$$I_k = - \operatorname{ld}(p_k) \tag{1}$$

Der Logarithmus hat hierbei ausschließlich rechenpraktische Vorzüge: man strebt ein additives Mengenmaß anstelle eines multiplikativen an und wählt den Logarithmus zur kleinsten Basis, also den dualen Logarithmus zur Basis zwei. Die Auswahl eines Zeichens aus einem Vorrat von zwei gleichwahrscheinlichen Zeichen, sagen wir 0 und 1, ist mit der Wahrscheinlichkeit ½ und somit dem Informationsgehalt 2 Bit verbunden (der Ausdruck „bit", binary digit, für die Einheit der Information geht auf John Tukey zurück). Der auf diese Weise formalisierte Hartleysche Informationsgehalt lässt sich begrifflich als ein Maß für den Neuigkeits- oder Überraschungswert eines Zeichens ansehen. Information kann also mit Neuigkeit oder Überraschung, letztlich aber mit Unwahrscheinlichkeit gleichgesetzt werden. Konzeptionell leistet ein solcher Informationsbegriff *de facto* nicht viel mehr als der Wahrscheinlichkeitsbegriff selbst.

Auf Basis der Grundidee von Hartley gelang es Claude Elwood Shannon Ende der 40er Jahre, eine beeindruckende formale Theorie der Information aufzubauen (Shannon & Weaver 1949), die dem nachrichten- und kommunikationstechnischen Zweck diente, die Bandbreiten von Übertragungskanälen je unterschiedlicher Quellen zu dimensionieren, die über unterschiedliche Alphabete oder Zeichenvorräte verfügen mit je unterschiedlichen Wahrscheinlichkeitsverteilungen. Ein zentrales Stück der nachrichtentechnischen Informations- und Kommunikationstheorie ist Shannons berühmte Formel über den mittleren Informationsgehalt einer Quelle bzw. den Erwartungswert des Informationsgehalts:

$$H = \langle I \rangle = \Sigma_k \, p_k \, I_k = - \Sigma_k \, p_k \operatorname{ld}(p_k) \tag{2}$$

Die Formel ist interessanterweise strukturell identisch mit der Boltzmannschen Entropieformel – jedenfalls bis auf ein Vorzeichen. Dies hat zu vielerlei Diskussionen Anlass gegeben. Shannon selbst hat seine Größe in Analogie zu Boltzmann mit dem griechischen Buchstaben H („Eta") versehen und als Informationsentropie bezeichnet. In der Tat lässt sich die Boltzmannsche Entropie als ein Maß für die Unkenntnis eines spezifischen Mikrozustandes in einem Makrozustand verstehen. Denn ein thermodynamischer Makrozustand ist ja in aller Regel kompatibel mit

vielen, ja sogar phantastisch vielen Mikrozuständen in dem Sinne, dass jeder dieser Mikrozustände die momentan *de facto* vorliegende Realisierung des Makro-
zustandes sein könnte. In diesem Sinne quantifiziert die Boltzmann–Entropie die
Unkenntnis des spezifischen Mikrozustandes in einem Makrozustand, die Boltzmannsche Entropie ist also eine bestimmte Art von Nicht-Information, Brillouin
hat umgekehrt Information als *Negentropie* bezeichnet (vgl. Lyre 2002, Kap. 2.1).

Die Hartley-Shannonsche Informationstheorie ist die wohl prominenteste Formalisierung des Informationsbegriffs als eines Maßbegriffs für Form, doch sie ist
beileibe nicht die einzige. Eine weitere, wirkungsmächtige Formalisierung von Information erfolgt im Rahmen der algorithmischen Informationstheorie. Diese geht
aber einen anderen Weg, hier geht es letztlich darum, Komplexität zu quantifizieren
(vgl. Mainzer 2007). Betrachten wir etwa die folgenden zwei Zahlenfolgen:

s1 = 1010101010101010101010101010101010101010

s2 = 0110111010000101101100011110010101110110

Die Sequenz s1 ist nach einem offenkundigen Muster gebildet, da sie einfach aus
einer abwechselnden Folge von Nullen und Einsen besteht. Demgegenüber lässt
sich für s2 kein Muster, oder jedenfalls nicht so ohne weiteres, angeben. s1 ist insofern weniger komplex als s2, als s1 sich auf eine kompakte Regel reduzieren lässt.
Denken wir uns die beiden Folgen in ähnlicher Weise weiter, ja vielleicht sogar ins
Unendliche fortgesetzt, so wird der Unterschied noch schlagender. Wollte man die
Folge s2 weitergeben, so müsste man, falls sich keine Regel oder kein Algorithmus
findet, nach der s2 gebildet wurde, die gesamte Folge übergeben. Die in s1 enthal-
tene Information lässt sich jedoch kompakt auf die Regel „schreibe so-und-so-oft
10" zurückführen. Hieran zeigt sich der Grundgedanke der algorithmischen Informationstheorie: der Informationsgehalt einer Zeichensequenz s ist als die Länge L
des kürzesten Programms (oder Turingmaschine) p_{min} definiert, das diese Sequenz
erzeugen kann:

$$I(s) = L(p_{min} \rightarrow s) \tag{3}$$

Das hat zur Folge, dass eine Zufallsfolge der Länge n den algorithmischen Informationsgehalt n hat. Eine Folge ist nach diesem Verständnis genau dann zufällig, wenn
sie nicht weiter komprimierbar ist, wenn also es keine kompaktere Art und Weise
gibt, die Folge zu übertragen, als sie ganz zu übertragen. Zufallsfolgen besitzen maximale Komplexität und sind nicht komprimierbar.

In der algorithmischen Komplexitäts- und Informationstheorie, die vor allem
auf Gregory J. Chaitin (1987) zurückgeht, lässt sich ein tiefliegendes Theorem, das
Chaitinsche Zufallstheorem, beweisen, das mit mehreren meta-mathematischen

Aussagen zusammenhängt. Das Zufallstheorem besagt, dass der algorithmische Informationsgehalt im Allgemeinen nicht berechenbar ist. Dies hängt damit zusammen, dass eine Turing-Maschine einer bestimmten Komplexität keine Sequenz mit größerer Komplexität erzeugen und deshalb auch nicht prüfen kann, ob eine längere Folge komprimierbar ist oder nicht. Hier kommt also ein interessantes, gegebenenfalls aber auch beunruhigendes Element der Subjektivität in die Theoriebildung hinein (siehe hierzu weiter unten).

Als Zwischenergebnis können wir festhalten, dass es durchaus unterschiedliche Arten und Weisen gibt, den Informationsbegriff zu formalisieren. Das aber heißt: es zeichnet sich kein einheitlicher Begriff von Information ab, sondern eine Pluralität der Informationsbegriffe. Für den Informations-Monisten oder auch Reduktionisten stellt dies offenkundig ein Problem dar: auf welches Informationsmaß soll er zurückgreifen? Ist Information begrifflich als (Un-)Wahrscheinlichkeit, Komplexität oder etwas Drittes anzusehen? Und selbst wenn sich hier eine Entscheidung herbeiführen ließe, so wäre dies nur dann ein Fortschritt, wenn sich hinsichtlich der neuen Begriffe größere konzeptionelle Klarheit gewinnen ließe; andernfalls hätte sich das Problem nur verschoben. Doch wie wir weiter unten sehen werden, ist beispielsweise der Wahrscheinlichkeitsbegriff hinsichtlich seiner Interpretation durchaus strittig, während umgekehrt der Begriff der Komplexität, wie etwa Ladyman et al. (2013) zeigen, auf vergleichbare Schwierigkeiten hinsichtlich seiner Formalisierung stößt wie der Informationsbegriff. Auch führen beide hier vorgestellte Informationstheorien auf Informationsmaße, die in einem gewissen Umfange kontext- und subjektabhängig sind, was zum nächsten Abschnitt überleitet.

22.4   Teil 3: Subjektivismus

In einem berühmt gewordenen Aufsatz zum quantenmechanischen Messproblem schreibt John Bell (1990):

„Here are some words which, however legitimate and necessary in application, have no place in a formulation with any pretension to physical precision: system, apparatus, environment, microscopic, macroscopic, reversible, irreversible, observable, information, measurement [...] Information? Whose Information? Information about what?"

Bell legt hier nicht weniger als eine schwarze Liste von Begriffen vor, die seiner Meinung nach in einer guten Physik nichts zu suchen haben. Sein Misstrauen gegenüber der Mehrheit der gelisteten Begriffe ist, dass es sich um epistemisch und subjektivistisch getränkte Begriffe handelt. Der Informationsbegriff steht dabei an oberster Stelle; denn Information kann nur sinnvoll definiert werden, wenn es jemanden gibt, der Empfänger von Information ist im Sinne eines epistemischen Subjekts. Würde man Information zu einem Grundbegriff der Physik machen, dann hielte – nach Ansicht John Bells und vieler anderer – ein unheilvoller Subjektivismus Einzug in die Physik. Die Physik sollte aber nach einem Objektivierungsideal streben: Sie fördert Mechanismen und Regelhaftigkeiten der Natur zu Tage (im Idealfalle Naturgesetze), die beobachterunabhängig und objektiv (oder doch wenigstens intersubjektiv) darstellbar und überprüfbar sind. Ein Subjektivismus physikalischer Grundbegriffe wäre mehr als problematisch.

Hierzu lässt sich wenigstens zweierlei anmerken. Erstens scheint es richtig zu sein, ja sogar entscheidend, den Begriff der Information vor dem Hintergrund eines Sender-Empfänger-Modells zu begreifen. Wo sinnvoll von Information die Rede ist, ist Information nur dasjenige, was von einem Empfänger aufgenommen und verarbeitet, vielleicht sogar verstanden werden kann. Der Empfänger ist dabei nicht notwendigerweise ein Mensch, aber doch wenigstens ein epistemisches Subjekt; d.h. ein Agent oder System, das eine Informations-Weiterverarbeitung im Rahmen und in Bezug auf ein größeres Netzwerk informationeller Bezüge, vielleicht sogar ein Überzeugungssystem, unterhält. Zwar gibt es Autoren, die versuchen, zu einem objektiven Begriff von Information zu gelangen, aber insofern diesen Autoren dann vermutlich eher ein rein strukturelles Informationsmaß vor Augen steht, ist es weitaus sinnvoller, die jeweils intendierten Maßbegriffe (Struktur, Wahrscheinlichkeit, Komplexität etc.) direkt zu benennen und zu verwenden. Eine darüber hinausgehende Einführung des Informationsbegriffs wäre nicht nur begrifflich überflüssig, sondern schlicht irreführend und falsch.

Eine zweite Anmerkung betrifft den mildernden Umstand, dass es tatsächlich aus guten Gründen strittig ist, inwieweit das Objektivierungsideal in der Physik überhaupt in Strenge umsetzbar ist – ob mit oder ohne Informationsbegriff. Die begrifflichen Zusammenhänge zwischen Shannonscher Information und Entropie in der Thermodynamik haben ihren gemeinsamen Ursprung in der Tatsache, dass beide Begriffe auf dem Begriff der Wahrscheinlichkeit und dessen Verwendung in der Thermodynamik fußen. Nicht minder irreduzibel ist der Wahrscheinlichkeitsbegriff in der Quantenmechanik. In den philosophischen Grundlagen der Wahrscheinlichkeitstheorie treten zwei gegensätzliche Auffassungen einander gegenüber: ein subjektiver und ein objektiver Wahrscheinlichkeitsbegriff. Im Bayesianismus, der bekanntesten Schule einer subjektivistischen Interpretation von

Wahrscheinlichkeit, werden Wahrscheinlichkeiten als subjektive Überzeugungsgrade eines epistemischen Subjekts aufgefasst. Die Überzeugungsgrade eines ideal rationalen Subjekts gehorchen dabei den Kolmogoroff-Axiomen der Wahrscheinlichkeitstheorie, ja mehr noch: die Axiome lassen sich unter dem Anspruch der Vermeidung eines sogenannten „Dutch books" (einer Kombination von Wetten, die, obwohl vom Subjekt als fair empfunden, in jedem Fall zu einem Verlust führt) sogar bayesianisch herleiten oder begründen.

Die objektivistische Wahrscheinlichkeitsinterpretation fußt auf der Annahme, dass es Dingen oder Ereignissen inhärent ist, sich so oder so zu verwirklichen. Betrachten wir etwa die Eigenschaft eines radioaktiven Atoms, zu einem bestimmten Zeitpunkt zu zerfallen. Alles, was wir quantitativ sagen können, ist, dass wir einem Ensemble von Atomen eine Wahrscheinlichkeitsverteilung zuordnen, die uns eine Halbwertszeit liefert, also diejenige Zeit, nach der die Hälfte der Atome im Durchschnitt zerfallen sein wird. Unsere Verwendung des Wahrscheinlichkeitsbegriffs beruht in diesem Falle nicht auf einer bloß subjektiven Beobachterunkenntnis über die genauen Zerfallsmechanismen der einzelnen Atome, sondern findet, nach objektivistischer Wahrscheinlichkeitsauffassung, ihre ontologische Begründung in der natürlichen, jedem einzelnen Atom innewohnenden Verwirklichungstendenz, nach Maßgabe seiner Halbwertszeit zu zerfallen. Die Zerfallseigenschaft ist insofern eine dispositionale Eigenschaft, die nur unter bestimmten Bedingungen manifest wird (im Gegensatz zu kategorialen Eigenschaften, die sich unter allen Umständen, also immer, manifestieren).

In den Grundlagen der Physik herrscht keine Einigkeit darüber, welche Wahrscheinlichkeitsinterpretation in welchen physikalischen Verwendungsfällen die jeweils angemessene ist. Tendenziell scheint eine subjektive Interpretation von Wahrscheinlichkeit in der Thermodynamik angemessen zu sein, insofern etwa die Verwendung von Makrozuständen im Formalismus der Statistischen Physik unserer bloß subjektiven Unkenntnis über den Mikrozustand entspringt. Demgegenüber steht die traditionelle Kopenhagener Interpretation der Quantenmechanik ebenso wie die spontane Kollapsinterpretation nach GRW einer objektivistischen Interpretation nahe, während Nicht-Kollaps-Theorien wie die Viele-Welten-Theorie und die Bohmsche Mechanik, die das quantenmechanische Messproblem eher aufzulösen als zu lösen beabsichtigen, eher mit einer subjektivistischen Interpretation verträglich sind (zur Philosophie der Quantenphysik vgl. Friebe et al. 2015).

Interessanterweise zieht ein subjektiver Wahrscheinlichkeitsbegriff nicht notwendig einen Subjektivismus in den Grundlagen der Physik nach sich. Es mag zulässig sein, einen subjektiven Wahrscheinlichkeitsbegriff zu verwenden, wenn diese Verwendung verträglich ist mit der ontologischen Annahme des Realismus, also krude gesagt der Annahme, dass die Welt in beobachter- und subjektunabhängiger

Weise existiert – auch wenn uns, etwa aus Gründen der epistemischen Limitiertheit, nicht immer sämtliche Informationen über den Realzustand zugänglich sind.
Andererseits muss eine subjektivistische Wahrscheinlichkeits-Interpretation von
der Voraussetzung ausgehen, dass das epistemische Subjekt und dessen Informa-
tionsstand für Zwecke der Weltbeschreibung mitberücksichtigt werden muss. Eine
besonders trickreiche Form von Subjektivismus tritt dann auf, wenn, wie in der
algorithmischen Informationstheorie, der Informationsgehalt bzw. die Komplexität
oder Kompressibilität einer Folge nicht allgemein berechenbar ist. Der algorithmische Informationsgehalt ist insofern notorisch kontext- und subjektabhängig.

22.5   Teil 4: Semantik

Bis hierher wurde der Informationsbegriff ohne Berücksichtigung semantischer
Fragen diskutiert, also ohne auf die *Bedeutung* von Information Bezug zu nehmen.
Dies ist charakteristisch für den *syntaktischen Aspekt* von Information, wie auch
Shannon und Weaver (1949) im Vorwort ihres Buches zur Shannonschen Informa-
tionstheorie betonen:

> „Das Wort Information wird in dieser Theorie in einem besonderen Sinn ver-
> wendet, der nicht mit dem gewöhnlichen Gebrauch verwechselt werden darf.
> Insbesondere darf Information nicht der Bedeutung gleichgesetzt werden. …
> Tatsächlich können zwei Nachrichten, von denen die eine von besonderer Be-
> deutung ist, während die andere bloßen Unsinn darstellt, in dem von uns ge-
> brauchten Sinn genau die gleiche Information enthalten. … Information in der
> Kommunikationstheorie bezieht sich nicht so sehr auf das, was gesagt wird,
> sondern mehr auf das, was gesagt werden könnte. Das heißt, Information ist
> ein Maß für die Freiheit der Wahl, wenn man eine Nachricht aus anderen aus-
> sucht."

In seiner Engführung auf den syntaktischen Aspekt von Information zeigt sich die
begriffliche Verkürzung und Limitiertheit des Hartley-Shannonschen Begriffs. Ein
vollumfänglicher Begriff von Information kann die semantische Komponente nicht
außen vor lassen. Bereits Charles William Morris, der Begründer der Semiotik, hat
darauf hingewiesen, dass für den Zeichenbegriff drei Dimensionen zu unterscheiden sind, die Morris (1938) als *syntaktische, semantische* und *pragmatische Dimension* bezeichnet. Auf den Informationsbegriff übertragen heißt das: die Syntax

regelt oder betrifft das Auftreten und die strukturellen Beziehungen informations-
tragender Vehikel, also der materiellen Träger von Information. Semantik betrifft
demgegenüber die Bedeutung bzw. den Gehalt der Vehikel, während die Pragmatik
schließlich die Funktionalität und Wirkungsweise der bedeutungshaften Vehikel
betrifft.

Warum reicht es nicht aus, sich auf den syntaktischen Aspekt von Information zu beschränken? Der Grund ist, dass wir Informations-Vorkommnisse darüber identifizieren und individuieren, welche Bedeutung sie tragen (semantische
Dimension) und welche Wirkung sie entfalten (pragmatische Dimension). Nicht
wenige Autoren stellen in diesem Sinne die von Morris eingeführte semiotische
Dreidimensionalität in Frage. Syntax, Semantik und Pragmatik lassen sich in Strenge nicht voneinander trennen. Dies passt, in der Umkehrung, zu unserer bisherigen
Kritik: eine rein syntaktisch verstandener Informationsbegriff kollabiert *de facto*
auf andere, semantik-freie Begriffe. Für einen vollumfänglichen Begriff von Information gilt dies nicht.

Wollte man den Begriff der Information zu einem Grundbegriff der Physik
machen, so müsste man speziell zeigen, inwieweit Fragen der Bedeutung und Semantik in der Physik eine Rolle spielen bzw. möglicherweise umgekehrt die Physik
Grundprobleme der Semantik letztlich klären kann. Es dürfte offensichtlich sein,
dass in beiderlei Hinsicht nach heutigem Kenntnisstand keine Veranlassung besteht, derart weitreichende Annahmen auch nur ansatzweise für plausibel zu halten. Fangen wir mit der zweiten Annahme an, inwiefern es der Physik gelingen
sollte, die Grundlagenprobleme der Semantik zu klären. Worin bestehen diese Probleme (in aller hier gebotenen Kürze)? Es handelt sich um nicht weniger als die Ge-
samtmenge der zentralen Fragestellungen der Sprachphilosophie und weiter Teile
der Philosophie des Geistes. Seit Frege gehört es zur herrschenden Meinung in
diesen Disziplinen, in Hinblick auf Bedeutung zwischen Extension und Intension
zu unterscheiden. Die Bedeutung eines Begriffs, etwa des Ausdrucks ‚Baum‘, liegt
zunächst darin, worauf sich dieser Begriff in der Welt bezieht. Als Extension oder
Referenz bezeichnet man die Menge aller Objekte in der Welt, die unter den Begriff
fallen, im Beispiel also die Menge aller konkreten Baum-Vorkommnisse. Wie Frege
an seinem berühmten Beispielsatz „Der Morgenstern ist der Abendstern“ deutlich
macht, geht Bedeutung über Referenz hinaus; denn in einer rein extensionalen
Semantik bliebe unverständlich, wieso wir diesen Satz als informativ empfinden,
obwohl doch die Ausdrücke ‚Morgenstern‘ und ‚Abendstern‘ sich auf denselben
Gegenstand in der Welt, den Planeten Venus, beziehen (also extensionsgleich sind).
Dies spricht nach Frege dafür, die Intension als weitere Bedeutungskomponente neben der Extension einzuführen, die den semantischen Gehalt eines Ausdrucks als

Art des Gegebenseins für das über den Ausdruck verfügende epistemische Subjekt zu erfassen sucht.

Vor dem Hintergrund dieser Unterscheidung ließe sich nun die gesamte Landschaft sprach- und geistphilosophischer Ansätze zur Semantik entfalten. Referentielle Semantiken wie etwa die kausale Theorie oder die Teleosemantik treten in Widerstreit mit nicht-referentiellen Semantiken wie etwa Sprach-Gebrauchstheorien oder Funktionale-Rollen-Semantik (spätestens hier wird die pragmatische Dimension sichtbar). Der Versuch, die hierbei auftretenden komplexen Fragen und Probleme wenigstens prinzipiell auf naturwissenschaftliches Vokabular zurückzuführen, entspricht der Programmvorstellung einer Naturalisierung von Semantik. Ein solches Programm konnte bislang nicht überzeugend durchgeführt werden (eine Vermutung über das vielleicht grundsätzliche Scheitern dieser Ansätze folgt im nächsten Absatz). Plausibler scheint allenfalls die Position des Interpretationismus, also der Annahme, die semantische Dimension beruhe letztlich auf einer äußeren Zuschreibung. Ahnvater einer solchen Position ist Daniel Dennett (1971). Doch auch diese Position, die in Strenge tatsächlich ein Anti-Realismus (genauer: Instrumentalismus) bezüglich Bedeutung und Gehalt ist, führt für alle praktischen Zwecke nicht auf einen Verzicht oder gar eine Elimination der semantischen Dimension. Im Gegenteil, nach Dennett lassen sich natürliche Systeme in der Welt dahingehend ordnen, ob die Ihnen bestangepasste Beschreibung eine rein physikalische, eine funktionale oder eine intentionale Zuschreibung ist. Systeme, die bedeutungshaft operieren, nennt er intentionale Systeme; und sie unterscheiden sich für alle praktischen Zwecke von denjenigen Systemen und Gegenstandsbereichen in der Welt, die rein physikalisch, ohne semantisches und intentionales Vokabular beschrieben werden können.

Nach dieser Ansicht klärt die Physik nur in einem sehr entfernten Sinne irgendwelche Grundlagenprobleme der Semantik, umgekehrt besteht der Witz physikalischer Weltbeschreibung für alle praktischen Zwecke gerade darin, semantisches Vokabular außen vor zu lassen. Physik ist geradezu dadurch charakterisiert, dass Fragen der Bedeutung und Semantik in ihr nicht auftreten. Der tiefere Grund für diese Trennung liegt letztlich darin, dass die Physik die Natur rein deskriptiv zu erfassen sucht, während auf der Ebene der funktionellen und intentionalen Beschreibung normative Elemente hinzutreten (Funktionen und Intentionen stehen unter Bedingungen der Erfüllung oder des dysfunktionalen oder fehlrepräsentationalen Misslingens).

22.6   Schlussbemerkung

Unsere Ausgangsfrage war, ob der Informationsbegriff fundamental, trivial oder mehrdeutig ist. Nur im ersteren Falle würde eine ernsthafte Chance bestehen, ihn als fundamentalen Einheits- und Vereinheitlichungsbegriff in den Naturwissenschaften fruchtbar zu machen. Ontologisch würde dies auf einen Informations-Monismus, wenigstens aber Informations-Reduktionismus hinauslaufen. In Teil 1 haben wir gesehen, dass derlei ambitiöse ontologische Programme an einer damit verbundenen, unplausiblen Hypostasierung von Information scheitern. Schränkt man sich auf den syntaktischen Aspekt von Information ein, was eine starke Einschränkung bedeutet, zeigte sich in Teil 2, dass auch in diesem Falle das Informationskonzept zu keiner begrifflichen Einheit, sondern auf eine Pluralität von Begriffen führt, die mit je unterschiedlichen Formalisierungsstrategien (syntaktischer) Information einhergehen. Anders gesagt: syntaktische Information kollabiert begrifflich wahlweise auf Struktur, Wahrscheinlichkeit, Komplexität, Kompressibilität etc. Da aber nicht nur unser alltagssprachlicher Gebrauch von Information es ferner nahelegt, den Informationsbegriff einerseits im Zusammenhang mit einem Sender-Empfänger-Modell und andererseits mit Fragen der Bedeutung zusammenzubringen, wurden in Teil 3 die unliebsamen Konsequenzen eines damit einhergehenden Subjektivismus für die Naturwissenschaft und in Teil 4 die für praktische Zwecke zum Teil drastischen Differenzen zwischen physikalischer Deskription und semantischer Zuschreibung aufgezeigt.

Fazit: Der Informationsbegriff verführt wie kaum ein anderer Begriff zu äquivoken Verwendungsweisen, die dann Ursprung erheblicher begrifflicher und sachlicher Verwirrungen sind. Seiner Leistungsfähigkeit ist in dieser Hinsicht deutlich Einhalt zu gebieten. Dies schmälert umgekehrt nicht die jeweiligen Stärken, die der syntaktisch eingeschränkte Begriff etwa der Hartley-Shannonschen Auftretens-Unwahrscheinlichkeit oder der algorithmischen Komplexität besitzt. Als Einheits- und Grundbegriff der Naturwissenschaften oder gar der Wissenschaften überhaupt ist dem Informationsbegriff jedoch eine Absage zu erteilen.

Literatur

Baeyer, Hans Christian von (2003). *Information as the New Language of Science*. London: Weidenfeld & Nicolson.

Bell, John S. (1990). Against 'Measurement'. *Physics World* 8: 33-40.

Chaitin, Gregory J. (1987). *Algorithmic Information Theory*. Cambridge: Cambridge University Press.

Dennett, Daniel (1971). Intentional Systems. *Journal of Philosophy* 68: 87–106.

Esfeld, Michael, Hg. (2012). *Philosophie der Physik*. Berlin: Suhrkamp.

Feynman, R. P. (1982). Simulating Physics with Computers. *International Journal of Theoretical Physics* 21(6/7): 467-488.

Friebe, Cord, M. Kuhlmann, H. Lyre, P. Näger, O. Passon & M. Stöckler (2015). *Philosophie der Quantenphysik*. Heidelberg: Springer Spektrum.

Friederich, Simon (2014). A philosophical look at the Higgs mechanism. *Journal for General Philosophy of Science* 45: 335-350.

Gleick, James (2011). *The Information: A History, a Theory, a Flood*. New York: Pantheon.

Hartley, Ralph V. L. (1928). Transmission of Information. *Bell System Technical Journal* 7: 535-563.

Ladyman, James, James Lambert & Karoline Wiesner (2013). What is a complex system? *European Journal for Philosophy of Science* 3: 33–67.

Leff, Harvey S. & Andrew F. Rex, Hg. (2002). *Maxwell's Demon 2: Entropy, Classical and Quantum Information, Computing*. Boca Raton: Taylor & Francis.

Lloyd, Seth (2004). *Programming the Universe*. New York: Knopf.

Lyre, Holger (1998). *Quantentheorie der Information*. Wien: Springer. (2. Aufl. Mentis, Paderborn, 2004).

Lyre, Holger (2002). *Informationstheorie: Eine philosophisch-naturwissenschaftliche Einführung*. München: Fink. (UTB 2289).

Lyre, Holger (2008). Does the Higgs Mechanism Exist? *International Studies in the Philosophy of Science* 22(2): 119–133.

Mainzer, Klaus (2007). *Thinking in Complexity: The Complex Dynamics of Matter, Mind and Mankind*. Berlin: Springer (5. Aufl.).

Morris, Charles (1938). *Foundations of the Theory of Science*. In: Neurath, O., R. Carnap und C. Morris, Hrsg.: *International Encyclopedia of Unified Science*. Chicago: The University of Chicago Press.

Norton, John (2013). All Shook Up: Fluctuations, Maxwell's Demon and the Thermodynamics of Computation. *Entropy* 15: 4432-4483.

Shannon, Claude E. & Warren Weaver (1949). *The Mathematical Theory of Communication*. University of Illinois Press, Urbana.

Smeenk, Chris (2006). The elusive Higgs mechanism. *Philosophy of Science* 73: 487–499.

Stonier, Tom (1990): *Information and the Internal Structure of the Universe. An Exploration into Information Physics*. Berlin: Springer.

Timpson, Christopher (2013). *Quantum Information Theory and the Foundations of Quantum Mechanics*. Oxford: Oxford University Press.

Weizsäcker, Carl Friedrich von (1985). *Aufbau der Physik*. München: Hanser.

Wiener, Norbert (1948). *Cybernetics or Control and Communication in the Animal and the Machine.* Cambridge, Mass.: MIT Press.

Wheeler, John A. (1989). *Information, Physics, Quantum: the Search for Links.* In: Kobayashi, S., H. Ezawa, Y. Murayama und S. Nomura, Hrsg.: *Proceedings of the 3rd International Symposium on the Foundations of Quantum Mechanics*, S. 354-368. Tokyo.

Zeilinger, Anton (2005). *Einsteins Schleier.* München: Goldmann.

Zuse, Konrad (1969). *Rechnender Raum.* Braunschweig: Vieweg.

IV
Ethische und politische Perspektiven

23. Handlung, Technologie und Verantwortung[1]

Julian Nida-Rümelin[2]

Abstract

Der Autor bietet eine philosophische Darstellung der Beziehung zwischen menschlichem Handeln und technischen Handlungsoptionen, wie beispielsweise Robotern und unserem Konzept sowie den Kriterien des Verantwortungsbegriffs. Er argumentiert, dass i) ein Akteur durch Intentionalität – der Fähigkeit, für Handlungen Gründe geben zu können - definiert wird, dass ii) diese Gründe normativ, objektiv und nicht-algorithmisch sind, und damit iii) autonome Roboter nicht als Akteure gelten, denen Verantwortung bezüglich ihrer Handlungen zugeschrieben werden kann. Der Autor sieht daher keine Notwendigkeit einer Änderung unseres Verantwortungsbegriffs in Anbetracht autonomer Roboter. Vielmehr, so der Autor, werden die Kriterien von Zuschreibung von Verantwortung komplexer.

---

1 Dieser Text basiert auf dem Vortrag, den Prof. Nida-Rümelin in englischer Sprache in Pisa am DIRPOLIS - dem Institut der Scuola Superiore Sant'Anna im Rahmen einer Tagung, veranstaltet im Zuge des Projekts *RoboLaw* am 29. November 2013 - gehalten hat. Die Übersetzung besorgte Dr. Fiorella Battaglia (LMU).

2 Julian Nida-Rümelin gehört neben Jürgen Habermas und Peter Sloterdijk zu den renommiertesten Philosophen in Deutschland. Er lehrt Philosophie und politische Theorie an der Universität München. Julian Nida-Rümelin ist Autor zahlreicher Bücher und Artikel sowie gefragter Kommentator zu ethisch, politischen und zeitgenössischen Themen. Sein Buch *Die Optimierungsfalle. Philosophie einer humanen Ökonomie* löste intensive Diskussionen in Unternehmen über die Rolle der Ethik in der ökonomischen Praxis aus. 2013 stieß er die Debatte zum Akademisierungswahn mit einem Interview in der Frankfurter Allgemeinen Sonntagszeitung an. Dazu erscheint im Oktober 2014 bei edition Körber-Stiftung der Essay „Der Akademisierungswahn – Zur Krise beruflicher und akademischer Bildung".

The author presents a philosophical account of the relationship between human agency, technological devices, e.g. autonomous robots, and our concept and criteria of responsibility. He argues that i) an agent is defined by having intentionalities qua giving and taking reasons for her actions, that ii) these reasons are normative, objective, and non-algorithmic, and that thus iii) autonomous robots are not real agents to which responsibility for actions can be ascribed. Therefore, the author sees no need to change the concept of responsibility in the face of autonomous robots. Instead, he considers the criteria of responsibility to have become more complex.

23.1   Einführung

Ich strebe die Entwicklung einer philosophischen Darstellung der Beziehung zwischen menschlichem Handeln, technischen Handlungsoptionen und unserem Konzept sowie den Kriterien des Verantwortungsbegriffs, an. Was verändert sich, wenn Technologien Teil menschlicher Handlungen werden? In dieser Frage gibt es zwei philosophische Lager: Das erste philosophische Lager, das sich vor allem auf Künstliche Intelligenz (KI) bezieht, geht davon aus, dass Computer und Roboter grundsätzlich wie menschliche Wesen handeln können. Es geht dabei weniger darum, ob Computer menschliches Denken imitieren können, sondern um die viel weitreichendere These, dass es zwischen Computern und Menschen keinen kategorischen Unterschied gebe. Ebenso ist die Idee nicht, dass sogenannte autonome Roboter wie menschliche Wesen handeln, sondern dass Menschen letztlich nichts anderes als Roboter, also algorithmisch agierende, von eindeutigen Input-Output-Relationen gesteuerte Entitäten, sind. Das andere, gegnerische philosophische Lager vertritt die Meinung, dass sich menschliches Handeln und menschliches Denken wesentlich von Software-basierten Prozessen unterscheiden. Um diesen Unterschied zu unterstreichen, greifen Vertreter dieses philosophischen Lagers wie ich auf einen konzeptuellen Raum der Begründung (das Geben und Nehmen von Gründen für Überzeugungen und Handlungen sowie für Emotionen oder emotionale Einstellungen) zurück, der an bestimmte mentale Eigenschaften gebunden ist (z.B. intentionale Zustände aufzuweisen), die Softwaresystemen und Robotern, wie humanoid auch immer, nicht zur Verfügung stehen.

   In unserem kulturellen Kontext sind die grundlegenden Begriffe, die jenes letztere philosophische Lager ausgearbeitet, bekannt und seit 2500 Jahren geläufig: Rationalität, Freiheit, (Willensfreiheit), Autonomie, Verantwortung, Verantwortung für Überzeugungen, für Handlung und Emotion. Nennen wir dies die „kanonische

Sicht". Die kanonische Sicht ist eingebettet in juristische Kriterien, wie man beispielsweise auch in § 20 des deutschen Strafgesetzbuches sehen kann.[3] Konfrontiert mit Neurophysiologen[4] und Robotern ergibt sich die Frage, ob wir diese Sichtweise ändern sollten. Sogar in der Rechtslehre sind beide philosophischen Parteien vertreten. Einige der Angehörigen wissen vielleicht nicht, zu welcher dieser Parteien sie gehören – allerdings gehören sie einer der beiden an.

Die kanonische Sicht, vor allem hinsichtlich der Rechtslehre, sollte in eine breitere kantianische Perspektive, entweder der Autonomie und Würde oder der Vernunft und Moral, als korrelierter Begriff eingebettet werden: In der kantianischen Auffassung entspricht Vernunft mehr oder weniger dem Begriff der Moral.

Ob erstere philosophische Partei an die Philosophie Humes gebunden ist, stellt eine schwierige philosophische Frage dar. Viele heutige Philosophen würden diese Frage bejahen. Die Kluft zwischen jenen, die den Begriff von Verantwortung auf nicht-menschliche Akteure erweitern, und jenen, die konstatieren, dass dies zum jetzigen Standpunkt der Dinge noch nicht möglich ist – dies werde ich an einem späteren Zeitpunkt noch diskutieren – stellt mehr oder weniger die Trennung zwischen Hume und Kant innerhalb der zeitgenössischen praktischen Philosophie dar.

Meine persönliche Ansicht geht über die kantianische Sicht hinaus. Selbstverständlich kann ich an dieser Stelle meine philosophische Position nicht vollständig darlegen. Dennoch möchte ich ein paar Bemerkungen machen: Immanuel Kant unterscheidet zwischen pragmatischen und moralischen Imperativen. Die pragmatischen Imperative leiten Handlungen, angeleitet vom Ziel des persönlichen Glücks. Moralische Imperative leiten Handlungen, die durch die Achtung vor dem moralischen Gesetz motiviert sind. Pragmatische Imperative leiten Handlungen heteronom, da Glück empirisch ist. Moralische Imperative leiten Handlungen autonom, denn der kategorische Imperativ, das moralische Gesetz, ist *a priori*. Eine weiterführende Überlegung aber zeigt, dass es die praktische Vernunft im Allgemeinen ist, die Autonomie konstituiert: Unabhängig von den verfolgten Zielen folgen rationale Akteure nicht der Humeschen Strategie der Wünsche-Überzeugungs-Optimierung. Rationale Akteure streben nach einer kohärenten Handlung, was einen mehr oder weniger weitgreifenden Abstand bezüglich gegenwärtiger Wünsche erfordert. Was Kant als Element moralisch-motivierter Handlung definiert, ist vielmehr ein universelles Merkmal vernünftiger Handlung. Ziele und Regeln, Werte und Normen sind verflochten, die Gesamtheit von Handlung kann durch subjektive Werte repräsentiert werden, sollten die einzelnen Handlungen kohärent zusammenhängen. Die Idee einer künstlichen *a priori* Begründung von

---

3   Ich verteidige die kanonische Sicht gegen ihre Kritiker in einer kleinen Reclam-Trilogie.
4   Henz et al. (2015).

Moral ist entweder nichtig (mangelnd an Inhalt, beispielsweise bezüglich der Ziele) oder, als eine strukturelle Eigenschaft von Handlung, allgemein, sei es moralisch oder anderweitig motiviert. Meine Ansicht von praktischer Vernunft lehnt diese Dichotomie ab. All unsere Rationalität ist an das Geben und Nehmen von Gründen gebunden.[5] Dies ist der primäre Fehler, der bei der üblichen Interpretation des Libet-Experiments, welches für manche gegen die Verantwortung menschlicher Akteure spricht, gemacht wird. Das Libet-Experiment ist so konzipiert, dass Gründe keine Rolle für die Entscheidung ,eine Hand zu bewegen' spielen können. Damit ist die Zuschreibung von Freiheit nicht in Gefahr und das Experiment kann nicht als Argument gegen die menschliche Freiheit aufgefasst werden (siehe Nida-Rümelin 2005, S. 45-105; Libet 1993 und Henz et al. 2015).

Gründe für Handlungen, Überzeugungen und emotionale Einstellungen bilden ein mehr oder weniger normatives System, welches propositionale und nicht-propositionale Einstellungen wie Überzeugungen, subjektiven Wert, moralische und nicht-moralische Beurteilung, sowie Emotionen, fasst.

Dieses *Geben und Nehmen von Gründen* ist konstitutiv für Interaktion, Kommunikation und menschliche Handlung im Allgemeinen. Es kann nicht in ein auf Algorithmen basiertes Verhaltenssystem übersetzt werden; Behaviorismus im Widerspruch zur Meta-Mathematik (siehe unten). In diesem Sinne sind Menschen keine Computer. Nichtsdestoweniger ist die Fähigkeit des Begründens nicht an spezifische empirische Eigenschaften der menschlichen Spezies gebunden. So kann es sein, dass auch nicht-menschliche Wesen in der Lage sind, zu argumentieren.

Es gibt Befunde innerhalb der Verhaltensbiologie von Tieren, die zeigen, dass man so etwas wie Meta-Kognition bei nicht-menschlichen Primaten finden kann. Die Frage ist nicht, ob eine Entität der menschlichen Spezies angehört; die Frage ist, ob diese Spezies fähig ist, sich von Gründen affizieren zu lassen und zu begründen. Insofern, als ein Wesen durch Gründe affiziert ist und fähig ist zu argumentieren, kann dessen Verhalten nicht als ein algorithmisch-basiertes System modelliert werden. Demnach ist das langjährige Ziel der aktuellen Robotik, ein Fußballteam aus Robotern zu bauen, welches alle menschlichen Fußballteams der Welt besiegen soll, etwas irreführend. Sogar das intelligenteste Roboter-Fußballteam der Welt könnte *prinzipiell* durch sein Algorithmus-basiertes Verhalten besiegt werden, da dieses *prinzipiell* berechnet und damit vorhergesehen werden kann. Dies bedeutet nicht, dass es unmöglich ist, ein Team aus Robotern zu bauen, welches ebenso gut ist wie ein Fußballteam, das gegen menschliche Fußballteams spielt. Abhängig von der Komplexität der epistemischen Situation könnte es sogar sein, dass ein derar-

---

5   Wir erklären Freiheit durch die Erläuterung der Praxis des Gründe Gebens und Gründe Nehmens und durch unser betroffen sein durch Gründe.

tiges Roboter-Team unbesiegbar ist, da menschliche Wesen nicht in der Lage sind, erfolgreiche Gegenstrategien zu entwickeln. Eine solche Unterlegenheit wäre aber lediglich der Effekt von epistemischen Einschränkungen. Im Prinzip kann jedes Roboter-Fußballteam durch nicht-algorithmische Strategien besiegt werden, da das Roboter-Team menschliche, nicht-algorithmische Handlung nicht berechnen kann.

Nun die Frage, die ich im Folgenden diskutiere: Welche Auswirkungen hat Technologie auf menschliche Handlung? Und welche Implikationen folgen im Zuge dessen auf unseren Begriff und die Kriterien von Verantwortung? Mein Interesse liegt vor allem in der Klärung der Frage, ob uns die Existenz von autonomen Robotern Gründe gibt, unseren Begriff der Verantwortung zu erweitern.

23.2 Was ist Handlung?

23.2.1 Handlungen als intentional kontrolliertes Verhalten

Was ist Handlung oder menschliche Handlung überhaupt? Handlung wird durch Intentionalität konstituiert. Durch Intention geleitetes und motiviertes Verhalten lässt sich als Handlung definieren. Alle und nur diese Bestandteile von Verhalten, welche durch Intention kontrolliert und motiviert sind, haben den Status einer Handlung.

Es scheint mir, dass viele optimistische Vertreter der naturalistischen „Partei" siebzig Jahre philosophischer Forschung bezüglich gerade dieser Frage außer Acht lassen. So gab es große Bemühungen, den Begriff einer Handlung genauer zu klären. Auf Elisabeth Anscombe[6] folgend gab es einige Autoren, die die Art der Intentionalität, die Handlung voraussetzt, diskutierten. Die gegenwärtige Handlungstheorie ist ein äußerst komplizierter und differenzierter Bereich der praktischen Philosophie.

Ich werde an dieser Stelle nur auf einzelne, ausgewählte Aspekte dieses Bereiches eingehen. Offensichtlich gibt es zwei sehr unterschiedliche Arten der Intentionalität innerhalb dieser Debatte – eine davon nennen wir für gewöhnlich *Entscheidungen*. Eine Entscheidung stellt eine sehr interessante Art der Intentionalität dar, da sie durch die Handlung selbst erst vollzogen wird, unabhängig von deren Folgen. Die andere Art der Intentionalität, beispielsweise Motivationen, differenziert sich

---

6 Siehe Anscombe (2000).

von ersterer. Motivationen werden lediglich durch die Folgen einer Handlung vollzogen, nicht aber durch die Handlung selbst.

Es ist äußerst interessant, wie empfindlich unser alltäglicher Sprachgebrauch gegenüber dieser Unterscheidung ist. Wenn ich beispielsweise sage (um ein politisch
inkorrektes Beispiel zu nennen) „Sie entschied sich Joe zu heiraten", wobei sie allerdings später herausfand, dass sie Joe nicht heiraten will, so würde sie niemals sagen
„Diese Entscheidung habe ich vor einem Jahr getroffen". Vielmehr würde sie sagen
„Ich *dachte,* ich hätte diese Entscheidung getroffen und darin aber geirrt (weil ich
meine Entscheidung geändert habe)". Intentionalität reicht nicht aus, um eine Entscheidung zuzuschreiben; der wichtigste Punkt ist, dass nach einer Entscheidung
nicht weiterführend *Pros* und *Kontras* überdacht werden. Eine Entscheidung been-
det den Deliberationsprozess.

Handlungen sind der Teil menschlichen Verhaltens, für welchen dem Menschen durch intentionale Kontrolle Verantwortung zugeschrieben wird. So ist Verantwortung nichts anderes als intentionale Kontrolle. Allerdings wird intentionale
Kontrolle von vernünftigen Akteure durch Gründe geleitet: Du hast dann Kontrolle
über das, was du tust, wenn du Gründe für dein Tun anführen kannst, und weiterführend begründen kannst, ob das was du tun willst richtig oder falsch ist.

23.2.2   Kontrolle als Selbstbild *vs.* Kontrolle als Zuschreibung

Diese Charakterisierung von menschlichem Handeln ist in der aktuellen Philosophie weitgehend akzeptiert. Allerdings sind Zweifel an der Charakterisierung einer
Interpretation dem Realismus zufolge keine Seltenheit. Skeptiker argumentieren,
dass diese Charakterisierung lediglich Teil eines Selbstbilds ist; sie bezieht sich auf
das, was wir selbst in uns sehen, wie wir uns als Akteure interpretieren. Sie behaupten weiterhin, dass dieses Selbstbild nicht auf objektiven Fakten innerhalb unserer
natürlichen Lebenswelt beruht. Ihres Erachtens sind Gründe lediglich Elemente
subjektiver Zustände. Gründe seien Teil einer Interpretation möglicher Ausgänge
unserer Handlung, leiten diese aber nicht (ebenso wenig unsere Überzeugungen
und Emotionen). Vielmehr rückten Gründe erst dann in den Vordergrund, nachdem wir uns in einer bestimmten Art und Weise verhalten haben. Sie seien Teil unseres Vorhabens, uns als kohärente Person darzustellen, die ihr eigenes Verhalten
kontrolliert. Dementsprechend sind Gründe immer *ex post:* Sie interpretieren, was
passiert ist, seien aber keineswegs Teil einer adäquaten kausalen Beschreibung dessen, *warum* etwas passiert ist. Tatsächlich seien Handlungen natürliche Geschehnisse und in diesem Sinne auch nicht von Verhalten zu unterscheiden. Handlungen

können daher im Prinzip durch die Naturwissenschaften erklärt (und beschrieben) werden; Gründe spielen keine Rolle in einer solchen Beschreibung oder Erklärung. Selbst wenn Gründe ein wesentlicher Teil unseres Selbstbilds sind, selbst wenn wir uns nicht von der Gewohnheit lösen können, unser Verhalten durch Gründe geleitet zu interpretieren, gibt es dennoch keine Gründe in der natürlichen Welt. Es gibt nur Ursachen, die grundsätzlich durch die Naturwissenschaften, einschließlich der Neurophysiologie, beschrieben werden können.

Radikalere Skeptiker fügen hinzu, dass die Idee schlüssiger Handlung ein kulturelles Konstrukt darstellt, dessen Wurzeln bis zu 300 Jahre zurückgreifen, in die Zeit der Europäischen Aufklärung. Ein Konstrukt, welches vielen Kulturen der heutigen Zeit noch fremd ist. Andere räumen ein, dass sich unser Gehirn vielleicht in dem Sinne weiterentwickelt hat, dass wir nicht anders können, als unser Verhalten durch Gründe geleitet zu interpretieren. Dennoch sollte dieser Charakterzug unserer Gehirnfunktion nicht realistisch interpretiert werden, argumentieren sie. Es ist die Zusammensetzung unseres Gehirns, welche diese Illusion erzeugt. Terminologisch scheint es, als hätten Freunde und Feinde praktischer Vernunft einen Kompromiss bezüglich des Begriffs „Zuschreibung" gefunden: Wir schreiben Menschen Gründe zu, wir schreiben Menschen Intelligenz zu, wir schreiben Menschen Wünsche zu, etc. Die Regeln der Zuschreibung intentionaler Zustände sind kompliziert und nur zum Teil durch die Sprachphilosophie analysiert.[7] Die Begrifflichkeit „zuschreiben" lässt es offen, ob hinter der Zuschreibung von intentionalen Zuständen Tatsachen stecken, ob Menschen *tatsächlich* intentionale Zustände haben und ob diese intentionalen Zustände *tatsächlich* eine (kausale) wichtige Rolle für menschliche Handlung spielen.

Beginnend mit Kant glaubten viele Philosophen, ein einheitliches Selbstbild auf der einen und ein deterministisches Weltbild auf der anderen Seite durch zwei – scheinbar zusammenpassende – Perspektiven zusammenzuführen: *Die Perspektive der ersten Person*, für welche intentionale Kontrolle unverzichtbar ist, und die *Perspektive der dritten Person*, für die die Zuschreibung von intentionalen Zuständen nichts weiter ist als (im besten Falle) ein heuristisches Werkzeug der Verhaltenserklärung. Während die handelnde Person zufrieden ist, wenn sie sich ihrer Beweggründe bewusst ist (gleichzeitig aber unzufrieden ist, wenn sie etwas getan hat, wovon sie nicht weiß, weshalb sie sich entschieden hat, dies zu tun), so kann die dritte Person nie zufrieden sein, wenn sie sich der (subjektiven) Gründe der handelnden Person bewusst ist. Die dritte Person entwickelt psychologische, neurophysiologische oder sogar physikalische Theorien, welche das fragliche Verhalten erklären. Die analytische Theorie zweier unterschiedlicher Sprach-Ebenen, welche

---

7   Siehe Brandom (1998).

unabhängig voneinander sind (Gilbert Ryle, auch Ludwig Wittgenstein), ist (vielleicht unwissentlich) eine Version der Sprachphilosophie des kantischen Kompatibilismus. Meine eigene Position unterscheidet sich von beiden dieser Versionen des Kompatibilismus.

23.2.3 Kontrolle als gradueller Begriff

Der größere Anteil der philosophischen Diskussion seit Kant hat eines zweier Extreme als Ausgangspunkt: Entweder ist menschliche Handlung durch Intention gesteuert oder sie wird durch außer-mentale Gründe gesteuert; entweder sind Menschen verantwortlich für das, was sie tun, oder sie sind es nicht. Diese Kontraposition resultierte in recht extremen philosophischen Paradigmen, vom Idealismus in der deutschen Philosophie an seinem Höhepunkt im 19. Jahrhundert auf der einen Seite bis hin zum Naturalismus von Quine in der analytischen Philosophie auf der anderen Seite.

Eine weit angemessenere Annäherung ist die des Gradualismus (und Pragmatismus). Den Ausgangspunkt stellt hier nicht eine philosophische Lehre dar, sondern er bezieht sich auf unsere alltägliche Verwendung von Zuschreibungen von Verantwortungen. Der *realistische* Pragmatist interpretiert jene Zuschreibung als ein Element realer Eigenschaften. Eine vernünftige Person zweifelt nicht an einer Überzeugung, wenn es keinen Grund für Zweifel gibt. „Der Vernünftige hat bestimmte Zweifel *nicht*", schreibt Wittgenstein in *Über Gewissheit* (1969/2005; siehe auch Wittgenstein 1953/2003). Unsere weit entwickelte Fähigkeit, die Stufen der Verantwortung eines Akteurs zu beurteilen und der sehr differenzierte normative Diskurs über Kriterien der Verantwortung sprechen beide sehr für diese Interpretation des Realismus. Dieser *normative Realismus* wird nicht von irgendeinem philosophischen Axiom deduziert. Vielmehr ist er ein Element unseres alltäglichen normativen Diskurses und unserer Praxis[8]. Individuen verlieren die Kontrolle über das, was sie tun, wenn sie betrunken sind. Dies ist offensichtlich ein gradueller Prozess: Das Gesetz definiert Kriterien in der Annahme, dass individuelle Verantwortung über eine bestimmte Grenze hinaus reduziert wird. Diese Kriterien verzichten nicht auf einen graduellen Begriff von Verantwortung. Je jünger Kinder sind, desto weniger Verantwortung tragen sie für ihre Handlungen. Ebenso unterminieren emotionale Instabilität oder manische Episoden Verantwortung.

---

8   Vgl. Nida-Rümelin 2016, S. 91-172.

23.2.4 Die Rolle von Kohärenz

All dies ist Teil unserer alltäglichen Interaktion; es ist Teil einer Lebensform, innerhalb welcher wir kooperieren. Wir sind nicht Skeptiker in unserem tagtäglichen Handeln und unserer Beurteilung bezüglich individueller Verantwortung. Ganz im Gegenteil, der Begriff und die Kriterien von Verantwortung spielen eine konstitutive Rolle für lebensweltliche Interaktion sowie lebensweltliches Beurteilen. Im Sinne von Max Weber ist eine autonom handelnde Person ein Idealtyp: Sie existiert nicht. Die idealerweise rationale und autonome Person entscheidet lediglich ein einziges Mal in ihrem Leben und stellt damit die Pluralität ihrer Gründe, welche sich in der Lebensform dieser Person äußern, auf. So gehören sowohl Fortschritte als auch Rückwürfe zu tatsächlichen Menschen. Sie wägen Gründe ab und verändern die relative Wichtigkeit über Jahre hinweg; einst berücksichtigte Bevorzugungen verändern sich und die epistemische Beschaffenheit, in welcher Entscheidungen getroffen werden, ist nicht optimal. Strukturelle Rationalität handelt von Kohärenz, praktisch ebenso wie theoretisch. Sie integriert Gründe, geleitete Überzeugungen und Handlungen, sodass das Ergebnis kohärent ist.

23.3 Drei Perspektiven

Zwei Anwälte des Deutschen Ethikrates argumentierten vor einer Weile, dass es zwei Perspektiven gibt.[9] Den Standpunkt der ersten Person und den der dritten Person. Die Perspektive der dritten Person werde durch die Beschreibung menschlicher Handlung in den Naturwissenschaften wiedergegeben, beispielsweise in der Neurophysiologie. Tatsächlich ist die neurophysiologische Beschreibung jedoch keine Personen-Sicht.

Vielmehr gibt es eine weitere Perspektive, die abhängig ist von dem Kontext, in welchem Gründe gegeben und genommen werden. *Du* bist Teil des *Gründe-Gebens und -Nehmens*. Dementsprechend ist es weniger eine Perspektive aus der dritten Person, vielmehr aber eine aus der zweiten Person: Du und ich, wir wissen, wie wir Gründe austauschen können, wir stimmen in einer Vielzahl praktischer und theoretischer Gründe überein, und dementsprechend verstehe ich dich. Wenn ein Mitglied einer gänzlich anderen Spezies, welche ebenfalls intelligent ist, das Verhalten von menschlichen Akteuren beschreiben würde, so könnte dieses weder die

---

9   Eine Dokumentation der Veranstaltung ist auch online zu finden auf:
     http://www.ethikrat.org/veranstaltungen/weitere-veranstaltungen/neuroimaging.

Perspektive der dritten noch die Perspektive der zweiten Person einnehmen. Es ist lediglich die Perspektive der Naturwissenschaften, welche ihm eine Beschreibung des Verhaltens ermöglicht. Die Art und Weise, wie wir mentale Zustände zuschreiben, wie beispielsweise die Beschreibung von intentionalen Zuständen oder Intentionalität, ist abhängig von der Zugehörigkeit zu einer gemeinsamen, geteilten Lebensform.

In einer Kultur, in der wir Gründe austauschen, könnte ein interkulturelles Problem auftreten. Von Kultur zu Kultur gibt es Invarianzen und wenn man nicht Teil jener Praxis ist, so kann man auch nicht deren Perspektive annehmen. So ist man beispielsweise nicht in der Lage, intentionale Zustände angemessen zu beschreiben oder zuschreiben zu können. Dies ist, meines Erachtens, ein sehr wichtiger Aspekt. Ich werde ihn den *pragmatischen Aspekt* nennen.

Ich betrachte Gründe als einen Grundbegriff und inzwischen stimmen mir hier einige Philosophen zu: Beispielsweise Scanlon (1998), der zuvor eine gegensätzliche Meinung vertrat, oder ähnlich Hilary Putnam (1990, 1992) in seinen neueren Arbeiten.

23.4   Gründe

„Grund" ist ein Grundbegriff. Er ist *normativ*. Und er ist, wie wir sehen werden, *objektiv* (und nicht subjektiv) und *nicht-algorithmisch*.

Gründe sprechen immer für Überzeugungen oder Handlungen oder emotionale Einstellungen. Es gibt keinen Grund ohne einen solchen normativen Inhalt. „Du *solltest* diese Ansicht vertreten, wenn du bestimmte Argumente hast". Dieses „sollte" ist ein epistemisches „sollte", kein moralisches. Im Zuge dessen scheint es mir sehr deutlich, dass die drei Begriffe von *Rationalität, Freiheit* und *Verantwortung* stark miteinander verbunden sind. Ich verzichte an dieser Stelle auf die Aussage, dass sie ebenso logisch verbunden sind, da ich nicht für die Dichotomie zwischen Analytischem und Synthetischem stehe. (siehe Nida-Rümelin 2001; 2005; 2011)

Dies bedeutet, dass man keine Verantwortung tragen kann, wenn man nicht rational ist und dass man nicht frei sein kann, wenn man keine Rationalität besitzt. Man kann nicht einfach eines dieser drei zentralen Konzepte menschlicher Anthropologie, die gar als Erbe des Projekts der *Aufklärung* angesehen werden können, auslassen, wie es so oft von Semi-Kompatibilisten, folgend der Tradition von Harry Frankfurt, getan wird. Diese akzeptieren, dass eine Person für etwas verantwortlich sein kann, bestreiten aber, dass eine Person frei ist.

23.5   Das Nicht-Algorithmische

Ich werde mich nun einem etwas komplizierterem Teil zuwenden, den ich bereits anfänglich adressiert habe: Gründe sind nicht-algorithmisch. Und, um ein wenig polemisch zu sein, wissen wir das seit den 1930ern. Bis heute hat niemand demonstrieren können, dass Alonso Church falsch lag bei seiner Darlegung, dass alle etwas komplexeren logischen Systeme nicht-algorithmisch funktionieren. Die Folge ist, dass Theoreme wie die Prädikatenlogik erster Stufe, einer der wohl grundlegendsten Teile der Logik, nicht durch algorithmische Prozesse, nicht durch eine Turing-Maschine, welche Zeilen des Beweises produziert, bewiesen werden können. Dies ist ein meta-mathematisches Ergebnis, welches niemand in Frage stellt. Deshalb finde ich es äußerst merkwürdig, dass ich unzählige Diskussionen mit Neurowissenschaftlern führen muss, welche das Gegenteil behaupten, wie beispielsweise, dass Begründungsprozesse herkömmliche, kausale Ketten sind. Wenn Logik ein essentieller Teil unseres Begründens ist, so haben wir schon lange bewiesen, dass Begründung nicht-algorithmisch ist.

Ist dies der Fall, und wir akzeptieren diejenige Auffassung kausaler Relationen, wonach kausale Relationen algorithmisch sind (wenn ich den aktuellen Stand der Tatsachen und aller relevanten Gesetze habe, so kann ich den nächsten Zustand der Tatsachen bestimmen), so wurde schon in den 1930er Jahren bewiesen, dass eine Begründung, in welcher logische Schlussfolgerungen auf dem Niveau der Prädikatenlogik erster Ordnung eine Rolle spielen, *kein kausaler Prozess ist*.[10]

23.6   Konsequentialismus

Diese Kritik einer begrenzten, naturalistischen Perspektive, der zufolge Begründung nichts weiter ist als ein kausaler, algorithmischer Prozess in unserem neurophysiologischen System, ist wichtig für eine weitere Debatte: Konsequentialismus *vs.* Deontologie in der Ethik, in der praktischen Philosophie und in der Rationalitätstheorie.

Wäre der Konsequentialismus wahr, so hätten wir einen Algorithmus, welcher uns sagen könnte, was rational ist. Dieser Algorithmus lautet wie folgt: Es gibt eine Wertfunktion, die Zuständen reelle Zahlen zuordnet. Ebenso gibt es eine Funktion,

---

10   Für weitere Informationen siehe Nida-Rümelin (2010).

welche Handlungen das „Sollen" zuschreibt, damit jene Wertfunktion maximiert wird. Dies ist eine algorithmische Art und Weise, wie wir entscheiden, was rational ist. Das Problem ist, dass das nicht adäquat ist. Dies liegt nicht daran, dass der Konsequentialismus die Komplexität moralischer Begründung nicht unterbringen kann. Konsequentialisten versuchen zu zeigen, wie man diese Komplexität durch eine andere Art der moralischen Begründung ersetzen kann. Ich halte diese Versuche für weniger überzeugend. J. J. Smart ist ein Beispiel (Smart und Williams 1973), auch Peter Singer (1993), oder sogar Richard Hare in seiner letzten Veröffentlichung über moralische Begründung – Moral Thinking (1981). Selbst wenn wir unsere Betrachtung auf moralische Aspekte beschränken, gibt es dennoch einige Aspekte, die gegen den Konsequentialismus sprechen. Das Hauptargument für mich allerdings ist, dass der Konsequentialismus ein zu einfaches Kriterium dafür ist, wie wir entscheiden, was wir machen sollten.

Wenn diese Analyse richtig ist, erklärt das, warum Konsequentialismus in der praktischen Philosophie und der Naturalismus als metaphysische Ansicht in der heutigen Philosophie oftmals kombiniert werden. Konsequentialismus reduziert die Komplexität des Gründe-Abwägens zu einer Maximierung der Wertfunktion. Er verwirft die Abwägung von Gründen und stellt ein simples und allumfassendes Maximierungsprinzip als Alternative dar. Allerdings ist es gerade diese Fähigkeit, Gründe abzuwägen, die den Kern einer menschlichen Lebensform darstellt. (Nida-Rümelin 2009)

23.7   Die intentionalistische Darstellung

Die Kombination von Konsequentialismus und Naturalismus bezüglich der Umformung oder Übersetzung des Begründungsprozesses in algorithmisch-kausale Prozesse funktioniert nicht. Die „intentionalistische Darstellung" von Handlung, wie ich sie nenne, definiert Handlung durch spezifische intentionale Zustände des Akteurs. Nachdem diese intentionalen Zustände von Gründen und Begründung umrahmt sind (Abwägen von Gründen), ist menschliche Handlung an die Komplexität von lebensweltlichen Gründen und Ansichten gebunden. Die intentionalistische Darstellung von Handlung ist eng verwandt mit der intentionalistischen Darstellung von Sinn.

Paul Grice hat eine paradigmatische Annäherung zur intentionalistischen Darstellung von Bedeutung geliefert. Das Problem hierbei war, dass viele seiner Anhänger meinten, die intentionalistische Darstellung von Bedeutung basiert auf einer einfachen (reduktiven) optimierenden Beschreibung von rationaler Handlung. In der Literatur wird behauptet, Paul Grice habe eine Theorie über die Bedeutung entwickelt, welche wiederum Teil der Entscheidungstheorie sei. Diese Interpretation wiederspricht allerdings einigen von Grice selbst veröffentlichten Artikeln und besonders der postumen Schrift *Aspects of Reason* (2005). Aber auch wenn Grice's intentionalistische Darstellung von Bedeutung in einen entscheidungstheoretischen Rahmen der Optimierung eingebettet werden könnte, sollte die intentionalistische Darstellung nicht abgelehnt werden.

Ich habe vorgeschlagen, diese auf paradigmatische Art und Weise neu auszurichten: Eine Person macht eine Aussage mit der Intention, dass der Adressat auf diese hin nachdenkt oder handelt (ohne dieses Denken oder Handeln allgemein zu bewirken).[11] Die intentionalistische Darstellung von Handlung basiert allgemein, so mein Verständnis, auf den konstitutiven Rollen von Gründen. Ähnlich basiert das intentionalistische Verständnis sprachlicher Bedeutung auf dem Austausch von Gründen (Gründe für Handlungen und Überzeugungen).

Betrachtet man diese intentionalistische Darstellung von Handlung und das intentionalistische Verständnis von Bedeutung, so kann man erkennen, dass es sich hier mehr oder weniger um die gleiche Theorie handelt. Etwas zu äußern bedeutet in gewisser Hinsicht auch zu handeln. Und diese Handlung hat Bedeutung insofern, als sie auf einem Prozess des Austauschs von Intentionen beruht.

Die Bedeutung einer Äußerung hängt von einem bestimmten kontrollierten Prozess des Austauschs von Intentionen durch das Geben und Nehmen von Gründen ab. Dies resultiert in einem Begriff menschlicher Verantwortung, welcher intentionalistisch und damit abhängig ist von Akteuren, die über Intentionen verfügen. Darüber hinaus ist er an Akteure, die Gründe abwägen und damit von ihnen affiziert sind, gebunden. Da die ganze Komplexität unserer Gründe nicht auf den Konsequentialismus reduziert werden kann, ist sie in diesem Sinne *deontologisch*, nicht aber lediglich im eng gefassten kantischen Sinne, und, äußerst wichtig, an den epistemischen Zustand der Person während der Handlung gebunden.

---

11   Siehe Nida-Rümelin (2009, 135-154).

23.8   Menschliche Verantwortung

Verantwortung wird durch zwei Faktoren eingeschränkt: Was eine Person weiß
und was sie kontrollieren kann. Reine Zufälle, welche einfach so passieren und daher nicht durch einen Akteur geleitet werden, können nicht dessen Verantwortung
beeinflussen. Ich bin mir sehr bewusst, dass diese Tatsache den Begriff der Haftung auf vielen Ebenen verändert. Es ist eine interessante Frage, wie die Praxis der
Haftung mit einer angemessenen Herangehensweise an Verantwortung kombiniert
werden kann.[12]

Ein angemessenes Verständnis von menschlicher Verantwortung ist intentionalistisch, non-konsequentialistisch, deontologisch und epistemisch. Es ist intentionalistisch insofern, als dass die Verantwortung eines Akteurs von seiner Fähigkeit,
seine eigenen Handlungen zu kontrollieren, abhängig ist; es ist non-konsequentialistisch, insoweit es keine Möglichkeit gibt, die Gesamtheit der Pluralität von Gründen auf ein Optimierungs-Prinzip zu reduzieren; es ist deontologisch, insofern
die Pflichten des Handelns von Pflichten bezüglich sozialer Rollen, Freiheiten und
Rechte, Verpflichtungen, etc. abhängig sind, die nicht auf adäquate Weise durch die
Optimierung von Zuständen realisiert werden können; es ist epistemisch insoweit,
als es *moral luck* nicht gibt: Geschehnisse, die ein Akteur nicht kontrollieren kann,
die allerdings trotzdem in die Konsequenzen einer Handlung intervenieren, fallen
nicht in den Verantwortungsbereich des Akteurs.

Diese Darstellung von Verantwortung nimmt den Akteur ernst. Es ist der Akteur selbst, der verantwortlich ist, nicht der Zufall (moralisches Glück); es ist der
Akteur, welcher verantwortlich ist, und nicht die sachlichen, vorhersehbaren Konsequenzen. So ist es beispielsweise notwendig, Gründe abzuwägen, um die Verantwortung eines Akteurs herauszuarbeiten.

23.9   Technologie

Inwiefern verändert Technologie menschliche Handlung? Handlung (*agency*)? Es
gibt mindestens drei Aspekte, die in diesem Zusammenhang eine wichtige Rolle spielen. Der Umgang mit Technologie verändert unsere Beurteilung von Risiko. Nehmen wir beispielsweise den Fall autonomer Roboter: Autonome Roboter
werden autonom genannt, da hinter ihren Handlungen kein menschlicher Akteur

---

12   Siehe Nida-Rümelin (2011).

steht. Ein autonomer Roboter ist in dem Sinne autonom, dass er dem Algorithmus entsprechend, welchen er durch sensorische Impulse empfängt, reagiert – dieser definiert sein Verhalten. Unsere Beurteilung von Risiko verschiebt sich: Ich weiß nicht immer, wie dieser autonome Roboter handeln wird. Wir können nicht *jede* erdenkliche Situation im Vorhinein testen. Wir verfügen über Wissen über die Software und können lediglich hoffen, dass diese in den meisten Situationen unseren Vorstellungen zufolge funktioniert. Ich weiß, dass wir diesen Roboter entworfen haben und damit auch eine bestimmte Vorstellung verfolgen darüber, wie dieser funktionieren soll. Dennoch haben wir keine direkte epistemische Kontrolle über das, was in allen möglichen Situationen passieren könnte. In der Terminologie der Entscheidungstheorie bezeichnet man dies nicht als eine Risikosituation, sondern als *Ungewissheit*. Je autonomer ein Roboter ist, desto mehr wird es zu einer Situation der Ungewissheit, nicht einer des Risikos. Die Frage, welche Kriterien wir verwenden, bleibt bestehen. In der traditionellen Entscheidungstheorie wechselt man von der zu erwartenden Nutzenmaximierung zu einem Maximin-Prinzip: Minimiere den größtmöglichen Schaden. Vielleicht ist es genau das, was wir bei der Benutzung von Robotern machen. Wir sollten versuchen, den größtmöglichen Schaden zu minimieren.

Ein weiterer, philosophisch sehr komplizierter, Aspekt ist der folgende: Nehmen wir die Analyse der philosophischen Handlungstheorie ernst, so ist es offensichtlich, dass das, was wir einer Person als Handlung – gegeben durch ihren epistemischen Zustand – zuschreiben, nicht präzise festgelegt ist.[13] Bewege ich intentional meine Hand, ist dies offensichtlich eine Handlung, welche mir zugeschrieben werden kann: Ich bewege meine Hand. Gegeben meinen epistemischen Zustand würde ich möglicherweise noch eine andere Handlung zuschreiben. Dass ich durch die Bewegung meiner Hand eine Wasserflasche fülle, da ich weiß, dass diese Bewegung der Hand Teil des Befüllens einer Wasserflasche ist. Ist das Wissen darüber, dass dieses Wasser vergiftet ist, Teil meines epistemischen Zustands, so befülle ich die Wasserflasche mit Gift. Ist das Wissen darüber, dass andere unter Umständen von diesem Wasser trinken werden, Teil meines epistemischen Zustands (juristisch ein sehr interessanter Punkt: *dolus eventualis*), so könnte uns dies Grund geben, eine derartige Aktion als Mord zu bezeichnen.

Wir sollten die Gesamtheit der Komplexität des Begründens innerhalb der Rechtslehre, der Philosophie und des alltäglichen Lebens ernst nehmen und nicht versuchen, diese durch eine einfache Optimierung zu substituieren. Der Gebrauch von Robotern verkompliziert die Situation. Er verändert zwar nicht den Begriff der Verantwortung, macht allerdings die Kriterien, auf deren Basis wir Handlung,

---

13   Siehe Anscombe (2000).

Verantwortung und Freiheit zuschreiben, schwerer greifbar. Mir scheint es, dass wir bisher noch keine angemessene Art und Weise gefunden haben, in derartigen Situationen Verantwortung zuzuschreiben.

Zusammenfassend kann man meine These dann wie folgt verstehen: Auch unter Berücksichtigung von Robotern bleibt der Begriff der Verantwortung an sich bestehen; es sind lediglich die Kriterien, die komplexer werden. Dies ist besonders im Fall autonomer Roboter interessant. Ich würde an dieser Stelle gerne erwähnen, dass sich meine anfängliche Anmerkung nicht nur auf Hardware beschränkt. Es ist nicht relevant, ob eine Entität durch herkömmliche Gehirnzellen oder durch anderes Material funktioniert. Der relevante Aspekt ist vielmehr, ob sie intentionale Zustände hat. Software und Roboter scheinen oftmals mentale Zustände zu haben. Zum jetzigen Standpunkt der Dinge können wir allerdings mit Sicherheit sagen, dass dem nicht so ist, dass sie keine mentalen Zustände haben. Wenn wir uns hier einig sind, so sind autonome Roboter keine Akteure, selbst wenn sie so aussehen oder handeln. Virtuelle Handlungen sollten keine Veranlassung dafür sein, unseren Begriff der Verantwortung zu modifizieren, wohl aber die Kriterien der Verantwortungszuschreibung so anzureichern, dass die neuen technologischen Möglichkeiten einbezogen werden können (Nida-Rümelin 2016).

Literatur

Anscombe, G.E.M. 2000. *Intention*. Cambridge, MA: Harvard University Press.
Brandom, R. 1998. *Making it Explicit. Reasoning, Representing, and Discursive Commitment.* Cambridge, MA: Harvard University Press.
Grice, P. 2005. *Aspects of Reason*. Oxford: Clarendon Press.
Hare, R.M. 1981. *Moral Thinking: Its Levels, Methods, and Point.* Oxford: Clarendon Press.
Henz, S., D.F. Kutz, J. Werner, W. Hürster, F.P. Kolb und J. Nida-Ruemelin. 2015. Stimulus-dependent deliberation process leading to a specific motor action demonstrated via a multi-channel EEG analysis. *Front. Hum. Neurosci.* 9:355. doi: 10.3389/ fnhum.2015.00355.
Libet, B. 1993. Unconscious Cerebral Initiative and the Role of Conscious Will in Voluntary Action. In *Neurophysiology of consciousness*, 269-306. Boston, MA: Birkhäuser Boston.
Nida-Rümelin, J. 1993. *Kritik des Konsequentialismus*. München: R. Oldenbourg Verlag.
Nida-Rümelin, J. 2001. *Strukturelle Rationalität*. Stuttgart: Reclam.
Nida-Rümelin, J. 2005. *Über menschliche Freiheit*. Stuttgart: Reclam.
Nida-Rümelin, J. 2009. *Philosophie und Lebensform*. Frankfurt am Main: Suhrkamp.
Nida-Rümelin, J. 2010. Reasons Against Naturalizing Epistemic Reasons: Normativity, Objectivity, Non-computability. In *Causality, Meaningful Complexity and Embodied Cognition*, hrsg. A. Carsetti, 203-2010. Dordrecht: Springer.

Nida-Rümelin, J. 2011. *Verantwortung*. Stuttgart: Reclam.
Nida-Rümelin, J. 2016. *Humanistische Reflexionen*. Frankfurt am Main: Suhrkamp.
Putnam, H. 1990. *Realism with a Human Face*. Cambridge, MA: Harvard University Press.
Putnam, H. 1992. *Renewing Philosophy*. Cambridge, MA: Harvard University Press.
Scanlon, T. 1998. *What We Owe to Each Other*. Cambridge, MA: Harvard University Press.
Singer, P. 1993. *Practical Ethics*. Cambridge: Cambridge University Press.
Smart, J.J.C. und B. Williams. 1973. *Utilitarism – For and Against*. Cambridge: Cambridge University Press.
Wittgenstein, L. 1969/2005. *Über Gewissheit*. Frankfurt am Main: Suhrkamp.
Wittgenstein, L. 1953/2003. *Philosophische Untersuchungen*. Frankfurt am Main: Suhrkamp.

24. Emergenz und Transdisziplinarität

Jürgen Mittelstrass[1]

Abstract

Emergenz und Transdisziplinarität sind Begriffe der neueren Wissenschaftstheorie. Emergenz bezieht sich auf Systemzustände, Transdisziplinarität auf die
methodische Ebene von Forschungsprinzipien. Während sich der Begriff des
Holismus darauf bezieht, daß sich der Zustand eines Gesamtsystems aus den
Zuständen seiner Teilsysteme und ihrer Wechselwirkungen ergibt, betont der
(starke) Emergentismus eine Nichtableitbarkeit der Systemeigenschaften aus
den Eigenschaften der Systemteile. Das gilt sowohl für die Phänomenebene als
auch für die Methoden- und Theorieebene, auf der es um das Verhältnis von
Disziplinarität und Transdisziplinarität im Erkenntnisprozeß geht.
Emergence and transdisciplinarity are concepts of modern philosophy of science.
Emergence refers to system states, transdisciplinarity to the methodological level
of research principles. While the notion of holism draws on the fact that the state
of the entire system results from the states of its subsystems and their interactions,
a (strong) emergentism emphasizes the non-derivability of the system properties
from the properties of the system parts. This applies both to the level of the phenom-
ena as well as to the methodological and theoretical level, at which the relationship
between disciplinarity and transdisciplinarity is situated.

---

1   Geb. 11.10.1936 in Düsseldorf. 1961 philosophische Promotion, 1968 Habilitation
    an der Universität Erlangen. 1970-2005 Ordinarius für Philosophie und
    Wissenschaftstheorie an der Universität Konstanz. 1989 Leibniz-Preis. 1997-1999
    Präsident der Allgemeinen Gesellschaft für Philosophie in Deutschland. 2002-
    2008 Präsident der Academia Europaea (London, UK). 2005-2015 Vorsitzender des
    Österreichischen Wissenschaftsrates.

Emergenz und Transdisziplinarität sind – zum Teil bereits modisch werdende – Begrifflichkeiten, mit denen versucht wird, wissenschaftliche Entwicklungen sowohl auf der Gegenstandsebene als auch auf der Theorieebene zu beschreiben bzw. zu erklären. Entsprechend im Folgenden einige kurze Bemerkungen vor allem begrifflicher Art – nichts wirklich Neues, aber wohl ein wenig Licht in manchmal recht dunkle begriffliche Verhältnisse Bringendes.

24.1   Begriffsgeschichten

Im Grunde geht es in der Rede von Emergenz und Transdisziplinarität, so fremd diese Zusammenstellung zunächst auch anmuten mag, um die ewig alten Fragen, was alt und was neu ist, wie alt das Alte und wie neu das Neue ist, und wie man aus dem Alten das Neue, das wiederum selbst im Laufe der Entwicklung das Alte wird, gewinnt. Oder anders, aus anderem Blickwinkel gefragt: Wie neu ist das Neue und wie alt ist das Alte? Das gilt sowohl von Entwicklungen, die wir nicht in der Hand haben – meist sprechen wir dann von natürlichen Entwicklungen – als auch von Entwicklungen, die wir in der Hand haben, die wir selbst steuern bzw. steuern können oder wollen. In diesem Falle tritt der Begriff des Fortschritts an die Stelle des Begriffs der Entwicklung. Zu den Entwicklungen, die wir nicht in der Hand haben, gehören z.B. biologische und geologische Entwicklungen, die in der Regel über große Zeiträume laufen, zu den Entwicklungen, die wir in der Hand haben, die meisten technischen und ökonomischen Entwicklungen, die zum Teil ebenfalls nicht von heute auf morgen erfolgen.

Manchmal verbindet sich auch beides miteinander, das, was wir nicht in der Hand haben, und das, was wir in der Hand haben, z.B. beim Thema Klima. Einst durfte man davon ausgehen, daß das Klima, etwa in Form von Kalt- und Warmluftfronten, für das uneingeschränkt Natürliche steht, heute wissen wir, daß der Mensch auch hier seine Hand mit im Spiele hat, daß es ein Natürliches, das uneingeschränkt oder streng Natürliche, nicht mehr gibt. Die Sphären des Natürlichen und des Nicht-Natürlichen, des Artifiziellen, sind immer weniger streng geschieden. Hier setzen Begriffe wie der der Emergenz an.

Begriffe haben ihre eigenen Karrieren. In der Wissenschaft werden sie in konkreten wissenschaftlichen Zusammenhängen gebildet und bleiben üblicherweise Teile dieser Zusammenhänge. Manchmal verlassen sie diese aber auch, vor allem dann, wenn andere, gemeint sind andere disziplinäre Zusammenhänge, sie für sich selbst entdecken. Das wiederum geschieht meist dort, wo es sich um unscharfe Begriffe handelt. Begriffliche Schärfe behindert inter- oder transdisziplinäre Anschlüsse, begriffliche Unschärfe fördert sie. Ein Beispiel dafür ist der Begriff des *Paradigmas*. Ursprünglich ein eher unscheinbarer, mit Teilen der Alltags- und Bildungssprache verbundener Begriff wird dieser von Thomas Kuhn in einem theoretischen Zusammenhang, einer Theorie der Wissenschaftsgeschichte bzw. in der Konzeption einer Theoriendynamik, zur Charakterisierung dessen, was er als ‚normale' Wissenschaft, Wissenschaft in einem Reifestadium, bezeichnet, hervorgehoben. Theoretische Paradigmen, so Kuhn, bestimmen den wissenschaftlichen Alltag; was sie genauer bedeuten – fundamentale Theorien, forschungsleitende Prinzipien, methodische Regeln etc. – bleibt weitgehend unklar. Eine kritische Analyse des verwendeten Paradigmenbegriffs weist mehr als 20 verschiedene Bedeutungen aus.[2] Offenkundig ist es eben diese Vieldeutigkeit, diese begriffliche Unschärfe, die die Übernahme dieses Begriffs in andere disziplinäre Bereiche, z.B. in die Literatur- und Rechtswissenschaften, wesentlich erleichtert hat; man mußte ihn sich nicht erst zurechtbiegen, um ihn auf je eigene disziplinäre Weise verwenden zu können.

Ein anderes Beispiel ist, wenn auch weit weniger vieldeutig und in die Nähe des Emergenzbegriffs führend, der in der Wissenschaftstheorie verwendete, aber auch in Biologie und Physik, in der sozialwissenschaftlichen Theoriebildung und in der Bestätigungstheorie Anwendung findende Begriff des *Holismus*. Mit ihm werden methodische Ansätze der Erklärung begrifflicher und empirischer Phänomene verstanden, die ihren Ausgangspunkt von einer ‚ganzheitlichen' Betrachtungsweise nehmen. Begrifflich bzw. methodologisch geht es dabei um die Unterscheidung zwischen der Teil-Ganzes-Beziehung und der Elementbeziehung, weil Ganzheiten zwar als Teilezusammenhang, nicht aber als bloße Summe ihrer Teile zu verstehen sind. Biologisch bezeichnet der Begriff des Holismus den Versuch, alle Lebensphänomene im Gegensatz zu den partikularen Auffassungen des Mechanismus und des Vitalismus aus einem ganzheitlichen ‚metabiologischen Prinzip' abzuleiten; physikalisch wird das Auftreten so genannter verschränkter Zustände in der Quantentheorie als Grundlage eines ontologischen Holismus betrachtet. Dieses Prinzip besagt, daß jedes physikalische System seine grundlegenden Eigenschaften unabhängig von anderen, von ihm getrennten Systemen besitzt. Bei zusammengesetzten Systemen ergibt sich der Zustand des Gesamtsystems aus Zuständen der Teilsysteme und deren Wechselwirkungen. In den Sozialwissenschaften bezeichnet ein methodologischer Holismus die Auffassung, daß gesellschaftliche Zusammenhänge nur in Begriffen sozialer Ganzheiten gedeutet und erklärt werden können.

Während biologische, quantenphysikalische und sozialwissenschaftliche Ausarbeitungen holistischer Vorstellungen den besonderen Deutungs- und Erklärungsanforderungen partieller Gegenstandsbereiche dienen sollen, geht es bei dem so

---

2 Vgl. Masterman (1970/1999, S. 61-66).

genannten *Bestätigungsholismus* um die bereichsübergreifende These, daß sich nur Theorien als ganze empirisch beurteilen lassen. Ähnlich wird im Rahmen eines so genannten *Bedeutungsholismus* oder *semantischen Holismus* argumentiert. Hier ergibt sich die Bedeutung einzelner Begriffe oder mit ihnen gebildeter Aussagen aus deren Wechselbeziehungen mit anderen Begriffen. Bedeutung tragen diese daher nicht isoliert, sondern erst im Zusammenhang umfassender Begriffssysteme. Dieser Holismus entsteht aus dem Bestätigungsholismus bei Hinzunahme der verifikationistischen Prämisse, daß die Bedingungen der empirischen Prüfung auch Aufschluß über Bedeutungen geben. Ein weiterer Grund ist die Feststellung, daß die Bedeutung wissenschaftlicher Begriffe nur im Zusammenhang der zugehörigen wissenschaftlichen Theorie verstanden wird und nicht durch schrittweise Kenntnis der einschlägigen Definitionen allein gewonnen werden kann. Nach dieser Kontexttheorie der Bedeutung gewinnt ein wissenschaftlicher Begriff erst durch seine theoretische Einbettung seinen spezifischen Gehalt.

Es sind holistische Ansätze dieser Art, die auf den Begriff der *Emergenz* führen, insofern es im Sinne des Bestätigungsholismus oder des semantischen Holismus die Systemeigenschaften sind, die Auskunft über das Verhalten eines Systems geben. Diese Eigenschaften, so wird gesagt, sind emergent.

24.2   Emergenz & Co.

Wie der Holismus bezieht sich auch der so genannte *Emergentismus* auf das Verhältnis von Eigenschaften, die das Ganze, und solche, die dessen Bestandteile betreffen, bzw. auf die Art und Weise, in der beide, die Ganzheitseigenschaften und die Teileeigenschaften, miteinander verknüpft sind. Emergenz besagt hier, daß die Eigenschaften von Ganzheiten aus den Eigenschaften ihrer Bestandteile und der Wechselwirkungen zwischen ihnen nicht erklärbar und insofern auch nicht vorhersehbar sind.[3] Für den Emergentismus stellt sich die Welt als eine Struktur hierarchisch organisierter Systeme dar, wobei die Eigenschaften von höherstufigen Systemen zwar durch die Eigenschaften ihrer Subsysteme festgelegt, aber doch wesentlich von diesen verschieden sind. Es treten jeweils andersartige Eigenschaften und Prozesse auf. Dabei wird zwischen einer starken und einer schwachen Emergenzthese unterschieden.

---

3   Das Folgende in direktem Anschluß an Carrier (2005). Zum Begriff der Emergenz vgl. ferner M. Carrier und J. Mittelstraß (1989, S. 127-130 bzw. 1991, S. 120-122) und K. Mainzer (2002, S. 84-85; 2007, S. 71-72).

Den Kern der *starken* Emergenzthese bildet eine Nichtableitbarkeits- oder Unerklärbarkeitsbehauptung der Systemeigenschaften aus den Eigenschaften der Systemteile. Eine emergente Eigenschaft ist unableitbar; ihr faktisches Auftreten ist insofern nicht zu erwarten und nicht vorhersehbar. *Schwache* Emergenz beschränkt sich dagegen auf die bloße Unterschiedlichkeit von System- und Teileeigenschaften und ist mit der theoretischen Erklärbarkeit der Systemeigenschaften verträglich. Das bedeutet: schwache Emergenz ist im wesentlichen ein Komplexitätsphänomen. Die klassische Formulierung einer starken Emergenzthese stammt von Ch. D. Broad (1925). Broad ging es um die angemessene Interpretation lebender Organismen. Sie sollte Organismen weder als bloße Maschinen noch als von einer besonderen lebenden Kraft bestimmt darstellen. Diese neovitalistische Auffassung wurde vor allem von H. Driesch (1909) vertreten, der die Lebewesen mit ‚Entelechien‘ ausgestattet sah, d.h. mit zielgerichteten biologischen Antrieben und Kräften. Broad suchte einen dritten Weg zwischen der mechanistischen und eben dieser vitalistischen Sicht des Lebens. Dieser Weg sollte durch eine Emergenzthese gefunden werden. Emergente Ganzheitseigenschaften sollten durch die andersartigen (qualitativen) Eigenschaften ihrer Teile der Sache nach festgelegt, jedoch nicht über diese erklärbar sein.

Der starke Emergentismus ist genauer durch die folgenden Bestimmungen charakterisiert: (1) Die Bedingung der *qualitativen Verschiedenheit*. Diese bezieht die Emergenzthese auf diejenigen Eigenschaften von Ganzheiten, die sich von den Eigenschaften ihrer Bestandteile wesentlich unterscheiden. (2) Die Bedingung der *Eigenschaftsdetermination*. Diese besagt, daß die Eigenschaften der Bestandteile hinreichen, um die betreffende Ganzheitseigenschaft hervorzubringen; diese hängt entsprechend nicht von zusätzlichen Einflußfaktoren ab. (3) Die Bedingung der *prinzipiellen Erklärungslücke*. Diese besagt, daß die Ganzheitseigenschaft durch die Eigenschaften der Teile einschließlich der zwischen ihnen bestehenden Wechselwirkungen nicht erklärt werden kann. Dabei ist die Existenz von emergenten Eigenschaften in diesem starken Sinne umstritten. Die einzigen Kandidaten im Rennen sind gegenwärtig *phänomenale* Eigenschaften (vgl. Stephan 2004). Der Punkt wäre dann, daß etwa im Rahmen der Philosophie des Geistes (*philosophy of mind*) im gegebenen neurologischen System nicht ableitbar und entsprechend nicht vorhersehbar wäre, daß in diesem System qualitative Erfahrungen etwa farblicher oder akustischer Art auftreten.

Für das im Zusammenhang mit dem Begriff der (wissenschaftlichen) Erklärung auftretende Thema Vorhersehbarkeit ist vor allem der temporale Aspekt der Emergenzthese interessant, d.h. für in Entwicklungen auftretende Ganzheitseigenschaften. Grenzen der Reduzierbarkeit (des Ganzen auf seine Teile) erweisen sich hier als Grenzen der Erklärbarkeit und der Vorhersehbarkeit. Dieser temporale Aspekt des

Neuen wird heute unter dem Begriff *creative advance of nature* behandelt und z.B. von K. Popper und J. C. Eccles vertreten (1977, S. 22-35; vgl. Čapek 1961, S. 333ff). Generell geht es im Begriff der starken und im Begriff der schwachen Emergenz um das Verstehen und Erklären mehr oder weniger spontaner Veränderungen in der Welt der Phänomene. Dieser Begriff in seiner doppelten Bedeutung läßt sich aber auch auf das Verstehen und Erklären mehr oder weniger spontaner Veränderungen in der Welt der Konzeptionen (Theorien) anwenden. Wenn wir von der Welt der (zu erklärenden) Phänomene sprechen, dann geht es z.B. um die Entstehung des Lebens aus dem Nicht-Leben, der reinen Materie, oder um die Entstehung mentaler Eigenschaften aus neuronalen Eigenschaften. Benachbarte Begriffe sind hier die Begriffe der Selbstorganisation und der Supervenienz. Der Begriff der *Selbstorganisation* bezieht sich auf das spontane Entstehen makroskopischer Strukturen aus mikroskopischen Strukturen unter bestimmten Randbedingungen (vgl. Carrier 1995) – der Akzent liegt hier im wesentlichen auf systeminternen Faktoren –, der Begriff der *Supervenienz* auf die Unterscheidung zweier Ebenen, etwa der Ebene des Physischen und des Psychischen, wobei die superveniente Ebene, hier das Psychische, als durch die subveniente Ebene, hier das Physische, determiniert gilt. Auf der sprachlichen Ebene bedeutet dies, daß die Semantik (semantische Unterschiede) supervenient relativ zur Syntax (syntaktischen Unterschieden) ist.[4]

Wenn wir im nunmehr abgeleiteten Sinne von der Welt der Konzeptionen (Theorien) sprechen, dann geht es einerseits um jene Konzeptionen, mit denen wir mehr oder weniger spontane Veränderungen in der Welt der Phänomene zu begreifen suchen, also z.B. um die erwähnten Begriffsbildungen von Selbstorganisation und Supervenienz, andererseits um unerwartete oder (in der Begrifflichkeit Kuhns) ‚revolutionäre' Veränderungen in eben diesen Konzeptionen, um das Auftreten neuartiger Theorien, die sich aus der bisherigen Theoriengeschichte nicht herleiten lassen. Als Beispiele können die Evolutionstheorie und die Relativitätstheorie dienen. Beide, die Welt der Theorien und die Welt der Phänomene, unter Emergenzgesichtspunkten oder anderen betrachtet, hängen natürlich miteinander zusammen. Theoriebildung wird nicht um der Theoriebildung willen betrieben, sondern um Phänomene (Sachverhalte) zu verstehen und zu erklären. Und doch ist beides nicht dasselbe. Vereinfacht gesagt: Im einen Falle geht es um die *Evolution der Phänomene* – man sagt dann auch, daß man es mit autopoietischen Verhältnissen zu tun hat –, im anderen Falle um die *Evolution der Gedanken*, die sich, wenn auch nicht ausschließlich, auf gegebene Verhältnisse, natürliche oder andere, beziehen. Schließlich kann auch die Welt der Artefakte Gegenstand der Theoriebildung sein.

---

4   Dazu M. Carrier und J. Mittelstraß (1989, S. 58-60, S. 212-215).

Hier sei im Folgenden von der Evolution der Gedanken, d.h. der Art und Weise, wie Theorien neuartige Theorien, Gedanken neuartige Gedanken generieren, und wie man einen derartigen, im übertragenen Sinne emergenten, das Neue durch alleinigen Rekurs auf das Alte nicht erklärbaren Prozeß fördern könnte, die Rede. Das Stichwort lautet: Transdisziplinarität.

24.3 Transdisziplinarität

Mit Emergenz, so haben wir gesehen, ist die Entstehung des Neuen in der Welt der Phänomene und der sie erklärenden Wissenschaft verstanden. Die Wissenschaft wiederum ist im wesentlichen disziplinär geordnet. Dabei wissen wir, daß das Neue in der Wissenschaft nur noch selten in ihren disziplinären Kernen, dort, wo auch das Lehrbuchwissen sitzt, entsteht, vielmehr vornehmlich an den disziplinären Rändern, zwischen den Disziplinen und über alles Disziplinäre hinweg. Wer auf das Neue in der Wissenschaft, auf Emergenz im Theoriesinne setzt, sollte folglich diesen Prozeß, der das Disziplinäre nicht obsolet werden läßt, aber auf eine neuartige Weise in die wissenschaftliche Pflicht nimmt, fördern. Das geschieht in einem besonderen Maße im Kontext der *Transdisziplinarität*, wobei sich dieser mittlerweile in der Welt der Wissenschaft etablierte Begriff gegenüber dem herkömmlichen Begriff der Interdisziplinarität dadurch abgrenzt, daß sich in der transdisziplinären Arbeit die Fächer und Disziplinen selbst verändern. Dazu einige kurze begriffliche Erläuterungen.[5]

Transdisziplinarität ist ein sowohl innerwissenschaftliches, die Ordnung des wissenschaftlichen Wissens und der wissenschaftlichen Forschung selbst betreffendes Prinzip als auch eine Forschungs- und Arbeitsform der Wissenschaft, wenn es darum geht, außerwissenschaftliche Probleme, z.B. Umwelt-, Energie- und Gesundheitsprobleme, zu lösen. In beiden Fällen ist Transdisziplinarität ein Forschungs- und Wissenschaftsprinzip, das dort wirksam wird, wo eine allein fachliche oder disziplinäre Definition von Problemlagen und Problemlösungen nicht möglich ist bzw., im Sinne der Suche nach dem wissenschaftlich Neuen, über derartige Definitionen hinausgeführt wird.

Nun treten reine Formen von Transdisziplinarität ebensowenig auf wie reine Formen von Disziplinarität oder Fachlichkeit. Wie Disziplinarität und Fachlichkeit ist auch Transdisziplinarität ein forschungsleitendes Prinzip bzw. eine idealtypische Form wissenschaftlicher Arbeit; Mischformen sind ihre Normalität. Wichtig ist al-

---

5   Vgl. Mittelstraß (2003).

lein, daß sich Wissenschaft und Forschung dessen bewußt sind und produktive Forschung nicht durch überholte, meist gewohnheitsmäßig vorgenommene Einschränkungen auf fachliche und disziplinäre Engführungen begrenzt wird. Fachliche und disziplinäre Kompetenzen bleiben damit, wie schon erwähnt, eine wesentliche Voraussetzung für transdisziplinär definierte Aufgaben und Arbeitsformen, aber sie allein reichen nicht mehr aus, um Forschungsaufgaben, die aus den klassischen Fächern und Disziplinen herauswachsen, erfolgreich zu bearbeiten.

Das muß in Zukunft zu neuen Organisationsformen führen, in denen die Grenzen zwischen den Fächern und Disziplinen blaß werden. Anders ausgedrückt: Transdisziplinarität ist erstens ein *integratives* Konzept. Sie löst Isolierungen, die sich in der Wissenschaftspraxis eingestellt haben, auf einer höheren methodischen Ebene auf, aber sie baut nicht an einem universalen Deutungs- und Erklärungsmuster. Transdisziplinarität hebt zweitens innerhalb eines historischen Konstitutionszusammenhanges der Fächer und Disziplinen Engführungen auf, wo diese ihre historische Erinnerung verloren und ihre problemlösende Kraft über allzu großer Spezialisierung eingebüßt haben, aber sie führt nicht in einen neuen fachlichen oder disziplinären Zusammenhang. Deshalb kann sie auch die Fächer und Disziplinen nicht ersetzen. Und Transdisziplinarität ist drittens ein wissenschaftliches Arbeits- und Organisationsprinzip, das problemorientiert über Fächer und Disziplinen hinausgreift, aber kein transwissenschaftliches Prinzip. Die Optik der Transdisziplinarität ist eine wissenschaftliche Optik, und sie ist auf eine Welt gerichtet, die, selbst mehr und mehr ein Werk des wissenschaftlichen und des technischen Verstandes, ein wissenschaftliches und technisches Wesen besitzt. Schließlich ist Transdisziplinarität viertens, und noch einmal, in erster Linie ein Forschungsprinzip, kein oder allenfalls in zweiter Linie, wenn nämlich auch die Theorien transdisziplinären Forschungsprogrammen folgen, ein Theorieprinzip.

Damit steht es auch im Dienste dessen, was, bezogen auf den Emergenzbegriff, als Evolution der Gedanken bezeichnet wurde. Emergenz im Theoretischen läßt sich nicht erzwingen – sie ist, wie im Falle der Evolution der Phänomene, aus der bisherigen Theoriegeschichte nicht deduzierbar –, aber sie läßt sich mittelbar fördern, indem man in der Wissenschaft die entsprechenden institutionellen und methodischen Voraussetzungen schafft. Oder anders gesagt: Transdisziplinäre Arbeitsformen fördern in der Wissenschaft nicht nur Problemlösungen bzw. stellen sich als eine zunehmend wichtiger werdende Voraussetzung für diese dar, sie schaffen auch einen Raum, in dem sich neue Wahrnehmungen, neue Problemkonstellationen, neue theoretische Einsichten zu bilden vermögen, in dem sich das Neue sowohl auf der Theorieebene als auch auf der Arbeitsebene leichter tut als in einem engen fachlichen oder disziplinären Rahmen. Emergenz im beschriebenen, schwachen oder starken Sinne ist dann immer noch ein Phänomen, das weder planbar noch vorhersehbar ist, aber durch fachliche und disziplinäre Gewohnheiten, die in der Regel stark ausgebildet sind, nicht behindert wird. Auch hier gilt, wie in anderen Verhältnissen auch: es hängt alles von den Umständen und von den Köpfen ab. Im übrigen – bevor der Gedanke der Emergenz selbst mythische Züge annimmt, das unerwartet Neue als irgendwie übernatürliches Geschehnis verstanden wird – ist immer alles anders als zuvor: die Lebensformen früher und heute, der Stand der Technik, der Medizin, der Freiheit und Unfreiheit ebenso, der Stand des Bildungs- und Rechtssystems früher und heute. Ob man das, auf das Unvorhersehbare, spontan im Empirischen wie im Theoretischen Auftretende achtend, als ,Emergenz', entsprechende Entwicklungen als ,emergent' bezeichnet oder nicht, ist nicht das Entscheidende, es sei denn, das Unvorhergesehene erweist sich gegenüber dem Bestehenden als so neu, so ,revolutionär', daß aus ihm das Unvorsehbare und in allen Aspekten Unvergleichbare wird. In der Welt der Gedanken ist das alles, wie die Wissenschaftsgeschichte lehrt, in der Regel wenig dramatisch, und in der Welt der Phänomene, wenn man sich weniger an den differenzierenden Blick der Erkenntnis- und Wissenschaftstheoretiker als vielmehr an die wissenschaftlichen Forschungsprogramme selbst hält, wohl auch.

Literatur

Broad, Ch. D. 1925. *The Mind and Its Place in Nature*. Routledge & Kegan Paul: London.
Čapek, M. 1961. *The Philosophical Impact of Contemporary Physics*. Princeton, NJ: Van Nostrand.
Carrier, M. 1995. Selbstorganisation. In *Enzyklopädie Philosophie und Wissenschaftstheorie* III, hrsg. J. Mittelstrass, 761-764. Metzler: Stuttgart/Weimar.
Carrier, M. 2005. emergent/Emergenz. In *Enzyklopädie Philosophie und Wissenschaftstheorie* II, 2. Aufl., hrsg. J. Mittelstrass, 313-314. Metzler: Stuttgart/Weimar.
Carrier, M. und J. Mittelstraß. 1989. *Geist, Gehirn, Verhalten. Das Leib-Seele-Problem und die Philosophie der Psychologie*. De Gruyter: Berlin/New York. (engl. [erweitert] *Mind, Brain, Behavior. The Mind-Body Problem and the Philosophy of Psychology*. DeGruyter: Berlin/New York, 1991.)
Driesch, H. 1909. *Philosophie des Organischen*, I-II. Leipzig: W. Engelmann.
Mainzer, K. 2002. *Zeit. Von der Urzeit zur Computerzeit*. 4. Aufl. München: C.H. Beck.
Mainzer, K. 2007. *Der kreative Zufall. Wie das Neue in die Welt kommt*. München: C.H. Beck.
Masterman, M. 1970/1999. The Nature of a Paradigm. In *Criticism and the Growth of Knowledge (Proceedings of the International Colloquium in the Philosophy of Science IV)*, hrsg. I. Lakatos und A Musgrave, 59-89. Cambridge: Cambridge University Press.
Mittelstraß, J. 2003. *Transdisziplinarität – wissenschaftliche Zukunft und institutionelle Wirklichkeit* (Konstanzer Universitätsreden 214). Konstanz: Universitätsverlag Konstanz.

Popper, K. R. und J. C. Eccles. 1977. *The Self and Its Brain*. New York: Springer.
Stephan, A. 2004. Phänomenale Eigenschaften, phänomenale Begriffe und die Grenzen Reduktiver Erklärung. In *Grenzen und Grenzüberschreitungen (XIX. Deutscher Kongress für Philosophie*, Bonn, 23.-27. September 2002). *Vorträge und Kolloquien*, hrsg. W. Hogrebe und J. Bromand, 404-416. Berlin: Akademie Verlag.

25. Überlegungen zu einer experimentellen Ethik

Christoph Lütge[1]

Abstract

In diesem Artikel werden Perspektiven aufgezeigt, wie Experimente der Sozialwissenschaften, Ökonomik und Psychologie für die Ethik fruchtbar werden können. Als Ergebnis wird ein neues Forschungsfeld „Experimentelle Ethik" vorgestellt. Forschungsfragen, praktische Implikationen sowie mögliche Kritikpunkte einer experimentellen Ethik werden diskutiert.

This article outlines some perspectives for an „Experimental Ethics" which makes use of experimental methods from economics and psychology. Central research questions, practical implications as well as possibilities for criticism are discussed.

In ihrer jahrtausendealten Geschichte hat die Philosophie immer wieder vor neuen Herausforderungen gestanden, seien es die Newtonsche Mechanik, Darwins Evolutionstheorie, die Quantenphysik oder die Entwicklungen im Bereich der künstlichen Intelligenz. Eine der jüngsten Herausforderungen ist die experimentelle Forschung in den Sozialwissenschaften, in der Ökonomik und Psychologie. Unter ihrem Einfluss ist in der Philosophie Mitte der 2000er Jahre eine Strömung entstan-

---

1   Prof. Dr. Christoph Lütge ist Inhaber des Peter Löscher-Stiftungslehrstuhls für Wirtschaftsethik an der TU München. Gastprofessuren in Venedig, Taipeh und Kyoto. Arbeitsgebiete: Wirtschafts- und Unternehmensethik, Allgemeine Ethik, CSR, Ethik und Internet. Wichtigste Buchveröffentlichungen: Order Ethics or Moral Surplus: What Holds a Society Together? (Lexington 2015), Ethik des Wettbewerbs (Beck 2014), Experimental Ethics (Mithrsg., Palgrave Macmillan 2014), Handbook of the Philosophical Foundations of Business Ethics (Hrsg., Springer 2013).

den, die sich *Experimentelle Philosophie* nennt und einige Kontroversen ausgelöst hat (vgl. etwa Grundmann et al. 2014). Eine der Intentionen dieser Richtung ist es, Grundfragen der Ethik und Moral einer stärkeren Berechenbarkeit näherzubringen und sich dabei von ‚gefühlten' Intuitionen zu lösen.

Meiner Ansicht nach hat sich diese Strömung jedoch zu sehr auf Probleme der theoretischen Philosophie konzentriert und solche der praktischen Philosophie (die m. E. für die experimentelle Methode sogar eher zugänglich sind) etwas vernachlässigt. Eine „Experimentelle Ethik" (Luetge et al. 2014) erscheint als Möglichkeit. Dieser Artikel kann zwar keinen kompletten Überblick über die Beiträge zu einer Experimentellen Ethik geben, wohl aber eine Einführung in die Möglichkeiten und Grenzen eines solchen Projekts.

Ich werde zunächst kurz die Entwicklung der Experimentellen Philosophie skizzieren, um dann zur Experimentellen Ethik überzugehen. Danach werde ich einige ihrer Vorläufer vorstellen. Die letzten drei Abschnitte beschäftigen sich mit den zukünftigen Chancen und Möglichkeiten, zentralen Forschungsproblemen sowie möglichen praktischen Anwendungen der Experimentellen Ethik.

25.1   Experimentelle Philosophie

Grundmann et al. (2014) haben bereits gute Überblicke über die (noch recht kurze) Geschichte und Entwicklung der Experimentellen Philosophie gegeben. Die Experimentelle Philosophie selbst sieht einen ihrer Ausgangspunkte darin, die Gewohnheit mancher Philosophen zu hinterfragen, sich auf Intuitionen zu berufen: Sind solche Intuitionen nur Intuitionen des jeweiligen Autors oder entsprechen sie tatsächlich Intuitionen, die in der Bevölkerung verbreitet sind? Experimentelle Philosophen (etwa Knobe 2007) karikieren nicht selten eine traditionelle Lehnstuhl-Philosophie, die sich nicht um empirisch vorhandene Meinungen und Intuitionen kümmere, sondern nur um die jeweils eigene Sichtweise und Interpretation. Die Experimentelle Philosophie erhebt dahingehend den Anspruch, die Intuitionen der ‚ordinary people' über hypothetische Fälle von philosophischem Interesse zu erheben und zur Grundlage ihrer Analysen zu machen. Ihre Vertreter haben einige interessante Effekte wie den sogenannten „Knobe-Effekt" (Knobe 2003) beschrieben, und sie haben auch, zumindest in manchen Fällen, originelle Einsichten in die moralischen Urteile der sogenannten gewöhnlichen Menschen gefunden. Inwieweit diese Ergebnisse tatsächlich relevant für die nicht-experimentelle Philosophie in einem weiteren Sinne sind, ist gegenwärtig umstritten und wird es sicherlich auch weiterhin bleiben. Es muss einschränkend hinzugefügt werden, dass nicht alle experimentellen Philosophen auch tatsächlich gut etablierte Methoden der bisherigen sozialwissenschaftlichen Experimentaldisziplinen verwenden, also etwa solche der Experimentellen Ökonomik oder der Experimentellen Psychologie.

Unabhängig von der verwendeten Methode gehören zu den bisher angesprochenen Teilbereichen und Problemen der Philosophie (ohne Anspruch auf Vollständigkeit) vor allem die Philosophie des Geistes (Schultz et al. 2011; Huebner et al. 2010), die Erkenntnistheorie (Buckwalter 2010; May et al. 2010), das Problem der Willensfreiheit (Nichols 2006; Weigel 2012; Feltz und Cokely 2009), die Metaphysik (etwa Probleme der Kausalität, vgl. Knobe 2009; Alicke und Rose 2012) und schließlich auch die Moralphilosophie (Kahane 2013; Knobe 2005; Greene 2012; Inbar et al. 2009; Strohminger et al. 2014), der ich mich im Folgenden zuwenden werde.

25.2 Experimentelle Ethik

Experimentelle Ethik ist als Begriff noch recht neu. Der erste Band mit diesem Titel dürfte Lütge et al. von 2014 sein. Darin findet sich u. a. eine Geschichte von Vorläufern der Experimentellen Ethik (Dworazik und Rusch 2014). Zum Teil verwenden andere Autoren den Begriff „Behavioural Ethics"[2], der allerdings sehr viel weiter gefasst ist als „Experimental Ethics" und insbesondere auch jegliche Art von empirischen Umfragen umfasst. Dagegen scheint es mir sinnvoller, den Begriff einer „Experimentellen" Ethik für Projekte und Ansätze zu reservieren, die tatsächlich – zumindest vorrangig oder in der Regel – *experimentelle* Methoden im engeren Sinne verwenden. Damit sind Methoden gemeint, die auf gezielte, kontrollierte Eingriffe setzen (vgl. Guala 2005), in Abgrenzung zu solchen, die nur in einem allgemeinen Sinn empirische Erhebungen ohne kontrollierte Eingriffe vornehmen. Das ist nicht als Abwertung gemeint, lediglich als (tentative) begriffliche Abgrenzung.

In der Tat scheint grundsätzlich die *Praktische* Philosophie, d.h. vor allem die Ethik, in ihren Fragestellungen empirischen und im Besonderen auch experimentellen Methoden gut zugänglich. Gerade experimentelle Arbeiten in den Sozialwissenschaften und in der Psychologie reichen thematisch oft in das Feld der Ethik hinein, etwa wenn es um Fragen von moralischer Motivation geht. Es erscheint daher natürlich, nach erfolgreichen Anwendungen der experimentellen Methode im Bereich der Ethik zu suchen.

---

2   Vgl. De Cremer und Tenbrunsel 2011, dort auch „behavioral business ethics".

Kwame Appiah, Philosoph an der New York University, hat als einer der ersten in seinem Buch „Ethische Experimente" (Appiah 2009), wenn auch eher metaphorisch als konkret-praktisch, ein entsprechendes Vorgehen in der Ethik gefordert. Er wirft vielen gegenwärtigen Paradigmen der Ethik vor, mögliche Kooperationen mit anderen, vor allem auch experimentellen, Disziplinen zu vernachlässigen. Gleichzeitig erinnert er daran, dass viele große Philosophen der Vergangenheit immer auch in einer anderen Wissenschaft geforscht haben, so etwa Descartes, Leibnitz oder Kant. Das ist natürlich nicht grundsätzlich neu oder unbekannt, dennoch ist es interessant, dass sich Appiah als Gegner vor allem die gegenwärtige analytische Ethik – und gerade nicht in erster Linie traditionellere, nicht-analytische Philosophen – vornimmt. Nach seiner Ansicht konzentrieren sich die Beiträge der analytischen Philosophie zur Ethik zu sehr auf Sprachanalyse und Gedankenexperimente, denen die Perspektive auf soziale und ökonomische Phänomene fehlt. Appiah verlangt, dass die Ethik sich stattdessen stärker den empirischen Einzelwissenschaften zuwenden und sich stärker mit Experimenten befassen sollte. Solche Experimente führt Appiah zwar nicht selbst durch, aber er bereitet mit seinen Arbeiten zumindest den Boden innerhalb der Ethik dafür.

25.3   Philosophische Vorläufer

Ich habe bereits betont, dass die Experimentelle Ethik zwar neu ist, dennoch aber Vorläufer hat. Auch andere Ansätze in der Ethik haben schon früher auf Methoden und Ergebnisse empirischer (allerdings nicht unbedingt experimenteller) Wissenschaften massiv zurückgegriffen.[3] Ich möchte hier auf zwei philosophische Ansätze hinweisen, die in diese Kategorie fallen: zum einen die Ethik auf naturalistischer Basis, zum anderen die Wirtschaftsethik mit ökonomischer Methode.

---

3   Vgl. Dworazik und Rusch 2014, die sich allerdings vor allem auf Arbeiten innerhalb der Psychologie und Sozialwissenschaften selbst konzentrieren.

25.3.1 Ethik auf naturalistischer Basis

Der Begriff Naturalismus wird in der Ethik oft mit großer Skepsis betrachtet, da G.E. Moore als ,ethischen Naturalismus' den Versuch bezeichnete, mit ungültigen Schlüssen Normen aus Fakten abzuleiten. Die Gefahr eines naturalistischen Fehlschlusses droht jedoch hier nicht, aus drei Gründen:

Zum einen wird nicht vom ethischen Naturalismus gesprochen, sondern, in schwächerer Form, nur von einer ,Ethik auf naturalistischer Basis'.

Zum zweiten wird nicht der Versuch unternommen, Normen direkt aus Fakten abzuleiten, sondern es wird akzeptiert, dass man immer auch mit einigen Basisnormen beginnen muss. Russell bemerkte einmal in anderem Zusammenhang: „auf dem kahlen Zweifel wachsen keine Gründe" (Russell 1912/1967, S. 133). Nur mit dem kahlen Zweifel kann eine Ethik auf naturalistischer Basis nicht beginnen. Sie behauptet jedoch, dass empirische Daten wie auch experimentelle Ergebnisse helfen können, den Möglichkeitsspielraum für Normen, die auch tatsächlich implementierbar sein sollen, einzuschränken.

Zum dritten verwende ich einen Begriff von Naturalismus, der sich nicht nur auf die Ergebnisse der Naturwissenschaften bezieht, sondern einen weiteren Begriff, der auch die Sozialwissenschaften und die Ökonomik mit einschließt (vgl. dazu Kitcher 1993 sowie auch Lütge 2001, Kap. 1 und Lütge 2004). Der Kern eines solchen Verständnisses von Naturalismus ist die Ablehnung einer ,prima philosophia', einer (idealtypischen) Form von traditioneller Philosophie, die die Einzelwissenschaften nicht ernst nimmt und sich ausschließlich auf vermeintliche Vernunftargumente, Diskurse und nicht-empirische Überlegungen stützt. Alle diese Elemente sind zweifellos wichtig für jede Art von Philosophie, dennoch kann und sollte auch die Philosophie Fortschritte in den Einzelwissenschaften – nicht nur nicht ignorieren, sondern – systematisch verarbeiten. Das gilt etwa für philosophische Ansätze wie die Neurophilosophie (Churchland 1989), die Evolutionäre Erkenntnistheorie (Campbell 1974, Vollmer 1975/1998) oder auch die Philosophie der Komplexität (Mainzer 2008). Die verwendeten Einzelwissenschaften dabei sind u.a. die Evolutionsbiologie, die Spieltheorie, die Institutionenökonomik, die Sozialpsychologie und andere. Es erscheint daher logisch und konsequent, jetzt auch den nächsten Schritt in die experimentelle Dimension zu gehen.

25.3.2 Wirtschaftsethik mit ökonomischer Methode

Die Wirtschaftsethik hat sich international etwa seit den späten siebziger Jahren, in Deutschland seit Mitte der achtziger Jahre als Disziplin etabliert. Sie ist nicht durchweg an philosophischen Fakultäten angesiedelt, gerade im angelsächsischen Raum sogar häufiger an den Business Schools. In Deutschland halten sich philosophische und ökonomische Institutionalisierung bei den gegenwärtig ca. 10-12 Lehrstühlen für Wirtschaftsethik in etwa die Waage (es ist dabei interessant festzustellen, dass die TU München unter den Exzellenz-Universitäten die einzige ist, die über einen Lehrstuhl für Wirtschaftsethik verfügt).

Sowohl im deutschen als auch im internationalen Raum lässt sich beobachten, dass eine rein philosophische Ausbildung für eine solche Position in der Regel nicht mehr ausreicht. Verglichen mit anderen Teilen der angewandten Ethik (etwa der Bio- und Medizinethik) hat sich die Einbeziehung einzelwissenschaftlicher Methoden und Ergebnisse in der Theoriediskussion der Wirtschaftsethik viel stärker durchgesetzt. Insbesondere die Vertreter der Ordnungsethik (etwa Karl Homann, Ingo Pies, Andreas Suchanek, Christoph Lütge, Markus Beckmann und andere) beziehen in ihren Arbeiten Methoden und Ergebnisse der Ökonomik, insbesondere der Institutionen- und Konstitutionenökonomik, seit vielen Jahren systematisch ein (vgl. Homann und Lütge 2013; Pies 2009a, 2009b; Suchanek 2001; Lütge 2014, 2015), sodass man geradezu von einer „Wirtschaftsethik auf ökonomischer Grundlage" sprechen kann.[4]

Die Ordnungsethik vertritt die Thesen, dass einerseits – auf theoretischer Ebene – in der Ökonomik als Disziplin selbst ethische Elemente und Prinzipien angelegt sind und dass andererseits – auf praktischer Ebene – das System der Ökonomie für ethische Zwecke einiges leistet, und zwar weit mehr, als viele Kritiker dieses Systems (etwa Chomsky 2000; Comte-Sponville 2009; Precht 2010) zur Kenntnis nehmen. So ist beispielsweise der Wettbewerb oft für alle Beteiligten von ethischem Wert, auch wenn die einzelnen Akteure nicht unbedingt ethisch oder gar altruistisch motiviert sein mögen (vgl. Lütge 2014). Darüber hinaus ist es das Projekt der Ordnungsethik, ethische Kategorien in ökonomischen Kategorien zu rekonstruieren und zu re-interpretieren (etwa Homann 2002; Lütge und Mukerji 2016): Pflicht beispielsweise wird als Ermutigung zum langfristigen (und nicht nur finanziellen) Investieren interpretiert, *phronesis* als ,weise' ökonomische und ethische Abwägung. Dieses und viele andere Beispiele zeigen, dass Ethik und Ökonomik

---

4   Im Unterschied zu einigen anderen Vertretern (etwa Suchanek 2001) vermeide ich es allerdings, explizit von einer „ökonomischen Ethik" zu sprechen, da ich der Ansicht bin, dass auch für die Philosophie hier noch eine wichtige Rolle zu spielen bleibt.

zwei Seiten *derselben* Medaille sind – und nicht zwei völlig getrennte Handlungs-
anforderungen.

Etwa seit den späten neunziger Jahren hat sich aber auch die Diskussionslage in
der Ökonomik selbst verändert. Mit den Arbeiten von Vernon Smith (2016 Gast
der TUM), Daniel Kahneman (die hierfür 2002 auch den Nobelpreis erhielten) und
Amos Tversky etablierte sich die Experimentelle Ökonomik als wesentliche neue
Strömung, die mittlerweile stark an Fahrt gewonnen hat. Es erscheint daher konsequent, dass die Wirtschaftsethik, soweit sie sich auf die Ökonomik stützt, auch
deren (wesentliche) Theorieentwicklungen nach- und mitvollzieht.

25.4   Chancen einer Experimentellen Ethik

Bei den Vorüberlegungen zu einer Experimentellen Ethik stellt sich unweigerlich
die Frage: Wozu kann eine Experimentelle Ethik dienen? Anders gefragt: Was kann
eine Experimentelle Ethik besser oder mehr leisten als eine traditionelle?

Diese Frage lässt sich hier nicht abschließend beantworten, aber es können
Richtungen aufgezeigt werden, in denen eine Antwort liegen könnte: Zweifellos
gibt es systematische Defizite in vielen traditionellen ethischen Argumentationen.
Die Ordnungsethik geht innerhalb der Wirtschaftsethik ähnlich vor, indem sie Un-
zulänglichkeiten in den Argumenten jener Wirtschaftsethiker aufzeigt, die schlicht
nach mehr moralischer Motivation und weniger Gier rufen, und empfiehlt stattdessen eine Untersuchung der Anreize und Dilemmasituationen, um nach institutionellen Verbesserungen zu suchen (vgl. Homann und Lütge 2013).

Ebenso sollte es möglich sein, systematische Defizite in traditionellen ethischen
Argumentationen herauszuarbeiten. Solche Defizite können etwa – zunächst in einem sehr allgemeinen Sinn – darin bestehen, dass systematisch die Rolle von em-
pirischem (und insbesondere von experimentellem) Wissen unterschätzt wird. Ein
anderes Defizit könnte darin bestehen, dass man vernachlässigt, wie schnell Moral
unter dem Druck der Anreize und Sachzwänge erodieren kann – und dies lässt sich
experimentell untersuchen.

Ich habe bereits erwähnt, dass hier noch kein vollständiges Programm einer
Experimentellen Ethik ausgebreitet werden kann. Folgende Punkte sollten jedoch
in einem solchen Programm bedacht werden:

1. Die Experimentelle Ethik sollte sich nicht nur als Kritik (an traditionellen Ansätzen) verstehen. Zum Teil werden hier im Zuge einer Überspitzung Zerrbilder
   von Philosophie gezeichnet, die als heuristische Ausgangspunkte ihre Berech-

tigung haben mögen, irgendwann aber auch wieder der Korrektur bedürfen. Schließlich sollte die Experimentelle Ethik selbst auch eine konstruktive Seite haben, in der sie ihre empirische und experimentelle Detailkenntnis zur Geltung bringen kann.

2. Ich schlage vor, die Experimentelle Ethik nicht mit einer ausgedehnten theoretischen Debatte beginnen zu lassen, sondern zunächst die Fruchtbarkeit dieses Ansatzes in seinen Anwendungen zu zeigen. Die theoretische Debatte sollte sich allmählich und begleitend entwickeln.

3. Systematische Defizite in ethischen Argumentationen sollten herausgearbeitet werden, insbesondere solche, die durch experimentelle Vorgehensweise angegangen und verbessert werden können.

4. Es muss auch das Verhältnis der Experimentellen Ethik zu anderen experimentellen Disziplin wie Experimenteller Ökonomik, Psychologie und auch zur Evolutionsbiologie geklärt werden. In welcher Hinsicht unterscheiden sich ihre experimentellen Methoden und Vorgehensweisen? Was macht die Besonderheit der Experimentellen Ethik aus?

5. Eine wichtige Aufgabe für die Experimentelle Ethik könnte bereits das Zusammentragen ethikrelevanter Ergebnisse aus verschiedenen Disziplinen sein – was in der Ethik noch ausbaufähig ist.

6. Schließlich könnte besonders die Angewandte Ethik von der Experimentellen Ethik profitieren: Gerade in der Medizinischen Ethik, Bioethik und auch in der Wirtschaftsethik (wenn wir sie eine Angewandte Ethik nennen wollen, vgl. Lütge 2014) dürften experimentelle Methoden und Ergebnisse Verbesserungen und Fortschritte bringen.

25.5   Mögliche Forschungsfragen einer Experimentellen Ethik

Für jedes neue Forschungsfeld ist es von zentraler Bedeutung, hinreichend originelle, substanzielle und nachhaltig interessante Forschungsfragen zu formulieren. Die folgende Liste enthält einige Vorschläge, welche ethischen Probleme für die experimentelle Ethik relevant sein könnten:

1. Ethische Bewertung und moralische Intuitionen:
   Wie bewerten Menschen tatsächlich unterschiedliche Handlungen, Folgen, Ergebnisse usw.? Welche gemeinsamen oder geteilten moralischen Intuitionen

lassen sich tatsächlich finden (und nicht nur theoretisch postulieren)? In welcher Weise verändern sich diese Bewertungen – und die damit verbundenen Handlungen – , wenn Menschen mit veränderten Anreizen, mit veränderten Spielregeln oder mit veränderten Strategien ihrer Interaktionspartner konfrontiert werden? Sind moralische Intuitionen in einer Gesellschaft ungleich verteilt, bleiben Sie unter allen Umständen stabil, können sie sich grundsätzlich verändern – was sich in Experimenten als Reaktion auf gewisse Erfahrungen modellieren lässt? Kann vielleicht das Konzept des Überlegungsgleichgewichts (John Rawls) in Experimenten nachgewiesen werden?

2. Moralische Motivation: Wie werden Menschen zu ethischen oder unethischen Handlungen motiviert?
Diese klassische Frage der Ethik gehen Vertreter der Experimentellen Ethik bereits neu an, etwa Schwitzgebel (2009, 2014), der u.a. auf der Basis von Daten über verschwundene Bibliotheksbestände argumentiert (Ethiker stehlen demnach genauso viele Bücher wie andere). Das sind nur erste Ansätze, die aber mit der Vorgehensweise der Ordnungsethik verwandt sind, welche seit langem argumentiert, dass mehr Wissen (auch über moralische Zusammenhänge) nicht ausreicht, um die Handlungen von Akteuren etwa im Gefangenendilemma systematisch und dauerhaft zu verändern. Moralische Motivation kann dort – wie, so die These, auch in vielen experimentellen Settings – nur dann aufrechterhalten werden, wenn Sanktionen zur Bestrafung der Trittbrettfahrer gegeben sind.

3. Sollen impliziert Können:
Das Prinzip „Sollen impliziert Können" spielt für jede Ethik auf naturalistischer Basis eine wesentliche, wenn nicht sogar die zentrale Rolle (vgl. Lütge und Vollmer 2004). Wenn wir von niemandem verlangen können (aber auch nicht verbieten müssen), dass sie 20 Meter hoch springt, dann muss im Grundsatz das Gleiche auch in sozialen und ökonomischen Zusammenhängen gelten: Wir können von Menschen nicht erwarten, dass sie systematisch gegen ihre eigenen Interessen verstoßen und sich dauerhaft selbst schaden.
Die Experimentelle Ethik kann diese Perspektive weiter verfeinern. Sie stellt Situationen und Situationstypen heraus, die über die klassische (institutionen-) ökonomische Perspektive hinausgehen, ohne deren grundsätzlichen Aussagen zu widersprechen. Aber die von Kahneman (2011) und anderen konstatierten „biases" und „mental traps" zeichnen nicht nur ein differenzierteres Bild menschlichen Handelns, sondern sind auch für die Gestaltung von Institutionen und den von ihnen ausgehenden Anreizen von Relevanz (vgl. etwa einige der Beiträge in Luetge et al. 2014). Sie erweitern somit die „Sollen impliziert Können"-Perspektive.

4. Welche Abwägungen und Kompromisse (‚trade-offs') nehmen Menschen bei ethischen Problemen unter welchen Bedingungen vor?

Auch in vielen traditionellen Ansätzen zur Ethik wird grundsätzlich in Rechnung gestellt, dass ethische trade-offs vorgenommen werden können. Die Experimentelle Ethik kann über solche trade-offs neue Erkenntnisse zusammentragen, einerseits über die Art und Weise ihres Zustandekommens, andererseits auch über ihren Inhalt hinsichtlich bestimmter Fragestellungen und Szenarien.

Die genannten möglichen Forschungsfragen stellen natürlich alles andere als eine vollständige oder abgeschlossene Liste dar. Aus meiner Sicht sind sie jedoch alle wert, weiterverfolgt zu werden. Dazu ist die Hilfe anderer experimenteller Disziplinen, etwa der experimentellen oder Verhaltensökonomik wie auch der entsprechenden psychologischen Ansätze, notwendig. Allerdings enthalten die Fragestellungen einer Experimentellen Ethik immer auch eine moralische bzw. ethische Dimension. Zwar grenzen Fragen und Untersuchungen der Experimentellen Ökonomik und Psychologie immer wieder auch an ethische Fragen an, aber nur die Experimentelle Ethik sieht diese moralische bzw. ethische Dimension als das zentrale Moment an. Eine genauere Abgrenzung wird erst im Verlauf der weiteren Entwicklung dieses Forschungsgebiets möglich sein.

25.6   Praktische Implikationen der Experimentellen Ethik

Da die Experimentelle Ethik noch am Beginn steht, mag es zwar – gerade da es sich um eine philosophische Disziplin handelt – noch etwas früh sein, bereits jetzt über Implikationen ihrer Ergebnisse nachzudenken. Man kann jedoch einige hypothetischer Konsequenzen aus Analogien zu anderen experimentellen Disziplinen skizzieren:

Bereits erwähnt wurde die Analogie zur Wirtschaftsethik: Gemäß der Experimentellen Ökonomik sollten wir bei Vorschlägen zu institutionellen Reformen einige systematische Vorurteile und Verzerrungen unserer Wahrnehmung sowie auch mentale ‚Fallen' berücksichtigen. Um nur drei zu nennen (vgl. Kahneman 2011): Es gibt systematische Verzerrungen

- zugunsten der Gegenwart und gegen die Zukunft,
- zugunsten von Sicherheit (Vermeidung von Verlusten) und gegen zusätzliche Gewinne sowie
- bei der Abschätzung von Zeithorizonten, bei Framing-Effekten u.a.

Auch bei der Sanktionierung von Verstößen gibt es entsprechende Effekte: Menschen sind bereit, Abweichler und Trittbrettfahrer zu bestrafen – vorausgesetzt, die

Kosten der Bestrafung sind nicht zu hoch. Menschen neigen dazu, die Rolle von Glück und Zufall in vielen Situationen zu unterschätzen, sowie dazu, aus einzelnen Fällen zu übergeneralisieren. Effekte dieser Art sollten auch beim Design von Institutionen berücksichtigt werden. Kahneman (2011) schlägt auch eine Reihe von Strategien für den Alltag vor, beispielsweise

• Optionen nicht zu früh auszuschließen,
• vor Entscheidungen systematisch zu versuchen, den Möglichkeitsspielraum auszuweiten.

In Bezug auf die Experimentelle Ökonomik schien es lange Zeit so, dass die praktischen Implikationen vor allem darin bestünden zu zeigen, der Mensch agiere und reagiere nicht nur wie ein homo oeconomicus. In dieser Weise wurden die Arbeiten von Kahneman und anderen lange Zeit rezipiert, in dieser Weise interpretieren auch nach wie vor viele Autoren selbst ihre Ergebnisse und gehen z.T. immer noch so weit zu behaupten, der ‚Markt' verderbe die ‚Moral'.[5]

Tatsächlich aber ist die Forschung in diesem Bereich seit Jahren deutlich differenzierter. Schon in früheren Jahren (Andreoni 1988; Yamagishi 1986, 1992; Yamagishi und Sato 1986), aber vor allem in letzter Zeit haben eine ganze Reihe von Studien gezeigt, dass moralische Motivation nicht einfach vorhanden ist, sondern – unter dem Druck von Anreizen – massiv erodieren, ja verschwinden kann (etwa Gürerk et al. 2006; Binmore 2010). Moralische Motivation ist danach nur ein Bestandteil eines größeren Bildes menschlicher Motivationen, Präferenzen und Gefühle. Auch die Experimentelle Ethik sollte daher die Bedingungen möglicher Erosion von Moral im Blick haben – und Implikationen nicht zu früh ziehen. Am Experimental Ethics Lab (EEL) der TU München, das ich 2011 eingerichtet habe, ist dies eine unserer wesentlichen Forschungsfragen; erste Ergebnisse finden sich unter anderem in Lütge et al. (2014), Levati et al. (2014), Rusch und Lütge (im Review) und Lütge und Rusch (2013). Beispielsweise lässt sich zeigen, dass nicht nur Verlierer bestrafen, sondern auch Gewinner (Jauernig et al. 2016). Ebenfalls sind experimentelle Ergebnisse und evolutionäre Befunde in hohem Maße kompatibel (etwa Rusch et al. 2014 sowie Dworazik/Rusch 2014): Unsere moralischen Begriffe und Kategorien sind demnach an die Situation der Vormoderne angepasst, sie hinken der sozialen und ökonomischen Entwicklung hinterher. Unsere moralische Erwartungshaltung lässt sich mit dem Begriff des „moralischen Mesokosmos" umschreiben: wie im naturwissenschaftlichen Mesokosmos versagen unsere Alltagskategorien, wenn wir sie über unsere Alltagswelt der mittleren Dimension hinaus anwenden. Wir können ganz gut mit Millimetern bis zu Kilometern rechnen, aber

---

5   Etwa Falk und Szech (2013), vgl. dazu als Kritik Lütge und Rusch (2013).

im ganz Großen und im ganz Kleinen versagt unsere Alltagsphysik: um zum Mond
zu fliegen, benötigt man schon die Relativitätstheorie, und in der Quantenphysik
gibt es Phänomene wie absoluten Zufall, die sich mit der Alltagsphysik nicht mehr
fassen lassen. Im Bereich der Moral, darauf deuten die Ergebnisse hin, ist es nicht
anders.[6] Experimentelle Studien von Joshua Greene und anderen bestätigen dies.[7]

25.7   Zu erwartende Kritikpunkte

Gerade neue Forschungsgebiete müssen sich auf Kritik gefasst machen, um Gegen-
argumente vorzubereiten. Die folgende Liste einiger aus meiner Sicht zu erwarten-
der Kritikpunkte ist dabei natürlich nur ein Anfang. Kritikpunkte werden sich vermutlich entweder gegen den experimentellen Ansatz allgemein richten oder gegen
den experimentellen Ansatz der Ethik im Besonderen.

Unter den allgemeinen Kritikern muss man sich auf jene gefasst machen, die
implizit davon ausgehen, dass Experimente in den Sozialwissenschaften (teilweise
auch in der Ökonomik) grundsätzlich fehl am Platze seien, da sie sich nicht mit der
vermeintlichen ‚realen Welt' beschäftigten. Soziologen aus unterschiedlichen La-
gern haben immer wieder experimentelle Methoden in dieser Weise kritisiert. Da
die experimentelle Methode jedoch in den letzten 15–20 Jahren massiv an Reputation gewonnen hat, sind diese Kritiker tendenziell auf dem Rückzug. Um ihnen zu
begegnen, ist es nach wie vor am sinnvollsten, 1) auf erfolgreiche experimentelle
Studien und Ansätze in anderen Disziplinen zu verweisen, und 2) darauf zu verweisen, dass der Vorwurf eines zu großen Idealismus gegenüber den experimentellen
Ansätzen nicht haltbar ist: Letztlich müssen alle Wissenschaften notwendig von
der Realität abstrahieren, wenn sie überhaupt Theorien und Modelle zur *Erklärung*
benutzen wollen – und nicht nur beschreiben.

Einige der Standard-Kritikpunkte gegen den experimentellen Ansatz in der
Ethik im Speziellen könnten sein:

1. Die Experimentelle Ethik begehe den naturalistischen Fehlschluss.
    Die kurze Antwort – ohne auf die Bibliotheken füllenden Arbeiten zu diesem
    Thema zu verweisen – ist, dass die Experimentelle Ethik keineswegs versucht,
    Normen aus Fakten abzuleiten. Stattdessen lässt sich ihre Arbeit im Rahmen

---

6   Zum Begriff des Mesokosmos vgl. Vollmer 1975/1998, 161ff., zu dem des „moralischen
     Mesokosmos" Lütge 2007, 121f.
7   Vgl. insbesondere Greene et al. 2001, Greene/Haidt 2002 sowie Greene 2012.

des Prinzips ‚Sollen impliziert Können‘ verstehen (Lütge und Vollmer 2004): Es geht um empirische Bedingungen implementierbarer ethischer Normen und Werturteile.

2. Die Begriffe „Experiment" und „Ethik" gemeinsam zu verwenden, klinge nach fragwürdigen Unternehmungen (wie dem – von manchen als fragwürdig angesehenen – Milgram-Experiment) und suggeriere, dass Menschen als Versuchskaninchen benutzt würden.

Das wäre ein Missverständnis. Natürlich sind in Experimenten Menschen als Subjekte involviert – aber das ist seit Jahrzehnten Standard in den experimentellen Sozialwissenschaften und wird von Ethikkomitees und anderen Gremien kontinuierlich begleitet und überprüft. Darüber hinaus beschäftigt sich aber die Experimentelle Ethik derzeit nicht mit ähnlich ‚invasiven‘ Versuchen wie dem Milgram-Experiment.

3. Was hat das mit Ethik zu tun?

Diese Kritik ist gelegentlich von traditioneller philosophischer Seite zu hören, so auch gegen einige Vorläufer der Experimentellen Ethik. Als Antwort kann man auf die offensichtlich ethischen Fragestellungen des Ansatzes verweisen, auf die untersuchten Forschungsprobleme und die entsprechenden philosophischen Begriffe (vgl. oben). Zu bestreiten, dass all dies für die Ethik relevant ist, hieße aber in der Tat, sich auf die Karikatur einer Lehnstuhlwissenschaft Philosophie zurückzuziehen, die wohl selbst hartnäckige Traditionalisten kaum vertreten wollten.

4. Experimentelle Ethik sei nichts anderes als (philosophische) Anthropologie.

Auch wenn sicherlich unterschiedliche Auffassungen darüber bestehen, was philosophische Anthropologie sein kann, sollte diese Kritik doch ernst genommen werden: Nein, experimentelle Ethik sollte sich nicht ausschließlich – nicht einmal in erster Linie und vorrangig – mit *individuellen* Akteuren beschäftigen. Stattdessen sollten die Bedingungen der sozialen Situation, die Randbedingungen, der Ordnungsrahmen und die davon ausgehenden Anreize eine wesentliche Rolle in den Experimenten spielen.

25.8 Konklusion

Abschließen möchte ich mit einer Bemerkung zum philosophischen Verständnis der Experimentellen Ethik: Gerade die Philosophie lässt Raum für unterschiedliche Methoden, darin liegt einer ihrer Vorteile gegenüber vielen Einzelwissenschaften. Diesen Vorteil sollte die Experimentelle Ethik auch nutzen – und dann ihre Fruchtbarkeit erweisen. Das Verständnis von Philosophie, das für mich dahinter steht, ist nicht mehr in erster Linie inhaltlich, etwa durch Begründungen, Vernunftargumente oder Sprachanalyse gekennzeichnet. Es ist eher ein organisatorisches, interdisziplinäres Verständnis, das eines ‚Interfaces der Wissenschaften' (Lütge 2003), mit dessen Hilfe unterschiedliche Disziplinen und unterschiedliche Methodologien zusammengebracht werden. Und auch dieses interdisziplinäre Verständnis steht nicht unverbunden neben der philosophischen Tradition, sondern hat seine Verankerung in ihr (vgl. Appiah 2009).

Literatur

Alicke, M. & D. Rose. 2012. Culpable Control and Deviant Causal Chains. *Personality and Social Psychology Compass* 6 (10): 723–735.

Andreoni, J. 1988. Why Free Ride? Strategies and Learning in Public Goods Experiments. *Journal of Public Economics* 37: 291–304.

Appiah, K. A. 2009. Ethische Experimente: Übungen zum guten Leben. München: Beck.

Binmore, K. 2010. Social Norms or Social Preferences?. *Mind and Society* 9: 139–158.

Buckwalter, W. 2010. Knowledge Isn't Closed on Saturday: A Study in Ordinary Language. *Review of Philosophy and Psychology* 1 (3): 395–406.

Campbell, D. T. 1974. Evolutionary Epistemology. In *The Philosophy of Karl Popper*, Vol. I, ed. Schlipp, Paul A., 413–459. Illinois: La Salle.

Chomsky, N. 2000. *Profit over People: Neoliberalismus und globale Weltordnung*. Hamburg: Europa Verlag.

Churchland, P. 1989. *Neurophilosophy – Toward a Unified Science of the Mind/Brain*. Bradford Book.

Comte-Sponville, A. 2009. *Kann Kapitalismus moralisch sein?* Zürich: Diogenes.

De Cremer, D. & A. E. Tenbrunsel. 2011. *Behavioral Business Ethics: Shaping an Emerging Field*. London: Routledge.

Dworazik, N. & H. Rusch. 2014. A brief history of Experimental Ethics. In *Experimental Ethics: Towards an Empirical Moral Philosophy*, Luetge, Rusch und Uhl, 38–56. Basingstoke: Palgrave Macmillan.

Falk, A. & N. Szech. 2013. Morals and Markets. *Science* 340: 707–711.

Feltz, A. & E. T. Cokely. 2009. Do Judgments About Freedom and Responsibility Depend on Who You Are? Personality Differences in Intuitions About Compatibilism and Incompatibilism. *Consciousness and Cognition* 18 (1): 342–350.

Greene, J. D., R. B. Sommerville, L. E. Nystrom, J. M. Darley, and J. D. Cohen (2001) 'An fMRI Investigation of Emotional Engagement in Moral Judgment', *Science*, 293, pp. 2105–2108.

Greene, J. D. and J. Haidt (2002) 'How (and Where) Does Moral Judgment Work?', *Trends in Cognitive Sciences*, 6, 12, pp. 517–523.

Greene, J. D. 2012. Reflection and Reasoning in Moral Judgment. *Cognitive Science* 36 (1): 163–177.

Grundmann, T., J. Horvath & J. Kipper. (Hrsg.). 2014. Die Experimentelle Philosophie in der Diskussion. Frankfurt: Suhrkamp.

Guala, F. 2005. *The Methodology of Experimental Economics*. Cambridge: Cambridge University Press.

Gürerk, Ö., B. Irlenbusch & B. Rockenbach. 2006. The Competitive Advantage of Sanctioning Institutions. *Science* 312: 108–111.

Homann, K. & C. Lütge. 2013. *Einführung in die Wirtschaftsethik* (3. Aufl.). Münster: LIT.

Huebner, B., M. Bruno & H. Sarkissian. 2010. What Does the Nation of China Think About Phenomenal States? *Review of Philosophy and Psychology* 1 (2): 225–243.

Inbar, Y., D. A. Pizarro, J. Knobe & Bloom, P. 2009. Disgust Sensitivity Predicts Intuitive Disapproval of Gays. *Emotion* 9 (3): 435–443.

Jauernig, J., M. Uhl & C. Luetge. 2016. Losers under suppression, winners under crossfire. An experiment on punishment after competition. Erscheint in: *Journal of Economic Psychology*.

Kahane, G. 2013. The Armchair and the Trolley: An Argument for Experimental Ethics. *Philosophical Studies* 162 (2): 421–445.

Kahneman, D. 2011. *Thinking, Fast and Slow*. New York: Farrar, Straus and Giroux.

Kitcher, P. 1993. *The Advancement of Science: Science without Legend, Objectivity without Illusions*. New York: Oxford University Press.

Knobe, J. 2003. Intentional Action and Side Effects in Ordinary Language. *Analysis* 63: 190–193.

Knobe, J. 2005. Ordinary Ethical Reasoning and the Ideal of 'Being Yourself'. *Philosophical Psychology* 18 (3): 327–340.

Knobe, J. 2007. Experimental Philosophy and Philosophical Significance. *Philosophical Explorations* 10 (2): 119–121.

Knobe, J. 2009. Folk Judgments of Causation. *Studies in History and Philosophy of Science Part A* 40 (2): 238–242.

Knobe, J. & S. Nichols, ed. 2008. *Experimental Philosophy*. New York: OUP.

Levati, M. V., M. Uhl & R. Zultan. 2014. Imperfect Recall and Time Inconsistencies: An Experimental Test of the Absentminded Driver 'Paradox'". *International Journal of Game Theory* 43: 65–88.

Lütge, C. 2001. Ökonomische Wissenschaftstheorie. Würzburg: Königshausen & Neumann.

Lütge, C. 2003. Philosophie als Interface der Wissenschaften. In *Kaltblütig: Philosophie von einem rationalen Standpunkt*, hrsg. W. Buschlinger und C. Lütge, 69–79. Stuttgart: Hirzel.

Lütge, C. 2004. Economics in Philosophy of Science: Can the Dismal Science Contribute Anything Interesting? *Synthese* 140 (3): 279–305.

Lütge, C. 2007. *Was hält eine Gesellschaft zusammen? Ethik im Zeitalter der Globalisierung*, Tübingen: Mohr Siebeck.

Lütge, C. 2014. *Ethik des Wettbewerbs: Über Konkurrenz und Moral*. München: Beck.

Luetge, C. 2015. *What Holds a Society Together? Order Ethics vs. Moral Surplus*. Lanham, Md.: Lexington.

Luetge, C. & N. Mukerji, ed. 2016. *Order Ethics: An Ethical Framework for the Social Market Economy*. Heidelberg/New York: Springer.

Luetge, C. & H. Rusch. 2013. The Systematic Place of Morals in Markets: Comment on Armin Falk & Nora Szech "Morals and Markets", *Science* 341 (6147): 714.

Luetge, C., H. Rusch & M. Uhl, ed. 2014. *Experimental Ethics: Towards an Empirical Moral Philosophy*. Basingstoke: Palgrave Macmillan.

Lütge, C. & G. Vollmer, Hrsg. 2004. *Fakten statt Normen? Zur Rolle einzelwissenschaftlicher Argumente in einer naturalistischen Ethik*. Baden-Baden: Nomos.

Mainzer, K. 2008. *Komplexität*. Paderborn: UTB.

May, J., W. Sinnott-Armstrong, J. G. Hull & A. Zimmerman. 2010. Practical Interests, Relevant Alternatives, and Knowledge Attributions: An Empirical Study. *Review of Philosophy and Psychology* 1 (2): 265–273.

Nichols, S. 2006. Folk Intuitions on Free Will. *Journal of Cognition and Culture* 6: 57–86.

Pies, I. 2009a. *Moral als Produktionsfaktor. Ordonomische Schriften zur Unternehmensethik*. Berlin: wvb.

Pies, I. 2009b. *Moral als Heuristik. Ordonomische Schriften zur Wirtschaftsethik*. Berlin: wvb.

Precht, R. 2010. *Die Kunst, kein Egoist zu sein. Warum wir gerne gut sein wollen und was uns davon abhält*. München: Goldmann.

Rusch, H. & C. Lütge. 2016. Spillovers From Coordination to Cooperation: Evidence for the Interdependence Hypothesis from the Lab. *Evolutionary Behavioral Sciences 10(1)*.

Rusch, H., C. Luetge & E. Voland. 2014. Evolutionäre und Experimentelle Ethik: eine neue Synthese in der Moralphilosophie? In *Bereichsethiken im interdisziplinären Dialog*, hrsg. M. Maring, 163–179. Karlsruhe: KIT Scientific Publishing.

Russell, B. 1967 (Orig. 1912). *Probleme der Philosophie*. Frankfurt: Suhrkamp.

Schultz, E., E. T. Cokely & A. Feltz. 2011. Persistent Bias in Expert Judgments About Free Will and Moral Responsibility: A Test of the Expertise Defense. *Consciousness and Cognition* 20 (4): 1722–1731.

Schwitzgebel, E. 2009. Do Ethicists Steal more Books? *Philosophical Psychology* 22: 711–725.

Schwitzgebel, E. 2014. The moral behavior of ethicists and the role of the philosopher. In *Experimental Ethics: Towards an Empirical Moral Philosophy*, Luetge, Rusch und Uhl, 59–64. Basingstoke: Palgrave Macmillan.

Strohminger, N., B. Caldwell, D. Cameron, J. Schaich Borg & W. Sinnott-Armstrong. 2014. Implicit Morality: A Methodological Survey. In *Experimental Ethics: Towards an Empirical Moral Philosophy*, Luetge, Rusch und Uhl, 133–156. Basingstoke: Palgrave Macmillan.

Smith, V. L. 2008. *Rationality in Economics: Constructivist and Ecological Forms*. Cambridge: CUP.

Suchanek, A. 2001. *Ökonomische Ethik*. Tübingen: Mohr Siebeck (UTB).

Vollmer, G. 1975, 1998. *Evolutionäre Erkenntnistheorie*. 7. Aufl. Stuttgart: Hirzel.

Weigel, C. 2012. Experimental Evidence for Free Will Revisionism. *Philosophical Explorations* 16 (1): 31–43.

Yamagishi, T. 1986. The provision of a sanctioning system as a public good. *Journal of Personality and Social Psychology* 51 (1): 110–116.

Yamagishi, T. 1992. Group Size and Provision of a Sanctioning System in a Social Dilemma. In *Social dilemmas. Theoretical issues and research findings*, hrsg. Liebrand, W. B. G., Messick, D. M. und Wilke, H. A. M, 267–287. Oxford: Pergamon Press (International series in experimental social psychology, 25).

Yamagishi, T. & K. Sato. 1986. Motivational bases of the public goods problem. *Journal of Personality and Social Psychology* 50 (1): 67–73.

26. Autonomie und Kontrolle in Big Data basierten Systemen

Sabine Thürmel[1]

Abstract

Autonomie und Kontrolle während der Analyse großer Datenmengen und ihrer Nutzung in Big Data basierten Systemen liegen im Fokus des Aufsatzes. Im Sinne von Klaus Mainzer, der sich immer wieder sehr engagiert für den verantwortlichen Umgang mit Daten und den klugen Einsatz von Big Data Ansätzen ausspricht, wird der Blick auf komplexe, adaptive Systeme gelenkt. Diese haben die Optimierung des persönlichen und sozialen Verhaltens zum Ziel. In zukünftigen proaktiven Systemen zur Stärkung der Gesundheit und des Wohlbefindens wird die Verhaltenssteuerung auf Basis von Big Data integriert sein. Einerseits bieten diese Systeme den Teilnehmern neue Formen des Wissens über sich selbst und der Selbstoptimierung. Andererseits beschränken die Steuerungsmechanismen, die in diese Systeme eingebaut sind, die Autonomie der Teilnehmer und unterwerfen sie der Systemsteuerung und -kontrolle. Daher ist es unabdingbar, dass ein seiner Verantwortung bewusster Innovationsprozess die Modellierung, die Realisierung und den Einsatz von Big Data Systemen begleitet.
The deployment of Big Data technologies forms an integral part of the latest generation in complex adaptive systems. In such systems, e.g. future proactive health and wellbeing systems, governance will already be embedded. Social engineering

---

1   Sabine Thürmel wurde an Technischen Universität München sowohl in Informatik (1989) wie auch in Technikphilosophie (2013, Betreuer: Klaus Mainzer) promoviert. Ihr Forschungsschwerpunkt liegt im Grenzgebiet Informatik und Philosophie, wobei sie von ihrer langjährigen Arbeit in der Wirtschaft profitiert. Aktuelle Veröffentlichungen finden sich zur Kollaboration mit technischen Agenten in soziotechnischen Systemen sowie zu verantwortungsvollen Innovationen in Big Data basierten Systemen.

restricts the autonomy of the participants. Thus a responsible innovation process guiding the modelling and employment of such systems is essential.

26.1   Einführung

Schon der Urvater aller Big Data Ansätze wusste, dass Daten ohne geeignete Prozesse keinen Sinn machen: Prozesse, um sie zu gewinnen, und Prozesse, um sie zu nutzen. Der US-amerikanische Marineoffizier und Hydrograph M.F. Maury veröffentlichte 1845 erste Wind- und Strömungskarten. Diese basierten auf alten Logbüchern und Karten. Ab 1847 wurde auf sein Betreiben hin mit meteorologischen Beobachtungen auf Marine und Handelsschiffen begonnen: „Every ship that navigates the high seas may henceforward be regarded as a floating observatory, a temple of science" (zitiert nach (Mayer-Schönberger und Cukier 2013, S. 75)). Maury, der *Pathfinder of the Seas*, nutzte diese Informationen, um 1854 die erste Tiefenkarte des Nordatlantiks zu erstellen. Sein Hauptwerk *The Physical Geography of the Sea* (1855) basierte auf mehr als 1.2 Millionen Datensätzen: „thus the young mariner would here find, at once, that he had already the experience of a thousand navigators to guide him" (zitiert nach (Mayer-Schönberger und Cukier 2013, S. 76)). Die Arbeit dieses frühen *Data Scientist* zeigt exemplarisch, dass Datenanalyse, Visualisierung und sinnvolle Nutzung Theorie und Praxis vereinen kann. So angewandt, bedeutet Big Data weder das „Ende der Theorie" (Anderson 2008) noch den bequemen Verzicht auf die eigene Urteilskraft und die Phronesis, die Klugheit im Handeln.

Maury arbeitete nach einem auf ihn zugeschnittenem Prinzip der Datengewinnung und Datenanalyse. Er nutzte die Erfahrung der Seeleute und ihre Messungen, um seine Karten zu erstellen. Dieser Prozess wurde zentral durch ihn gesteuert und ausschließlich von ihm kontrolliert. Die Seefahrer konnten seine Karten wiederum eigenständig im praktischen Einsatz nutzen. Wie steht es in heutigen Big Data Systemen um Autonomie und Kontrolle bei der Datenanalyse und bei der Datennutzung? Die folgenden Ausführungen wollen zum einen erste Hinweise geben, wie die Datenanalyse auf Basis von Parallelarbeit und kollaborativen Ansätzen unterstützt werden kann, wenn - wie zu Maurys Zeiten – lokal vorhandene Informationen und Ressourcen genutzt werden. Zum anderen wird aufgezeigt, dass in Big Data basierten Systemen, deren Ziel die proaktive Gestaltung der Zukunft ist, die Autonomie der Nutzer durch das System selbst eingeschränkt wird.

26.2   Autonomie und Kontrolle bei der Datenanalyse

Heute ist Big Data durch die vier V charakterisiert: *„volume, velocity, variety, and veracity".* *Volume, Variety* und *Veracity* waren bereits Kennzeichen von Maurys Big Data. Unter seiner zentralen Kontrolle und unter Mithilfe von Marine und Handel wurden die Karten erstellt und kontinuierlich verbessert. Heutzutage jedoch fallen die zu verarbeitenden Daten in Echtzeit an. *Velocity* erfordert neue Ansätze, der Informationsflut Herr zu werden: *Crowd Sourcing,* die Auslagerung von Teilaufgaben an beliebige Internetnutzer, und *Citizen Science,* der Einsatz sogenannter Bürgerwissenschaftler zur Kategorisierung von Flora, Fauna und fernen Galaxien, erinnern an Maurys Zuarbeiter. Es gibt auch schon erste Ansätze, bei der Datenanalyse sogenannte Multiagentensysteme einzusetzen. Hier wird auf die Kooperation von Softwareagenten gesetzt: ein Multiagentensystem (MAS) ist ein lose gekoppeltes Netz von Softwareagenten, die interagieren, um Probleme zu lösen, deren Bewältigung über die Fähigkeiten und das Wissen der individuellen Agenten hinausgehen. Es existieren heute bereits eine Vielzahl von Ansätzen zur Modellierung und Realisierung von Softwareagenten. Insbesondere die von Mainzer (1997) als „schwache Agententechnologien" bezeichneten Systeme, in denen Softwareagenten eigenständig vorgegebene Ziele zu erreichen suchen, Information austauschen und sich an Veränderungen im Netz anpassen, sind bereits in vielerlei Anwendungen erprobt. Heutige Softwareagenten können sich dezentral organisieren und so programmiert werden, dass sie eigenständig entscheiden, wann und wie sie die vorgegebenen Ziele zu erreichen suchen. Die Charakteristika, die Mainzer (1997) unter dem Oberbegriff „starke Agententechnologie" zusammenfasste, stellen jedoch auch heute noch Herausforderungen für Forschung und industrielle Nutzung dar. Es sind dies das Verfolgen eigener Ziele, der Besitz einer eigenen Motivationsstruktur und die auf diesen Fähigkeiten basierende Lernfähigkeit. Sobald starke Agententechnologien zum Einsatz kommen werden, wird eigenständiges Handeln der Softwareagenten erfolgen. In Zusammenhang mit Big Data können Multiagentensysteme zum einen für die dezentral organisierte Datenanalyse genutzt werden. Zum anderen kann das Verhalten der individuellen Agenten selbst *on the fly* durch Big Data verbessert werden (Ghose 2013). Nicht zuletzt kann die Perspektive von der Fokussierung auf Transaktionen und Datenströme durch den Blick auf Ökosysteme abgelöst werden „to see how structures emerge, proliferate, and morph into other structures" (Zeng und Lusch 2013, S. 4). So werden schnappschussartig gewonnene Daten und Korrelationen in einen größeren Kontext eingebettet. Die dynamische Organisation von sozialen Netzwerken, wirtschaftlichen Märkten und soziotechnischen Systemen kann durch „big data with an ecosystem lens" (Zeng und Lusch 2013, S. 4) in den Blick genommen werden. Auf diese Weise kann die Datenanalyse heute auf

Basis von Parallelarbeit und kollaborativen Ansätzen unterstützt werden, wenn –
wie zu Maurys Zeiten – lokal vorhandene Informationen und Ressourcen genutzt
werden. Meist jedoch erfolgt die Verarbeitung der Echtzeitdaten durch spezielle
Datenanalyseprogramme, die nicht auf kooperativer Parallelarbeit basieren und
(nur) das Aufzeigen statistisch relevanter Korrelationen zum Ziel haben.

26.3   Autonomie und Kontrolle bei der Datennutzung

Auf Big Data basierte Korrelationen können zum einen als Startpunkt eigenen
Denkens und Handelns genutzt werden: nicht nur Nachrichtendienste und Finanz-
dienstleister nutzen Echtzeitbenachrichtigungen nach Auswertung von Twitter-
nachrichten, um schneller als die Konkurrenz auf neue Entwicklungen aufmerksam
zu werden. Auch Wissenschaftler und Ingenieure finden in Big Data Korrelationen
Anregungen, die sie eine andere Sicht auf ihr Thema gewinnen lassen. Wie May-
er-Schönberg zusammenfasst, „lassen sich also mit Hilfe von Big Data nicht bloß
bereits vorgefasste Hypothesen bestätigen, sondern automatisiert neue Hypothesen
generieren und evaluieren" (Mayer-Schönberger 2015, S. 15). Hierbei bedeutet „bestätigen" und „evaluieren" nicht exakte Ursachenforschung, sondern eine Pragmatik im Vorgehen. Das „Ende des Ursachenmonopols" (Mayer-Schönberger 2015, S.
15) kann die Adaptionsgeschwindigkeit an sich ändernde Umstände auf Basis einer
auf statistischen Korrelationen beruhenden Sicht auf die Wirklichkeit erhöhen. Vor
dem damit potenziell einhergehenden „Missbrauchs von Big-Data Korrelationen
für kausale Zwecke" (Mayer-Schönberger 2015, S. 19) warnt insbesondere Klaus
Mainzer immer wieder eindringlich. Er spricht sich u.a. in (Mainzer 2013), (Mainzer 2014) und (Mainzer 2015) sehr engagiert für den verantwortlichen Umgang
mit Daten und den klugen Einsatz von Big Data Ansätzen aus: „Algorithmen sind
mächtig und hilfreich. Aber alleine sind sie blind" (Mainzer 2015, S. 65).

Big Data Ansätze können für die Optimierung des persönlichen und sozialen
Verhaltens genutzt werden: Big Personal Data und Big Social Data sind nicht ferne
Forschungsziele, sondern heute bereits Realität. Erste Big Personal Data Anwendungen stehen heute schon jedermann dank *Smartwatches* und *Gesundheitsapps*
zur Verfügung. Das quantifizierbare Selbst dient als Grundlage für die kontinuierliche Selbstverbesserung (Selke 2014). Big Social Data werden nicht nur für die
Trendanalyse eingesetzt, sondern können auch zur Optimierung sozialer Systeme
genutzt werden (Pentland 2014). Heutige verteilte Systeme der Gesundheitsüber-
wachung leiten Information über die Vitalparameter weiter und lassen die Ärzte
bei Bedarf direkt mit den Patienten interagieren. In zukünftigen proaktiven Syste-

men zur Stärkung der Gesundheit und des Wohlbefindens wird die Verhaltenssteuerung auf Basis von Big Data integriert sein: „proactive P4 medicine: predictive, preventive, personalized and participatory …has two major objectives: to quantify wellness and to demystify disease. P4 medicine has striking implications for society – including the ability to turn around the ever-escalating costs of healthcare" (Hood und Flores 2012, S.1). Diese Form des *Social Engineering* zielt auf die Anpassung an individuelle Ziele und auf in der Systemebene vorgegebenen Zwecke ab. Konformität und nicht individuelle Entscheidung und Entwicklung ist hier das Ziel. Teilnahme, aber nicht Teilhabe werden gefördert. Dieses Vorgehen führt zum Paradox der Partizipation und dem Paradox der Autonomie: einerseits bieten Big Data basierte Systeme den Teilnehmern neue Formen der Wissens über sich selbst und der Selbstoptimierung. Andererseits beschränken die Steuerungsmechanismen, die in diese Systeme eingebaut sind, die Autonomie der Teilnehmer und unterwerfen sie der Systemsteuerung und -kontrolle.

Die Einschränkung der menschlichen Autonomie und die Kontrolle in soziotechnischen Systemen kann noch genauer charakterisiert werden. Mit Gransche et al. (2014) kann zwischen drei Stufen der Kontrolle in technischen Systemen unterschieden werden: operative Kontrolle, strategische Kontrolle und normative Kontrolle. Dieser Ebene können die folgenden Stufen der Autonomie zugeordnet werden: Kontrolle der Mittelwahl und des Mitteleinsatzes zur operativen Kontrolle, Entscheidung über Strategien der Zweckerfüllung zur strategische Kontrolle und Anerkennung/Ablehnung/Setzung der Zwecke auf Ebene der normativen Kontrolle. In einer Vielzahl technischer Systeme erfolgt schon die operative und strategische Kontrolle durch das System selbst. Cyberphysikalische Systeme stehen exemplarisch für diesen Trend (Mainzer 2010, S.207ff). In heutigen soziotechnischen Systemen finden sich Softwareagenten und Roboter als Interaktionspartner (Thürmel 2015). Die Delegation der Mittelwahl, operative Strategieentscheidung, Aushandlung von Zielprioritäten bei der Verwaltung knapper Ressourcen kann durch technische Systeme autonom erfolgen. Die Zwecksetzung ist i.A. wirtschaftlich begründet und nur zum Teil einer normativen Kontrolle unterworfen. Die Autonomie der Nutzer, so auch die Patientenautonomie, wird eingeschränkt. Dieser Wandel von Interaktion und Kontrolle ist Herausforderung für Big Data basierte Systeme. Daher ist es unabdingbar, dass ein seiner Verantwortung bewusster Innovationsprozess die Modellierung, die Realisierung und den Einsatz von Big Data Systemen begleitet: „Responsible innovation is a collective commitment of care for the future through responsive stewardship of science and innovation in the present." (Owen et al. 2013, S. 36).

Value sensitive Governance, d.h. wertorientiertes Design im Sinne von Friedman et al. (2006) sowie van den Hooven (2007) und Einsatz der Systeme unter

Berücksichtigung eines verbindlichen wertorientierten Verhaltenskodex sind unabdingbar. Die Werte, denen ein Big Data basiertes System verpflichtet ist, können kulturelle Unterschiede aufweisen. Jedoch sind klare Zuständigkeiten, Fairness für alle Beteiligten und Transparenz unabdingbar, gerade weil diese Systeme die Autonomie der Nutzer und Betroffenen beschränken.

Literatur

Anderson, Chris. 2008. The End of Theory: The Data Deluge Makes the Scientific Method Obsolete *Wired Magazine 16.07.* http://archive.wired.com/science/discoveries/magazine/16-07/pb_theory. Zugegriffen: 20.4.2015.

Friedman, Bayta, Peter Kahn und Alan Borning. 2006. Value Sensitive Design and information systems. In *Human-Computer Interaction and Management Information Systems: Foundations,* hrsg. Ping Zhang und Dennis Galletta, 348–372, New York: ME Sharpe.

Ghose, Aditya. 2013. Agents in the Era of Big Data: What the "End of Theory" Might Mean for Agent Systems *Proceedings of the 16th International Conference on the Principles and Practice of Multi-Agent Systems (PRIMA 2013),* 1-4, Springer LNCS: Heidelberg.

Gransche, Bruno, Erduana Shala, Christoph Hubig, Suzana Alpsancar und Sebastian Harrach. 2014. Wandel von Autonomie und Kontrolle durch neue Mensch-Technik Interaktionen. Grundsatzfragen autonomieorientierter Mensch-Technik-Verhältnisse, Stuttgart: Frauenhofer Verlag.

Hood, Leroy und Mauricio Flores. 2012. A personal view on systems medicine and the emergence of proactive P4 medicine: predictive, preventive, personalized and participatory. *New Biotechnology.* doi:10.1016/j.nbt.2012.03.004.

van den Hoven, Jeroen. 2007. ITC and Value Sensitive Design. In The Information Society: Innovations, Legitimacy, Ethics and Democracy , IFIP International Federation for Information Processing, Volume 233, hrsg. P. Goujon, S. Lavelle, P. Duquenoy, K. Kimppa, und V.Laurent, 67-72, Boston: Springer.

Mainzer, Klaus. 1997. Künstliches Leben und virtuelle Agenten. Zur digitalen Evolution intelligenter Netzwelten *Telepolis 16.1.1997,* http://www.heise.de/tp/r4/artikel/6/6212/1.html. Zugegriffen: 20.4.2015.

Mainzer, Klaus. 2010. *Leben als Maschine?: Von der Systembiologie zur Robotik und Künstlichen Intelligenz.* Paderborn: Mentis Verlag.

Mainzer, Klaus. 2013. *Die Berechnung der Welt: Von der Weltformel zu Big Data.* München: C.H. Beck.

Mainzer, Klaus. 2014. BIG DATA und die neue Weltordnung, *viernull-magazin 02/2014:* 28-32.

Mainzer, Klaus. 2015. Algorithmen sind mächtig und hilfreich. Aber alleine sind sie blind. *Faszination Forschung. TUM Research Highlights 15/14:* 64-75.

Mayer-Schönberger Victor und Kenneth Cukier. 2013. *BIG DATA, a revolution that will transform how we live, work and think.* London: John Murray.

Mayer-Schönberger Victor. 2015. Was ist Big Data? Zur Beschleunigung des menschlichen Erkenntnisprozesses *APuZ 11-12/2015*, 14-19.

Owen Richard, Jack Stilgoe, Phil Macnaghten, Mike Gorman, Erik Fisher und Dave Gustion. 2013. A Framework for Responsible Innovation. In *Responsible Innovation: Managing the responsible emergence of science and innovation in society*, hrsg. Richard Owen, John Bessant und Maggy Heintz, 27-50, Chichester, UK: John Wiley & Sons.

Pentland Alex. 2014. Social Physics: How Good Ideas Spread – The Lessons from a New Science. Melbourne,London: Scribe.

Selke, Stefan. 2014. Lifelogging: Wie die digitale Selbstvermessung unsere Gesellschaft verändert. Berlin: Econ.

Thürmel, Sabine. 2015. The Participatory Turn - A Multi-Dimensional Gradual Agency Concept for Human and Nonhuman Actors. In *Collective Agency and Cooperation in Natural and Artificial Systems. Explanation, Implementation and Simulation*, Philosophical Studies Series 122, hrsg. Cathrin Misselhorn, 45-62, Berlin: Springer.

Zeng, Daniel und Robert Lusch. 2013. Big Data Analytics: Perspective Shifting from Transactions to Ecosystems, IEEE Intelligent systems 28(2): 2-5.

27. Was kann man aus Jonathan Franzens Roman "Freedom" über Energie und Nachhaltigkeit lernen?

Thomas Hamacher[1]

Abstract

Gute Romane sind unbestritten Seismographen ihrer Zeit. Sie kondensieren Entwicklungen und Ideen in Personen und Handlungen, die gerade in ihrer Besonderheit Repräsentanten für ein Allgemeines sind. Romane werfen einen Blick auf das „wirkliche" Leben und ergänzen damit die „nackten" Zahlen der „unpersönlichen" Statistiken. Als ein Beleg für diese Aussage soll der Roman „Freedom" von Jonathan Franzen als Beispiel analysiert werden, in dem das Thema Energie eine große Nebenrolle spielt. Daraus lässt sich dann die Frage ableiten, ob eine Beschäftigung mit Literatur eine sinnvolle Erweiterung von Systemanalysen darstellt.

1 Prof. Hamacher (*1964) forscht auf dem Gebiet der Energie- und Systemanalyse. Schwerpunkte der Forschung sind städtische Energiesysteme, Integration Erneuerbarer Energien in das Stromnetz und innovative nukleare System, insbesondere Fusion. Weitere Schwerpunkte der Arbeit sind Methoden und Grundlagen von Energiemodellen. Nach seinem Studium der Physik in Bonn, Aachen und an der Columbia University in New York promovierte Prof. Hamacher zum Thema „Baryonische B-Zerfälle" an der Universität Hamburg. Seit 1996 arbeitete Prof. Hamacher am Max-Planck-Institut für Plasmaphysik, zuletzt als Leiter der Gruppe für Energie- und Systemstudien. Von 2010 bis 2013 war Prof. Hamacher kommissarischer Leiter des Lehrstuhls für Energiewirtschaft und Anwendungstechnik. Im September 2013 wurde er als Full Professor für Erneuerbare und Nachhaltige Energiesysteme berufen. Zugleich ist er Direktor der Munich School of Engineering (MSE). Prof. Hamacher ist Mitglied des Wissenschaftszentrums Umwelt (WZU) der Universität Augsburg.

Good novels are seismographs of their time. They portray developments and ideas in people and actions, which are representatives of something general precisely due to their distinctiveness. Novels take a look at the "real" life and thus complement the "naked" numbers of "impersonal" statistics. As evidence for this statement, the novel "Freedom" by Jonathan Franzen shall be analyzed as an example, in which the topic of energy plays an important supporting role. In view of this the question is asked whether the study of literature constitutes a meaningful extension of system analysis.

27.1   Einleitung

Gute Romane sind unbestritten Seismographen ihrer Zeit. Sie kondensieren Entwicklungen und Ideen in Personen und Handlungen, die gerade in ihrer Besonderheit Repräsentanten für ein Allgemeines sind. Gute Romane erlauben sich Diskussionen zu führen, die sich aus der Enge der öffentlichen politischen Diskussion lösen und Themen aufgreifen, die letztlich entscheidend sind, aber immer wieder ausgespart werden. Technische und wirtschaftliche Aspekte stehen dabei, wenn überhaupt erwähnt, eher im Hintergrund. Damit sind Romane wichtige Zeugen für eine Zeit. Romane werfen einen Blick auf das „wirkliche" Leben und ergänzen damit die „nackten" Zahlen der „unpersönlichen" Statistiken. Als ein Beleg für diese Aussage soll im Folgenden der Roman „Freedom" von Jonathan Franzen als Beispiel analysiert werden, indem das Thema Energie eine große Nebenrolle spielt. Daraus lässt sich dann die Frage ableiten, ob eine Beschäftigung mit Literatur, letztlich natürlich auch mit anderen Produkten des Literatur- und Kunstbetriebes, eine sinnvolle Erweiterung von Systemanalysen darstellt.

Die Systemanalyse ist wissenschaftstheoretisch eine der umstrittensten Wissenschaftsdisziplinen. Leider nimmt die Disziplin selber diese Herausforderung nicht wirklich an und diskutiert nicht, wieweit die Zukunft der sozio-technischen Systeme berechenbar ist. Der Fluss an Aufträgen aus Politik und Entscheidungsträgern ist zu konstant und lässt keinen Raum für Selbstzweifel. Auch in dieser Hinsicht soll der Beitrag gelesen werden, nämlich inwieweit die jeweils beschriebene Alltagswelt hilft die Komplexität von Systemen zu reduzieren und dabei die Bedeutung einzelner Systemkomponenten, wie z.B. hier der Autos, herauszustellen und „berechenbarer" zu machen.

Der Roman „Freedom" von Jonathan Franzen (2010) spielt in den ersten zehn Jahren des 21. Jahrhunderts. Dabei gehen diverse Rückblenden deutlich weiter zurück. Diese zehn Jahre haben insbesondere die USA durch Katastrophen und

Kriege geführt, die die Welt verändert haben. Am Anfang stand der 11. September, gefolgt von den Kriegen in Afghanistan und Irak, gefolgt von Finanzkrise und dem Zusammenbruch der Lehmann Brothers. Am Ende stand als Lichtblick der erste schwarze Präsident der USA, der die Menschen mit dem Slogan „Yes We Can" für sich gewonnen hat. Die drei „Helden" des Romans sind selber keine Protagonisten der Entwicklung. Sie sind höchstens Beobachter.

Was hat der Roman mit Energie und dem Streben nach einer nachhaltigen Entwicklung zu tun bzw. mit dem Thema, dass die Forschung an der TU München im Energiebereich betrifft. Hier eine erste Antwort:

1. Zum ersten nutzt der Roman das Thema Energie- und Umweltschutz, hier insbesondere den Vogelschutz, als Hintergrund der Handlung. Dabei geht es um so umstrittene Themen wie *Mountaintop Removal Mining* im Kohlebergbau bzw. die beginnende intensive *Shale Gas*-Ausbeutung.

2. Der Roman spielt quasi immer wieder am Walden-See. Das Buch „Walden oder Leben in den Wäldern" von Henry David Thoreau (2004) ist die Beschreibung eines „Aussteigers", der am eigenen Leibe zeigen will, dass ein anderer Lebensstil möglich ist und dem Menschen mehr Freiheit und insbesondere Zeit bietet. Das Buch ist sicherlich ein Meilenstein in der Lebensstildiskussion. Die Diskussion heute neu Aufzugreifen ist mehr als überfällig.

3. Dann geht der Roman aber noch einen Schritt weiter. Wichtiger als die Frage „wie soll der Mensch leben", wird die Frage „wie viele Menschen sollen auf der Erde leben". Mit dem Stichwort „Überbevölkerung" wird die radikalste Begrenzung des menschlichen Einflusses auf die Umwelt diskutiert. Das Thema ist alt und wurde von Robert Thomas Malthus (1798) schon am Ende des achtzehnten Jahrhunderts entdeckt. Exponentielles Wachstum bringt früher oder später jede Ressource an ihre Grenzen. Das gleiche Kalkül wurde in den siebziger Jahren in der Studie „Grenzen des Wachstums" (Meadows et al. 1972) aufgegriffen. Die Begrenzung des Bevölkerungswachstums in China und in Ansätzen auch in Indien hat das Thema dann zum Tabu werden lassen. In den Zeiten der „Überalterung" scheint das Thema für die Industrieländer ohnehin keine Relevanz mehr zu haben.

Das Papier ist wie folgt organisiert: Zuerst wird ein ganz grober Überblick über die Personen und Handlungen des Romans gegeben. Dieser Teil wird mit einer kurzen Beschreibung des Autors abgerundet. Dann werden drei Themen aus dem Roman nacheinander aufgegriffen: 1) Die neue „fossile" Ära in den USA mit *Mountaintop Removal Mining* und *Shale Gas* 2) Walden oder das einfache Leben 3) die Begrenzung der Weltbevölkerung als radikalste Form der Nachhaltigkeit.

Daraus werden dann Schlussfolgerungen für die heutige Energie- und Nachhaltigkeitsforschung abgeleitet, bzw. inwieweit die Ansätze aus dem Roman in die Forschung integriert werden könnten. Zum Abschluss wird die Eingangsfrage wieder aufgegriffen und ein Ausblick auf eine Verknüpfung von Literaturwissenschaft und Energie- und Nachhaltigkeitsuntersuchungen gegeben.

27.2 Ein schneller Überblick über die Personen und Handlungen des Romans und den Autor

Der Roman beschreibt die Entwicklung der Liebe zwischen Patty Emerson, später Berglund, und Walter Berglund. Die beiden kommen aus ganz verschiedenen Familien. Patty aus einer Juristen-Politiker Familie in New York und Walter kommt aus Minnesota. Sein Vater ist ein einfacher Mann. Er besitzt ein kleines Motel. Die beinahe prototypische Zeichnung der Familienverhältnisse mit Demokraten auf der einen und Republikanern auf der anderen Seite kehrt im Roman an verschiedenen Stellen wieder auf. Patty und Walter passen aber jeweils nicht in ihr familiäres Umfeld und haben darin sicher eine der stärksten Gemeinsamkeiten.

Die beiden lernen sich am College an der University of Minnesota über Walters Freund Richard Katz kennen. Patty hat dabei ein Auge auf Richard geworfen. Deswegen dauert es etwas bevor sie zusammenkommen. Aber dann geht alles schnell. Sie gründen eine Familie, setzen zwei Kinder in die Welt. Bis es nach Jahren zwischen Richard und Patty zu einer kurzen Affäre kommt. Fürs erste bleibt der Ehebruch unentdeckt. Trotzdem führt die Tat schrittweise zu Veränderungen, die die Berglunds aus ihrem bürgerlichen Leben in St. Paul nach Washington führt. Richard produziert angeregt durch den Ehebruch mit Patty die Platte „Nameless Lake", die ihn zu einem erfolgreichen Musiker macht. Walter wird durch den Erfolg seines alten Freundes angestachelt, gibt seine Stelle bei 3M auf und geht dann wie weiter unten noch diskutiert nach Washington. Dort stellt er sich auf ganz besondere Weise in den Dienst des Vogelschutzes und bekämpft nebenbei die Überbevölkerung. Dann kommt der Ehebruch doch noch an den Tag, weil Patty alles, in Folge einer Therapie, in einem Buch festgehalten hat. Richard sorgt dafür, dass Walter das Buch in die Hände bekommt. Danach kommt es zum Bruch zwischen Walter und Patty. Walter tröstet sich mit seiner jungen hübschen indischen Assistentin Lalitha, die aber wenig später bei einem Verkehrsunfall ums Leben kommt. Er geht dann wieder zurück nach Minnesota und arbeitet wieder im Naturschutz.

Patty geht nach New York erst zu ihrem Geliebten Richard. Bleibt dort aber nur kurz. Dann geht sie zurück zu ihren Eltern, bzw. lebt allein in New Work. Sie arbeitet als Lehrerin.

Nach sechs Jahren kehrt sie zu ihrem Mann zurück und gemeinsam ziehen sie nach New York. Dies mutet fast wie ein Märchen an, aber am Ende ist die Liebe zwischen zwei Menschen eben doch für ein ganzes Leben bestimmt. Aber dieser Frage möchte ich hier natürlich nicht nachgehen.

Der Roman ist ein Familienroman der reinsten Sorte mit all den Spielarten der Verstehens und Missverstehens zwischen Ehepaaren, Eltern und Kindern und insbesondere auch Geschwistern untereinander. Daneben werden auch sehr prototypische Sozialtypen gezeichnet. Der intellektuelle demokratisch wählende Städter gegen den SUV fahrenden republikanisch wählenden Nicht-Ostküstler usw. Die Symbolik der Autos wäre eine eigene kleine Untersuchung wert.

Jonathan Franzen ist einer der bekanntesten amerikanischen Schriftsteller der Gegenwart. Er wurde 1959 in Western Springs bei Chicago geboren. Auf der Titelseite des Nachrichtenmagazin Time wurde er im August 2010 als „Great American Novelist" bezeichnet. Neben Freedom wurde sein Roman „The Corrections" bekannt. Bei beiden Romanen steht das Thema Familie im Mittelpunkt.

27.3 Die neue fossile Ära in den USA: *Mountaintop Removal Mining* und *Shale Gas*

Die ersten zehn Jahre des 21. Jahrhunderts haben die USA insbesondere was die Energieversorgung angeht vor zwei ganz große Herausforderungen gestellt:

1. Nach den Anschlägen am 11. September 2001 wurde es immer klarer, dass das politische Engagement im mittleren Osten zur Sicherung insbesondere der Erdölreserven eine große Bürde darstellt. Seitdem die englische Flotte Anfang des zwanzigsten Jahrhunderts von Kohle auf Erdöl umgestellt hatte, war die Sicherung der Erdölquellen im mittleren Osten eine ganz besondere geopolitische Aufgabe erst Großbritanniens und dann der USA.

2. Der starke Anstieg des Ölpreises, der nicht zuletzt als Ergebnis des rasanten weltweiten Wirtschaftswachstums erfolgte, hat insbesondere die Menschen in den USA getroffen. Dabei haben die hohen Energiepreise einen Druck geschaffen, der dann zum Ausbruch der Finanzkrise führte. Die USA war noch mehr als Europa an niedrige Energiepreise gewöhnt.

Damit war der Hunger auf eine neue, am besten heimische, Energiequelle als Ersatz groß. Es ist kein Zufall, dass der Roman dieses Thema als Grundmotiv aufgreift. Walter Berglund hat seinen Anteil an der amerikanischen Energiewende. Wie schon erwähnt, wird Walters Ehrgeiz durch Richards Erfolg angestachelt. Nach seinem Studium hatte er erst bei 3M im Umweltbereich gearbeitet. Er selber lebt recht grün und fährt selbst im kalten Winter mit dem Fahrrad. Die Arbeit mutet etwas bieder an im Angesicht seiner Fähigkeiten und auch der Ansprüche, die er als Student hatte. Auf der Suche nach neuen Herausforderungen geht er zuerst zur amerikanischen Naturschutzorganisation „Nature Conservancy". Hier erklimmt er schnell die Karriereleiter und wird für das Bundesland der zuständige Geschäftsführer. Dann wechselt er aber nach Washington und arbeitet dort als Geschäftsführer des „Cerulean Mountain Trust". Der Name leitet sich dabei vom „Cerulean Warbler", dem deutschen Pappelwaldsänger, ab. Die Idee des Trustes ist es ein möglichst großes zusammenhängendes Waldgebiet zu kaufen, um damit wieder geeignete Brut- und Lebensbedingungen für Zugvögel bereitzustellen. Der eigentliche Kopf hinter dem Trust heißt Vin Haven und ist gut mit der Familie des Präsidenten Bush befreundet. Dabei wird der Ankauf des Landes nicht zuletzt darüber finanziert, dass erst mal ein Teil des Landes dem Mountain-Top-Mining geöffnet wird und erst nach der Ausbeutung wieder in seinen Naturzustand hergestellt wird. Tatsächlich stellt sich dann heraus, dass der Landkauf auch unter dem Aspekt der späteren Ausbeutung von *Shale Gas* erfolgt. Während der erste Kompromiss Walter von Anfang an klar war, erfährt er von der zweiten Finte erst im Lauf seiner Arbeit.

Mountain Top Removal Mining wird in einigen Teilen der USA als Verfahren zum Kohlebergbau genutzt. Dabei wird die Bergspitze meist weggesprengt, um dann leichter an die Kohlevorkommen zu gelangen. Die Auswirkungen auf die Landschaft und Ökologie können erheblich sein. Diese insbesondere in den Appalachen eingesetzte Form des Kohlebergbaus ist deswegen auch in den USA sehr umstritten.

Die Verbindung von Kohlebergbau und Naturschutz klingt aus Walters Mund recht plausibel. Sie ist aber insbesondere unter dem Aspekt der späteren Nutzung des Landes zum Abbau von *Shale Gas* nur noch zynisch. Der ganze Vorgang zeigt die Zerrissenheit einer Gesellschaft, die durchaus sehr stolz auf ihre Natur ist, die aber das Land braucht um einen unermesslichen Energiehunger zu stillen. Hier zeigt sich dann auch eine Parallele zur deutschen Energiewende, die auch nur gelingt, wenn wir erheblich in die Gestaltung der Landschaft eingreifen, und dies tun wir mit dem gleichen Argument: dem Umweltschutz. Das scheinbar so skurrile amerikanische Beispiel hält uns hier einen Spiegel vor und was in den USA schnell wie eine hässliche Fratze wirkt, ist auch bei uns nicht gleich ein schöner Jüngling. Dieser Punkt ist von großer Bedeutung für die Weiterentwicklung des Energiesystems und verlangt die Entwicklung neuer Forschungszweige wie der Energiegeographie.

Zurück zum amerikanischen Energiehunger. Es hat in Deutschland einige Zeit gebraucht, bevor man die amerikanische Energiewende so richtig wahrgenommen hat. Heute, nachdem die USA zu einem der größten Erdölproduzenden in der Welt aufgestiegen ist und die Entwicklung des Ölpreises spannender als jeder Krimi geworden ist, hat dies auch der Letzte in Deutschland verstanden. Dabei muss man anerkennen, die amerikanische Energiewende verändert die Welt, die deutsche muss noch zeigen, ob sie dies langfristig schafft.

Als Seismograph hat das Buch von Franzen dieses sich ankündigende Erdbeben aufgezeichnet und eben nicht nur das, es zeigt eine Gesellschaft, die weder bereit noch in der Lage ist, auf einen hohen Energieverbrauch zu verzichten. Leider sind die hierfür verantwortlichen Organisationen wie die IEA in Paris nicht so schnell und präzise wie Franzen, sie reagieren immer etwas zeitverzögert. Allein deswegen sollten wir mehr ausländische Romane lesen!

27.4   Walden oder das Leben in den Wäldern

Jetzt wenden wir uns dem „einfachen Leben" als einer Lösung zum oben angesprochenen Energieproblem zu. Wenn die Menschen viel weniger Energie verbrauchen würden, wenn sie wie Walter mit dem Fahrrad zur Arbeit führen, dann wäre das Problem wohl nicht annähernd so drückend. Diese Diskussion wird insbesondere durch viele Verweise auf das Buch „Walden" des Schriftstellers Henry David Thoreau geführt. Damit wird eine ganz besondere amerikanische Tradition, die ein Leben mit und in der Natur mit einfachsten Mitteln zum Ideal erhebt, aufgegriffen. In Energie- und Umweltdiskussionen wird die Forderung nach einem neuen, einfachen Lebensstil immer wieder aufgestellt und oftmals auch als einziger realistischer Ansatz beschrieben. Heute wird die Diskussion oft mit dem Begriff der Suffizienz geführt (Princen 2005; Schneidewind und Zahrnt 2013; Paech und Paech 2011).

Walden beeindruckt durch den stringenten Beweis, dass ein einfacher und bescheidener Lebensstil uns insbesondere Zeit gibt. Thoreau verbringt immer nur wenige Stunden damit das Lebensnotwendige zu beschaffen und viel Zeit damit zu lesen und die Natur zu genießen. Da Thoreau das beschriebene Leben tatsächlich geführt hat, stellt sich nicht die Frage nach der möglichen Umsetzung. Thoreau verbringt eineinhalb Jahre am Walden See in einer ganz kleinen und bescheidenen Hütte. Es ist so gerade der Gegensatz eines Lebensstils, der Bedeutung und Wichtigkeit in der Tatsache findet, dass man keine oder nur sehr wenig Zeit hat, wie er heute eher gegenwärtig ist.

An die Stelle des Walden Sees, der in Concord in der Nähe von Boston in den USA liegt, tritt hier ein kleiner See in Minnesota in der Nähe von Walters erstem Wohnort Hibbins. Die Familie von Walters Mutter hat ein kleines Ferienhaus am See. Als er das kleine Ferienhaus renovieren will, damit es später vermietet werden kann, da nur so ein Verkauf verhindert werden kann, lernt er die besondere Schönheit der Natur kennen. Später vererbt die Mutter das Haus allein an ihren Sohn Walter, da die anderen Brüder das kleine Juwel nicht verdient haben und nicht zu schätzen wissen.

Das kleine Ferienhaus am See wird zum zentralen Ort für die Schlüsselereignisse des Romans. Hier geht Patty mit Richard fremd, hier werden die erfolgreichen Lieder von Richard gedichtet, hierhin wendet sich Walter nachdem er seine Frau verlassen hat und hier kommen Walter und Patty wieder zusammen. Aber hier bleiben sie eben nicht leben, nachdem sie wieder zusammen sind. Sondern sie ziehen nach erfolgreicher Versöhnung wieder nach New York und wandeln ihr kleines Anwesen am See in ein kleines Vogelschutzgebiet um.

Damit wendet sich Franzen ganz klar gegen Thoreau und macht deutlich, dass heute ein Rückzug in die Wälder eben nicht mehr gelingt. Dies wird in dem Buch umso klarer, als in der Zeit des amerikanischen Baubooms die Häuser bis an den See kommen und insbesondere die zahlreichen Katzen das ehemalige Vogelparadies empfindlich stören.

Für eine grundlegende Diskussion über Nachhaltigkeit ist eine Rückbesinnung auf Thoreau von großer Bedeutung. Wie schon angedeutet wird hier insbesondere über die Verfügbarkeit der Zeit diskutiert und genau diese Diskussion klammern wir immer wieder gerne aus. Sicherlich hat der technische Fortschritt die Lebensarbeitszeit deutlich verkürzt und damit in den letzten zweihundert Jahren Raum für Bildung, für alle möglichen Freizeitaktivitäten und natürlich auch den Altersruhestand geschaffen. Aber die ständige Steigerung des Lebensstandards war wichtiger als ein Zugewinn an Zeit. Wären wir beim Lebensstandard der frühen siebziger Jahre stehen geblieben und hätten jeden Fortschritt in der Produktivität in Freizeit umgewandelt, dann wären wir heute sicher eher bei einer Zwanzig-Stunden-Woche und nicht bei der Rückkehr der Vierzig-Stunden-Woche. Daneben hätten wir einen deutlich reduzierten Ressourcenverbrauch.

Die Diskussion über das einfache Leben ist in der abendländischen, aber auch in asiatischen Debatten eines der ganz großen Themen. Ausgangspunkte sind dabei in Teilen in der kynischen und stoischen Philosophie zu finden bzw. in der christlichen Tradition insbesondere im Lukasevangelium. In der christlichen Tradition stellt das Aufkommen der Bettelorden im Mittelalter einen besonderen Höhepunkt dar. Heerscharen von jungen Menschen folgten dieser Bewegung, ohne dass sie den ganz großen Durchbruch in der Kirche erlangen konnte. Der heilige Franziskus und der heilige Dominikus führten durch ihre Ordensgründungen zu einer neuen Reformbewegung in der Kirche. In der Folge brachten diese so herausragende Persönlichkeiten wie Thomas von Aquin und Bonaventura hervor. Beim Blick auf die Kirche in Deutschland heute ist oft nicht viel von diesem Armutsgedanken hängen geblieben. Es soll sogar Bischöfe geben, die nicht wie von Aposteln verlangt (Lk 10,1-13) zu Fuß ohne Schuhe laufen, sondern mit großen deutschen Luxusautos fahren.

„Walden" greift diese Tradition nicht explizit auf. Es verweist nur in gleicher Weise auf den Gewinn an Freiheit durch den Verzicht auf Eigentum und Konsum. Walden hat aber die Diskussion wieder in Gang gebracht und Anhänger gefunden. Tolstoi und Gandhi seien hier als besonders prominente Beispiele genannt. Dabei haben weder Thoreau noch Tolstoi noch Gandhi den Wunsch nach starkem wirtschaftlichem Wachstum in ihren Ländern verhindern können. In Indien wird Gandhi sicher bis auf den heutigen Tag von vielen Menschen verehrt, sein Lebensstil aber von den wenigsten freiwillig geteilt.

Der Roman greift dabei weniger das einfache Leben per se auf, obwohl Walter Anklänge daran zeigt. Viel eindrücklicher ist aber die Diskussion der Zeit und der Bedeutung von bezahlter Arbeit. Die Heldin Patty, die eine ehrgeizige Sportlerin war, betont immer wieder, dass sie eine Entscheidung nie bereut hat, nämlich Mutter zu werden. Deswegen ist sie zu Hause geblieben und hat sich sehr traditionell um ihre Kinder gekümmert, während ihre erfolgreichen Eltern genau dafür keine Zeit hatten. Wieder ist es die Zeit, die wir heute in unserem ökonomischen Geist lieber dem Wachstum des Bruttosozialproduktes als unseren Kindern zur Verfügung stellen. Und dies gilt natürlich für Männer wie für Frauen. Hier erhält die Zeitdiskussion von Thoreau und Franzen eine enorme politische Bedeutung.

Eine Diskussion über nachhaltige Entwicklung wird zu einer Diskussion über Zeit und wie wir Zeit verwenden. Diese Diskussion ist mehr als überfällig. Eine andere Verwendung unserer Zeit bedeutet sicher erstmal auch Verluste von Einkünften, da weniger Zeit für die Erwerbstätigkeit aufgebracht wird. Aber erst in der Gegenüberstellung von gewonnener Frei-Zeit und Verzicht auf Konsum und Güter kann eine neue Balance gefunden werden.

27.5   Überbevölkerung

Aber das Buch geht in seinen Betrachtungen noch wesentlich weiter. Der radikalste Weg zur Erreichung der nachhaltigen Entwicklung gelingt durch einen drastischen Rückgang der Weltbevölkerung. Die Verbreitung dieser Idee ist die wahre Leidenschaft und der größte Widerspruch unseres Helden Walter. Schon als Student versucht er seinen Freund Richard zu überreden mit seinen Liedtexten die jungen Menschen von einem kinderlosen Leben zu überzeugen. Trotzdem bekommt seine Frau, kaum dass sie verheiratet sind, zwei Kinder.

Bei seiner Arbeit in Washington greift er den Kampf wieder auf. Nachdem ihm klar wird, dass dem Trust nicht an Natur- und Umweltschutz gelegen ist, stürzt er sich ganz auf freie Mittel, die ihm zur Verfügung stehen, um damit ein Sommercamp für junge Menschen zu gestalten, die nicht zuletzt durch die Lieder von Richard von dieser neuen kinderlosen Lebensform überzeugt werden sollen.

Der Roman nimmt hier wiederholt Bezug auf die „Grenzen des Wachstums", den Bericht des Club of Rome, der durch die Volkswagenstiftung finanziert wurde und seine besondere Glaubwürdigkeit nicht zuletzt durch die damals noch ungewöhnliche Nutzung von Computern gewann. Die Grenzen exponentiellen Wachstums hatte aber schon Malthus mehr als hundert Jahre zuvor in seinen Arbeiten festgestellt.

Franzen macht klar, dass dieser Gedanke insbesondere im religiösen Umfeld der USA heute keine Freunde finden wird. Und auch in Deutschland werden steigende Geburtenraten gefordert und sinkende mit allen möglichen Mitteln bekämpft. Aus Walters Sicht müsste das Kinderkriegen bestraft, die Kinderlosigkeit belohnt werden, vielleicht durch die Einführung von Kindersteuern statt Kindergeld. Für die TU München könnte es bedeuten, dass deutsche bzw. europäische Studenten hohe Gebühren zahlen müssten, während Studenten aus bevölkerungsreichen Ländern umsonst studieren dürften und obendrein einen sicheren Platz im Studentenwohnheim hätten. Aus der Logik unseres Helden wäre dies mehr als vernünftig.

Auf alle Fälle sieht es erstmal nach einer Lösung aus. Eine Milliarde Menschen kann die Erde mit den Mitteln der heutigen Technologie sicher unterstützen und dabei ausreichend Nahrungsmittel und Ressourcen aus nachwachsenden Rohstoffen bereitstellen. Bei zehn Milliarden Menschen wird es schnell eng. Die jüngsten Abschätzungen legen nahe, dass es in diesem Jahrhundert noch erhebliches Wachstum geben kann und auch eine Zahl von 12,3 Milliarden Menschen nicht unrealistisch ist (Corey und Brook 2014). Schlüssel wird dabei die Entwicklung in Afrika sein.

Aber der Roman macht hier eben in der Person Walters das Dilemma klar. Auch für ihn sind Kinder die Basis seines Lebensglückes, obwohl ihm seine Vernunft die

Kinderlosigkeit rät. Damit ist jenseits jeder Familien- oder Einwanderungspolitik die zentrale Herausforderung auf dem Tisch: wie kann ein Mensch sein privates Glück in Familie und eben auch Kindern finden ohne die Überbevölkerung zu stärken. Der Roman gibt hier keine Antwort, sondern zeigt nur Walters Entscheidung.

27.6   Schlussfolgerungen für die Forschung

Aus diesen Betrachtungen lassen sich schon einige Schlussfolgerungen für die Forschung ziehen, die dann aber in einem abschließenden Teil noch um eine methodische Forderung erweitert werden sollen.

27.6.1. Die neue fossile Herausforderung

Die Entwicklungen in den USA haben, wie schon angedeutet, die Energiewelt auf den Kopf gestellt. Dabei waren es teilweise eher mittelständische Unternehmen, die durch die Bohr- und Fracking-Technologien dem fossilen Zeitalter ein zweites Leben eingehaucht haben. Die Folgen davon können sehr Vielfältig sein: die Gefährdung von Klimazielen, wobei insbesondere in den USA durch die vermehrte Nutzung von Erdgas anstelle von Kohle das Umgekehrte zu beobachten ist; die Veränderung von Randbedingungen, die die USA als Industriestandort attraktiv machen; der Zusammenbruch der politischen Systeme, die sich durch den Export von teuren fossilen Energieträgern finanzieren usw. Auf alle Fälle stellen sich hier viele Fragen, die wir bisher nicht gut beantworten können.

Die Erforschung der neuen „fossilen Renaissance" ist wichtig und darf nicht durch voreilige Forderungen, sich doch ganz auf erneuerbare Energien zu konzentrieren, verwässert werden. Insbesondere sollte die alte Idee des Methanzeitalters als wichtige Brücke in eine neue Energiewelt wieder aufgenommen werden. Sollten mehr nicht-konventionelle Erdgaslagerstätten nutzbar sein, dann liegt hier sicher eine große Chance. Erdgas könnte weltweit als Brückentechnologie die intensive Nutzung von Kohle ablösen und damit zu einer deutlichen Reduktion der klimaschädlichen Treibhausgasemissionen führen.

27.6.2. Energie und Landschaft

Der Zusammenhang von Energie und Landschaft wird zu einer der zentralen Herausforderungen, eben nicht nur in den USA. Die Wissenschaft allein kann hier keine finale Lösung präsentieren, aber hilfreiche Handreichungen vorbereiten. Landschaftsarchitekten und -ökologen sind hier gefragt, die neue Wissenschaft der Energiegeographie sollte helfen. Was verstehen wir unter der Landschaft. Der amerikanische Philosoph und Thoreau-Freund Emerson schreibt dazu in seinem Buch „Natur": „In der Wildnis finde ich etwas Wertvolleres und Verwandteres als auf den Straßen und in den Dörfern. In der ruhigen Landschaft, und besonders in der weit entfernten Linie am Horizont, erblickt der Mensch etwas, das so schön ist wie seine eigene Natur."[2] Die Veränderung der Landschaft auch durch erneuerbare Energien muss zu einem Thema werden, das mit Mut in einer breiten gesellschaftlichen Diskussion geführt werden muss. Die Landschaft ist auch heute in den meisten Industrieländern schon das Ergebnis von Land- und Forstwirtschaft. Die neuen Formen der Energiewandlung, seien es fossile Technologien oder erneuerbare, werden durch oft weit sichtbare Technologien noch zusätzliche Bestandteile liefern, die erst fremd in der Landschaft sind. Hier ist ein Dialog zwischen Bevölkerung und Planern notwendig, die „Notwendigkeiten" der neuen Energiewelt mit dem Wunsch nach einer „schönen" Landschaft verbinden.

27.6.3. Zeit und Nachhaltigkeit

Die Verkehrs- und zunehmend auch die Energieforschung stützen sich auf Informationen über den Zeitverlauf der menschlichen Tätigkeit. In der Verkehrsforschung ist uns dies klar, in der Energieforschung noch nicht. Doch die die Anpassung der zeitlichen Nachfrage an die fluktuierenden erneuerbaren Energien verlangt auch eine viel besseres Verständnis der Zeitbudgets der Menschen. Der Verbrauch soll an das Erzeugungsangebot angepasst werden. Dabei werden prädiktive Regelungen entwickelt, die wissen müssen, wann wir wo sind und was wir dann machen. Google und Co werden uns natürlich dabei helfen. Zeit spielt also eine Rolle.

Dies um die Betrachtung von Zeitbudgets an sich zu erweitern ist eine erste Basis, um verschiedene Lebensstile zu diskutieren, zu vergleichen und energetisch zu bewerten. Ingenieure werden keine Vorgaben für den Lebensstil machen, sie

---

2   Natur (Nature), zitiert nach der deutschen Übersetzung von Harald Kiczka, Zürich 1988, S. 17

sollten aber klar machen, was welcher Lebensstil für die Erde bedeutet. Der Weg über die Zeitbudgets ist hier mehr als elegant und kann uns helfen die Diskussion insgesamt etwas nüchterner zu führen. Die Diskussion muss dabei nicht auf das Thema Energie beschränkt bleiben, wie der Roman zeigt, kann damit auch das wichtige Thema der Kinderbetreuung noch einmal geführt werden.

27.6.4. Bevölkerung und Nachhaltigkeit

Es wird wohl kaum gelingen eine Tragfähigkeit des Ökosystems Erde für den Menschen zu bestimmen, sprich wir können nicht eine maximale Bevölkerungszahl nennen, die die Erde nicht überschreiten darf ohne unweigerlich zerstört zu werden. Auch eine Reduktion der Weltbevölkerung auf eine Milliarde Menschen ist eher unrealistisch. Und trotzdem kann die Größe Weltbevölkerung stärker in den Fokus genommen werden, egal ob wir über Energie, über Wasser, über Ernährung oder auch Gesundheit reden. Die Erforschung der Demographie muss bei allen Nachhaltigkeitsuntersuchungen im Zentrum stehen. Stattdessen nehmen wir hier meist die Bevölkerungsdaten der UN unkritisch auf und nutzen sie. Hier ist mehr Sorgfalt und eigene Forschung gefragt. Dabei sind Themen wie Migration von großer Bedeutung, wie sehen Gesellschaften in Zukunft aus, welche Sprachen, welche Religionen sind vorherrschend.

Insgesamt muss das Thema expliziter nicht zuletzt in Energiestudien Einzug erhalten.

27.7. Die Bedeutung der Literatur für die Energiesystemanalyse

Die Analyse von Systemen hat eine lange Tradition und ihren Ursprung in Biologie und Regelungstechnik. Als wichtiger Wegbereiter kann hier das Buch „General System Theory" von Ludwig von Bertalanffy genannt werden. Im Energiebereich bildete sich dann zunehmend eine eigene Energiesystemanalyse heraus, die eigene Werkzeuge und Methoden entwickelt hat. Basis für diese Untersuchungen sind zahlreiche Energiestatistiken und Energieberichte. Die Daten werden dabei teilweise von staatlichen bzw. internationalen Institutionen gesammelt und herausgegeben oder von Firmen, die sich teilweise auf die Bereitstellung dieser Daten spezialisiert haben.

Die Auswertung eines Romans ist in diesen Methoden nicht vorgesehen. Die zugegeben noch oberflächliche Auswertung des Romans „Freedom" von Jonathan Franzen zeigt aber, dass eine Beschäftigung mit der Literatur in mehrfacher Hinsicht hilfreich sein kann: a) das Thema *Shale Gas* spielte hier schon eine Rolle als es in Deutschland noch nicht wirklich präsent war, hier kann die Literatur als Frühindikator gute Beobachter dienen. b) Die Diskussionen und Fragestellungen gehen über die „üblichen" Diskussionen hinaus. Im Roman ist dies die Diskussion über die Bevölkerungspolitik. Und c) mit Walden wird eine amerikanische Tradition angesprochen, die durchaus einen wichtigen Beitrag zur Lösung beitragen kann, wenn sie richtig verstanden wird.

Die Entwicklung eines Verfahrens zur Nutzung der Literatur setzt sicherlich den Einsatz von „Beobachterseminaren" in Literaturinstituten voraus. Hier werden ohnehin oft „Neuerscheinungsseminare" durchgeführt, die einen Blick auf die neusten Romane werfen. Damit könnten sicherlich schnell die relevanten Neuerscheinungen identifiziert werden. Problematischer ist die Methode zur Auswertung der Romane. Was ist das Ergebnis von Beobachtungen, was sind Übertreibungen usw. Hier hat die Literaturwissenschaft sicher ein breites Portfolio an Herangehensweisen. Sie müssen nur in den Fragestellungen auf die Bedürfnisse der Energiesystemanalyse angepasst werden.

Literatur

Bradshaw, Corey J. A. und Barry W. Brook. 2014. Human population reduction is not a quick fix for environmental problems. *PNAS* 111 (46): 16610-16615.
Franzen, Jonathan. 2010. *Freedom*. New York, NY: Farrar, Straus and Giroux.
Malthus, Thomas Robert. 1798. An Essay on the Principle of Population As It Affects the Future Improvement of Society with Remarks on the Speculations of Mr. Godwin, M. Condorcet and Other Writers (1 ed.). London: J. Johnson in St Paul's Church-yard.
Meadows, Donella H., Dennis L. Meadows, Jorgen Randers und William W. Behrens III. 1972. *The Limits to Growth*. New York: Universe Books.
Paech, Niko und Björn Paech. 2011. Suffizienz plus Subsistenz ergibt ökonomische Souveränität. Stadt und Postwachstumsökonomie. *Politische Ökologie* 124: 54-60.
Princen, Thomas. 2005. *The Logic of Sufficiency*. Cambridg, MA: MIT Press.
Schneidewind, Uwe und Angelika Zahrnt. 2013. Damit gutes Leben einfacher wird: Perspektiven einer Suffizienzpolitik. München: Oekom Verlag.
Thoreau, Henry David. 2004. *Walden oder das Leben in den Wäldern*. Zürich: Diogenes.

28. Von STS zu STSE angesichts des Atomunfalls in Japan

Naoshi Yamawaki[1]

Abstract

Dieser Beitrag zielt darauf ab, einen neuen Ansatz für die integrativen Wissenschaften aufzustellen, und zwar in Hinblick auf die Folgen des Atomunfalls in Japan. Der unerwartete Atomunfall in Fukushima fordert uns dazu heraus, nicht nur die unverantwortlichen Systeme von Tepco sowie des Wirtschaftsministeriums in Japan gründlich zu kritisieren, sondern auch die Ideologie von ‚Atoms for Peace' in Frage zu stellen. Dabei geht es darum, eine neue öffentliche Ethik, die der modernen progressiven Geschichtsauffassung fehlt, zu entwickeln und die Grenze der sogenannten STS zu überwinden.

This paper aims at putting forward a new viewpoint for integrative science in face of the aftermath of the nuclear accident in Fukushima. Indeed, this catastrophic accident requires us not only to criticize the irresponsible systems of TEPCO and METI in Japan, but also to question the ideology of Atoms for Peace. Furthermore, it is indispensable for us to develop a new public ethics, which the modern progressive viewpoint of history has been ignoring, and to overcome the limit of STS (Science, Technology and Society).

---

1   Naoshi Yamawaki ist Professor em. für Sozialphilosophie an der Universität Tokyo und Dekan an der Seisa Universität in Japan. Zahlreiche Veröffentlichungen, zuletzt: *Glocal Public Philosophy – Toward Peaceful and Just Societies in the Age of Globalization*, Lit Verlag 2016, "Philosophie und Ethik für die integrale Gesellschaftstheorie und die nachhaltige Gesellschaft", in: Daisenonßji e. V und der Deutschen Akademie der Technikwissenschaften-acatech (Hg.), *Integration als globale Herausforderung der Menschheit*, J.H Rölle, 2012, 173-189, "Demokratie und civil society in Japan aus Sicht der *public philosophy*," in Foljanty-Jost, Gesine, and Hüstebeck, Momoyo (eds.) *Bürger und Staat in Japan*, Halle: Universitätsverlag Halle-Wittenberg, 2013, 19-40.

Der vorliegende Beitrag beschäftigt sich mit den Integrativen Wissenschaften[2] und zwar in der Absicht einen neuen Ansatz vorzuschlagen, der für die Überwindung der Krise der modernen technischen Gesellschaft unentbehrlich zu sein scheint.

28.1.   Überlegungen zum Ereignis vom 11. März 2011

Um dieses Thema erörtern zu können, will ich zunächst, da ich in Japan geboren, aufgewachsen und dort als Philosoph tätig bin, auf das Ereignis vom 11. März 2011, nämlich die große Katastrophe Ostjapans, zu sprechen kommen. Wie weitgehend bekannt ist, besteht dieses Ereignis aus drei Katastrophen, einem Erdbeben, einem Tsunami und einem Atomunfall. Während das Erdbeben und der Tsunami zu Naturkatastrophen gehören, stellt m.e. der Atomunfall eindeutig eine von Menschen verursachte Katastrophe dar. Man kann nicht genau voraussagen, wann, wo und in welchem Maße ein großes Erdbeben oder ein großer Tsunami geschehen wird. Beim Atomunfall geht es hingegen um eine Unzulänglichkeit einer von Menschen hergestellten Anlage, auch wenn der Unfall durch eine Naturkatastrophe wie dem Tsunami verursacht worden ist. Im Folgenden will ich auf diesen Punkt aus der Sicht der Vorhersagefähigkeit eingehen.

28.1.1 Das Ereignis des Tsunamis

Japan ist ein Land mit häufigen Erdbeben. Daher wird jährlich eine große Summe aus dem Staatsbudget für die Erdbebenvorhersage zur Verfügung gestellt. Es steht jedoch fest, dass man trotzdem weder das Erdbeben von Ostjapan noch das von Kobe vom 16. Januar 1995 vorhersagen konnte. Wahrscheinlichkeitsvoraussagen großer Erdbeben in bestimmten Gebieten innerhalb einiger Jahre werden zwar manchmal berichtet, es darf aber nicht übersehen werden, dass es fast keine Vorhersage von Kobe gab. Was das Erdbeben vom 11. März 2011 betrifft, haben nur wenige Wissenschaftler behauptet, dass sich ein großes Erdbeben der Stärke 9, dessen Epizentrum vor der Sanriku-Küste liegt, ereignen würde. Stattdessen herrschte damals die Meinung, dass sich ein solch starkes Erdbeben, wenn überhaupt, in der Tokai-Region, die in der Mitte Japans liegt, ereignen würde. In Anbetracht die-

---

2   Seit zehn Jahren leiten Klaus Mainzer und ich gemeinsam die Deutsch-Japanische Gesellschaft für Integrative Wissenschaft.

ser Tatsache setzt sich unter Seismologen immer mehr die Ansicht durch, dass die computerbasierte Erdbebenvorhersage im Grunde genommen unmöglich sei, und dass es keinen Sinn mache, eine solche Vorhersage aus dem Staatsbudget großzügig zu unterstützen.[3] Man sollte also diese Investition sparen und damit das Geld für Maßnahmen eines Katastrophenschutzes verwenden. Die Vorhersage der Naturkatastrophe durch Computersimulation sei zwar in jenem Sinne sinnvoll, dass sie die Menschen in den betreffenden Gebieten risikobewusster mache. Dieselbe Vorhersage weise aber gleichzeitig eine gegenteilige Wirkung auf: Sie könnte die Menschen, die nicht in den betreffenden Gebieten wohnen, beruhigen und gleichzeitig ihr Risikobewusstsein mindern (das betrifft das jüngste große Erdbeben in Kumamoto).

In Bezug auf den Tsunami hat man aufgrund der Daten der letzten 100 Jahre behauptet, dass ein Tsunami mit einer Wellenhöhe von über 10 m die japanischen Küsten nicht treffen würde. Wellenbrecher wurden dementsprechend errichtet, was jedoch bedauerliche Folgen mit sich gebracht hat. Hinweise in Handbüchern oder bei Katastrophenübungen für den Tsunamifall, wie zum Beispiel: Es ist nicht gefährlich, wenn man sich schon im zweiten Stock befindet, waren nutzlos. Viele Menschen, die sogar im dritten Stock ihre Zuflucht suchten, sind in Wirklichkeit ums Leben gekommen. Eine persönliche Erfahrung, die ich vor 10 Jahren angesichts eines Besuches des Taro-Bezirks der Stadt Miyako und eines dort errichteten Wellenbrecher mit einer Höhe von ca. 10 m machte, bestätigt diese tragischen Fehleinschätzungen. Die Stadt war stolz auf ihn, der den Tsunami im Jahr 1960 überstanden hat, welcher durch das große Chile-Erdbeben verursacht worden war. Dieser Wellenbrecher hat jedoch die Bewohner unaufmerksam hinsichtlich neuer Tsunami gemacht. Im März 2011 hat der Tsunami mit einer Höhe von über 10 m diesen Wellenbrecher zerbrochen und zahlreiche Opfer gefordert, die wegen des Wellenbrechers unbesorgt geblieben waren.

In Hinblick auf diese Tragödie gibt es heute einen Vorschlag von der Präfektur Miyagi Wellenbrecher mit einer Höhe von 15 m zu errichten. Viele Bewohner widersprechen jedoch diesem Plan, weil die Möglichkeit, dass sich ein Tsunami mit einer Höhe von über 15 m ereignet, sowieso nicht ausgeschlossen sei, und weil solche riesigen Wellenbrecher den Anblick der schönen Küsten beeinträchtigen würden. Sie behaupten darüber hinaus, es sei viel wichtiger, erdbebensichere Gebäude zu bauen, Fluchtwege festzulegen und, da es einen zeitlichen Unterschied zwischen Erdbeben und Tsunami gibt, einen besseren Katastrophenplan zu erstellen, der eine schnelle Evakuierung für die Bewohner ermöglicht. Die Stimmen gegen den Bau

---

3   Vgl. Geller (2011). Das jüngste große Erdbeben in Kumamoto hat sicher diese Ansicht bestätigt.

riesiger Wellenbrecher, die der Landschaft schaden, beherrschen die öffentliche Meinung. Die Entscheidung darf auf jeden Fall weder dem Computer noch der Wissenschaft vorbehalten sein, sondern muss eine Politik treffen, welche ihr Gewicht auf die Stimme der Bewohner legt. Es geht hier also nicht um Wissenschaft, sondern um Ethik im weiteren Sinne.

28.1.2   Das Ereignis Atomunfall

Ich möchte nun den Atomunfall von Fukushima behandeln. Dieser Unfall ist durch einen Tsunami verursacht worden. Die Betreiberfirma Tepco und das Wirtschaftsministerium haben das Risiko einer durch einen riesigen Tsunami verursachten Kernschmelze aus Gedankenlosigkeit nicht thematisiert. Dies führte zu dem großen Chaos bzw. Durcheinander an der Unfallstelle unmittelbar nach der Katastrophe. Nicht wenige Menschen haben aber auf die Möglichkeit hingewiesen, dass nicht nur die Stromversorgung des Atomkraftwerkes, sondern auch die Notstromversorgung mit Dieselmotoren unter der Erde im Falle eines großen Erdbebens oder eines riesigen Tsunamis versagen könne. Wenn man an das Gemeinwohl denkt, sollte man eigentlich auch mittels Simulation den Fall der Zerstörung der Stromversorgung untersuchen, welche Schäden dort entstehen können und welche Maßnahmen getroffen werden sollten. Die leitenden Kräfte von Tepco und des Wirtschaftsministeriums haben nie diese Intention verfolgt, sondern waren stets nur auf der Suche nach Gewinn. Sie haben nach dem Unfall behauptet, dass die Situation unvorstellbar gewesen sei. In ihrer Ausrede hat sich ihre „nicht gemeinnützige Natur" direkt widergespiegelt. Ich möchte hier auch ein Beispiel anführen, das zeigt, dass ein Computersystem bei diesem Atomunfall nutzlos war. Speedi (System for Prediction of Environmental Emergency Dose Information) wurde mit einem Budget von 11,3 Milliarden Yen (entspricht ungefähr 85 Millionen Euro) entwickelt. Es hat für Kritik gesorgt, als die Daten von Speedi eine Weile nach dem Atomunfall nicht veröffentlicht wurden. Der Grund für die Nichtveröffentlichung war, dass man die Möglichkeit nicht ausschließen konnte, dass der Wind kurzfristig seine Richtung wechseln wird. Die von der Windrichtung abhängigen Daten von Speedi können also keine Evakuierungsrichtung bestimmen und die Menschen nur verwirren. Man kam daher zu dem Urteil, dass Speedi nichts anderes als ein Anemometer oder Windmesser ist, sodass das Budget für die Erhaltung und Entwicklung dieses Computersystems im folgenden Jahr in hohem Maße reduziert worden ist.

Man wendet jetzt versuchsweise verschiedene Roboter für Entsorgungsarbeiten nach dem Atomunfall an, sie arbeiten aber nicht so gut wie erwartet. Sie sind zwar sehr nützlich für die Beseitigung der Trümmer des Atomkraftwerkes oder des Atommülls im Kernreaktor und mögen außerdem eine große Rolle bei der künftigen Reaktorstilllegung spielen. Dennoch ist es schwer zu sagen, ob die Roboter für die Beseitigung radioaktiv verseuchten Wassers oder für die Messung der Strahlungsdosis im Reaktorkern gut funktionieren. Sie sind außerdem nicht in der Lage, Cäsium-134, welches in den betroffenen Gebieten immer noch nicht abnimmt, nachzuweisen.

Ein weiterer Hinweis sei erwähnt, wonach die Meinungen über die Gefahr niedriger Strahlenbelastung sowie die diesbezügliche Politik unter Medizinern, Physikern und Technologen diametral auseinander gehen. Die einen behaupten, dass eine Strahlenbelastung, die 100 mSv pro Jahr nicht übersteigt, die Gesundheit des menschlichen Körpers wenig gefährde.[4] Andere vertreten die Meinung, dass eine jährliche Strahlenbelastung von nur 1 mSv für Kinder und Schwangere schädlich sei. Haben die ersteren recht, brauchen die Menschen, die außerhalb der schon bestimmten Gefahrengebiete wohnen, nicht evakuiert zu werden. Wenn aber die letzteren Recht haben, muss die Regierung mit Staatsbudget die Risikogebiete neu bestimmen und die dort wohnenden Menschen evakuieren lassen. Eine Computersimulation kann in diesem komplexen Fall nicht beweisen, welche Meinung richtig ist. Wenn wir zu dem Urteil gelangen, dass eine Beweisführung nicht möglich ist, so sollten wir ethische Kriterien in die Abwägung einbeziehen. Es geht dabei sowohl um die Pflichtenethik, nach der eine Evakuierung aufgrund des Vorsorgeprinzips und des Sozialrechts stattfinden soll, als auch um kluge Überlegung (phronesis), nach welcher die Evakuierung mit den daraus entstehenden Risiken (z. B. innere Instabilität) verglichen werden soll.

28.2   Notwendige Integration der Wissenschaften inklusive der Ethik

Was lässt sich aus den oben genannten Fallbeispielen integrativ-wissenschaftlich schlussfolgern?

Es ist zunächst klar, dass die Analysen und Maßnahmen bezüglich des Atomunfalls eine integrative Forschung von Sozial-, Natur- und Ingenieurwissenschaften inklusive der Ethik (im weiteren Sinne) fordern. Die integrative Forschung ist in

---

4   Zu dieser Meinung vgl. Yamashita (2011).

Wirklichkeit in Japan, wo die Zusammenarbeit zwischen Philosophen, Naturwissenschaftler und Ingenieure sehr selten gemacht wird, schwer durchführbar.
Es gibt außerdem keinerlei Anzeichen dafür, dass eine solche Organisation wie
die Ethik-Kommission der deutschen Regierung in Japan gegründet wird. M.E.
ist Japan diesbezüglich vergleichbar mit Frankreich, einem von der Atomenergie
stark abhängigen Land. Im Kontrast zu Deutschland revidiert Frankreich seine
Atomenergiepolitik nicht und beginnt infolge auch nicht mit einer integrativen
Forschung. Die japanische Regierung behauptet deshalb mit lauter Stimme, dass
Deutschland, obwohl ab März 2022 kein Atomkraftwerk mehr in Betrieb sein dürfe, dennoch Atomenergie im Notfall aus Frankreich importieren könne. Man dürfe
daher Japan nicht, da es aus Inseln besteht, mit Deutschland vergleichen.[5]

Wie dem auch sei, ich würde darauf hinweisen, dass das Ereignis des Atomunfalls in Japan nicht allein für die Regierung, sondern auch für das System der
Wissenschaften gilt, welches vom bisherigen Spezialismus gefesselt ist. Der briti-
sche Schriftsteller und Physiker, Charles Percy Snow, hat einmal in seinem Werk
The two cultures and a second look (1967) kritisiert, dass Geistes- und Naturwissenschaftler sich nicht miteinander verständigen wollen. Die Kluft zwischen den
beiden Seiten fällt heute noch deutlicher auf, während die breiten Probleme vom
Atomunfall bis hin zur Atomwaffe von Natur- Ingenieur-, und Sozialwissenschaft-
lern und Philosophen (Ethikern) eine fachübergreifende Zusammenarbeit fordern.

Die Propaganda: „Atomenergie ist eine traumhafte alternative Energie gegen-
über der fossilen Energie" hat ihren Ursprung in der Rede des US-Präsidenten
Dwight David Eisenhower vor der UN-Versammlung im Jahr 1953 sowie in der
Gründung der IAEA im Jahr 1957. Nach dem Zweiten Weltkrieg in Japan, den
Opfernd und Zerstörungen durch die Nuklearschläge, hat das Volk inklusive politisch linker Physiker die zivile Nutzung der Atomenergie im Allgemeinen als positiv betrachtet, und bei der Weltausstellung 1970 in Osaka ist der Strom aus dem
Atomkraftwerk in der Präfektur Fukui mit Applaus begrüßt worden. Das Problem
der Beseitigung radioaktiver Abfälle war jedoch außer Acht gelassen worden und
der Sicherheitsmythos, dass Atomkraftwerke auch in Japan, wo sich Erdbeben sehr
oft ereignen, sicher sind, wurde zu Unrecht betont. Es gab aber natürlich einige
Wissenschaftler, die konsequenterweise gegen zivile Nutzung der Atomenergie waren, und zwar aufgrund der Auffassung, dass die Atomkrafttechnologie im Grunde
gefährlich und insgesamt schädlich sei, auch wenn es nützliche Seiten gebe. Der
im Jahre 2000 verstorbene Chemiker, Jinzaburo Takagi, kann z.B. als repräsenta-

---

5   Vgl. Japan's energy policy under Basic Act on Energy Policy. On April 11.2014. http://
     www.enecho.meti.go.jp/en/category/others/basic_plan/pdf/4th_strategic_energy_plan.
     pdf.

tiv für solche Wissenschaftler benannt werden.[6] Trotz der Atomunfälle von Three Mile Island und Tschernobyl und obwohl der Schnelle Brüter Monju schon ein paar Monate nach Beginn der Stromerzeugung in 1995 wegen Natrium-Austritts außer Betrieb gesetzt worden ist, haben der Sicherheitsmythos sowie der Mythos des nuklearen Kreislaufs weiter Japan beherrscht, bis der Atomunfall in Fukushima geschah. Die gegenüber der Atomkrafttechnologie kritischen Atomtechniker wurden in universitären Instituten aufs Abstellgleis geparkt und mussten hinnehmen, als ewiger Assistent zu arbeiten. Die Forschungen über Atomunfälle wurden missachtet, und man hat sich einen großen Atomunfall nicht vorgestellt, was schließlich zu dem großen Chaos nach dem Atomunfall in Fukushima geführt hat.

Viele Wissenschaftsphilosophen und STS (Science, Technology and Society)-Forscher in Japan haben nun angefangen darüber nachzudenken, warum sie keine kritische Rolle in dieser Situation spielen konnten. Meiner Überzeugung nach sollten STS-Studien heute die Ethik, die sich mit Werten beschäftigt, in ihren Forschungsansatz integrieren und ihren Namen von STS in STSE ändern. Die Notwendigkeit dafür wird klar, wenn man versteht, warum sich die deutsche Kommission, die für den Atomausstieg plädiert hat, „Ethik"-Kommission nennt. Unter den STS-Forschern verbreitet sich jetzt immer mehr der Begriff „Trans-Science"-Problem, eine Fragestellung, für deren Lösung man sich zwar an die Wissenschaft wenden kann, auf die man jedoch keine befriedigende Antwort bekommen kann, wenn man sich nur auf die Wissenschaft verlässt. Dieser Begriff soll Anfang der 1970er Jahre zum ersten Mal von dem amerikanischen Kernphysiker Alvin M. Weinberg geprägt worden sein, der an dem Manhattan-Projekt beteiligt war (Weinberg 1972). Er hatte jedoch noch die Perspektive, dass die Trans-Science-Fragen durch den wissenschaftlichen und technischen Fortschritt schließlich gelöst werden sollten. In diesem Zusammenhang bezieht sich Trans-Science in seinem Sinne nur wenig auf die Ethik.[7] Angesichts der gegenwärtigen Situation, in der der Plan des nuklearen Kreislaufs fast fehlgeschlagen ist, und die Entwicklung des Thorium-basierten Atomkraftwerkes, von dem Weinberg geträumt hatte, in Schwierigkeiten geraten ist, ist es unentbehrlich, eine gründliche ethische Erörterung sowie eine Entscheidung für oder gegen Atomenergie aufgrund der wissenschaftlichen, technischen und sozialwissenschaftlichen Kenntnisse vorzunehmen.

---

6   Zur kurzen Information über Jinsaburo Takagi, vgl. de.wikipedia.org/wiki/Jinzabur%C5 %8D_Takagi
7   Weinberg hatte seinen Traum von Thorium Flüssigsalzreaktoren. Vgl. Martin (2012).

28.3 Rehabilitierung der öffentlichen Ethik, die der modernen progressistischen Geschichtsauffassung fehlt.

Nun möchte ich im Zusammenhang mit diesem Thema einen Blick auf die Ideengeschichte werfen.

Francis Bacon, der heute als Vater der STS-Studien bezeichnet werden könnte, hat Anfang des 17. Jahrhunderts ein wichtiges Wissenschaftsprojekt entworfen. Das Projekt bestand darin, eine Wissenschaft zu gründen, die für die Menschen so nützlich sein kann wie die drei wichtigsten Entdeckungen in der Renaissance, nämlich der Kompass, der Typendruck und das Schießpulver, und eine Wende der Naturanschauung zu schaffen. Gerade Bacon hat davon geträumt, mit der Technik in die Natur einzugreifen und deren Kräfte zu ändern, um das menschliche Wohlergehen zu verbessern (1963). Er behauptete daher, dass die teleologische Naturanschauung, die als Voraussetzung der Scholastik gilt, sofort beseitigt werden sollte, weil sie nicht zur Verbesserung des menschlichen Lebens oder dem Fortschritt der Gesellschaft beitragen kann. Er schlug stattdessen die Gründung eines Wissenssystems vor, welches das, was bisher als unmöglich betrachtet worden ist, ermöglicht, nämlich das Wohlergehen der Menschen zu fördern . Wr können sagen, dieser Gedanke stützt sich auf eine utilitaristische öffentliche Ethik.[8]

Dieses Projekt von Bacon für die moderne Zivilisation wurde von den Aufklärern im 18. Jahrhundert in Frankreich enthusiastisch begrüßt. Marquis de Condorcet hat davon ausgehend eine grandiose Theorie der Zivilisation entwickelt. Er hat in seinem Werk *Esquisse d'un tableau historique des progrès de l'esprit humain* in großem Stil erläutert, wie die Menschheit mit der Technik und den naturwissenschaftlichen Kenntnissen von Unwissenheit und Wahn zur Zivilisation fortgeschritten sei. Er hat dort die menschliche Fähigkeit als unbegrenzt und unerschöpflich betrachtet (Condorcet 1976). Dieser Traditionsstrang eines „Fortschritts der Menschheit durch Wissenschaft" wurde im 19. Jahrhundert von Henri de Saint-Simon und Auguste de Comte übernommen und weiter entwickelt. Saint-Simon hat Bacon für „den ersten Propheten in der politischen Forschung über das Industriesystem" angesehen und seine Organisation zeitweilig als Bacon-Gesellschaft bezeichnet. Comte, Schüler von Saint-Simon, hat auch Bacon als bahnbrechenden Denker seiner eigenen Wissenschaftskonzeption betrachtet. Seine progressistische Auffassung der Geschichte von der theologischen über die metaphysische bis zur empirischen Stufe wurde vor allem ausgehend von der Bacon´schen Idee entwickelt. Das Bacon-Projekt hat sich in der französischen Aufklärung auf diese

---

8   Zur Kritik an Bacons Projekt vgl. Jonas (1979) sowie Spaemann und Löw (1985).

Weise in Verbindung mit der Idee (Ideologie): „Sozialer Fortschritt mit naturwissenschaftlichen Kenntnissen" entwickelt. Womöglich könnte man in dieser Ideengeschichte eine Antwort auf die Frage finden, warum die Atomkrafttechnologie in Frankreich, einem von der Atomenergie stark abhängenden Land, philosophisch nur selten kritisiert wird?

Karl Marx im 19. Jahrhundert und zahlreiche Marxisten im 20. Jahrhundert haben diese progressistische Auffassung weitestgehend übernommen. Marx hat behauptet, das Proletariat, nämlich eine „Klasse, die zu der Bürgergesellschaft gehört und zugleich nicht gehört", werde eine Revolution vollziehen und dies realisiere eine emanzipierte Gesellschaft, in der die Entwicklung der Freiheit jedes Menschen die Voraussetzung für die Freiheit aller Menschen sei. Marx hat dabei großes Gewicht auf die Entwicklung der Produktionskraft gelegt und die Verwirklichung der Industriegesellschaft als Fortschritt der Zivilisation betrachtet (Marx und Engels 2012). Der wissenschaftlich-technische Fortschritt als solcher war kein Gegenstand der Kritik. Nach dem Zweiten Weltkrieg gab es in Japan eine heute nicht mehr ernstzunehmende marxistische Behauptung des „Japanischen Rates gegen Atom- und Wasserstoffbomben", wonach die Atombombe der Sowjetunion als Gegenmaßnahme gegen die USA anzuerkennen sei. Selbst der berühmte politisch-linke Physiker Mitsuo Taketani, der einen Standpunkt gegen die Sowjetunion einnahm, war nicht gegen die friedliche Nutzung der Atomenergie. Man sagt, diese Einstellung sei der Grund für den Zwist zwischen ihm und dem oben genannten Jinzaburo Takagi.[9]

Die „progressistische Geschichtsauffassung" von Bacon bis zum Marxismus hat nunmehr infolge der zwei großen Weltkriege sowie angesichts der Probleme der Massenvernichtungswaffen und der Atomkrafttechnologie ihre Grundlage verloren. Wir müssen jetzt zweifellos den Mythos aufgeben, dass alle wissenschaftlich-technischen Fortschritte die Wohlfahrt der Menschheit mit sich bringen würden. Es macht allerdings keinen Sinn, die positiven Möglichkeiten einer technologischen Zivilisation zu verneinen. Insofern würde ich die entartungstheoretische Geschichtsauffassung von Nietzsche oder Spengler keineswegs vertreten. Um die Krise der atombasierten Zivilisation zu überwinden, reicht es nicht, nur zur technischen Vernunft oder dem naturwissenschaftlichen Denken zu greifen. Wir müssen nun eine neue praktische Vernunft konzipieren, die in Form einer öffentlichen Ethik verstanden werden sollte.[10] Diese öffentliche Ethik beschäftigt sich mit der Trans-Science-Frage, für deren Lösung man sich zwar an die Wissenschaft wenden kann, auf die man jedoch keine Antwort bekommen kann, wenn

---

9   Dazu Kato (2013).
10   Dazu Yamawaki (2015).

man sich nur auf die Wissenschaft verlässt. Ich möchte diesbezüglich meine Ansicht von der abstrakten Techniktheorie sowie der Kritik der Moderne etwa durch
Heidegger (1962) scharf abgrenzen. Wie der hervorragende US-amerikanische
Technikphilosoph, Andrew Feenberg, betont hat, vermag die abstrakte und spekulative Auffassung Heideggers, wonach das Wesentliche der Technik als Gestell
betrachtet werden sollte, es nicht, zwischen Elektrogerät und Atomkraftwerk oder
zwischen Agrartechnik und Holocaust zu unterscheiden (Feenberg 1999; 2004).
Heideggers Auffassung zeigt außerdem keinen Weg zur praktischen Philosophie
oder der Ethik, die die negative Seite der Neuzeit überwinden könnte.

28.4   Zur STSE-Methodik

Bereits auf der letzten Deutsch-Japanischen Tagung habe ich drei zentrale Perspektiven einer inter- bzw. transdisziplinären Gesellschaftstheorie, die den Spezialismus
der Wissenschaften überwinden kann und in dem Sinne als Projekt der Post-Spezi-
alisierung zu bezeichnen ist, wie folgt zusammengefasst:

1. Empirische Untersuchung der gegenwärtigen sozialen Realität, die sich auf die
   Frage bezieht, was wir wissen müssen.
2. Normative Theorie bezüglich der gegenwärtigen und zukünftigen Gesellschaft,
   die sich auf die Frage bezieht, was wir tun sollen.
3. Sozialpolitische Forschung bezüglich der zukünftigen Verwirklichung der
   Norm, die sich auf die Frage bezieht, was wir tun können (Yamawaki 2012, insbesondere S. 178).

Aus dieser dreifachen Perspektive der STSE-Forschung nach dem 11. März 2011
ergeben sich die folgenden konkreten Themen:

Bezüglich des ersten Punktes sollten wir auf die folgenden Ereignisse zurück-
blicken: die Entdeckung der Atomenergie durch Otto Hahn, Fritz Strassman und
Lise Meitner; das Manhattan-Projekt, welches in den USA von John Robert Op-
penheimer geleitet wurde, und die Atombombenabwürfe auf Hiroshima und Na-
gasaki als dessen Folge.[11] Die Erklärung „Atoms for Peace" des US-Präsidenten
Eisenhower und die Gründung der IAEA; die Verbreitung der Atomkraftwerke in
den hochindustrialisierten Ländern; die Protestbewegung in den 1970er Jahren bis
zum gegenwärtigen Atomausstieg in Deutschland, während sich Frankreich zu einem stark von der Atomenergie abhängigen Land entwickelt hat; die Geschichte

---

11   Zu diesem Thema, vgl. Rose (2012) und Baggott (2011).

von der japanweiten Verbreitung der Atomkraftwerke über den Atomunfall in Fukushima bis hin zu der heutigen chaotischen Situation. All diese Ereignisse gehören zu einer großen unvollendeten Geschichte der Moderne und diese Geschichte sollte aus der Sicht der Wissenschaftsgeschichte sowie der Geschichte des sozialen Denkens interpretiert werden. Verschiedene gegenwärtige Probleme im Kontext der Atomkrafttechnologie unterschiedlicher Länder sollten darüber hinaus sozialwissenschaftlich untersucht und behandelt werden.

Der zweite Punkt ist mit den folgenden Themen verbunden: die gerechte Risikoverteilung bezüglich der zivilen Atomkrafttechnologie; Verantwortung für die zunehmenden Atomabfälle gegenüber den nachfolgenden Generationen; die Frage nach der umweltbezogenen Gerechtigkeit im Zusammenhang mit den öffentlichen Gelder, mit dem sich arme Gebiete finanziell bereichern können; die Sicherung der Menschenrechte (vor allem der sozialen Rechte) der Bewohner der betroffenen Gebiete nach dem Atomunfall. Diese Themen sollten Diskussion könnten verschiedene Normen (Utilitarismus, Liberalismus von John Rawls, Kommunitarismus, Katholizismus, Neo-Aristotelismus, Diskursethik, Verantwortungsethik und die aus verschiedenen Religionen stammenden Normen) vermutlich in Konflikt stehen.

Zum dritten Punkt gehören Themen wie die zukünftige Energiepolitik, vor allem die Politik alternativer Energien im Falle des Atomausstiegs, Kosten für Reaktorstilllegung, Methoden der endgültigen Beseitigung oder die Zwischenlagerung der Atomabfälle. Diese Themen müssen hinsichtlich von Wissenschafts- und Technologiepolitik ernsthaft und umfassend diskutiert werden. Die computerbasierte Forschung der komplexen Systeme sowie Computersimulationen spielen dabei selbstverständlich eine wichtige Rolle.

Literatur

Bacon, F. 1963. *Advancement of Learning*. London: Everyman's Library.
Baggott, J. 2011. *The First War of Physics: The Secret History of the Atom Bomb 1939-49*. Reprint. New York, NY: Pegasus Books.
Condorcet, M. J. A. N. d. C. 1976. *Entwurf einer historischen Darstellung der Fortschritte des menschlichen Geistes*. Frankfurt a.M.: Suhrkamp.
Ethik-Kommission Sichere Energieversorgung. 2011. *Deutschlands Energiewende – Ein Gemeinschaftswerk für die Zukunft (Abschlussbericht)*. Berlin: Presse- und Informationsamt der Bundesregierung.
Feenberg, A. 1999. Questioning Technology. Abingdon: Routledge.
Feenberg, A. 2004. Modernity and Technology: Heidegger and Marcuse. Abingdon: Routledge.

Geller, R. J. 2011. Shake-up time for Japanese seismology. *Nature* 472 (7344): 407-409.

Heidegger, M. 1962. *Die Technik und die Kehre*. Pfullingen: Neske Verlag.

Jonas, H. 1979. *Das Prinzip der Verantwortung: Versuch einer Ethik für die technologische Zivilisation*. Frankfurt: Insel.

Kato, T. 2013. Sozialismus in Japan: Die Logik gegen die Atombombe und für Atomkraftwerke. Tokio: Iwanami, S. 45-207 (aus dem Japanischen).

Marx, K. und F. Engels. 2012. *Das Kommunistische Manifest*. München: Beck.

Martin, R. 2012. *Superfuel: Thorium, the Green Energy Source for the Future*. Macmillan Science.

Rose, R. 2012. *The Making of Atomic Bomb*. 25th Anniversary Edition. New York, NY: Simons and Schuster.

Snow, C. P. 1967. *Die zwei Kulturen: Literarische und naturwissenschaftliche Intelligenz*. Stuttgart: Klett.

Spaemann, R. und R. Löw. 1985. *Die Frage Wozu? Geschichte und Wiederentdeckung des teleologischen Denkens*. München: Pieper.

Weinberg, A. M. 1972. Science and Trans-Science. *Minerva* 10 (2): 209-222.

Yamashita, S. 2011. Studying the Fukushima Aftermath People are Suffering from Radiophobia. *Spiegel Online International*, 19. August 2011. http://www.spiegel.de/international/world/studying-the-fukushima-aftermath-people-are-suffering-from-radiophobia-a-780810.html.

Yamawaki, N. 2012. Philosophie und Ethik für die integrale Gesellschaftstheorie und die nachhaltige Gesellschaft. In *Integration als globale Herausforderung der Menschheit*, hrsg. Daisenonßji e.V. und Deutsche Akademie der Technikwissenschaften, 173-189. Dettelbach: J. H. Röll.

Yamawaki, N. (Hg.). 2015. *Science, Technology and Social Ethics* Tokio: University of Tokyo Press (aus dem Japanischen).

The manufacturer's authorised representative in the EU is Springer
Nature Customer Service Centre GmbH, Europaplatz 3, 69115 Heidelberg,
Germany. If you have any concerns regarding our products, please
contact ProductSafety@springernature.com

Printed and bound by CPI Group (UK) Ltd, Croydon, CR0 4YY
27/04/2026
02097564-0009